INTERNATIONALISATION AND GLOBALISATION IN MATHEMATICS AND SCIENCE EDUCATION

Internationalisation and Globalisation in Mathematics and Science Education

Edited by

Bill Atweh
Curtin University of Technology, Perth, Australia

Angela Calabrese Barton
Columbia University, New York, USA

Marcelo C. Borba
State University of São Paulo - Rio Claro, Brazil

Noel Gough
La Trobe University, Melbourne, Australia

Christine Keitel
Freie University Berlin, Berlin, Germany

Catherine Vistro-Yu
Ateneo de Manila University, Quezon City, Phillipines

and

Renuka Vithal
University of KwaZulu-Natal, Durban, South Africa

Springer

A C.I.P. Catalogue record for this book is available from the Library of Congress.

ISBN 978-1-4020-5907-0 (HB)
ISBN 978-1-4020-5908-7 (e-book)

Published by Springer,
P.O. Box 17, 3300 AA Dordrecht, The Netherlands.

www.springer.com

Printed on acid-free paper

This project is supported by the Mathematics Education Research Group of Australasia (MERGA)

TABLE OF CONTENTS

Preface ix

About the Editors xi

About the Contributors xv

SECTION 1 THEORETICAL PERSPECTIVES

Chapter 1 Mathematical Literacy and Globalisation 3
Ole Skovsmose

Chapter 2 Epistemological Issues in the Internationalization and Globalization of Mathematics Education 19
Paul Ernest

Chapter 3 All around the World: Science Education, Constructivism, and Globalization 39
Noel Gough

Chapter 4 Geophilosophy, Rhizomes and Mosquitoes: Becoming Nomadic in Global Science Education Research 57
Noel Gough

Chapter 5 Science Education and Contemporary Times: Finding our way through the Challenges 79
Lyn Carter

Chapter 6 Social (In)Justice and International Collaborations in Mathematics Education 95
Bill Atweh and Christine Keitel

Chapter 7 Globalisation, Ethics and Mathematics Education 113
Jim Neyland

Chapter 8 The Politics and Practices of Equity, (E)Quality and Globalisation in Science Education: Epriences from both sides of the Indian Ocean 129
Annette Gough

SECTION 2 ISSUES IN GLOBALISATION AND INTERNATIONALISATION

Chapter 9 Context or Culture: Can TIMSS and Pisa Teach us about what Determines Educational Achievement in Science? 151
Peter Fensham

Chapter 10 Quixote's Science: Public Heresy/Private Apostasy 173
Paul Dowling

Chapter 11 The Potentialities of (Ethno) Mathematics Education: An Interview with Ubiratan D'ambrosio 199
Ubiratan D'Ambrosio and Maria do Carmo Domite

Chapter 12 Ethnomathematics in the Global Episteme: Quo Vadis? 209
Ferdinand Rivera and Joanne Rossi Becker

Chapter 13 POP: A Study of the Ethnomathematics of Globalization Using the Sacred Mayan Mat Pattern 227
Milton Rosa and Daniel Clark Orey

Chapter 14 Internationalisation as an Orientation for Learning and Teaching in Mathematics 247
Anna Reid and Peter Petocz

Chapter 15 Contributions from Cross-National Comparative Studies to the Internationalization of Mathematics Education: Studies of Chinese and U.S. Classrooms 269
Jinfa Cai and Frank Lester

Chapter 16 International Professional Development as a Form of Globalisation 285
Hedy Moscovici and Gary Varrella

Chapter 17 Doing Surveys in Different Cultures: Difficulties and Differences – A Case from China and Australia 303
Zhongjun Cao, Helen Forgasz and Alan Bishop

Chapter 18 The Benefits and Challenges for Social Justice in International Exchanges in Mathematics and Science Education 321
Catherine Vistro-Yu and Kathryn Irwin

Chapter 19 Globalisation, Technology, and the Adult Learner of Mathematics 343
Gail FitzSimons

SECTION 3 PERSPECTIVES FROM DIFFERENT COUNTRIES

Chapter 20 Balancing Globalisation and Local Identity in the Reform of Education in Romania 365
Mihaela Singer

Chapter 21 Voices from the South: Dialogical Relationships and Collaboration in Mathematics Education 383
Mónica Villarreal, Marcelo C. Borba and Cristina Esteley

Chapter 22 Globalization and its Effects in Mathematics and Science Education in Mexico: Implications and Challenges for Diverse populations 403
Edith Cisneros-Cohernour, Juan Carlos Mijangos Noh, María Elena and Barrera Bustillos

Chapter 23 In Between the Global and the Local: The Politics of Mathematics Education Reform in a Globalized Society 421
Paola Valero

Chapter 24 Singapore and Brunei Darussalam: Internationalisation and Globalisation through Practices and a Bilateral Mathematics Study 441
Khoon Yoong Wong, Berinderjeet Kaur, Phong Lee Koay and Jamilah Binti Hj Mohd Yusof

Chapter 25 Lesson Study (Jyugyo Kenkyu) from Japan to South Africa: A Science and Mathematics Intervention Program for Secondary School Teachers 465
Loyiso Jita, Jacobus Maree, and Thembi Ndlalane

Chapter 26 The Post-Mao Junior Secondary School Chemistry Curriculum in the People's Republic of China: A Case Study in the Internationalization of Science Education 487
Bing Wei and Gregory Thomas

Chapter 27 Globalisation/Localisation in Mathematics Education: Perception, Realism and Outcomes of an Australian Presence in Asia 509
Beth Southwell, Oudone Phanalasy and Michael Singh

Author Index 525

Subject Index 535

PREFACE

The past 100 years have witnessed a rapid increase of international contacts and collaborations between academics around the globe in the form of conferences, publications, courses for international students, exchanges of curricula and professional development programs and, of course, a multitude of cross-national comparative studies and other projects. In spite of their prevalence, ethical implications, and possible economic and political consequences, these international activities have rarely been subject to explicit research and critique.

Moreover, these interactions occur within a wider context of the globalisation of every aspect of our personal, social and academic lives. Mathematics and science education might be two of the most globalised subjects of the school curriculum under the masks of objectivity, valuelessness, universality of their respective "truths" and their perceived relationship to the economic development aspirations of every nation. These assumptions are often inadvertently carried over to the disciplines of mathematics and science education themselves, including teacher education, curriculum development, professional development and research. This volume is a contribution towards putting these assumptions under our collective critical gaze.

At the same time that trends of globalisation might be providing increasing opportunities for our academic work, consultancies and publications, it is also leading to an ever-increasing gap between the *haves* and *have-nots*, between the rich and the poor. Although these increasing differentiations are found within each country, they are more prominent along the south-north and west-east divides. This edited collection of diverse works is intended to maintain our vigilance about the prevalence of these patterns in our attempts to promote the international standing of our professions.

In calling for proposals for contributions to this volume, the editors identified the following aims for the collection:

- Develop theoretical frameworks of the phenomena of internationalisation and globalisation and identify related ethical, moral, political and economic issues facing international collaborations in mathematics and science education.
- Provide a venue for the publication of results of international comparisons of cultural differences and similarities rather than merely of achievements and outcomes.

- Provide a forum for critical discussion of the various models of international projects and collaborations.
- Provide a representation of the different voices and interests from around the world rather than a consensus on issues.

The call for expressions of interests for authoring chapters was widely circulated around the world using electronic lists, international conference attendance and the editors' personal contacts and networks, inviting academics in both mathematics and science education to consider sharing their experiences and learnings through authoring chapters related to the above aims. In particular, the call for chapters targeted a variety of authors with varying levels of accomplishment on the international scene and authors from non-English speaking backgrounds. To achieve this variety of voices several mechanisms were put in place. Firstly, the composition of the editorial team itself represented a wide geographical spread from Latin and North America, Europe and Africa, and East Asia and Australia. Secondly, the editors deliberately encouraged joint authorship between less experienced and more experienced authors, and from English speaking and non-English speaking backgrounds. Thirdly, a multi-loop peer review and editing process generated iterative writing and constructive comments from a variety of critical friends. Chapters were reviewed by at least three peers from within the community of authors and editors.

As a result the authors – all of whom are or have been involved in some bilateral, regional or multinational projects – represent voices from a wide range of nations including Argentina, Armenia, Australia, Brazil, Brunei Darussalam, China (including Hong Kong and Macau), Denmark, Germany, Israel, Laos, Mexico, New Zealand, the Philippines, Romania, Singapore, South Africa, the United Kingdom and the United States of America. Projects relating to Colombia and Japan are also reported upon.

This book is sponsored by the Mathematics Education Research Group of Australasia (MERGA). Previous volumes in this series include *Research and Supervision in Mathematics and Science Education* published in 1998 and *Sociocultural Research on Mathematic Education: An International Perspective* published in 2001.

The Editors

ABOUT THE EDITORS

Bill Atweh is an Associate Professor in mathematics education at the Curtin University of Technology, Perth, Australia. His research interests include socio-cultural factors in mathematics education including gender and socioeconomic background; globalisation, post school pathways and action research. He is the Vice President of Publication of the Mathematics Education Research Group of Australasia. He has conducted research and professional development activities in Brazil, Colombia, Cuba, Mexico, Korea, Philippines and Vietnam. His previous editorial experiences include *Action research in practice: Partnerships for social justice in education* (Routledge) *Research and supervision in mathematics and science education* (Erlbaum) and *Sociocultural research on mathematics education: An international perspective* (Erlbaum). Email: b.atweh@curtin.edu.au

Marcelo C. Borba is a professor of the graduate program in mathematics education and of the mathematics department of UNESP (State University of Sao Paulo), campus of Rio Claro, Brazil. He is a member of the editorial board of *Educational Studies in Mathematics* and a consultant for several journals and funding agencies both in Brazil and abroad. He is the editor of *Boletim de Educação Matemática* (BOLEMA). He gave talks in countries such as Canada, Denmark, Mozambique, New Zealand and United States. He has been a member of the program committee of several international conferences. He wrote several books, book chapters and papers published in Portuguese and in English, and he is the editor of a collection of books in Brazil, which includes ten books to date. Email: mborba@rc.unesp.br

Angela Calabrese Barton is an expert in urban science education and issues of equity and diversity. She received her PhD in Curriculum, Teaching and Educational Policy from Michigan State University in 1995. Her research focuses on issues of equity and social justice in science education, with a particular emphasis on the urban context (Calabrese Barton, 2002, 2003). Her work has been published in *Educational Researcher, American Education Research Journal, Educational Policy and Practice, the Journal of Research in Science Teaching, Science Education, Curriculum Inquiry*, among others. Her most recent book, *Teaching Science for Social Justice* (Teachers College Press), won the 2003 AESA Critics Choice Award. Her other recent book, *Re/thinking Scientific Literacy* won the 2005 AERA Division K award for Exemplary Research. She has also been awarded the Early Career Research Award, National Association for Research in

Science Teaching, 2000; Kappa Delta Pi Research Award (Teaching and Teacher Education), American Education Research Association, Division K, 1999; Early Career Award, National Science Foundation, 1998–2003; National Academy of Education Spencer Fellow, 1996–1998; and the Outstanding Dissertation Award, Michigan State University, Department of Teacher Ed, College of Education, 1995. Email: acb@msu.edu

Noel Gough is a Foundation Professor of Outdoor and Environmental Education at La Trobe University, Victoria, Australia. He is the author of *Blueprints for Greening Schools* (1992), *Laboratories in Fiction: Science Education and Popular Media* (1993), and numerous journal articles. He is coeditor (with William Doll) of *Curriculum Visions* (2002), which has also been translated into simplified and traditional Chinese, and the founding editor of *Transnational Curriculum Inquiry: the Journal of the International Association for the Advancement of Curriculum Studies.* His research interests, which he has pursued in Australia, North America, and southern Africa, include poststructuralist and postcolonialist analyses of curriculum change, with particular reference to environmental education, science education, internationalisation and inclusivity. In 1997 he received the inaugural Australian Museum Eureka Prize for Environmental Education Research. Email: n.gough@latrobe.edu.au

Christine Keitel is a Professor for Mathematics Education at Freie University Berlin and Vice-President (Deputy Vice Chancellor) of teaching and research. She was the director of the Basic Components of Mathematics Education for Teachers (BACOMET) project during 1989–1993; director of the NATO Research Workshop during 1993–1994; and member of the Steering Committee of the OECD project Future Perspectives of Science, Mathematics and Technology Education for 1989–1995. She was an Expert Consultant for the and for the TIMSS Video and Curriculum Analysis Project in 1993–1995 and a member of editorial boards of several journals for curriculum and mathematics education and on the Advisory Board of Kluwer's Mathematics Education Library. She has been founding member of the National Coordinator and Convenor/President of International Organisation of Women and Mathematics Education (IOWME) 1988–1996, Vice President, Newsletter Editor and President of the Commission Internationale pour l'Etude et l'Amélioration de l'Enseignement des Mathématiques (CIEAEM) 1992–2004, and member of the International committee of the International Organsiation of Psychology and Mathematics Education (PME) 1988–1992. Email: keitel@zedat.fu-berlin.de

Catherine Vistro-Yu, a Professor in the Mathematics Department of the Ateneo de Manila University, The Philippines, is a mathematics educator. She teaches mathematics and mathematics education courses to classroom teachers of both the primary and secondary levels. Her research interests lie mostly in the area of mathematics teacher education and children's understanding of mathematics although she has studied other important concerns as well, such as the use of technology in mathematics education and mathematics teachers' beliefs. In the last decade, she has been actively engaged in large-scale mathematics education projects and programs in her country that address issues of curriculum, teacher competencies in mathematics, and student achievement, among others. Catherine's international network has provided her with valuable opportunities for significant collaboration with colleagues in Asia through the SEACME and EARCOME, and with other foreign colleagues through ICME. Email: cvistro-yu@ateneo.edu

Renuka Vithal, is a Professor in mathematics education and Dean of the Faculty of Education at the newly merged University of KwaZulu-Natal, Durban, South Africa. She has published both nationally and internationally and written widely in these fields including articles in journals and conference proceedings, books, and chapters in books. She serves as reviewer and is a member of the editorial boards of several national and international journals. Her recent publications include *In search of a pedagogy of conflict and dialogue for mathematics education* (2003, Kluwer). She co-edited with Prof Jill Adler and Prof Christine Keitel a volume titled *Researching Mathematics Education in South Africa: Perspectives, practices and possibility.* (2005, HSRC, Pretoria). She has also been the South African project leader for an international study on Learners' perspectives of grade eight mathematics classrooms. Prof. Vithal has served as education expert in the South African National Commission of UNESCO and in the executive and as chair of the Southern African Association for Research in Mathematics, Science and Technology Education. She is an institutional auditor for the Council on Higher Education; and is a member of the South African National Committee for the International Mathematical Union representing the Association for Mathematics Education of South Africa. Email: vithalr@ukzn.ac.za

ABOUT THE CONTRIBUTORS

María Elena Barrera Bustillos has a Masters degree in Higher Education and a Teaching Certificate from the Universidad Autónoma de Yucatan. She is also a specialist in the governance of educational institutions by the Instituto Nacional de Administración Púbilca (INAP), and a certified accreditation specialist by the Unión de Universidades de América Latina. Her main research interests are in the areas of Administration and Educational Policy, where she has conducted research on institutional evaluation and accreditation, the evaluation of advising, student services, and leadership, among other topics. In addition, she has a strong interest in curriculum program development and evaluation. She is a member of several evaluation technical committees at the national and state levels in Mexico. Currently, she is the Dean of the College of Education at the Universidad Autónoma de Yucatan. Email: bbustil@uady.mx

Alan Bishop is an Emeritus Professor of Education in the Faculty of Education, Monash University. For 23 years he was a university lecturer at the University of Cambridge before going to Monash University, Australia in 1992 as Professor of Education. He was President of the Mathematical Association, UK; was on the Royal Society's Mathematics Education committee; and was the UK National Representative on the International Commission for Mathematics Instruction, advising Government agencies on policies regarding mathematics education. His research interests cover various aspects of mathematics education. He conducted a research survey to advise the influential Cockcroft Committee in the UK which changed policy regarding mathematics education in many countries. He advises UNESCO on mathematics education matters and contributed the module on Numeracy for UNESCO's resource material on Science and Technology Education. Email: alan.bishop@education.monash.edu.au

Jinfa Cai is a Professor of Mathematics and Education and the Director of Secondary Mathematics Education at the University of Delaware. He is interested in how students learn mathematics and solve problems, and how teachers can provide and create learning environments so that students can make sense of mathematics. He has explored these questions in various educational contexts, both within and across nations. He was a 1996 National Academy of Education Spencer Fellow. In 2002, he received an International Research Award and a Teaching Excellence Award from the University of Delaware. He has been serving on the Editorial Boards of *Journal for Research in Mathematics Education, Mathematics Education Research Journal, and Zentralblatt fuer Didaktik der Mathematik.* He has been a visiting professor in various institutions, including Harvard University in 2000–2001. Email: jcai@math.udel.edu

Zhongjun Cao received his PhD in Mathematics Education in the Faculty of Education, Monash University, Australia in 2004. His thesis investigated students' attitudes towards mathematics in China and Australia. Previously he started his PhD study, he worked as a Lecturer in the Department of Mathematics, Henan University, China for ten years. His research covers a variety of topics in mathematics education and in tertiary education, which include learning approaches, assessment, attitudes and values of teachers and students, and persistence of tertiary students. He is currently working at the Victoria University, Australia. Email: zhongjun_cao@yahoo.com.au

Lyn Carter has published widely in the areas of globalisation and science education in a number of prominent journals and books including *Journal for Research in Science Teaching* and Springers' *International Handbook Globalisation and Education Policy Research.* She has also published in *Science Education.* Her other research interests include the use of postcolonialist theory and sustainability science as counter discourses to globalisation. Lyn lectures in science and technology education to undergraduate primary and secondary teacher education students in the Trescowthick School of Education on the Melbourne Campus of the Australian catholic University. She also lectures in postgraduate education particularly in the areas of research methodologies and contemporary issues in curriculum. Email: l.carter@patrick.acu.edu.au

Edith Cisneros-Cohernour is a Professor at the Universidad Autónoma de Yucatan, Mexico. A former Fulbright fellow, she received her PhD in Higher Education Administration and Evaluation from the University of Illinois at Urbana-Champaign on 2001. From 1994–2001, she was also affiliated with the National Transition Alliance for Youth with Disabilities, the Bureau of Educational Research, and the Center for Instructional Research and Curriculum Evaluation of the University of Illinois at Urbana-Champaign. Her areas of research interest are evaluation, professional development, organizational learning and the ethical aspects of research and evaluation. Among her publications are *Situational Evaluation of Teaching* (2000), co-authored with Robert E. Stake; *Strategies for Effective Instruction: Mexican American Mothers and Everyday Instruction* (1999), co-authored with Robert P. Moreno; *Probative, Dialectic, and Moral Reasoning in Program Evaluation*, co-authored with Migotsky et al (1997); *Communities of Practice* with Benzie, Mavers and Somekh (2005); and *Influencia del Contexto Sociocultural en el Liderazgo Escolar* en Mexico con Bastarrachea (2006). Email: cchacon@tunku.uady.mx

Ubiratan D'Ambrosio is an Emeritus Professor of Mathematics, State University of Campinas/UNICAMP, São Paulo, Brazil. Previously, he was Associate Professor of Mathematics and Graduate Chairman, State University of New York/SUNY at Buffalo (1968–1972); Professor of Mathematics and Director of the Institute of Mathematics, Statistics and Computer Science (1972–1980) of UNICAMP, Brazil; Chief of the Unit of Curriculum of the Organization of American States, Washington DC (1980–1982); Pro-Rector [Vice-President] for University Development, (1982–1990) of UNICAMP, Brazil. Currently, he is a Professor at the PUCSP/Pontifícia Universidade Católica de São Paulo, and guest professor at the USP/Universidade de São Paulo and the UNESP/Universidade Estadual Paulista and the President of the Brazilian Society of History of Mathematics/SBHMat. Recently he received a citation from the American Association for the Advancement of Science for "imaginative and effective leadership in Latin American Mathematics Education and in efforts towards international cooperation." (1983) as well as the Kenneth O. May Medal in the History of Mathematics granted by the International Commission on History of Mathematics (2001), and the Felix Klein Medal of Mathematics Education granted by the International Commission of Mathematics Instruction/ICMI (2005). Email: http://vello.sites.uol.com.br/ubi.htm

Maria Do Carmo Domite has been a mathematics educator at the Faculty of Education of the University of Sao Paulo since 1997. Her main research interests are in areas of problem posing, Ethnomathematics and Indigenous education. She obtained a Master of Arts in Mathematics Education from the University of Georgia, USA in 1985 and her Doctor of Philosophy from the University of Campinas in 2004. Email: mcdomite@usp.br

Paul Dowling is a sociologist and former teacher of mathematics. His research over the past twenty years has involved the development of an organisational language for the analysis of pedagogic texts, sites and technologies. He is author of *The Sociology of Mathematics Education: Mathematical Myths/Pedagogic Texts* and Sociology as Method: Departures from the forensics of culture, text and knowledge and co-author of the best seller, *Doing Research/Reading Research: A mode of interrogation for education.* His work (represented on his website at http://homepage.mac.com/paulcdowling/ioe/) engages a wide range of theoretical perspectives (from Foucault to Baudrillard to Bernstein to Douglas and back) and empirical settings (ranging from a bus journey in Rajasthan to a *Monument* in Trafalgar Square to the A-Bomb Dome in Hiroshima, as well as mathematics and science texts and edutainment sites). He believes that educational research properly *interrogates* rather than directly informs professional educational practice and policy. He is based at the Institute of Education, University of London, but also spends several months each year in Japan. Email: p.dowling@ioe.ac.uk

Paul Ernest studied mathematics, logic and philosophy at Sussex and London universities in the UK, where he obtained his BSc, MSc and PhD degrees. He became a qualified teacher in the 1970s teaching school mathematics in London. He subsequently held lecturing positions in the universities of Cambridge and the West Indies. Paul Ernest is currently Emeritus Professor of the Philosophy of Mathematics Education at Exeter University, UK, as well as visiting professor at the universities of Trondheim and Oslo. He directs the specialist masters and doctoral programmes in mathematics education at Exeter that attract students from almost every continent. He is well known internationally for his research and conference contributions and he has published over 200 papers, chapters and books ranging across the field of mathematics education, as well as contributions to the philosophy of mathematics. His main research interests concern fundamental questions about the nature of mathematics and how it relates to teaching, learning and society. He is currently working on a semiotic theory of mathematics and education. He is best known for his work on philosophical aspects of mathematics education and he edits the international web-based Philosophy of Mathematics Education Journal, located at <http://www.ex.ac.uk/~PErnest/> His books include *The Philosophy of Mathematics Education* (Falmer, 1991) and *Social Constructivism as a Philosophy of Mathematics* (SUNY Press, 1998). Email: p.ernest@ex.ac.uk

Cristina Esteley is a mathematics education Professor for pre-service mathematics teachers at the Pedagogy and Human Sciences Institute of Villa María University in Córdoba, Argentina. She received her Masters in Education from The City College of the City University of New York. She has co-supervised Masters students and has served as director of a number of Córdóba Agency Science projects for in-service mathematics teachers. She has taught mathematics at the secondary and university levels, co-authored several journal articles, and served as consultant for secondary schools in Córdoba. Email: rguillet@arnet.com.ar

Peter Fensham is an Emeritus Professor of Education at Monash University, Australia where he developed a strong research group in science education. In his recent book, *Defining an Identity*, he discusses the emergence of science education as an international field of research. Now attached as Adjunct Professor to the School of Mathematics, Science and Technology Education at the Queensland University of Technology, he has been involved in large international assessment projects for science education for more than a decade, and their implications for the science curriculum of schooling. Email: p.fensham@qut.edu.au

Gail Fitzsimons was a teacher of mathematics, statistics, and numeracy subjects to adult students of further and vocational education in community, industry, and institutional settings for 20 years. Gail was awarded an Australian Research Council Post-Doctoral Research Fellowship, 2003–2006, for a project entitled: *Adult Numeracy and New Learning Technologies: An Evaluative Framework*. In 2002 Gail published a revised version of her doctoral thesis as a monograph entitled: *What counts as mathematics? Technologies of power in adult and vocational education*, through Kluwer Academic Publishers. She has also edited and contributed chapters to numerous books, as well as acting as a reviewer for Springer. She is an associate editor for the *Mathematics Education Research Journal*, an editorial panel member for the *Australian Senior Mathematics Journal*, and was inaugural editor of the electronic journal *Adults Learning Mathematics – An International Journal*. Email: gail.fitzsimons@education.monash.edu.au

Helen Forgasz is an Associate Professor in the Faculty of Education, Monash University, Australia. Before embarking on an academic career, Helen was a secondary teacher of mathematics, physics and computing for ten years. On completion of her PhD, Helen was awarded a prestigious Australian Research Council Australian Postdoctoral Research Fellowship and worked for three years on a research project that was an extension of her doctoral work. She then worked as a Research Fellow before taking up a lectureship in mathematics education at Deakin University, a position she held for three years. Helen's research interests relate to all levels of mathematics education – primary, secondary, and tertiary – and include equity issues with a focus on gender issues, beliefs and attitudes, learning environments, and computer use. Email: helen.forgasz@education.monash.edu.au

Annette Gough is a Professor of Science and Sustainability Education and Head of the School of Education at RMIT University, Melbourne, Australia. Her research interests include feminist, poststructuralist and postcolonialist analyses of curriculum policy, design and development in environmental and science education in Australia, South Africa, Canada, Korea and globally. She has recently completed an ARC research project on Improving Middle Years Mathematics and Science (with Russell Tytler and Susie Groves). She is the author of *Education and the Environment: Policy, Trends and the Problems of Marginalisation* (ACER Press) and of numerous book chapters and journal articles. She is a member of the editorial boards of the *Australian Journal of Environmental Education*, the *Canadian Journal of Environmental Education*, *EIU Journal*, *Environmental Education Research*, the *Eurasian Journal of Science, Mathematics and Technology Education*, the *Journal of Biological Education*, and the *Southern African Journal of Environmental Education*. She is a past president and life fellow of the Australian Association for Environmental Education, a previous vice president of the Science Teachers Association of Victoria and was Victorian Environmental Educator of the Year in 2000. Email: annette.gough@rmit.edu.au

Kathryn Irwin, from the University of Auckland, New Zealand, has investigated the mathematical thinking of students from 4 through 15 and successful teaching methods for improving their mathematical concepts. Her early research was on the concepts of compensation and co-variation in young children. Later work resulted in several publications on older students' understanding of decimal fractions and the use of contexts to overcome common misconceptions. This has been followed by several evaluations of a teaching style that encourages older students to develop mental strategies that help them develop an understanding of how numbers can be decomposed for easy mental calculation, processes that lead to algebraic thinking. With Catherine Vistro-Yu, she has published two papers on students' concepts of linear measurement and effective ways of teaching measurement. Email: k.irwin@auckland.ac.nz

Loyiso Jita obtained his PhD in Curriculum, Teaching and Educational Policy at Michigan State University, USA. He is a senior lecturer in the Department of Curriculum Studies at the University of Pretoria, where he is also a former director of the Joint Centre for Science, Mathematics and Technology education. His research interests are in the areas of science and mathematics education, curriculum reform and instructional leadership. His recent publications on science and mathematics teachers' identities have appeared in the South African journal *Perspectives in Education*. Dr Jita is currently working on several funded research projects on school reform and classroom change in science and mathematics education. Email: loyiso.jita@up.ac.za

Jamilah Binti Hj Mohd Yusof is a senior lecturer and Head of the Department of Science and Mathematics Education, Sultan Hassanal Bolkiah Institute of Education, Universiti Brunei Darussalam. She taught mathematics in secondary and primary schools for 9 years before she joined Universiti Brunei Darussalam in 1989 as a lecturer. She now teaches primary mathematics education courses and has been involved in the education of primary school teachers and in-service education in Brunei. Her interests in mathematics education include creativity in mathematical word problems and mathematical errors in fractions work among primary school pupils. Email: jmy_yusof@yahoo.co.uk

Berinderjeet Kaur is an Associate Professor of Mathematics Education at the National Institute of Education in Singapore. She began her career as a secondary school mathematics teacher. She taught in secondary schools for 8 years before joining the National Institute of Education in 1988. Since then, she has been actively involved in the education of teachers and heads of departments. Her primary research interests are in the area of comparative studies and she has been involved in numerous international studies of mathematics education. As the President of the Association of Mathematics Educators from 2004–2008, she has also been actively involved in the professional development of mathematics teachers in Singapore and is the founding chairperson of the Mathematics Teachers Conferences that started in 2005. Email: berinderjeet.kaur@nie.edu.sg

Phong Lee Koay is an Associate Professor of Mathematics Education at the National Institute of Education in Singapore. She has been involved in the training of mathematics teachers in Malaysia, Brunei Darussalam, and Singapore. She is an author of a series of primary mathematics textbooks used in Singapore. Her interests in mathematics education include the use of investigative approach and technology in teaching elementary mathematics and middle school mathematics. Email: phonglee.koay@nie.edu.sg

Frank Lester is the Chancellor's Professor of Education and Professor of Mathematics Education and Cognitive Science at Indiana University. His primary research interests lie in the areas of mathematical problem solving and metacognition, especially problem-solving instruction. From 1991 to 1996 he was the editor of the *Journal for Research in Mathematics Education*, following a four-year term as editor of that journal's monograph series. He also serves as consulting editor for several other research journals. From 1999–2002, he served on the Board of Directors of the National Council of Teacher of Mathematics. He is editor of the soon-to-be published *Second Handbook of Research in Mathematics Teaching and Learning Education*. He has been on the faculty at Indiana University since 1972. Email: lester@indiana.edu

Prof. Kobus Maree is a professor in the faculty of education (University of Pretoria (UP)). A triple doctorate, he is internationally recognised for his work in e.g. career counselling. His research focuses on optimising the achievement of disadvantaged learners and providing cost-effective career facilitation and maths education to all persons. As the author or co-author of more than 40 books and chapters in books and 90 articles in accrediated scholarly journals, and recipient of numerous awards for his research, he is frequently interviewed on radio and television. He was a finalist in the National Science and Technology Forum awards in 2006, and he received the Exceptional Academic Achiever Award at UP from 2004–2009. Prof. Maree was elected as a member of the South African Academy for Science and Arts in 2003 and elected as a member of the Academy of Science of South Africa (ASSAf) in 2006. He has a C rating from the NRF (he was recently invited to reapply for rating). He is the editor-in-chief of *Perspectives in Education,* consulting editor of *Gifted Education International,* and a member of the Editorial Boards of six more scholarly journals.

Hedy Moscovici is an Associate Professor and Director of the Center for Science Teacher Education at the California State University – Dominguez Hills. Born and raised in Socialist Romania, she received her Bachelors and Masters degrees in Biology/General Science and Parasitology/Microbiology from the Hebrew University in Jerusalem, Israel. She has taught science and mathematics in secondary schools in Jerusalem and earned her PhD in Science Education at Florida State University. Hedy's research interests focus on challenges and dilemmas in the infusion of inquiry science and problem/scenario-based mathematics in urban schools in relation to critical pedagogy and cultural pluralism. In addition, she is also involved in curricular changes and professional development activities in countries that are coming out of communism/socialism (e.g., Romania, Armenia). Email: hmoscovici@csudh.edu

Thembi Ndlalane is a senior lecturer at the University of Pretoria in South Africa. She is a graduate from Leeds University in England specialising in Science Education. She is a former director for the Science Education Project. She has taught science at secondary school level for 20 years. She has been working in collaboration with Hiroshima and Naruto Universities in Japan for six years on the project attempting to improve the quality of teaching science and mathematics in one of the provinces of South Africa. She has just completed her PhD. Her research interest is on teacher networks/clusters and the opportunities provided to teachers to work as peers in improving their pedagogical content knowledge. Email: tndlalane@postino.up.ac.za

Jim Neyland is currently the Director of Postgraduate Programmes in the Faculty of Education, Victoria University, New Zealand. His main research interests are in the philosophy of education, curriculum theory, and mathematics education. He has been a high school teacher, and a pre-service lecturer. He has also played a leadership role in curriculum development in mathematics at the national level, and is on the editorial board for the journal *Curriculum Matters*. He is the editor and co-writer of *Mathematics Education: A Handbook for Teachers* published by the National Council for Teachers of Mathematics. He has lectured in the mathematics department at Victoria University, and now lectures in the school of education studies. Email: jim.neyland@vuw.ac.nz

Daniel Clark Orey is a Senior Fulbright Specialist to Katmandu University, Nepal and recent CNPq Fellow at the Universidade Federal de Ouro Preto, Brasil. Professor Orey is currently the Coordinator and Principle Investigator of the Algorithm Collection Project, Coordinator of the Trilha da Matemática de Ouro Preto and the Coordinator of Luso-Brazilian Studies Group at California State University, Sacramento, where he is Professor of Multicultural and Mathematics Education in the College of Education and an instructor in the Department of Learning Skills. He is the former Director of Professional Development and the Center for Teaching and Learning at California State University, Sacramento. He earned his doctorate in Curriculum and Instruction in Multicultural Education with an emphasis in mathematics and technology education from the University of New Mexico in 1988. His Mellon-Tinker funded field research took him to Highland Maya Guatemala and to Puebla, México. He is a founding board member, and serves as Vice President for North America (1996 – present) and General Secretary (1995) of the Sociedade Internacional para Estudos da Criança. In 1998, he served as a J. William Fulbright Scholar to the Pontifícia Universidade Católica de Campinas in Brazil. This chapter was written while a visiting researcher at the Universidade Federal de Ouro Preto with support from CNPq (Conselho Nacional de Desenvolvimento Científico e Tecnológico). Contact: http://www.csus.edu/indiv/o/oreyd/ http://www.csus.edu/indiv/o/oreyd/

Peter Petocz is an Associate Professor in the Department of Statistics at Macquarie University, Sydney. As well as his work as a professional statistician, he has a long-standing interest in mathematics and statistics pedagogy, both in practical terms and as a research field. He is the author of a range of learning materials, (textbooks, video packages and computer-based materials) and has been recently recognised with a national teaching award. In collaboration with Anna Reid, he has undertaken joint research over a period of several years in topics including music, sustainability, statistics and mathematics. Through the intersection of qualitative and quantitative research paradigms to explore learning and teaching in higher education, they bring research strength to their studies. Email: Peter.Petocz@mq.edu.au

Oudone Phanalasy currently works as a lecturer in mathematics at the National University of Laos, with an emphasis in discrete mathematics. He has been an assistant and a lecturer since 1981, following his graduation from the Pedagogical University of Vientiane, Laos with a BSc degree in mathematics. He holds a Graduate Diploma of Science (1996) and a MSc by research (1999), both from

Northern Territory University, Australia. His fields of interest include graph theory and mathematics education. His publications include the mathematics textbooks for the national secondary curriculum, and discrete mathematics texts for the university level. He has served as the government technical counterpart for a variety of national curriculum development and training projects, together with international consultants and international and bilateral agencies, including UNESCO, UNICEF, ADB, and Sida/SAREC (Swedish Government funding). He is currently conducting research in the field of graph theory. Email: ophanalasy@yahoo.com

Anna Reid is an Associate Professor in the Institute for Higher Education Research and Development at Macquarie University, Sydney, where she has a particular responsibility for the integration and enhancement of international perspectives with the curriculum. Her role encompasses research development and research and applications within tertiary learning and teaching environments. Her research focuses on the professional formation of students through their university studies and has been oriented across a range of disciplines such as music, design, law and environmental education. In collaboration with Peter Petocz, she has undertaken joint research over a period of several years in topics including music, sustainability, statistics and mathematics. Through the intersection of qualitative and quantitative research paradigms to explore learning and teaching in higher education, they bring research strength to their studies. Email: Anna.Reid@mq.edu.au

Ferdinand Rivera is an assistant professor of mathematics education in the Department of Mathematics at San José State University, San José California, USA where he teaches mathematics and mathematics education courses. A recipient of an National Science Foundation Career grant, he is currently working in an urban middle school classroom conducting a longitudinal research on students algebraic thinking from 6th to 8th grade. His primary research interests are algebraic thinking at the middle school level and technology in mathematical learning. Also, trained in the cultural studies in education, his other research interest involves postmodern theorizing in relation to mathematics and curriculum theory in general. Email: FerdinandRivera@math.sjsu.edu

Milton Rosa is a mathematics teacher from Brazil who teaches algebra and geometry at Encina High School, in Sacramento, California. From 1988 to 1999, he taught mathematics in public middle, high and technical schools in Amparo, São Paulo, Brazil. In 1999, he was invited to come to California to participate in the international mathematics visiting teacher exchange program sponsored by the California Department of Education. He earned his Masters in Curriculum and Instruction, with an emphasis in mathematics education from California State University in Sacramento. He has written several articles and books in Portuguese, Spanish and English languages. His research fields are ethnomathematics and modeling. He is also interested in the connection between the acquisition of a second language and the acquisition of a mathematical knowledge for immigrant students. Email: milrosa@hotmail.com and MRosa@sanjuan.edu

Joanne Rossi Becker is a Professor of mathematics education in the Department of Mathematics at San José State University, San José, California, USA, where she has taught for over 20 years. At SJSU Joanne teaches courses for prospective elementary and secondary teachers of mathematics, including problem solving courses, methods of teaching mathematics, and supervision of secondary student teachers. For many years she has directed professional development programs for in-service teachers that focus on enhancing teachers' content knowledge in mathematics and developing teacher pedagogical content knowledge through classroom coaching, lesson study, and critical examination of students' performance assessments. Her research interests include gender and mathematics, teacher professional development, and early algebraic thinking. Email: becker@math.sjsu.edu

Florence Mihaela Singer is a senior researcher at the Institute for Educational Sciences, Bucharest, Romania. As head of the Experts Group, head of the Curriculum Component, and then as president of the National Curriculum Council, she was one of the coordinators of the process of designing and implementing the Romanian National Curriculum for grades 1–12. She has published more than 150 articles and books concerning mathematics learning, curriculum development, and human cognition. During the last ten years, she has worked as international education consultant in Romania (within the World Bank education programs), Republic of Moldova, Tajikistan, and the United States. Some of her mathematics textbooks have been translated from Romanian in German, Hungarian, Georgian and Russian. Her ongoing research is focused on the interaction between complexity and abstraction in knowledge building at the higher education level. Email: mikisinger@gmail.com

Juan Carlos Mijangos Noh has a Masters degree in Social Anthropology from the University of Yucatan, México. He received his PhD in Educational Sciences from the University of La Havana, Cuba in 2002. He is a member of the National Research System in Mexico and a member of the American Educational Research Association. Recently, he published a book about popular education and articles related to teacher education in Mexico. He has conducted research about Mayan students' culture in the Yucatan Peninsula and its influence on the educational processes of these students, particularly on the use of Mayan language in elementary schools. He has been professor for the University of Quintana Roo, and the Normal School Rodolfo Menéndez de la Peña. Currently, he is a research Professor at the College of Education of the Universidad Autonoma de Yucatan. Email: jc.mijangos@uady.mx

Michael Singh is a Professor of Education at the University of Western Sydney where he works with colleagues across the University in undertaking research, consultancies and research-based teaching on continuity and change in education within a framework that foregrounds issues of social justice, social, multi-cultural and ecological diversity. Previously, he was Professor of Language and Culture at RMIT University and was Head of RMIT Language and International Studies, which proved to be very innovative and highly successful under his leadership.

At RMIT he worked to establish the Globalism Institute which has now established an outstanding reputation for investigating issues of globalisation and cultural diversity. He is now undertaking a project funded by the Australian Research Council with Fazal Rizvi (University of Illinois) investigating the uses of international education by students from India and China, exploring what this means for reworking the curriculum and pedagogy of Australian education. He has also investigated innovations in the global business of English language teaching with Peter Kell (University of Wollongong) and Ambigapathy Pandian (Universiti Sains Malaysia). Email: m.j.singh@uws.edu.au

Ole Skovsmose has a special interest in critical mathematics education. Recently he has published *Travelling Through Education*, which investigates the notions of mathematics in action, students' foreground, globalisation, ghettoising with particle reference to mathematics education. He is professor at Aalborg University, Department of Education, Learning and Philosophy. He is member of the editorial boards of *Nordic Studies in Mathematics Education*, *Bolema* (a Brazilian journal in Portuguese), *For the Learning of Mathematics*, *Mathematics Education Research Journal*, *African Journal of Mathematics, Science and Technology Education*, *Adults Learning Mathematics Journal*, *Mathematics Education Library* (Springer). Together with Alan Bishop and Thomas Popkewitz he is the editor of *Critical Essays in Education* (Sense Publisher). He has been co-director of The Centre for Research of Learning Mathematics, a co-operative project between Roskilde University Centre, Aalborg University and The Danish University of Education. He has participated in conferences and given lectures about mathematics education in many different countries, including Australia, Austria, Brazil, Canada, Colombia, Germany, Norway, Sweden, USA, England, Hungary, Iceland, South Africa, Greece, Portugal, Spain, and Denmark. Email: osk@learning.aau.dk

Beth Southwell has a long history in teaching undergraduates and graduates in mathematics education at the University of Western Sydney. She has been a consultant to national and state governments and other organizations in curriculum development and a range of other aspects of mathematics education. She also has an extensive experience in working with educational systems in overseas countries. She has been honoured by professional organizations in which she has been active with a Fellowship, three Life Memberships and a Queen's Silver Jubilee Medal. Her research and professional publications are numerous and she has been a consistent and constant presenter at state, national and international conferences on mathematics education. Email: b.southwell@uws.edu.au

Gregory Thomas completed his undergraduate studies in science education at James Cook University of North Queensland and began a high school teaching career in 1988. He taught secondary chemistry, biology and science for 10 years. In 1996 he received a National Excellence in Teaching award in recognition of his exemplary classroom practice, particularly in the area of developing students' cognition and metacognition. He completed a Masters of Educational Studies at Monash University, Australia in 1992 and a PhD at the Queensland University of Technology in 1999. He is currently Professor and Head of the Department of Mathematics, Science, Social Sciences and Technology at the Hong Kong Institute of Education, a Visiting Professor at South China Normal University, and a member of the Hong Kong Research Grants Council Panel for Humanities, Social Science and Business Studies. Email: gpthomas@ied.edu.hk

Paola Valero is an Associate Professor in the Department of Education, Learning and Philosophy, Aalborg University. Her initial background was in Linguistics and Political Science. Since 1990 she has been doing research in the area of mathematics education, with particular emphasis on the political dimension of mathematics teaching and learning, and of mathematics teacher education. Her research integrates sociological and political analysis of mathematics in different institutional settings, and different aspects of mathematical learning and teaching. She has published several papers in books, journals and conferences proceedings. Email: paola@learning.aau.dk

Gary Varrella is an Associate Professor at Washington State University in Extension Education, working in Spokane County as the 4-H Youth Development Educator. He has taught high school science and agriculture. He earned his PhD in Science Education from the University of Iowa. His research relates to 4-H youth development, professional development, the development of teacher expertise, curriculum, and evaluation. Dr. Varrella has worked extensively in the republic of Armenia and Azerbaijan with school and universities conducting teacher enhancement, program and curriculum development, and evaluative activities. He has held positions at universities in California, Iowa, Ohio, Virginia, and Washington State. Gary has worked extensively in the Republics of Armenia and Azerbaijan with schools and universities conducting teacher enhancement, program and curriculum development, and evaluative activities. Email: gvarrella@spokanecounty.org, and gfvarrella@comcast.net

Monica Villarreal is a calculus professor at the Faculty of Agronomy of Cordoba University. She concluded her doctorate in mathematics education at UNESP, Rio Claro. She has supervised Masters students and has directed various research projects in Argentina. She is researcher of the Argentinean National Council of Scientific and Technological Researches (CONICET). She is a consultant of BOLEMA, one of the most important mathematics education journals in Brazil, and of *Revista de Educacion Matematica*, a journal in Argentina. Email: mvilla@agro.uncor.edu

Bing Wei is an assistant professor of science education in the Faculty of Education, University of Macau. Before he moved to Macau in the early 2006 he taught in Guangzhou University, China for more than ten years. He was educated at Beijing Normal University and The University of Hong Kong. He taught chemistry in a secondary school prior to working in universities. His research interests include social contexts of science curriculum, scientific literacy, history and philosophy of science and science teaching, and science teacher development. His recent international publications appear in *Science Education*, *International Journal of Science*

Education, *Research in Science Education*, and *Science Education International*. He is also the author of a book in Chinese on science curriculum development. Email: bingwei@umac.mo

Khoon Yoong Wong is an Associate Professor and Head of the Mathematics and Mathematics Education Academic Group at the National Institute of Education, Nanyang Technological University, Singapore. He was a mathematics teacher in Malaysia and a mathematics educator at Curtin University of Technology, Murdoch University, and Universiti Brunei Darussalam. He has participated in the revision of the national mathematics curriculum in Malaysia, Brunei Darussalam, and Singapore. Now he teaches mathematics education courses and his special interests are in mathematics teacher education and the use of multi-modal strategies and ICT to teach school mathematics. Email: khoonyoong.wong@nie.edu.sg

SECTION 1

THEORETICAL PERSPECTIVES

1

MATHEMATICAL LITERACY AND GLOBALISATION

Ole Skovsmose

Department of Education, Learning and Philosophy, Aalborg University, Fiberstraede 10, DK-9220 Aalborg East, Denmark. E-mail: osk@learning.aau.dk

Abstract: If mathematics and power are interrelated in a globalised world, what does that mean for a mathematical literacy to be either functional or critical? The discussion of this question is organised in three steps:

First, different processes of globalisation are outlined. The thesis of indifference – that mathematics is a pure science without any socio-political or technological significance – is contrasted with the thesis of significance – that mathematics in action can operate in powerful ways, and power can be exercised though mathematics in action

Second, the processes of constructing, operating, consuming and marginalising are analysed. Here mathematics is operating, and mathematical literacy might be either functional or critical: (1) Processes of construction include advanced systems of knowledge and techniques, by means of which technology, in the broadest interpretation of the term, is maintained and further developed. (2) Processes of operation refer to work practices and job functions where mathematics may operate, although without surfacing in the situation. (3) Processes of consuming refer to situations in which one is addressed as a receiver of goods, information, services, obligations, etc. (4) Processes of marginalising turn out to be an aspect of globalisation, governed by a neo-liberal economy, which is far from being inclusive

Third, as conclusion, I get to the aporia, which questions the very distinction: functional-critical. On the one hand, I find this distinction important with respect to mathematical literacy. On the other hand, the distinction is vague, maybe illusive. Being both important and vague-illusive indicates the aporia we have to deal with, with respect to any critical mathematics education

Keywords: Mathematical literacy, globalisation, ghettoising, uncertainity

'Mathematical literacy' is far from being a well-defined term. Theconcept can be related to notions like empowerment, autonomy and 'learning for democracy'.[1] Talking about empowerment also brings us to talk about disempowerment, and

[1] See, for instance, Jablonka (2003).

B. Atweh et al. (eds.), Internationalisation and Globalisation in Mathematics and Science Education, 3–18.

one could consider to what extent 'mathematical literacy' could connote, say, 'regimentation' and 'indoctrination'.

Michael Apple (1992) has distinguished between two types of literacy being either 'functional' or 'critical'.[2] One could see functional literacy as first of all defined through competencies that a person might have in order to fulfil a particular job function. Working conditions and political issues are not challenged through a functional literacy, while a critical literacy addresses exactly such themes. Such a literacy is also included in what Paulo Freire has referred to as a 'conscientizaçao': a deeper reading of the world as being open to change.

A critical mathematical literacy includes a capacity to read a given situation, including its expression in numbers, as being open to change. Reading the world drawing on mathematical resources means, according to Eric Gutstein (2003), to use mathematics to 'understand relations of power, resource inequities, and disparate opportunities between different social groups and to understand explicit discrimination based on race, class, gender, language, and other differences. Further, it means to dissect and deconstruct media and other forms of representation and to use mathematics to examine these various phenomena in one's immediate life and in the broader social world and to identify relationships and make connections between them' (Gutstein, 2003, p. 45).

Notions like 'functional' and 'critical' might, however, assume very different meanings depending on what context we are considering. What could they mean with respect to 15 year old students in a provincial town in Denmark? To immigrant students in Denmark? To students from a Mexican minority community in a USA metropolis? To students from an Indian community in Brazil? To students from Palestine? To students living in a war zone? To students from an impoverished province of India, just discovered by an international company as a site for production of electronic equipment? And what does it mean for students living in a well-off neighbourhood? One could also think of students in elementary education, or of university students, or of people who do not have the opportunity to go to school. The distinction functional-critical could have very different interpretations depending on the context of the learner. Furthermore, even with reference to a particular practice, it might be difficult to point out what observations and what phenomena signify that we are dealing with either a critical or a functional learning.

As a consequence, we should not have great expectations about reaching a conceptual clarification with respect to the functional-critical distinction. Nevertheless, I want to address the following question: If mathematics and power are interrelated in a globalised world, what does that mean for a mathematical literacy to be either functional or critical? The discussion of this question will be organised in three steps: First, I will make some comments on globalisation and on the power-dimension of mathematics (Sects. 1 and 2). Second, I will refer to four groups of

[2] Instead of 'critical', I have previously talked about 'reflective' knowledge with respect to mathematics. This refers to a competence in evaluating how mathematics is used or could be used. Reflections could address both simple and complex uses of mathematics. See Skovsmose (1994).

people, constructors, operators, consumers and 'disposables' with respect to whom a mathematical literacy might be either functional or critical (Sects. 3–6). Third, as a conclusion, I arrive at an aporia which accompanies critical mathematics education and questions the very distinction 'functional-critical' (Sect. 7).

1. Setting the Scene: Globalisation

Globalisation refers to processes that have been defined and elaborated in very many different ways.[3] Let me, however, limit myself to the following six points.

First, it is generally agreed that processes of globalisation are facilitated by *information and communication technologies*. In theorising technology,[4] a principal issue is to what extent social development is determined by a technological development; but with respect to globalisation, technological impacts seem to be taken for granted. Manuel Castells (1996, 1997 ,1998) has carefully analysed the 'informational age' and the 'network society'. And it appears that the very networking to a large extent is constructed, not by stone and bricks, but by 'packages': those electronic units, easy to install, which establish new procedures, routines and forms of communication.

Second, it appears that globalisation is betrothed with a *free-growing capitalism*. Thus, Beck (2000) talks about a 'disorganised capitalism', which could sound misleading, as 'disorganised' might indicate a lack of power and efficiency. But if being 'disorganised' indicates that the growth of capitalism is operating through a new more powerful dynamic and that it is getting out of control (if not getting in control), then the word is well-chosen. Globalisation refers to an opening up of new markets.

Third, the processes of globalisation do not follow any simple predictable route. Determinism assumes the existence of some patterns of social development. I find, however, that processes of social development exceed in complexity what any 'logic' might be able to grasp. In particular, I find that processes of globalisation include so many interrelated factors that any possible pattern gets lost in complexities. This idea is also included in the notions of 'risk society' and 'world risk society' as developed by Beck (1992, 1999). Elsewhere, I have talked about social development as 'happenings', emphasising that the capacity to grasp what is taking place, is not granted to the people participating in the situation.[5] In this sense I find indeterminism a basic challenge to any social theorising addressing processes of globalisation.

Fourth, globalisation includes *distribution and redistribution of 'goods' and 'bads'*. The liberal aspect of the globalised economy can be illustrated by the movements of supply chains, i.e. the chains leading from raw material to the final commodity. The direction of a supply chain can, nowadays, be changed according

[3] See, for instance, Bauman (1998), Beck (2000), and Hardt and Negri (2004).
[4] See, for instance, Ihde (1993).
[5] See Skovsmose (2005b).

to emerging priorities. It can be taken as a given that a 'company belongs to the people who invest in it – not to its employees, suppliers, not the locality in which it is situated' (Albert J. Dunlap, quoted after Bauman, 1998, p. 6). The meaning of this statement is clear: A company is a freely moving entity, and to big companies borders are no restriction. The demands for profit could imply that production becomes located in areas where cheap labour is available, and 'cheap labour' not only refers to the level of salary, but also to the level of security measures to be taken. The permanent possibility of moving the company, the production and the capital is a defining element of a globalised capitalism.[6] Goods are produced and distributed on a global scale, and the production of goods is accompanied by a production of 'bads', that might be in the form of pollution and damage to the environment or to the people involved in the production.

Fifth, poverty accompanies free-growing capitalism, and globalisation turns into *ghettoising*, which also includes huge areas of Europe, USA and parts of their biggest metropolis. Ghettoised people are immobilised people. As Bauman emphasises: 'Ghettoes and prisons are two varieties of the strategies for "tying the undesirable to the ground" of confinement and immobilization' (Bauman, 2001, p. 120). In case we consider ghettos as a reservoir for extra labour force, the erection of the 'modern' hyperghetto seems irrational.[7] This ghetto does not serve as any reservoir, and certainly not as a reservoir for consumers who could help to speed up informational capitalism. The hyperghetto, operates as a dumping ground for people who have no role to play in globalised capitalism. Bauman refers to Loïc Wacquant who observes that 'whereas the ghetto in its classic form acted partly as a protective shield against brutal racial exclusion, the hyperghetto has lost its positive role of collective buffer, making it a deadly machinery for naked social relegation' (see Bauman, 2001, p. 122). Some of the immense favelas rapidly growing around cities like São Paulo and Rio de Janeiro might serve as illustrations. The film *Cidade de Deus* (*City of God*) might give an impression of what 'naked social relegation' could mean.

Sixth, globalisation could be *armed*. While the First and Second World Wars were between two, more or less equally strong powers, the wars of today, as in the time of colonisation, are between incongruent enemies. Armed globalisation tries to control minorities, located at strategic positions, close to oil pipelines for instance. Regions without any apparent strategic significance can, however, be ignored. Thus, the genocides that took place in Rwanda and Sudan were, from the perspective of a free-growing capitalism, without significance.

Certainly other processes of globalisation could be enumerated, but let me raise a different question: How could one judge such processes? Let me refer to just two alternatives. The first position, *globalism*, celebrates the new worldwide market,

[6] Different techniques, like ranking of 'risk countries', facilitate companies' judgments of where to locate different supply chains, and where to allocate investments.

[7] Contrasting the hyperghetto one could think of the 'classic' ghetto as exemplified by the Jewish communities that maintained a cultural homogeneity that served as a protection against an often hostile environment.

which seems to become established through globalisation. It assumes that the free market can solve social problems, and, consequently, a free market, expanded to global dimensions, represents the definitive problem solver. Globalism embraces neo-liberalism. One could, however, make a particular point of relating other concerns to a programme of globalisation. This brings us to a second alternative: One could try to *think globally* in being concerned about justice and equality on a global scale. My own perspective is the latter one. (And this perspective, I assume, has already become reflected in the way I have briefly highlighted six aspects of globalisation.)

2. Mathematics and Power

In these initial comments about globalisation, I made no explicit mention of knowledge or of mathematics. But the knowledge-power issue can easily surface. Daniel Bell (1980) has suggested that knowledge and information are strategic resources. As capital and labour previously have been basic to any theory of value, so knowledge and information are ready to take over this position.[8] Such formulations provide a quantitative way of relating power and knowledge, while Michel Foucault (1977, 1989, 1994) has related power and knowledge through qualitative analyses.

How then to see the relationship between a particular form of knowledge, namely mathematics, and power? One could negate the existence of such relationships and claim the *thesis of indifference*. Thus, a knowledge theory of value has not been specific about mathematics.[9] And related to this, social theorising, as expressed in more general considerations about the emergence of the network society and processes of globalisation, does not make any substantial remarks about mathematics. [10] A related expression of the thesis of indifference may follow from Foucault's studies. He scrutinised the relationship between power and knowledge through his studies of madness, punishment, jails, schooling, and sexuality. None of these accounts, however, tried to uncover the impact of the content of the scientific revolution and of how mathematics, as a technology of power, influences (if not co-fabricates) technological and socio-political development. I realise that Foucault could not be expected to investigate each and any relevant topic, so I do not fault him for this omission. But I find it problematic if research priorities applied by Foucault become paradigmatic for what to address in every knowledge-power analysis. In particular, I find it problematic if one does not find it relevant to analyse mathematics from a knowledge-power perspective. Finally, we should

[8] So the function of production, Q, previously defined as a function of two variables $Q = Q(C, L)$, where Q denotes output, Ccapital input, and L labour input, may take the form $Q = Q(C, L, S)$, where S refers to communication and/or business services. See, for instance, Tomlinson (2001).

[9] Thus, the notions of knowledge and information are not further analysed by Bell (1980). In fact both notions operate as 'dummies' in his theory of value. So do they in Castells (1996, 1997, 1998).

[10] See, for instance, Bauman (1998); Beck (2000); Castells (1996, 1997, 1998); Archibugi and Lundvall (Eds.) (2001), and Hardt and Negri (2004).

not forget that the thesis of indifference has received eager support from within the mathematical research community; for instance a programmatic statement was made by G. H. Hardy (1967) about the pureness of pure mathematics: Mathematics has never had any social impact!

I find the thesis of indifference problematic, and I have argued in favour of the *thesis of significance*: mathematics interacts with power, and this interaction has a political, technological and economic significance.[11]Let me recapitulate just three points of this argumentation. First, I relate mathematics to action. Previously, mathematics has been thought of as the language of science, and this idea was accompanied by a conception of language as a descriptive tool. If we, instead, consider language from the perspective of speech act theory and discourse theory, we capture the point that language 'forms the world', and 'forms actions in the world'. This inspired me to consider how mathematics in action provides ways of seeing, doing, organising, constructing, processing, deciding, etc. Second, when I talk about mathematics as interesting from the knowledge-power perspective, I have a broad concept of mathematics in mind. My notion of mathematics is not limited to mathematics curriculum at any level. I see mathematics as also including all forms of techniques operating in technological enterprises, in engineering, in economics, in banking. I would even doubt that one could hope to find simple unifying characteristics of mathematics. This concept-stretching – and I am well aware of this – facilitates my arguments against the thesis of insignificance. Third, I see the mathematics-power relationship illustrated though several more detailed considerations about mathematics in action.[12]

The thesis of significance suggests that mathematics in action can operate in powerful ways, and power can be exercised through mathematics in action. This thesis brings a particular significance to the discussion of mathematical literacy. The functional-critical distinction refers to two different ways of addressing a mathematics-power interaction. I will try to address this interaction with respect to a simplified grouping referring to constructors, operators, consumers and 'disposables', who might come to deal with the mathematics-power relationship in different ways. Naturally such a grouping represents a gross analytical simplification. Nevertheless, it makes it possible for me to address the content of a mathematical literacy in a more specific way.

3. Constructors

Processes of construction include advanced systems of knowledge and techniques, by means of which technology, in the broadest interpretation of the term, is maintained and further developed.[13] It is the task of universities and other institutions of further education to provide competences relevant for *constructors*, and any

[11] See, for instance, Skovsmose (1994) for a discussion of the formatting power of mathematics, and Skovsmose (2005b) for a discussion of the 'apparatus of reason' and of 'mathematics in action'.

[12] See, for instance, Skovsmose (2005b).

education of engineers, economists, computer scientists, pharmacists, etc. includes mathematics.

One example of how mathematics is part of the processes of construction is the very construction of the computer. Thus, the principles of the functioning of a computer, including its limits for computation, were grasped even before the construction of the first computer. The computer itself, both hardware and software, includes materialised forms of mathematical algorithms. And new inventions related to computers draw on mathematics. One example is found in the development of cryptography. This is an old technique, and an ongoing refinement has taken place, although within the frame of tradition. Thus, a particular selected book (one copy kept by the sender and one copy kept by the receiver) or an elaborated mechanical procedure, such as the Enigma-machinery used by the Germans during the Second World War, was used for coding and decoding. A radical new approach was only discovered through a rethinking of certain properties of mathematical functions, the so called trap-door function the inverse of which is difficult to determine, and the observation that no algorithm was likely to be identified for the factorisation of a number n, in the case that n was a multiple of two large prime numbers of, say, about 50 digits each. In fact, the time needed for identifying the factorisation of such a number n, using the computer facilities now available, is millions of years. A surprising observation, considering that it only takes about two lines of the size of the present book to write a number of 100 digits. The whole new insight in mathematics-based cryptography, then, was condensed into packages, which could turn into goods for sale, and be installed in each and every computer. Cryptography is essential in the globalised economy (not to mention warfare).[14]

The emergence of the new cryptography illustrates the more general observation: mathematics provides a form of technological freedom by establishing a 'technological imagination'. Numerous technological devices could not have been identified and produced without the use of a sophisticated mathematics-based technological imagination. The technical feasibility of e-mailing and electronic networking, fundamental to globalisation, could never be conceptualised through common sense.

Mathematics provides the possibilities for 'hypothetical reasoning', which refers to analysing the consequences of an imaginary scenario. By means of mathematics we are able to investigate particular details of a not-yet-realised design. Thus, mathematics constitutes an important instrument for carrying out detailed thought experiments. However, mathematics also places severe limitations on hypothetical reasoning, as any technological design has implications not identified by hypothetical reasoning (and which may not be possible at all to identify by such reasoning). As a consequence, some of the implications of a realised

[13] By technology I include not only its 'machinery' but also the organisation, the know how and the procedures for design and decision making.

[14] See Skovsmose and Yasukawa (2004).

design might be very different from the calculated implications of the mathematically described hypothetical situation. Nevertheless, hypothetical reasoning is one important element in the whole process of construction. One tries to see what a construction might include before it is ever constructed. But mathematics-based hypothetical reasoning may overlook even principal consequences of technological initiatives.

We are not used to thinking of mathematicians and highly qualified technicians as being in need of empowerment. They do not seem short of mathematical literacy. However, one can still consider the distinction between functional and critical mathematical literacy: What does this distinction mean with respect to constructors?

A worrying element in the history of science is the linking of science with ethically questionable projects. One could think of the involvement of science in the Nazi war machinery. This could not function without a titanic scientific involvement, including all kinds of scientific activities such as the investigation of differential equations in order to predict the range of artillery and trajectory rockets.[15] One factor facilitating an ethics-blind and 'functional' application of science and mathematics has to do with the *disaggregation of tasks*. Such a disaggregation is common in processes of construction: one complex task becomes split up into very many, still very challenging tasks. These tasks are then positioned within particular research programmes. In this way they take on a 'purified' significance, or by taking over a classic significance having to do with, for instance, the solutions of partial differential equations. Problems related to cryptography easily become decomposed into a variety of theoretical tasks. Disaggregation means a relocation of tasks into alternative discourses that might filter away socio-political and ethical concerns.

Processes of disaggregation and relocation might eliminate reflective and critical considerations. At universities and technical schools, one can observe a strong tendency to eliminate critical issues through a disaggregation of the curriculum into units to be taught and learnt and tested separately. One could learn mathematical techniques in one course, and apply them in different courses to address problems with technological and economic implications that never are elucidated. Here we experience a 'blind' feeding of mathematics into the processes of construction. To me this exemplifies what functionality could mean with respect to advanced mathematical literacy.

One could think of alternatives: How to organise an education for future constructors in such a way that reflective elements are included in their (mathematics) education? The approach of project work in university mathematics education can be seen as one attempt to prevent from disaggregation in educational matters, and instead to provide a more holistic approach, which makes reflection possible. [16] I do not claim that such project work has proved successful, but it illus-

[15] See Mehrtens (1993).

[16] See Vithal, Christiansen and Skovsmose (1995).

trates what it could mean to be concerned with critical and not just with functional elements in an advanced mathematical literacy for construction.

4. Operators

Bringing technology into operation in work practices and job functions includes many different elements, one being mathematics. However, mathematics might not surface in the situation. The *operator* might not be aware of the mathematical content of the procedures he or she performs.

Tine Wedege (2002) observed how the person responsible for loading an airplane has to take into account how well balanced the plane is before take-off. The luggage has to be loaded into different compartments in such a way that the balancing factor remains within a certain interval. The actual calculation of this factor is done by a computer according to algorithms that only the engineer may know. However, the person responsible for loading has to provide inputs to the computer, and it has to be judged what could be done in case the balancing factor appears too close to the upper or lower security limits. A reloading could be needed. Such an alternative could, however, cause other problems, for instance, the take-off could be delayed. It could also be considered if the remaining luggage could be loaded and stored in such a way that the balancing factor does not conflict with security limits. The person in change of the loading is operating within a system, constructed on the basis of a deeper insight into airplane stability. This insight, however, becomes operational through the construction of a decision-making system.

This is one particular example of mathematics in operation. There are many other mathematics-based systems brought into operation in all kinds of job functions: ticket reservations in the travel industry; procedures for buying and selling houses; banking or any kind of financial operation, etc. Insurance companies would nowadays be unable to operate without having access to adequate systems and programmes. Taxi driving has been reorganised through computer based systems that make it possible to minimise the distance driven by a car with no passengers, while navigation systems point out shortest routes when necessary. Modern farming is dependent on systems for monitoring the feeding and growth of animals. In medicine a huge variety of systems and equipment are highly computerised, and mathematics for medicine has emerged as a university study. One should not forget that modern warfare is now a computerised operation. Mathematics is part of very many processes of operation. [17]

In all such different areas routines can be established. Thus, a model for airline booking provides a grand scale of 'routinisation'. This is simply one of the basic reasons for the success of a booking model and other similar schemes of management. Furthermore, mathematically based actions could provide an 'authorisation'. It is possible to refer to some calculations (which 'obviously' cannot be different) for carrying out certain tasks or for justifying some decisions. However,

[17] See, for instance, FitzSimons (2002).

sometimes an 'authorisation' with references to numbers could mean a pseudo-authorisation, where reference to numbers only serves to obscure as 'objective' a decision based on other factors.

How could one look at mathematical literacy with respect to an operator? What would the 'functional-critical' distinction refer to in this context? Let us consider adult mathematics education. One could try to provide adults with knowledge and techniques that make it easier for them to enter the labour market. Or one could enhance the proficiency of employees with respect to certain job functions. This seems straightforward when we consider the interests of the company which has employed the people in question. If an insurance company should fund the costs of any further education for their staff, the education should be directed towards well-specified aims related to the functionality of the staff, such as being able to operate with a new system of providing instant offers, based on inputs that the consumer could provide on the spot. But what could it mean to be critical in this case? To take up a critical position with respect to how the system is operating? Could the system provide an offer which, at a closer look, would be insufficient considering other information about the customer, not included as numbered inputs to the system? Or would being critical mean to give extra thought about what it means to bring such systems into operation? Thus, the introduction of this new system might be the first step in establishing a do-it-yourself system, which is intended to make a reduction in the size of an insurance company's staff possible. Or could 'being critical' also mean to consider what it could mean to operate within a system, which in the end could be operated by a staff in, say, India, where low-paid highly qualified operators could be hired? Are such considerations part of a critical mathematical literacy for operators?

One characteristic of the school mathematics tradition is the overwhelming number of exercises which the students have to do. During primary and secondary school, the total number of exercises easily exceeds 10,000. This extreme preoccupation with exercises might, at fist glance, seem pathological. However, a readiness to follow orders and do so in a careful way, could be a 'functional' quality for being an operator. From the perspectives of 'adaptability' and 'functionality', the mathematics offered in their secondary or tertiary education might appear to be adequate.

But what, then, could a critical mathematical literacy mean with respect to operators? Here I consider again the notion of routinisation and authorisation. Routines can be built on numbers, and the routine of loading an airplane is just one example. One point here concerns the readiness to consider the reliability of certain numbers. How accurately do they represent a situation? Could they include some miscalculations or misjudgements? An authorisation could include a decision and a justification of this based on numbers. Here the issue of responsibility enters: Would it be possible to base a decision on possible implications of these numbers, which might be reliable, more or less? A concern for developing students' awareness of issues of reliability and responsibility could be one suggestion for what a critical mathematical literacy for operators might include.[18]

5. Consumers

Statements from experts are expressed each and every day on television and in the newspapers. Numbers and figures concerning elections, the economy, exchange rates, casualties, and investments are mixed with advertising of any number of 'special offers'. Somebody must be listening to all this. I will call these people *consumers*. I use 'consumers' in a broad and slightly ironic interpretation; the expression 'citizen' might be more appropriate. However, as citizens we are in many situations constituted as consumers, and it is this role I want to discuss here.

Most directly we are addressed as possible consumers, when all kinds of offers are presented to us. And while products have increased in variety, prices have increased in complexity. A product need not be anything tangible, but could be a service in terms of, say, an insurance offer. And prices turn into a complexity of conditions for payment including rents and terms. In other situations we are constituted as consumers, but in a broader sense: We look at news, we receive information, ideas, priorities, 'life styles', opinions, entertainment; we listen to opinions, arguments, justifications, questionable legitimations, and decisions. All such things also have to be consumed by somebody.

A report published by Teknologirådet (1995) discusses the increasing use of computer-based models in political decision-making. The report refers to 60 models, covering areas such as economics, environment, traffic, fishing, defence, and population. It emphasises that this extended use of mathematical models may erode conditions for democratic life: Who constructs the models? What aspects of reality are included in the models? Who has access to the models? Who is able to control the models? If such questions are not adequately clarified, traditional democratic values may be hampered. The report emphasises, in particular, that models related to traffic and environmental issues, such as the construction of a bridge, are often used in support of decisions which cannot be changed. In several cases it appears that models are used in order to legitimate *de facto* decisions, as a model-construction may provide numbers and figures that justify already made decisions. So, mathematics operates in the space between establishing justification and dubious forms of legitimation of decisions and actions. As consumers or as citizens we are constantly facing justifications and legitimisations for decisions based on complex models.

What could a mathematical literacy mean with respect to all such forms of consuming? What does it mean for this literacy to be functional? Most directly it could mean that one is able to read all such information. One could consider citizenship from a 'receiving' or consuming perspective. A citizen should be able

[18] A detailed discussion of reliability and responsibility in relation to classroom practice is discussed in Alrø and Skovsmose (2002).

to receive information from 'authorities'. If the citizen were not able to read information put into numbers, then society would not be able to operate. There is much discussion as to what functionality could mean in this aspect. Consumers' mathematics has been developed from a highly pragmatic perspective. This pragmatism has dominated many textbooks with elaborated examples of mathematics in daily-life situations. Much effort has been devoted to ensuring a functional mathematical literacy through mathematics education.

One could, however, also consider citizenship from a different standpoint. As a citizen one should not only be able to receive from authorities as a functional receptive consumer, but also to 'talk back' to authorities.[19] This brings us to the notion of critical citizenship. Could a critical citizenship be supported by the development of critical mathematical literacy? And what could such literacy mean in this context? Here there are no lack of attempts and examples (see, for instance, Frankenstein, 1989, 1998; Gutstein, 2003), which give meaning to the critical dimension of a mathematical literacy.

6. 'Disposables'

Processes of globalisation are brutal, and some groups of people seem not to be necessary for these processes. A steady growth of favela-like neighbourhoods gloomily testifies that free-growing globalised capitalism is not an inclusive economy. Instead it marginalises in great measures people as being '*disposables*'.

People from a favela will see and know about the nearby affluence, although far out of reach. Things, however, can be stolen, and many other 'perverted' forms of economic connections are established between marginalised groups and globalised capitalism, such as drug dealing. Nor should we forget the less profitable business of selling sunglasses, lighters and other items possible to carry around along the streets where cars come to a stop, or at least slow down, being trapped by traffic jams. A different form of relationship is exemplified by many groups of Brazilian Indians, who maintain traditions of their own, and who do not define themselves with reference to what they miss in a relationship with the globalised world. But certainly they feel threatened, as their environments turn into sites for exploitation.

From the perspective of globalised capitalism, both groups, however, might appear to be peripheral. Marginalisation seems consequential, and new forms of apartheid emerge,[20] not strictly related to racial categories, although often with a strong correlation to racial factors. (Thus, people living in Brazilian favela maintain a high overrepresentation of blacks and coloured people compared to other communities in Brazil.) The principal point of the new globalised apartheid is to isolate groups that neither provide potential markets for the globalised economy, nor provide resources for production, but who could turn into a disturbing factor. So, globalisation keeps processes of ghettoising in full swing.

---

[19] See, for instance, Nowotny, Scott and Gibbons (2001).

[20] See Hardt (2004) for comments about the new apartheid.

What could mathematical literacy mean in this case? One could first remark that from a neo-liberal perspective, there is no real incentive to invest in education of the 'disposables'. When decisions are based on input-output considerations, such education might appear irrelevant. The situation looks different, however, if we consider education as a human right. What then could a mathematics education contain? And what would the difference between a functional and a critical mathematical literacy mean in this situation?

With reference to the problem of marginalisation, it is important to acknowledge people's right to have the opportunity to operate functionally. Here one could raise a critical issue with respect to several educational practices inspired by ethnomathematical programmes. There, mathematics educators have paid special attention to the students' backgrounds in order to establish a learning which is sensitive to their cultural roots. And, certainly, I find this important. However, I also find it important to consider students' *foregrounds*, and this means considering what might be their aspirations and hopes in life. It might well turn out that some of these aspirations are better served by a mathematics education which brings the students into an adequate position for further education, while a strict ethnomathematical approach might limit some of their possibilities.[21]

This brings me back to the issue of education of constructors. My concern about the critical aspect of mathematical literacy does not exclude the need for functionality. My point is not to outline being functional and being critical as two mutually exclusive qualities. An expertise presupposes a functionality, but this does not reduce the need for a critical dimension of any competence. I do not see any point in contrasting functional and critical aspects of a mathematical literacy referring to those becoming constructors or to the situation of people who might tend to be marginalised through processes of globalisation.

Often it has been emphasised that functional competencies have to be developed before any critique could become established. How can one reflect on ethical issues of, say, nanotechnology if one has no idea about the matter? For me this is a most problematic assumption. I do not operate with any assumption about any order with respect to functionality and critique. This applies to expertise. There is no point in postponing reflective elements in university education to the last year of the study. The same is the case with respect to mathematical literacy of potentially marginalised groups of people. This point is clearly illustrated by Freire's approach: learning to read and write can be strongly facilitated by providing scope for critical reflection.[22] This brings about the insight that the development of functional and critical competencies need not be separated in different educational processes, when we consider mathematical literacy.

---

[21] For a further discussion of foreground, see Skovsmose (2005a).

[22] See also Powell (2002).

7. Uncertainty

Mathematics is part of the processes of construction, operation, consuming and marginalisation. In such processes one can experience a mathematics-power interaction. This means that it becomes important to consider what it could mean for a mathematical literacy to be functional or critical (or both), and that the analysis of this issue could bring about different insights which respect to the different groupings one might consider.

As already mentioned, one could have very different positions with respect to the processes of globalisation. By affirming globalism, one could embrace a neo-liberal perspective and see globalisation as an important step in expanding the free market, which provides a sound basis for solving social problems of all kinds. One could, however, be concerned about this liberalism being included in globalisation. One could try to think globally and consider to what extent justice and equality could come to make up part of the processes of globalisation.

Similar alternative positions can also be related to mathematics in action. On the one hand, one could consider mathematics in action as representing an ultimately valuable thing and celebrate the free usage of mathematics. In mathematics education research, this neo-liberal movement is voiced by the claim that, as a mathematics educator, one must serve as an ambassador of mathematics. Any illumination through mathematics is necessarily healthy and sound. According to this liberalism, a functional mathematical literacy is sufficient. There is no need to invent a distinction between being functional and critical with respect to mathematical literacy.

On the other hand, one could also find that mathematics in action cannot simply be considered an ultimate good, as the processes of technological imagination, hypothetical reasoning, routinisation and authorisation may include both attractive and problematic features. This position does not mean the abolition of mathematics-based technologies, but it represents an uncertainty with respect to how mathematics in action might operate. I consider mathematics in action, similar to any action, as being in need of critical evaluation. Such a position makes it important for mathematics education to become critical also with respect to the content and socio-political functions of mathematics.

However, we have to face one more profound uncertainty. As indicated in the introduction, it is not an easy task to point out what the distinction between functional and critical could mean in a given mathematics educational context. The distinction is not placed on any firm ground.

In several contexts, I have referred to the aporia which is included in any attempt to develop a critical mathematics education.[23] The aporia represents a basic dilemma with respect to some notions and distinctions. In general, an aporia represents a situation where rationality seems in danger of committing suicide. In this context, I see a dilemma with respect to the distinction between functional and critical. On the

[23] See, for instance, Skovsmose (2005b)

one hand, I find this distinction important. It refers to mathematical literacy, which could be either functional or critical. It refers to processes of construction, operating, consuming, marginalising which also could be addressed functionally or critically. On the other hand, the distinction is difficult to maintain. It is vague, maybe illusive. When one tries to analytically grasp it and address educational practices, it turns out to be difficult to handle. Being both important and vague/illusive indicates the aporia we have to deal with, with respect to mathematics in action and mathematical literacy.

Acknowledgements

I want to thank Miriam Godoy Penteado for critical comments and suggestions for improving preliminary versions of this paper, and Gail FitzSimons for completing a careful language revision. The paper results from the research project 'Learning from diversity', funded by The Danish Research Council for Humanities and Aalborg University. The paper represents a further development of my paper 'Ghettoising and Globalisation: A Challenge for Mathematics Education, *Proceedings for XI Inter-American Conference on Mathematics Education*, 13–17 July, Blumenau, Brazil (on disk).

References

Alrø, H., & Skovsmose, O. (2002). *Dialogue and learning in mathematics education: Intention, reflection, critique*. Dordrecht: Kluwer.

Archibugi, D., & Lundvall, B. -Å. (Eds.) (2001). *The globalizing learning economy*. Oxford: Oxford University Press.

Apple, M. (1992). Do the standards go far enough? Power, policy and practice in mathematics Education. *Journal for Research in Mathematics Education*, *23*, 412–431.

Bauman, Z. (1998). *Globalization: The human consequences*. Cambridge: Polity Press.

Bauman, Z. (2001). *Community: Seeking safety in an insecure world*. Cambridge: Polity Press.

Beck, U. (1992). *Risk society: Towards a new modernity*. London: SAGE Publications.

Beck, U. (1999): *World risk society*. Cambridge: Polity Press.

Beck, U. (2000). *What is globalization?* Cambridge: Polity Press.

Bell, D. (1980). The social framework of the information society. In T. Forrester (Ed.), *The microelectronics revolution* (pp. 500–549). Oxford: Blackwell.

Castells, M. (1996). *The information age: Economy, society and culture. volume I: The rise of the network society*. Oxford: Blackwell Publishers.

Castells, M. (1997). *The information age: Economy, society and culture. volume II, The power of identity*. Oxford: Blackwell Publishers.

Castells, M. (1998). *The information age: Economy, society and culture. volume III, End of millennium*. Oxford: Blackwell Publishers.

FitzSimons, G. E. (2002). *What counts as mathematics? Technologies of power in adult and vocational education*. Dordrecht: Kluwer.

Foucault, M. (1977). *Discipline and punish: The birth of the prison*. Harmondsworth: Penguin Books. (First French edition 1975.)

Foucault, M. (1989). *The archeology of knowledge*. London: Routledge. (First French edition 1969.)

Foucault, M. (1994). *The order of things: An archaeology of the human sciences*. New York: Vintage Books. (First French edition 1966.)

Frankenstein, M. (1989). *Relearning mathematics: A different third R – Radical Maths*. London: Free Association Books.

Frankenstein, M. (1998). Reading the world with maths: Goals for a critical mathematical literacy curriculum. In P. Gates (Ed.), *Proceedings of the first international mathematics education and society conference* (pp. 180–189). Nottingham: Centre for the study of Mathematics Education, Nottingham University.

Gutstein, E. (2003). Teaching and learning mathematics for social Justice in an urban, latino school. *Journal for Research in Mathematics Education, 34*(1), 37–73.

Hardt, M., & Negri, A. (2004). *Multitude*. New York: The Penguin Press.

Hardy, G. H. (1967). *A mathematician's apology*. With a Foreword by C. P. Snow. Cambridge: Cambridge University Press. (1st edition 1940.)

Ihde, D. (1993). *Philosophy of technology: An introduction*. New York: Paragon House Publishers.

Jablonka, E. (2003). Mathematical literacy. In A. J. Bishop, M. A. Clements, C. Keitel, J. Kilpatrick & F. K. S. Leung (Eds.), *Second international handbook of mathematics education* (pp. 75–102). Dordrecht: Kluwer.

Mehrtens, H. (1993). The social system of mathematics and national socialism: A survey. In S. Restivo, J. P. van Bendegem & R. Fisher, R. (Eds.), *Math worlds: Philosophical and social studies of mathematics and mathematics education* (pp. 219–246). Albany: State University of New York Press.

Nowotny, H., Scott, P., & Gibbons, M. (2001). *Re-Thinking science: Knowledge and the public in an age of uncertainty*. Cambridge: Polity Press.

Powell, A. (2002): Ethnomathematics and the challenges of racism in mathematics education. In P. Valero & O. Skovsmose (Eds.), *Proceedings of the third international mathematics education and society conference* (pp. 15–28). Copenhagen, Roskilde and Aalborg: Centre for Research in Learning Mathematics, Danish University of Education, Roskilde University and Aalborg University.

Skovsmose, O. (1994). *Towards a philosophy of critical mathematical education*. Dordrect: Kluwer.

Skovsmose, O. (2005a). Foregrounds and politics of learning obstacles. *For the Learning of Mathematics, 25*(1), 4–10.

Skovsmose, O. (2005b). *Travelling through education: Uncertainty, mathematics, responsibility*. Rotterdam: Sense Publishers.

Skovsmose, O., & Yasukawa, K. (2004). Formatting power of 'mathematics in a package': A challenge for social theorising? *Philosophy of Mathematics Education Journal*. (http://www.ex.ac.uk/~PErnest/pome18/contents.htm)

Teknologirådet (1995): *Magt og modeller: Om den stigende anvendelse af edb-modeller i de politiske beslutninger*. Copenhagen: Teknologirådet.

Tomlinson, M. (2001). New roles for business services in economic growth. In D. Archibugi & B. -Å. Lundvall (Eds.), *The globalizing learning economy* (pp. 97–107). Oxford: Oxford University Press.

Vithal, R., Christiansen, I. M., & Skovsmose, O. (1995). Project work in univeristy mathematics education: A danish experience: Aalborg university. *Educational Studies in Mathematics, 29*(2), 199–223.

Wedege, T. (2002). Numeracy as a basic qualification in semi-skilled jobs. *For the Learning of Mathematics, 22*(3), 23–28.

2

EPISTEMOLOGICAL ISSUES IN THE INTERNATIONALIZATION AND GLOBALIZATION OF MATHEMATICS EDUCATION

Paul Ernest
University of Exeter, UK p.ernest@ex.ac.uk

Abstract: This chapter discusses some of the universalizing forces at work in the globalization and internationalization of mathematics education. Wikipedia is used a both a definitional source for the concepts of globalization and internationalization, as well as exemplifying the Anglophone and eurocentric domination of the knowledge economy worldwide. This web based encyclopedia also exemplifies mode 2 knowledge production outside of the academy, which is related to the place of ethnomathematics in society. The distinction between mode 1 and 2 knowledge production is used to critique the ideological discourse of mathematics which asserts that it is universal and sustains economic and social activity, and is an Anglophone academic production. It is argued that the role of mathematics is inseparable from the dominant background ideology of capitalism-consumerism, through which it helps to sustain the economic supremacy of the developed countries of the North. However, some possibilities for countering these effects via the development of critical mathematical literacy in learners and citizens are also indicated

Keywords: epistemology, knowledge production, knowledge economy, ideology, critical mathematical literacy, ethnomathematics, mathematized identities, spreadsheet metaphor, cultural difference

1. Introduction

In this chapter I explore a number of themes broadly associated with the globalization and internationalization of mathematics and mathematics education. In particular, I look at epistemological presuppositions as they relate to these domains and the associated ideological discourses and values, both overt and covert.

B. Atweh et al. (eds.), Internationalisation and Globalisation in Mathematics and Science Education, 19–38.

An interesting case study in the globalisation and internationalization of knowledge is that of Wikipedia, the online encyclopaedia. For example this is an entry from Wikipedia (2005) on globalisation and internationalization.

Globalization (or globalisation) in its literal sense is a social change, an increased connectivity among societies and their elements due to transculturation [the phenomenon of merging and converging cultures]; the explosive evolution of transport and communication technologies to facilitate international cultural and economic exchange. The term is applied in various social, cultural, commercial and economic contexts.

'Globalization' can mean:

- The formation of a global village – closer contact between different parts of the world, with increasing possibilities of personal exchange, mutual understanding and friendship between 'world citizens',
- Economic globalization – more freedom of trade and increasing relations among members of an industry in different parts of the world, with a corresponding erosion of national sovereignty in the economic sphere.
- The negative effects of for-profit multinational corporations – the use of substantial and sophisticated legal and financial means to circumvent the bounds of local laws and standards, in order to leverage the labour and services of unequally-developed regions against each other.

It shares a number of characteristics with internationalization and is used interchangeably, although some prefer to use globalization to emphasize the erosion of the nation state or national boundaries.

This definition describes and defines (and indeed embodies, more on this below) the dichotomy inherent in the concept of globalization. Namely the increased networking and connectivity between peoples and knowledge, on the one hand, and the imposition of hierarchy and potentially exploitative power relations on the other hand.

Wikipedia originates in Florida, USA, but is of anonymous multi-authorship, with anyone worldwide with internet access being able not only to access it freely but also to add and edit the text and associated links as and when they wish. This means that the text lacks authority. It is not vouched for by established authorities or august sponsoring institutions such as traditional universities or publishing houses. But the readership operates like any learned academy or learning 'conversation' (Ernest 1998) with different voices counterbalancing each other against prejudices, by (potentially) editing out one sided, polemical or self-promoting additions. In addition, for any entry it is possible to view the history of its edits to see how past and current versions of an article have been written, edited, and revised, and by whom.

This history is loosely analogous to a mathematical proof, showing the stages of reasoning through which the final assertion is derived. However, unlike a proof, interventions or additions need not be solidly grounded in logic or shared assumptions. Furthermore this process never arrives at a terminal stage, like a proved theorem. For social constructivist philosophies of mathematics this is not such a strong disanalogy since it is claimed that proofs and theorems proved never reach final state, but are always tentative and subject to revision (Ernest 1998; Hersh 1997; Lakatos 1978; Tymoczko 1986).

English is the original language of the encyclopaedia, but parallel versions are available in almost 200 other languages and 8 of them from Portuguese to Japanese have over 100,000 articles each. Thus Wikipedia is an example of the globalization, internationalization and localization of knowledge that was impossible to imagine even a decade ago.

The democratic and open features of Wikipedia whereby any reader can edit and add to it must of course be seen against the backdrop of the 'digital divide', the socio-economic and knowledge gap between communities that have access to computers and the internet and those who do not. Furthermore, although 'global' in the sense of its outreach, access and potential authorship, it is North American based and biased. This is manifested in a number of ways including the past emphasis on US interest centred news and anniversaries on its daily homepage (corrected since 2006), the use of Americanized spellings, and the US location and ownership of the project administration which makes policy decisions and intervenes to resolve editorial conflicts.

Wikipedia is a prime example of the globalization and internationalization of knowledge. It has an implicit epistemology and associated values which are asymmetric and ethnocentric. Knowledge is presented as neutral and balanced, as if emerging from some unspecified idealized and objectified origin, moving outward from this origin, and downwards in localization. Knowledge is never presented as originating in peripheral or distributed cultural contexts. Overall the epistemology and its ideological underpinning inadvertently expresses a rationalistic, Americanized and Eurocentric viewpoint.

Whereas internationalization is the adaptation of products for potential use virtually everywhere Wikipedia describes localization as means of adapting products such as publications and software for non-native environments, especially nations and cultures other than those of product origin. Localization is the addition of special features for use in a specific locale. This description is in the voice of the producer, market-maker or corporate manager. Localization is conceptualized in top-down mode, from the perspective of the supplier or producer rather than in bottom-up mode, reflecting the interests of the consumer, user or participant. Customs, aesthetics, values and other cultural aspects of the context are seen as obstacles to be surmounted for sales or flavourings to be captured in products, rather than as integral and intrinsically valuable features of a community and the local cultural context.

One of the central dimensions of globalization is the role of knowledge, and in particular its commodification and exploitation in the global knowledge economy

(Peters 2002). The knowledge economy differs from the traditional economy in several key respects (Skyrme 2004):

1. Information and knowledge are not depleted through use;
2. The effect of location is neutralized through the use of information and communication technologies;
3. National laws, barriers and taxes are difficult to apply in the globalised knowledge economy;
4. Pricing and value depend heavily on context, and the same information or knowledge can have vastly different value to different people at different times;
5. Human capital, i.e., knowledge and competencies, is a key component of value.

This chapter and book concern the impact and relevance of globalization, internationalization and the knowledge economy for mathematics education. So how do these changes impact on mathematics education? First, it should be noted that the term 'mathematics education' is ambiguous, for it refers both to a set of practices, encompassing mathematics teaching, teacher education, and research training, as well as to a field of knowledge with its own terms, concepts, problems, theories and literature (and indeed its own practices). There are a number of ways in which the effects of globalization and internationalization can and do occur in the domains of mathematics education. Four of the central dimensions of potential and actual impact are:

1. International marketing of mathematics curricula and higher education, including recruitment of international students, distance learning programmes and the international franchising of courses in mathematics education;
2. Mobility of knowledge workers in education, including international recruitment of mathematics teachers, university lecturers, teacher educators and researchers and more generally the international mobility of mathematics education faculty for employment and consultancy;
3. International collaboration in the organisation, development and dissemination of research and knowledge through projects (e.g. TIMSS, PISA) sometimes paid for by international funding agencies (e.g., World Bank, UNESCO), as well as conferences (e.g., ICME, PME, CERME) and organising bodies (e.g., ICMI, CIEAM) bringing together researchers from many countries;
4. Global domination in the production, warranting and regulation of research and knowledge in mathematics education by Anglophone Western countries, via the controlling interests of the agencies and institutions mentioned in 3 and the leading international journals and publications in mathematics education (Ernest 2006).

In addition, underneath the issues and problems of mathematics education lies an epistemological fact, namely the universality of mathematics in the modern world. Irrespective of whether this universality is seen as an essential feature of mathematical knowledge, or whether mathematics and its universality are viewed as products of human relations and activities past and present, mathematics is globally ubiquitous. In this respect mathematics is unique, and this privileges mathematics education on the international scene. Although its position is almost rivalled by science, medicine, computing or English, unlike these subjects mathematics is taught universally from the beginning of schooling, its symbolism is universal, and its uses underpin the functioning of all modern societies.

All of these themes concern knowledge: the production and transfer of knowledge, knowledge workers, and the social institutions of knowledge production and dissemination. One of the challenges for epistemology is to reconceptualise knowledge to take account of its new relationships in the market economy and the discourses of globalisation and internationalization. Knowledge can no longer be seen as isolated from its contexts of production, dissemination and use, and these contextual factors inevitably raise social, cultural, political, economic, and ethical issues too. There are also the highly current problems of knowledge transfer, the problematic notions of the knowledge or learning society and the knowledge economy. Even the concept of knowledge itself is a problematic one, from a philosophical perspective. Modernism and postmodernism are associated with very different conceptions of knowledge. Modernist perspectives see knowledge as objective, abstract, depersonalized, value-neutral and unproblematically transferable between persons and groups. In contrast, postmodernism views knowledge as socially and culturally embedded, value-laden, and not transferable across contexts without significant transformations and shifts in meaning. An analysis of knowledge types and modes of knowledge production can help to resolve some of the dilemmas and problems raised by these issues.

2. Knowledge Types and Knowledge Production

Contemporary epistemology tends to focus on propositional knowledge although attention has also been given to the explicit-tacit distinction ('knowing that' versus 'knowing how', e.g., Ryle 1949, Ernest 1999). However there is an ancient tradition in philosophy going back to Aristotle that makes a more fine grained set of distinctions.

Aristotle (1953) distinguishes three domains of knowledge and states of knowing. These are *Theoria*, *Techne* and *Praxis*. *Theoria* concerns abstract and universal knowledge, termed *Episteme*, that is applicable to all circumstances. *Techne* is taught knowledge, something like technique, originally concerning the farm, household, and everyday needs for making and composing objects, including poetry. In Aristotle's day this was considered to be a low form of knowledge. The modern word Technology derives from this kind of know-how, and only recently has it has become associated with tools and machines.

Praxis corresponds to what one might call practical wisdom. It is knowledge acquired through the process of doing. However unlike *Techne* is has a central ethical dimension, it is reasoned and right action with human good for its object.

Although *praxis* is a term taken up by Marxist theorists, and indeed corresponds in some respects to Habermas' (1971) third knowledge paradigm type, namely Critical Theory with its emancipatory interest, it barely figures in current epistemology (outside of philosophy of education). However, *Techne* has lost its negative connotations. Although for a long time described as applied knowledge – and this attribution relegates it to secondary status (the shadow of *Theoria*) – it has recently come into its own in a postmodern analysis of knowledge production into two types.

Gibbons et al. (1994) describe the production of knowledge in postmodernity as falling into two modes. Mode 1 knowledge production comes from a disciplinary community and its outcomes are those intellectual products produced and consumed inside traditional research-oriented universities. It is the kind of knowledge that is associated with research degrees such as the PhD. The legitimacy of such knowledge is determined by universities, the academics working within the knowledge area, and the academic journals that disseminate the knowledge. This corresponds to Aristotle's *Theoria* and *Episteme*, and is the subject of traditional epistemology.

In contrast, mode 2 knowledge is the identification and solution of practical problems in the day-to-day life of its practitioners and organizations, rather than centering on the academic interests of a discipline or community. Mode 2 knowledge is characterized by a set of attributes concerned with problem-solving around a particular application and context.

1. The different knowledge and skills of the practitioners are drawn together solely for the purpose of solving a socially (including industrially, commercially and technologically) motivated problem, and hence are integrated and transdisciplinary rather than confined to a single academic discipline.
2. The trajectory follows the problem-solving activity, and the context, conditions and even the research team may change over time as determined by the course of the project.
3. Knowledge production is carried out in an extensive range of formal and informal organizations including but extending well beyond universities.
4. The focus on socially motivated problems means there is social accountability and reflexivity built in from the outset of the project.

The key point made by Gibbons et al. (1994) is that the 'know how' generated by mode 2 practices is neither superior nor inferior to mode 1 university-based knowledge. It is simply different. There are different sets of intellectual and social practices required by mode 2 production compared with those likely to emerge in mode 1 knowledge production.

Mode 2 knowledge production may be newly recognised in postmodernity, but it is not a new phenomenon. It is a great irony that all knowledge production originates historically in mode 2 activities. Only with the development of specialized professions and academies focusing on the creation and validation of knowledge have these activities been appropriated and transformed into the mode 1 knowledge production institutions that now claim the ownership of epistemology and its products and processes.

Aristotle recognised *Techne* nearly 2500 years ago. Although modernist epistemology, since Descartes, has focused on explicit propositional knowledge, other traditions have acknowledged broader forms of knowledge production. The shift of emphasis onto knowledge production as opposed to knowledge itself raises issues to which epistemology has traditionally been blind, namely those of power and social context. Ever since ancient times the 'knower' as recognised by philosophy has been a member of an exclusive and elite group. Whether it was free citizens in Ancient Greece, the learned priesthood in the middle ages, or professional academics in modernity, these groups have arrogated to themselves and their institutions the powers of discernment of truth (knowledge discovery) and knowledge ratification (justification), and trustworthiness.

Mode 2 knowledge production refocuses the emphasis on the functionality of knowledge in context. Its underlying philosophy is thus closer to that of pragmatism in the tradition of Peirce, James, Dewey and more recently Rorty (1979). There is also a redolence of an instrumentalist philosophy of science, as found in, for example, Duhem, Poincaré and Quine (Losee 1980), according to which theories are viewed as tools as opposed to representations of reality. From these perspectives, knowledge is viewed functionally, as a means of solving problems, rather than as truth, something eternal existing in its own right. So this fits with postmodernism, since it rejects the metanarrative of certainty that modernism presumed to validate all knowledge (Lyotard 1984). Postmodernism see this fictive metanarrative as serving to safeguard the privileged knowledge-makers and their institutions. Instead, its own and differing perspective reconnects knowledge and knowing with their socio-cultural origins, namely human problem solving activities.

The acknowledgement of the key roles of power and social context in knowledge production also admits a third axis, that of economics or money (Foucault 1972). Foucault argues that to refuse to see the inextricable intertwining of power, knowledge and economics in discursive practices – the cultural milieus in which we humans all operate – is to be deceived by the purist and self-serving ideologies of modernism and its cultural elites. But as philosophy and the academy wake up to the fact that they do not own or control knowledge production, especially mode 2 knowledge production, we see that there is a powerful and ubiquitous knowledge market and globalised knowledge economy already exploiting the cultural embeddedness of all knowledge products and production.

3. Mathematical Knowledge and Mathematics in Society

Elsewhere (Ernest 1991, 1998) I have described the modernist paradigm of mathematical knowledge as pure, certain, perfect, unchanging and other-worldly. I will not reiterate my characterization of this absolutist view of mathematics and the grounds for rejecting it here, except to refer to some of the range of scholars who have contributed to this critique (Davis and Hersh 1980, Hersh 1997, Tymoczko 1986, Lakatos 1976, Wittgenstein 1956, Restivo 1992).

Despite this philosophical critique there is widespread acceptance of the modernist view and its corollary that everyday, applied or socially embedded mathematical usage is underpinned by the knowledge created by mathematicians. In other words, it is claimed that academic or research mathematics is what drives the social applications of mathematics in such areas as education, government, commerce and industry, as well as applications in science, engineering, etc. Historically, this is an inversion of the true state of affairs. In the Middle East clay tokens and pictograms inscribed on clay tablets were developed as communicative signs for quantity to facilitate trade and commerce within those contexts (Schmandt-Besserat 1978). This led to the development of written language and mathematics. Five thousand years ago in ancient Mesopotamia it was the need for scribes to tax and regulate commerce, as well as to perform ritual functions, in the well-organised kingly states, that led to the setting up of scribal schools in which mathematical methods and problems were systematized. This led to the founding of the academic discipline of mathematics.

> the creation of mathematics in Sumer was specifically a product of that school institution which was able to create knowledge, to create the tools whereby to formulate and transmit knowledge, and to systematize knowledge.
>
> (Høyrup 1980:45)

Although on and off since this beginning pure mathematics has taken on a life of its own through being internally driven either within this tradition (e.g., scribal problem posing and solving) or outside it (e.g., the ancient Greeks' separation of pure geometry from practical 'logistic'), practical mathematics has had a vitally important life outside of the academy. Even today the highly mathematical studies of accountancy, actuarial studies, management science and information technology applications are mostly undertaken within professional or commercial institutions outside of mode 1 knowledge production sites, and with little immediate input from academic mathematics. These areas have evolved as mode 2 knowledge production sites, and even where they have become formalized and professionally certifying, it is typically outside of traditional universities.

Nevertheless, the received, and in my opinion, mistaken view is that academic mathematics drives its more commercial, practical or popular applications. This ignores the fact that a two-way formative dialectical relationship exists between mathematics as practised inside and outside the academy. For example, overweight and underweight bales of goods are understood to have given rise to the plus and minus signs in medieval Italy, so important for history of mathematics. However it was the acceptance of negative roots to equations that finally forced the recognition of the negative integers as numbers within the discipline of mathematics. The same holds true of imaginary numbers. Now, of course, as well as playing a central role in pure mathematics, negative and imaginary numbers play an essential role in such areas as banking and electrical engineering.

The point I wish to make here is that although undoubtedly much of modern research mathematics derives from Type 1 knowledge production, there is a much broader category of mathematical knowledge that comes from Type 2 knowledge production. In accountancy, actuarial studies, management science, information technology applications, advanced engineering, economic modelling, and other areas new knowledge and techniques are developed and exploited solely for the purposes of solving practice-specific problems. Many of these applications are not viewed or described as mathematical, but are seen as belonging to and characterized by the domain of application.

Beyond these current commercial and industrial domains, there are informal, often more traditional problems and contexts, in which knowledge production also takes place and has taken place historically. Going back in time, the current view accepted by many scholars is that oral proto-mathematics and ethnomathematics have developed in all human cultures. Thus the word 'tik' meaning finger, digit, one, and 'pal' meaning two have been identified by linguistic theorists in the conjectured proto-language out of which all human languages developed, tens of thousands of years ago (Lambek 1996). In the great ancient high civilisations of the past, including the Chinese, Indian, Sumerian, Egyptian and Mayan cultures, written mathematics developed as a discipline connected with accountancy and central administration, as noted above. In the millennia that followed elements of this knowledge were transformed and elaborated by the Greek, Indian, and Arabian cultures and in the past 700 years by the Renaissance and modern European cultures, although developments also continued elsewhere such as in China and India (Joseph 1991). Thus in a discussion of globalization and internationalization in mathematics education it is ironic to note the global and international roots of mathematics which is now so often presented as a European creation and export.

Outside of the roots and trajectory of the development of academic mathematics, the broad and living informal cultural presence of ethnomathematics in most times and places is a further expression of both type 2 knowledge production and of its usage. The unique and universal characteristic of human beings is that we all have and make cultures, and every culture includes elements we can label as mathematical. Ethnomathematics should be understood in a broad sense referring to activities such as ciphering, measuring, classifying, ordering, inferring, and modelling in a wide variety of socio-cultural groups (D'Ambrosio 1985). It also includes the basket and rug designs, sand and body patterns, quipu, etc. made by various groups and relying on a sense of the possibilities of symmetry and form. Examples of socio-cultural groups include the different peoples studied in Africa by Gerdes (1986) and Zaslavsky (1973) and in the Americas by Ascher (1991) as well as the street mathematics (Nunes 1992) and shopping mathematics (Lave 1988) of modern urban life. Ethnomathematics is not about the exotic conceptions of 'primitive peoples'. The power of the concept of ethnomathematics is to challenge the notion that mathematics is only produced by mathematicians.

Ethnomathematics as informal and folk mathematics is an intrinsic part of most people's cultural activities, but academic mathematicians have appropriated, decontextualised (and recontextualized), elaborated and reified mathematical knowledge, until it has a life of its own. A common strategy in both analytical philosophy and mathematics is to factor out the origins of knowledge and only consider its final rationalized form. In keeping with this strategy, mathematical knowledge is seen from both modernist and the popular perspectives it has generated to be a pure substance that reflects the structure of a superhuman and timeless realm (Plato's World of Forms, Cantor's paradise), thus denying its ethnomathematical origins. This is a major historical and philosophical falsification. Identifying mathematics with academic, research or school mathematics has the result of deprecating and rendering invisible culturally distributed mathematics.

However, despite the importance of the concept, there are a number of theoretical problems associated with ethnomathematics. For example, whose viewpoint defines mathematics and ethnomathematics? Some multiculturalist views of ethnomathematics sees mathematics in cultural activities such as basket and rug designs, sand and body patterns, pebble games and ritual dances. Such a perception may only be possible for those enculturated into the Western academic mathematical worldview, and who have developed a 'mathematical gaze'. It can be seen as a form of intellectual cultural-imperialism, which abstractly factors out one component from a concretely given practice, whose purpose is some overall form of material production, or perhaps of religious observance.

There is a debate currently underway in the mathematics education community about ethnomathematics, and some scholars argue that it has become a romantic liberal 'shibboleth' (Brown and Dowling 1988). Ethnomathematics, as the authentic proto-mathematical modes of thinking of peoples of traditional cultures has come to symbolize their dignity, wisdom and authenticity. However, this is problematic for a number of reasons. First, although there are phenomena we might label ethnomathematical, there is no unified body of mathematical knowledge, and no social institution of ethnomathematics. Ethnomathematical modes of thought are an intrinsic part of a variety of situated practices whose focus and purpose is likewise concretely situated. The very gaze which identifies a part of the practice or its product as mathematics has been constituted through the specific abstracted modes of an academic or school mathematical training. Second, if ethnomathematics means culturally embedded mathematics, then the greatest site of ethnomathematics is in the variety of practices in industrial societies, from computer game sub-cultures and shopping to gambling and 'loan-sharking'. Third, prioritizing ethnomathematics in education does not address the problem of the connection of academic and school mathematics with power, and its potential role in empowering learners to take more control over their own lives. Thus conventional achievements in mathematics open up doors of opportunity for all students. So certification in school or academic mathematics is important. Having a good grasp of social mathematics, including the mathematics and statistics used to support political arguments in society, is necessary for critical citizens to be able to participate fully in modern society and assert their rights. That mathematics is largely determined by the academy and school.

Of course most uses and applications of mathematical knowledge, whoever's gaze defines it, does not constitute knowledge production. Gerdes (1985) distinguishes between the creation of new ethnomathematical knowledge, and the 'frozen' mathematics embodied in most unselfconscious ethnomathematical practices which may be creative in terms of other products like baskets, patterns, etc. Only the creation of new ethnomathematical knowledge is a candidate for characterization as mode 2 knowledge production.

A similar analysis is possible concerning mathematics in industrialized society. The mathematization of modern society and modern life has been growing exponentially, so that now virtually the whole range of human activities and institutions are conceptualised and regulated numerically, including sport, popular media, health, education, government, politics, business, commercial production, and science. Thus, for example, spectator sport is understood by its audiences in highly and largely quantified terms (e.g., the viewing of sports on television involves the absorption of many screenfulls of digital information) as well as being regulated by financial considerations. The initial conceptualisation and installation of such systems is an outcome of mode 2 knowledge production, but the utilization of these systems and consumption of their products is not.

In modernity the scientific worldview has come to dominate the shared conceptions held widely in society and by individuals. This worldview prioritizes what are perceived as objective, tangible, real and factual over the subjective, imaginary or experienced reality, and over values, beliefs and feelings. This perspective rests on a Newtonian realist worldview, etched deep into the public consciousness as an underpinning 'root metaphor' (Pepper 1948), even though the modern science of relativity and quanta has shown it to be untenable. In the late- or post-modern era this viewpoint has developed further, and I wish to claim that a new 'root metaphor' has come to dominate, namely that of the accountant's balance-sheet. From this perspective the ultimate reality is the world of money, finance, and other associated quantifiables. Elements of such a critique are present in Critical Theory and the work of Marcuse (1964); Young (1979); Skovsmose (1994); Restivo et al. (1993). The way this scheme and its mechanisms work is as follows. Many aspects of modern society are regulated by deeply embedded complex numerical/algebraic systems, e.g., supermarket checkout tills with automated bill production, stock control etc.; tax systems; welfare benefit systems; industrial, agricultural and educational subsidy systems; voting systems; stock market systems. These automated systems carry out complex tasks of information capture, policy implementation and resource allocation. Niss (1983) named this the formatting power of mathematics and Skovsmose (1994) terms the systems involved which are embedded in society the 'realized abstractions'. Such systems are all the outcomes of mode 2 knowledge production.

Thus complex mathematics is used to regulate many aspects of our lives, e.g., our finances, banking and bank accounts, but with very little human scrutiny and intervention, once the systems are in place. There are two overall effects: first, most of contemporary industrialized society is regulated (and subject to surveillance) by embedded and part-hidden complex mathematical-based systems ('black boxes'). These are automated through the penetration of computers and information and communication technologies into all levels of industry, commerce, bureaucracy and institutional regulation. The computer penetration of society was only possible because the politicians', bureaucrats' and business leaders' systems of exchange, government, control and surveillance were already quantified, just as they were in a more rudimentary form 5000 years ago at the beginning of mathematics.

Second, individuals' conceptualisations of their lives and the world about them is through a highly quantified framework. The requirement for efficient workers and employees to regulate material production profitably, motivated the structuring and control of space and time, and for workers' self-identities to be constructed and constituted through this structured space-time-economics frame (Foucault 1970, 1976). Thus we understand our lives through the conceptual meshes of the clock, calendar, work timetables, travel planning and timetables, finances and currencies, insurance, pensions, tax, measurements of weight, length, area and volume, graphical and geometric representations, and so on. This positions individuals as regulated subjects and workers in an information controlling society and state, as beings in a quantified universe, and as consumers in post-modern consumerist society.

One of the most important ways that this is achieved is through the universal teaching and learning of mathematics from a very early age and throughout the school years. The central and universal role of arithmetic in schooling provides the symbolic tools for quantified thought, including the ability to conceptualize situations quantitatively. The high penetration of everyday life, the media and other dimensions of culture cultivates and reinforces the development of quantified identities in modern citizens. This is now so widespread and universal that it is not only taken for granted and invisible, but is also seen to be necessary and inevitable, despite being socially constructed.

My claim is that the overt role of academic mathematics, the outcome of mode 1 knowledge production which we recognise as mathematics *per se*, in this state of affairs is minimal. It is management science, information technology applications, accountancy, actuarial studies and economics and other fields of mode 2 knowledge production, which are the roots of and inform this massive mathematization on the social scale. Underpinning this, at both the societal and individual levels, is the balance-sheet metaphor, for economic or market value is the common unit in which virtually all of the activities and products of contemporary life are measured and regulated.

This means that although there is an information revolution taking place, increased mathematical knowledge is not needed by most of the population to cope with their new roles as regulated subjects, workers and consumers. More mathematics skills beyond the basic are not needed by the general populace in industrialized societies to 'cope' with and serve these changes, as opposed to critically mastering them. Most governments are happy to limit the populace's roles to service rather than mastery, even in the personal domain. A corollary is that contrary to the widespread myth, national success in international studies of mathematical achievement is not the creator of economic success unless having compliant subjects and consumers is needed.

It is ironic to watch the rise and fall of the industrial powers of nations and hear the media and politicians attributing the success to the national methods of teaching mathematics (and science and technology) employed in these countries. Such attributions have regularly been heard in the UK over the past few decades. It looks like a form of phallocentrism, the hero worship of the powerful. In the 1970s and 1980s much attention was directed at the educational methods of USSR, attention which has now died away since its political and economic collapse. Shortly after, Germany was claimed to be an exemplary country which had got its economics and mathematics teaching right. These claims ceased after reunification and the consequent economic rebalancing. In the mid 1990s adulation of the methods of Pacific Rim countries was at its height. But with their fall from economic grace after the meltdown of their stock markets and currencies, this died down somewhat, although the tiny dictatorship in Singapore is held up as an example to emulate. It is likely that the politicians will tell us to emulate India and China next, in view of their rapid economic growth.

Although advanced mathematical skills are not widely needed for economic development, there is of course a need for a small highly skilled technical elite. These are the persons who design and control the information systems and mechanisms, as well as a group of specialist technicians to service or programme them. These need to be present in all industrialized societies. But this group represent a tiny minority within society and their very special needs should not determine the goals of mathematics education for all. In addition, it is not academic mathematics which underpins the information revolution. It is instead a collection of technical mathematised subjects and practices which are largely institutionalised and taught, or acquired in practice, outside of the academy. In other words, it is the mode 2 knowledge producers who contribute this important role in society. Most of the public do not need advanced mathematical understanding for economic reasons, and the minority who do apply mathematics acquire much of their useful knowledge in institutions outside of academia or schooling. This has been termed the 'relevance paradox', because of the "simultaneous objective relevance and subjective irrelevance of mathematics" in society (Niss 1994, p. 371). Society is ever increasingly mathematised, but this functions at a level invisible to most of its members.

I have offered a simple sketch of the role of mathematics in society to illustrate the growing importance of mode 2 knowledge production in the numerate areas that underpin the functioning of modern industrial society. I argue that the teaching and learning of mathematics generally produces subjects with quantified identities and outlooks, as well as through an advanced training route, a small highly trained elite, who develop and maintain the quantified social systems and structures. Such impacts are the results of the globalization and internationalization of mathematics education worldwide.

Historically, European colonial powers imposed their educational systems on their colonies, with mathematics and the language of the rulers at the heart of the curriculum. The British did this in India and throughout the Empire, just as France did in Algeria and its colonies. Missionary schools likewise provided a mixture of religious education and the elementary education in the style of their mother countries. In the move to post-colonial times, this has changed as educational interventions have been provided to developing countries as a dimension of aid, sponsored by the World Bank, US Aid and other similar sources. In all of these cases the educational systems, methods and values from more powerful nations were imposed on weaker or less developed ones. Because these recent impositions are seen as charitable gifts, aid, or development support, until recently it has not been possible for the recipient countries to question the values, intentions and methods of the imposed education. But there is growing awareness among recipient countries that together with the educational support comes a hidden curriculum in the form of views of knowledge, values, and ideologies. An unreflective imposed mathematics (or science) curriculum is often based on decontextualised algorithms and methods, includes contexts and examples from an alien culture (the donor culture), and takes no account of local customs or cultural practices. Thus, for example, development aid in the 1960s and 1970s coincided with the New Maths in Western countries and led to such debacles as the teaching of set theory to primary school children in Brazil and Papua New Guinea. This was often neither useful nor comprehensible to the students (Howson 1973).

Sometimes the export of mathematics teaching methods is well intentioned but still culturally insensitive. For example, the teaching of problem solving and discovery learning is sometimes promulgated with missionary zeal to countries with strong traditions of the teacher as an unquestioned authority, as in some Islamic and Far Eastern cultures. Such approaches applied in a superficial way in the source countries were unsuccessful, and not surprisingly fared even worse in countries where the received values were dissonant with the approach. Some recent large scale survey research conducted in Sweden (Bentley, 2003) suggests that there are optimal class sizes for different teaching approaches in mathematics, with large classes (over 30 pupils) performing better with traditional direct instruction methods, and small classes performing better with progressive methods (cooperative group work, problem solving and practical projects). If these results are generalizable, the utilitisation of progressive teaching methods in schools and countries where large classes are the norm not surprisingly leads to disappointing results.

However, great caution must be exercised against characterizing the cultures of developing countries in a monolithic and over-simplified way. As we know from the developed world, there is more cultural variation within countries than across them. For example, the most powerful and persistent findings on differences in educational achievement concern class and socio-economic status (TGAT 1988). Poorer persons in virtually all countries perform significantly worse than their better off peers in mathematics (and across the curriculum). In part this is due to access to educational and home resources (worse schools, higher pupil-teacher ratios, less books, less computer access, etc). But it is also very likely in large part due to cultural differences, and the differential cultural capital that different students groups bring with them (Bourdieu and Passeron 1977). This accounts for the reason that certain ethnic groups with low socio-economic status (recent immigrants from far Eastern Countries to the USA) outperform others (African-American and Native-American groups). Another longstanding difference in educational achievement, namely that between rural and urban students may also be reducible to socio-economic factors.

Developing countries typically have a very broad spectrum of wealth and socio-economic status, with the rich and the relatively wealthy urban middle classes at the top, and the rural poor at the bottom. Often there are also large migrant populations in cities living in barrios and ghettos, the new urban poor, also near the bottom of the wealth spectrum. There is often a very striking contrast between the cultures and identities of the rich and urban middle classes, on the one hand, and rural poor, on the other hand. The urban middle classes in developing countries such as Brazil, Egypt, India, Pakistan, South Africa and Thailand, for example, will very likely be culturally closer to the educated classes in developed countries than to the rural poor in their own countries. The latter are more likely to adhere to the traditional customs, beliefs and cultural practices. The former are more likely to have Westernized identities involving the consumption of goods and services (cars, appliances, fashion clothing, makeup, fast-food, travel, computers and the internet, satellite television, popular music, etc.). Such identities will often be constructed as composites, sometimes in tension, between traditional values, beliefs and observances and a Westernized consumerist identity. Such tensions can be problematic, as is expressed by Middle Eastern women forced to wear the hijab (headscarf) in public in their own countries, but who remove them in the privacy of their own homes or when visiting the West. In contrast, Islamic girls in France who wish to wear their hijab to school as an expression of their cultural identity and religious commitment have been denied this right by law.

However, such tensions can also be productive. For example, much of the most vibrant and prized new fiction in the English language has emerged from the English Diaspora, citizens of ex-colonies or oppressed peoples who combine immigrant or postcolonial identities with Westernized sensitivities, including Chinua Achuebe, John M. Coetzee, Nadine Gordimer, Toni Morrison, V. S. Naipaul, Salman Rushdie, Wole Soyinka, Derek Walcott.

Although modern audio-visual media perhaps have the greatest impact, the influence of Western curriculum and pedagogies on the teaching and learning of mathematics may make a significant contribution to the construction of Westernized identities in developing countries. Where mathematics is presented in a Eurocentric way by means of historical references and examples drawn from daily life it encourages learners to admire and valorise the West and its lifestyles. This very likely contributes to the development of aspirations towards Western consumerist lifestyles in learner identities. In extreme cases it might lead to rejection of one's own culture and even self-loathing. At present there is little data in this area, it being both technically and theoretically difficult and politically sensitive an area to research. But the export of Western mathematics education methods, even when judged successful in terms of achievement outcomes, may well have other negative impacts.

I have stressed some of the potentially negative impacts of the globalization and internationalization of the teaching and learning of mathematics emanating from the West. My stress on the negative is deliberate, to offset the widespread (e.g., both official and public) perception that developmental aid in education is an unqualified benefit to recipient countries. Instead, my emphasis is to caution that the importation or imposition of an ill-thought through mathematics curriculum is very likely problematic. There must be a careful determination of local needs, possibilities and resources. Well informed local expertise is required to make sound judgements about local needs, as well as about how educational reforms can be implemented and disseminated (Broomes 1981; Broomes and Kuperus 1983). The problems of the latter are already well evidenced within countries such as the United Kingdom where centralised curriculum innovations are unintentionally transformed and degraded in the process of dissemination and implementation as they are passed down through a system of hierarchical training sessions Ernest (1991).

Above, I discuss how the perceived role of mathematics in society and the received aims for the teaching and learning of mathematics can lead to its deployment as a force of the subjectification of learners and citizens. In addition, the globalization and internationalization of the mathematics curriculum can also intensify such effects. There is, fortunately, another aspect of education that can counter this; the vital emancipatory role of education, and the role of mathematics in critical citizenship (Ernest 1991; Frankenstein 1983, 1989; Powell and Frankenstein 1997; Skovsmose 1994.).

Critical mathematical education aims to empower learners as individuals and citizens-in-society, by developing mathematical confidence and power both to overcome barriers to higher education and employment and thus increasing economic self-determination; and to foster critical awareness and democratic citizenship via mathematics. This includes critically understanding the uses of mathematics in society: to identify, interpret, evaluate and critique the mathematics embedded in social, commercial and political systems and claims, from advertisements to government and interest-group pronouncements.

Unless schooling helps learners to develop the knowledge and understanding to identify the prevalent mathematizations of modern society, and the confidence to question and critique them, they cannot be in full control of their own lives, nor can they become properly informed and participating citizens. Instead they may be manipulated by commercial, political or religious interest groups, or become cynical and irrational in their attitudes to social, political, medical and scientific issues.

Once mathematics becomes a 'thinking tool' for viewing the world critically, it contributes to the political and social empowerment of the learner, and ideally, should contribute to the promotion of social justice and a better life for all. The larger aim of a critical mathematics education is social change towards a more just and egalitarian society via the empowerment of the citizenry.

However, such political aims for mathematics teaching addressing contentious social issues in the classroom are clearly controversial. To date I know of no country where such aims and approaches have had official sanction in mainstream education, because they involve a critical stance towards the social *status quo*, whatever the political orientation of the government, left, centrist, or right.

Where such approaches have been successfully implemented, it has been through local or grass roots initiatives, typically with marginal or disempowered groups. The rural children of the Italian School of Barbiana (1970) took control of their own education in a powerful questioning way, utilizing mathematics as a tool for critical analysis. Mellin-Olsen (1987) reports the inspiring project work in mathematics of low attaining pupils in Norway. Frankenstein (1989) developed a critical numeracy course for adults returning to education in Boston. This is one of the few areas in developed countries where critical mathematics education is addressed, as the title *Numeracy for Empowerment and Democracy?* of the 2001 conference of the Adults Learning Mathematics conference shows (Ostergaard Johansen and Wedege 2002)

Perhaps the best known approaches to critical education have emerged in developing countries. Paulo Freire (1972) working with landless peasants in Brazil developed a programme of education with the aim of achieving critical consciousness or 'conscientisation' "a permanent critical approach to reality in order to discover it and discover the myths that deceive us and help to maintain the oppressing dehumanizing structures." (Dale et al. 1976, p. 225). Julius Nyerere initiated the Tanzanian programme of *Education for Self-Reliance* "to prepare people for their responsibilities as free workers and citizens in a free and democratic society, albeit a largely rural society. They have to be able to think for themselves, to make judgements on all the issues affecting them" (Lister 1974, p. 97). D'Ambrosio in Brazil and Gerdes in Mozambique have both developed and promoted ethno-mathematics as a means of locally based critical mathematics education. They have also had a major impact on mathematics education research worldwide in raising awareness of the diversity of localized mathematical thinking and its philosophical and political significance.

Thus there are, if not opposing forces, oppositely motivated movements which act to counter the globalizing and internationalizing forces in the mathematics curriculum. However, these are relatively minor on the global scale and may pass as unobserved or irrelevant by globalizing and internationalizing forces.

4. Conclusion

Overall, I have discussed some of the universalizing forces at work in the globalization and internationalization of mathematics education. The ideological discourse of mathematics suggests that it is universal and sustains economic and social activity, and emanates from academic European sources (including North America). This is used as a widespread justification for the internationalization and globalization of mathematics curricula emanating from the West. I have argued that all of the assumptions involved are problematic, and used the distinction between mode 1 and

2 knowledge production to challenge some of them. Nevertheless, the widespread acceptance of this ideological discourse helps sustain the economic supremacy of the developed countries of the North. It also helps to maintain the power and economic differentials within these countries. While indicating some possibilities for trying to counter these effects (through critical mathematical literacy) I have also acknowledged that role of mathematics is inseparable from the dominant background ideology of capitalism-consumerism. This argument can be extended to show how features of the ideological discourse of mathematics education also works to maintain the dominance of the field by Anglophone Western universities Ernest (2006).

References

Aristotle (1953). *The ethics of Aristotle* (The Nichomachean Ethics, J. A. K. Thomson, Trans.). London: Penguin Classics.

Ascher, M. (1991). *Ethnomathematics: A multicultural view of mathematical ideas, pacific grove.* California: Brooks/Cole.

Bentley, P. O. (2003). *Mathematics teachers and their teaching: A survey study*, Göteborg Studies in Educational Sciences 191. Sweden: Gothenburg University.

Bourdieu, P., & Passeron, J. C. (1977). *Reproduction in education, society and culture.* London: Sage.

Broomes, D. (1981). Goals of mathematics for rural development. In R. Morris (Ed.), *Studies in mathematics education.* Paris: UNESCO, *2*, 41–59.

Broomes, D., & Kuperus, P. K. (1983). Problems of defining the mathematics curriculum in rural communities. In M. Zweng, T. Green, J. Kilpatrick, H. Pollak & M. Suydam (Eds.), *Proceedings of fourth international congress on mathematical education* (pp. 708–711). Boston: Birkhauser.

Brown, A., & Dowling, P. (1988). *Towards a critical alternative to internationalism and monoculturalism in mathematics education*, London: Institute of Education, University of London.

D'Ambrosio, U. (1985). Ethnomathematics and its place in the history and pedagogy of mathematics. *For the Learning of Mathematics*, *5*(1), 44–48.

Dale, R., Esland, G., & MacDonald, M., (Eds.). (1976). *Schooling and capitalism.* London: Routledge and Kegan Paul.

Davis, P. J., & Hersh, R. (1980). *The mathematical experience.* London: Penguin Books.

Ernest, P. (1991). *The philosophy of mathematics education.* London: Falmer Press.

Ernest, P. (1998). *Social constructivism as a philosophy of mathematics.* Albany, NY: State University of New York Press.

Ernest, P. (1999). Forms of knowledge in mathematics and mathematics education: Philosophical and rhetorical perspectives. *Educational Studies in Mathematics*, *38*(1–3), 67–83.

Ernest, P. (2006). Globalization, Ideology and research in mathematics education. In R. Vithal, M. Setati & C. Malcolm, (Eds.), *Methodologies for researching mathematics, science and technological education in societies in transition.* Durban, South Africa: UNESCO-SARMSTE, University of KwaZulu-Natal.

Foucault, M. (1970). *The order of things: An archaeology of the human sciences.* London: Tavistock.

Foucault, M. (1972). *The archaeology of knowledge.* London: Tavistock.

Foucault, M. (1976). *Discipline and punish.* Harmondsworth: Penguin.

Frankenstein, M. (1983). Critical mathematics education: An application of Paulo Freire's epistemology. *Journal of Education*, *165*(4), 315–339.

Frankenstein, M. (1989). *Relearning mathematics: A different third r – radical maths.* London: Free Association Books.

Freire, P. (1972). *Pedagogy of the oppressed.* London: Penguin Books.

Gerdes, P. (1985). Conditions and strategies for emancipatory mathematics education in underdeveloped countries. *For the Learning of Mathematics*, *5*(1), 15–20.

Gerdes, P. (1986). How to recognise hidden geometrical thinking, *For the Learning of Mathematics*, *6*(2), 10–12 & 17.

Gibbons, M., Limoges, C., Nowotny, H., Schwartzman, S., Scott, P., & Trow, M. (1994). *The new production of knowledge.* London: Sage.

Habermas, J. (1971). *Knowledge and human interests.* London: Heinemann.

Hersh, R. (1997) *What is mathematics, really?* London: Jonathon Cape.

Howson, A. G., (Ed.). (1973). *Developments in mathematical education.* Cambridge: Cambridge University Press.

Høyrup, J. (1980). Influences of institutionalized mathematics teaching on the development and organisation of mathematical thought in the pre-modern period, Bielefeld. Extract reprinted in J. Fauvel and J. Gray, (Eds.), *The History of Mathematics: A reader* (pp. 43–45). London: Macmillan, 1987.

Joseph, G. G. (1991). *The crest of the peacock: Non-European roots of mathematics.* London: Penguin Books.

Lakatos, I. (1976). *Proofs and refutations: The logic of mathematical discovery.* Cambridge: Cambridge University Press.

Lakatos, I. (1978). *Philosophical papers* (2 vols.). Cambridge: Cambridge University Press.

Lambek, J. (1996). Number words and language origins. *The Mathematical Intelligencer*, 18(4), 69–72.

Lave, J. (1988). *Cognition in practice: Mind, mathematics and culture in everyday life.* Cambridge: Cambridge University Press.

Lister, I., (Ed.). (1974). *Deschooling.* Cambridge: Cambridge University Press.

Losee, J. (1980). *A historical introduction to the philosophy of science* (2nd ed.). Oxford: Oxford University Press.

Lyotard, J. F. (1984). *The postmodern condition: A report on knowledge.* Manchester: Manchester University Press.

Marcuse, H. (1964). *One dimensional man.* London: Routledge and Kegan Paul.

Mellin-Olsen, S. (1987). *The politics of mathematics education*, Dordrecht: Reidel.

Niss, M. (1983). Mathematics education for the 'Automatical Society'. In R. Schaper, (Ed.), *Hochschuldidaktik der Mathematik* (Proceedings of a conference held at Kassel 4–6 October 1983) (pp. 43–61). Alsbach-Bergstrasse, Germany: Leuchtturm-Verlag.

Niss, M. (1994). Mathematics in society. In R. Biehler, R. W. Scholz, R. Straesser, & B. Winkelmann, (Eds.), *The didactics of mathematics as a scientific discipline* (pp. 367–378). Dordrecht: Kluwer.

Nunes, T. (1992). Ethnomathematics and everyday cognition. In D. A. Grouws, (Ed.), *Handbook of research on mathematics teaching and learning* (pp. 557–574). New York: Macmillan.

Ostergaard Johansen, L., & Wedege, T. (Eds.). (2002). *Numeracy for empowerment and democracy?* (Proceedings of the 8th international conference of Adults Learning Mathematics at Roskilde, Denmark, June 2001). Roskilde, Denmark: ALM.

Pepper, S. C. (1948). *World hypotheses: A study in evidence.* Berkeley, CA: University of California Press.

Peters, M. (2002). Education policy research and the global knowledge economy, *Educational Philosophy and Theory*, *34*(1), 91–102.

Powell, A., & Frankenstein, M. (Eds.). (1997). *Ethnomathematics: Challenging eurocentrism in mathematics education.* Albany, NY: SUNY Press.

Restivo, S. (1992). *Mathematics in society and history.* Dordrecht: Kluwer.

Restivo, S., Van Bendegem, J. P., & Fischer, R. (Eds.). (1993). *Math worlds: Philosophical and social studies of mathematics and mathematics education.* Albany, NY: SUNY Press.

Rorty, R. (1979). *Philosophy and the mirror of nature.* Princeton, NJ: Princeton University Press.

Ryle, G. (1949). *The concept of mind.* London: Hutchinson.

Schmandt-Besserat, D. (1978). The earliest precursor of writing. *Scientific American, 238*(6), 50–58.

School of Barbiana (1970). *Letter to a teacher.* Harmondsworth: Penguin Books.

Skovsmose, O. (1994). *Towards a philosophy of critical mathematics education.* Dordrecht: Kluwer.

Skyrme, D. (2004). *Global knowledge economy*, http://www.skyrme.com/insights/21gke.htm, Webpage accessed on 26 April 2004.

TGAT (Task Group on Assessment and Testing) (1988). *A report.* London: Department of Education and Science.

Tymoczko, T. (Ed.). (1986). *New directions in the philosophy of mathematics.* Boston: Birkhauser.

Wikipedia (2005). 'Globalization' consulted on 31 December 2004 at <http://en.wikipedia.org/wiki/Globalization.>

Wittgenstein, L. (1956). *Remarks on the foundations of mathematics*, revised edition. Cambridge, MA: Massachusetts Institute of Technology Press, 1978.

Young, R. M. (1979). Why are figures so significant? The role and the critique of quantification. In J. Irvine, I. Miles, I. & J. Evans, (Eds.), *Demystifying social statistics* (pp. 63–74). London: Pluto Press.

Zaslavsky, C. (1973). *Africa counts.* Boston: Prindle, Weber and Schmidt.

3

ALL AROUND THE WORLD SCIENCE EDUCATION, CONSTRUCTIVISM, AND GLOBALISATION

Noel Gough
La Trobe University, Victoria, Australia

Abstract: This chapter explores a number of challenges, uncertainties, and opportunities facing science education as new and complex global processes affect the ways in which knowledge is produced and circulated. Major themes of the chapter include the difficulties of implementing Western science education programs in cross-cultural and/or multicultural settings and the extent to which the doctrine of constructivism resolves issues of cultural difference, even for those science educators who are particularly attentive to the cultural contexts of science and science education. It is argued that although Western science educators cannot speak from outside their own Eurocentrism, asking questions about the globalization of science education as a cultural practice might help to make both the limits and strengths of Western science's knowledge traditions more visible.

Keywords: constructivism, globalization, Eurocentrism

> Well, it's not just you
> And it's not just me
> This is all around the world
>
> —Paul Simon (1986) 'All around the world; or, the myth of fingerprints'

My purpose in this chapter is to explore some of the challenges, uncertainties, and opportunities that face science educators as new and complex global processes shape the activities of knowledge production. These activities include whatever we might understand by 'science education' (such as formal school programs and informal learning via popular media) as well as the ways in which the people we call 'scientists' go about their work (these are, of course, interrelated activities: science education is influenced by what scientists do – or at least by what science

B. Atweh et al. (eds.), Internationalisation and Globalisation in Mathematics and Science Education, 39–55.

educators *believe* that scientists do – and most adult scientists have experienced science education in some form). In the first part of this chapter I will focus on issues of 'cultural blindness' that might accompany attempts to implement Western science education programs in cross-cultural and/or multicultural settings (e.g. in non-Western countries or in culturally diverse communities in the West). I will then consider the appropriateness of privileging 'constructivist' views of learning as a response to apprehensions of cultural difference in science education. During the past two decades, constructivism has become something of a new orthodoxy of Western science and mathematics education and my purpose here is to demonstrate that the limits of its applicability in non-Western cultural contexts also draw attention to its limitations as a theoretical framework for science education policy and research in Western societies.

1. Globalising Western science; or, the Myth of (no) Fingerprints

Until relatively recently in human history, the social activities that have produced distinctive forms of knowledge have for the most part been localised. The knowledges generated by these activities have thus borne what Sandra Harding (1994) calls the idiosyncratic 'cultural fingerprints' (p. 304) of the times and places in which they were constructed. The knowledge that the English word 'science' usually signifies is no exception, since it was uniquely coproduced with industrial capitalism in seventeenth century northwestern Europe. The internationalisation of what we now call 'modern Western science'[1] was enabled by the colonisation of other places in which the conditions of its formation (including its symbiotic relationship with industrialisation) were reproduced.

The global reach of European and American imperialism gives Western science the *appearance* of universal truth and rationality, and it is often assumed to be a form of knowledge that lacks the cultural fingerprints that seem much more conspicuous in knowledge systems that have retained their ties to specific localities, such as the 'Blackfoot physics' described by David Peat (1997) and comparable knowledges of nature produced by other aboriginal societies. This occlusion of the cultural determinants of Western science contributes to what Harding (1993) calls an increasingly visible form of 'scientific illiteracy', namely, 'the Eurocentrism or androcentrism of many scientists, policymakers, and other highly educated citizens that severely limits public understanding of science as a fully social process':

> In particular, there are few aspects of the 'best' science educations that enable anyone to grasp how nature-as-an-object-of-knowledge is always

[1] I realise that this formulation – 'modern *Western* science' rather than just 'science' or 'modern science' – introduces a problematic 'West versus the rest' dualism and might appear to overlook the historical influences of other cultures (e.g., Islamic, Indian, Chinese, etc.) on its evolution. However, I also want to emphasise that I am referring to science as it was produced in Europe during a particular historical period and to those of its cultural characteristics that have endured to dominate Western (and many non-Western) understandings of science as a result of Euro-American imperialism.

> cultural. . . These elite science educations rarely expose students to systematic analyses of the social origins, traditions, meanings, practices, institutions, technologies, uses, and consequences of the natural sciences that ensure the fully historical character of the results of scientific research (p. 1).

Over the last few decades, various processes of political, economic, and cultural globalisation – including the increasing volume of traffic in trade, travel, and telecommunications networks crisscrossing the world – have helped to make some multicultural perspectives on 'nature-as-an-object-of-knowledge' more visible, such as those that have been popularised as the 'wisdom of the elders' (Knudtson & Suzuki, 1992) or 'tribal wisdom' (Maybury-Lewis, 1991). The publication in English of studies in Islamic science (e.g. Sardar, 1989) and other postcolonial perspectives on the antecedents and effects of modern Western science (e.g. Third World Network, 1988; Petitjean, Jami & Moulin, 1992; Sardar, 1988) has raised further questions about the interrelationships of science and culture. However, economic globalisation is also – simultaneously and contradictorily – encouraging cultural homogenisation *and* the commodification of cultural difference within a transnational common market of knowledge and information that remains dominated by Western science, technology, and capital.

Scepticism about the universality of Western science provokes a variety of responses from scientists and science educators. Aggressive (and well-publicised) defenders of an imperialist position include scientists such as Paul Gross and Norman Levitt (1994) who heap scorn and derision on any sociologists, feminists, postcolonialists, and poststructuralists who have the temerity to question the andro-centric, Eurocentric, and capitalist determinants of scientific knowledge production.[2] Although I am sure that many science educators take a similar position to Gross and Levitt,[3] I prefer to attend to the less obvious – and thus perhaps more insidious – forms of imperialism manifested by science educators whose ideological standpoints appear to be closer to my own. It is for this reason that, in the remainder of this chapter, I will focus a good deal of my critical attention on a paper in which William Cobern (1996) explicitly calls for science education researchers 'to use a constructivist model of learning to both support the need for, and facilitate, investigations of how science education can be formulated from different cultural perspectives' (p. 296). Cobern (1996) claims to reject 'an acultural view of science' (p. 295) and criticizes colleagues who assume a 'cultural deficit' in scientific

[2] Gross and Levitt (1994) give the impression that the academic left's 'quarrels with science' are chiefly the result of ignorance, scholarly incompetence, irrationality and/or ideological prejudice, an impression they underscore with a litany of personal abuse: for example, they refer to Sandra Harding's 'megalomania' (p. 132), Donna Haraway's 'delusions of adequacy' (p. 134), and Katherine Hayles's 'mathematical subliteracy' (p.104) for whose work 'the word *crackpot* unkindly leaps to mind' (p. 103, emphasis in original).

[3] I have neither sought nor sighted published examples of science educators engaging in the kind of attack on critics of Western science mounted by Gross and Levitt, although I have heard these authors quoted or referred to with approval and even admiration by some science education researchers and teachers at academic and professional conferences and other gatherings.

understanding in 'non-western and traditional cultures' (p. 296) – positions that I support unequivocally. His paper – especially when read in conjunction with other documents produced under the auspices of the Scientific Literacy and Cultural Studies Project (SLCSP)[4] that he directs – convinces me that his respect for non-Western and traditional cultures is sincere. Nevertheless, I will argue that for all of his undeniably good intentions, Cobern falls short of rejecting an acultural view of science. Moreover, what Cobern (1996) appears to mean by 'making science curricula authentically sensitive to culture and authentically scientific' (p. 295) is using constructivism to make Western scientific imperialism universally user-friendly. I will address the issue of constructivism in the next section of this essay. In this section I will focus on the reluctance of many Western science educators to fully accept the implications of confirming the proposition that 'nature-as-an-object-of-knowledge is always cultural' and the rhetorical strategies they use to persuade learners that the world Western scientists imagine and represent is 'real' and that the knowledge they produce is universal.

For example, one way in which Western scientists privilege their discipline is to stipulatively define its uniqueness. The physicist Paul Davies deploys this strategy in his response to the question, 'Can Western science have all the answers?'

> From the point of view of the new physics, there is no other science. A construct of Western rationalism, using the language of mathematics, science lays claim to the status of universal truth regardless of cultural context (quoted in Slattery, 1995, p. 15).

By explicitly locating the position from which he speaks within the knowledge system produced by the members of his own disciplinary community, Davies makes it difficult to dispute Western science's claim to universal truth because, by his stipulation, 'there is no other science' to contradict it. From this standpoint, one can understand Blackfoot physics as wise and efficacious local knowledge – but it cannot be 'science'.

Cobern (1996) adopts a similar tactic to Davies by defining what counts as 'science' in terms of cultural exclusion:

> If 'science' is taken to mean the casual study of nature by simple observation, then of course all cultures in all times have had their own science. There is, however, adequate reason to distinguish this view of science from modern science (p. 307).

The distinction Cobern makes here is difficult to sustain in the light of evidence that the 'study of nature' was performed by some 'not modern' cultures in ways that cannot be diminished by terms such as 'casual' and 'simple'. For example, as David Turnbull (1991) points out, people from south-east Asia began to systematically colonise and transform the islands of the south-west Pacific some ten thousand years

[4] The SLCSP is funded by a grant from the (US) National Science Foundation. Details of SLCSP reports and publications are available via the Project's home page at <http://www.wmich.edu/slcsp/slcsp.htm>

before what is Eurocentrically described as the 'birth of civilisation' is alleged to have taken place in the Mediterranean basin. The Micronesian navigators combined knowledge of sea currents, marine life, weather, winds and star patterns to produce a sophisticated and complex body of natural knowledge which, combined with their proficiency in constructing large sea-going canoes, enabled them to transport substantial numbers of people and materials over great distances in hazardous conditions. They were thus able to seek out new islands across vast expanses of open ocean and to establish enduring cultures throughout the Pacific by rendering the islands habitable through the introduction of new plants and animals. Although the knowledge system constructed by these people did not involve the use of either writing or mathematics – and it is thus easy to stipulate that it is not 'modern science' – it is patronising and indefensible to describe it as 'the casual study of nature by simple observation'.

If the knowledge produced by Western scientists is only 'consumed' in cultural sites dominated by Western science, then their claim to its universality could be understood as a relatively harmless conceit. But we are increasingly seeing attempts to generate global knowledge in areas such as health (necessitated, in part, by the global traffic in drugs and disease) and environment (for example, global climate change) which draw attention to the cultural biases and limits of Western science. For example, as Brian Wynne (1994) reports, the models of climate change devised by the Intergovernmental Panel on Climate Change (IPCC) up to the early 1990s equated global warming mainly with carbon emissions (and ignored factors such as cloud behaviour, and biological processes such as marine algal fixing of atmospheric carbon and natural methane production), and yet were understood by many Western scientists as producing universally warranted conclusions. From a non-Western standpoint, these same IPCC models can be seen to reflect the interests of developed countries in obscuring the exploitation, domination, and social and political inequities underlying global environmental degradation. But if global warming is understood as a problem for *all* of the world's peoples, then we need to find ways in which *all* of the world's knowledge systems – Western, Blackfoot, Islam, whatever – can jointly produce appropriate understandings and responses. I will not presume to suggest (indeed, I cannot imagine) what a Blackfoot or Islamic contribution to such jointly produced knowledge might be, but I am prepared to assert that a coexistence of knowledge systems is unlikely to be facilitated by the adherents of any one system arbitrarily privileging their own criteria for distinguishing it (uniquely) as 'modern science' and thereby laying claim to producing 'universal truth regardless of cultural context'.

This claim to the universality of Western science is usually advanced by drawing attention to its supposed power to produce ahistorical and transcultural generalisations, exemplified by Michel Serres' (1982) ironic assertion that 'entropy increases in a closed system, regardless of the latitude and whatever the ruling class' (p. 106). Cobern (1996) deploys a similar strategy (without any obvious irony) when he notes that science textbooks from around the globe are 'strikingly similar' and asserts that 'one expects a discussion of the observed phenomenon known as photosynthesis to appear in all basic biology textbooks regardless of cultural location' (p. 299); he adds that 'it makes sense that an isolated scientific concept (e.g. photosynthesis) is *a*cultural' (p. 299, his emphasis). But even if we agree that photosynthesis *can* be 'observed' (as distinct from induced from other observations), this only 'makes sense' if we assume that the concepts Western scientists invent to represent natural phenomena are, as Richard Rorty (1979) puts it, 'transparent to the real' (p. 368) To assert that photosynthesis (or entropy) is '*a*cultural' is to naturalise the social construction of scientific knowledge. This is not to deny that there are observable phenomena that Western science represents in terms of the process of photosynthesis. What is at stake here is not *belief* in the real but confidence in its representation – in Rorty's (1979) words, 'to deny the power to "describe" reality is not to deny reality' (p. 375). Furthermore,

> we need to make a distinction between the claim that the world is out there and the claim that the truth is out there. To say that the world is out there, that it is not our creation, is to say, with common sense, that most things in space and time are the effects of causes which do not include human mental states. To say that truth is not out there is simply to say that where there are no sentences there is no truth, that sentences are elements of human languages, and that human languages are human creations... The world is out there, but descriptions of the world are not (Rorty, 1989, p. 5).

Thus, the *concept* of photosynthesis, like the concepts of 'entropy' and 'closed systems', cannot be 'acultural'. Whatever it is that a leaf does independently of 'human mental states', its representation as 'photosynthesis' is clearly a human invention. If it is true that a discussion of photosynthesis appears 'in all basic biology textbooks regardless of cultural location', then this could be taken as testimony to the power of a particular ruling class to impose its meanings universally rather than an expression of the universal meaningfulness of the concept of photosynthesis.[5]

In addition to photosynthesis, Cobern (1996) refers to 'phenomena such as motion, force, life and gravity' (p. 304) as if they signified transcultural 'realities' rather than constructs of Western science. This allows Cobern (1996) to make the contradictory assertions that 'science content is science content regardless of culture to be sure, but... communicated science, which includes science education, is inculturated' (p. 300). As with photosynthesis, even if we agree that motion, force, life and gravity *can* be 'observed', these observations still have to be made by *culturally located* humans who must also construct, with the cultural materials at hand, the

[5] Reading Cobern's article in its entirety leads me to believe that he is well aware of the distinction between the *real* and a *representation* of the real that I am making here, but his references to photosynthesis can be interpreted as contradicting such an awareness. To say that 'it makes sense that an isolated scientific concept (e.g. photosynthesis) is *a*cultural' carries many cultural assumptions, including the assumption that it is sensible to conceptually isolate and name a hypothesis about the processes that relate foliage to energy conversions. I would have no disagreement with Cobern if he had written that *from a Western cultural standpoint* it 'makes sense that an isolated scientific concept (e.g. photosynthesis)' *appears to be* '*a*cultural'.

representations which enable them to communicate their observations – to produce the testimonies to experience that we call 'facts'. In other words, if what Cobern means by 'science content' is exemplified by photosynthesis, motion, force, life and gravity, then it is always already 'communicated science' and 'inculturated'.

Like many thoughtful science education researchers, Cobern appears to be struggling to reconcile a realist ontology with the view that scientific knowledge is socially constructed. Indeed, within the discourses of science education research, it is not difficult to find such unequivocal statements as: 'The objects of science are not the phenomena of nature but constructs that are advanced by the scientific community to interpret nature' (Driver et al., 1994, p. 5). But in the discourses of science education policy and practice we tend to find a different story. For example, a draft version of *A National Statement on Science for Australian Schools* (Australian Education Council 1991, p. 4) explicitly recognised the social and cultural dimensions of scientific activity, but asserted nevertheless that the truth claims of scientists should be privileged by the special qualities of the method used to produce them: '*Although* science is socially constructed, the processes and principles of science *still* enable scientific knowledge to be developed which is generally reliable, useful and well accepted' (my emphases). It is worth considering what might be implied by the terms 'although' and 'still' here. Are the authors suggesting that the social construction of knowledge diminishes its reliability, usefulness and acceptability? If so, are they implying that it is possible to imagine knowledge that is *not* socially constructed and, if so, who – or what – is in a position to make such a judgment? The deferential 'although' suggests that the authors are apologising for science being socially constructed, but then they reassure the reader that, nevertheless ('still'), this troublesome complication can be overcome by applying 'the processes and principles of science' – as if social constructedness were a curable disease. This rhetorical ploy reasserts the privileged status of scientific knowledge by insinuating that its method transcends (or in principle *can* transcend) social construction.[6]

2. Globalising Western Science Education; or, a Myth of Constructivism

According to Cobern (1996, p. 301), constructivist thought supplies 'a view of learning that is *transferable across*, and *appropriate for*, different cultural environments' (his emphasis). His confidence in the cross-cultural applicability of constructivism underlies his argument that 'science education research and curriculum

[6] Later versions of the Australian Education Council's *Statement* are less grudging in their affirmation of science as a social construction. However, although these later versions might be politically (and philosophically) more 'correct', I suspect that many policy-makers, teachers, and researchers remain attracted to the draft position. For example, in affirming their support for Rom Harré's (1986) realist ontology, Driver et al. (1994) adopt the position 'that scientific progress has an empirical basis, *even though* it is socially constructed and validated' (p. 6, my emphasis).

development efforts in non-western countries can benefit by adopting a constructivist view of science and science learning' (p. 295). For Cobern (1996), constructivism 'suggests a conceptualisation of scientific knowledge in which it is reasonable to expect culture-specific understandings of science' (p. 304). By way of example, Cobern (1996) argues that we should not expect Nigerian students and students in Western countries to understand science in exactly the same way and emphasizes that this does not mean that the Nigerian understandings will be unscientific: 'Rather, their scientific viewpoint will reflect their Nigerian worldview... the problem in non-western science education is not to make it more scientific, but to make it less culturally western' (pp. 304–5). Although I can support Cobern's aspirations up to a point,[7] I do not share his confidence that constructivism provides any impetus for science educators to accept the cultural specificity of the knowledge constructed in the name of Western science. As I have pointed out above, Cobern's own commitment to constructivism does not prevent him from assuming that such scientific constructs as photosynthesis are 'acultural'. Nor does Cobern seem to recognize that constructivism is itself a construct of Western science education research and, therefore, that a constructivist theory of learning is not necessarily a universal truth.

Given that it is not always clear what Western science educators have in mind when they invoke the term constructivism, the idea that it might provide a transcultural model of learning seems a somewhat tenuous hope. Paul Cobb (1994) notes that constructivism is often reduced to the mantra-like slogan that 'students construct their own knowledge', and points to the difficulties that arise if we apply this theory reflexively and try to explain 'how so many mathematics and science educators have individually constructed this supposedly indubitable proposition' (p. 4). Clive Sutton (1992) draws attention to 'the unfortunate blurring of the distinction between personal and social constructivism... Most writers with a background in science teaching remain stubbornly psychological rather than sociological, and it is *personal constructs* that they have in mind, not *social constructs'* (p. 108, emphasis in original). For example, Kenneth Tobin (1990) writes of 'social constructivist perspectives on the reform of science education' but his essay is concerned with the social context in which learners' deploy personal constructs rather than with the implications for science education of the social construction of reality and its representations.

Rosalind Driver et al. (1994) assert that 'the core commitment of a constructivist position, [is] that knowledge is not transmitted directly from one knower to another, but is actively built up by the learner' (p. 5). Driver and her colleagues clearly recognise the need to go beyond psychologistic and individualistic forms of constructivism and claim that their position on learning science is informed 'by a view of scientific knowledge as socially constructed and by a perspective on the

---

[7] My support for Cobern's goal of making science 'less culturally western' is tempered by my concern that this formulation may assume that there is some universal, 'acultural' core or essence of science that is distorted by the 'noise' of Western culture – and that 'the problem in non-western science education' is therefore to deliver science without that noise. I would prefer to see the 'problem' as one of making the cultural specificities of *all* sciences more explicit.

learning of science as knowledge construction involving both individual and social processes' (p. 5). However, although these authors argue that 'the view of scientific knowledge as socially constructed and validated has important implications for science education', these seem to relate principally to the efficacy of the procedures for inducting neophytes into a knowledge production system:

> Once such knowledge has been constructed and agreed on within the scientific community, it becomes part of the 'taken-for-granted' way of seeing things within that community.
> ... learning science involves being initiated into scientific ways of knowing. Scientific entities and ideas, which are constructed, validated, and communicated through the cultural institutions of science, are unlikely to be discovered by individuals through their own empirical enquiry; learning science thus involves being initiated into the ideas and practices of the scientific community and making these ideas meaningful at an individual level (Driver et al., 1994, p. 6).

Although I can accept this argument up to a point, the reference to '*the* scientific community' seems to suggest that these authors are assuming a monoculture of science that is tacitly Western. I am also troubled by the word 'initiated'. It is surely defensible to help learners to make personal sense of the Western scientific 'way of seeing things', but *initiating* them into Western scientific 'ways of knowing' could be understood as precluding or limiting their access to other ways of knowing. It is one thing for scientists to take their own constructions for granted, but quite another for science educators to insist that learners who have not yet chosen science as their vocation should do likewise. Even if the connotations of 'initiated' are acceptable, I would want to add the proviso that learners should simultaneously be 'initiated' into methods of exposing the historically and culturally specific determinants of 'scientific ways of knowing' and of the means by which 'scientific entities and ideas... are constructed [and] validated'. This is partly a matter of historical reinterpretation – of understanding that, say, the apparent fruitfulness of Newtonian mechanics was very largely determined by an androcentric and Eurocentric scientific community (see, for example, Jansen, 1990) – and partly a matter of contemporary cultural critique, such as recognising the extent to which ethnocentrism and biological determinism pervade the educational philosophies of John Dewey, Jean Piaget, and Paulo Freire, whose work continues to inform constructivist educational reforms in many nations (see, for example, Bowers, 2005, who refers to constructivism as 'the Trojan horse of Western imperialism', p. 79).

However, neither historical reinterpretations nor contemporary cultural criticisms of Western science are necessary attributes of constructivist science education. Science education informed by constructivism does not necessarily problematise the cultural construction of scientific knowledge but, rather, attempts to use knowledge of learners' personal constructs to generate more effective strategies for persuading students to adopt Western scientists' social constructions. By way of illustrating this assertion, I will consider two of the three examples used by Richard

Gunstone (1988) to introduce constructivist research on students' 'interpretations of natural phenomena' (NB. I have retained Gunstone's numbering of these examples for later reference):

> *Example 2...*
> A physics graduate in a one-year course of teacher training was in a group shown a bell jar containing a partially inflated balloon. When asked to predict what would happen to the balloon when air was evacuated from the bell jar, he answered 'The balloon will float'. His reason: 'Because gravity will be reduced'...
>
> *Example 3...*
> Large samples of science and physics students from each of the ages 13 to 17 years were given questions about a ball thrown in the air. The questions asked whether the force on the ball was up, down or zero for three positions shown on diagrams – ball rising, ball at highest point, ball falling. The most common response at all five age levels was 'up, zero, down'. This response, which embraces the belief that a force is needed in the direction of motion to maintain that motion, was given by about half of the 16 and 17-year-old physics students (p. 74).

The first point to note about these examples is that the 'natural phenomena' that students are being asked to interpret are all highly contrived or abstracted. Indeed, to say that 'a bell jar containing a partially inflated balloon' is intended to demonstrate a 'natural' phenomenon is a little like saying that animals in zoos display 'natural' behaviours. Nevertheless, Gunstone (1988) uses this example to illustrate two constructivist research findings:

> [Students' ideas/beliefs] can be remarkably unaffected by traditional forms of instruction. ... a tertiary physics graduate... apparently continues to interpret the world around him via a belief that gravity is an atmosphere-related phenomenon (i.e. without air there is no gravity)....
>
> Some students can hold the scientists' interpretations given in instruction together with a conflicting view already present before instruction. The science interpretation is often used to answer questions in science tests, and the conflicting view retained to interpret the world. This is illustrated... by example 2 (where the graduate involved could readily answer questions requiring Newton's Law of Gravitation) and by example 3 (where some 50 per cent of senior students holding the force-needed-in-direction-of-motion belief could successfully solve standard $F = ma$ problems) (p. 75).

I have already quoted Cobern's (1996) reference to 'phenomena such as motion, force, life and gravity' (p. 304) and here Gunstone provides another example of the tendency of science educators to naturalise what is socially constructed by referring to a representation as a phenomenon. Yet, as Katherine Hayles (1993) notes, 'gravity, like any other concept, is always and inevitably a representation' (p. 33). Within communities of working scientists, the conflation of a phenomenon and its representation may be a relatively harmless linguistic short cut. Projected beyond these communities, such conflations create the impression that the world Western scientists imagine and represent is 'real' and that the knowledge they produce is universal. Thus, in Gunstone's account, gravity is accorded the status of natural phenomenon and Newton's Law of Gravitation is the privileged explanation. Alternative representations of the phenomena to which 'gravity' and 'gravitation' refer are not considered – and constructivism does not necessarily invite them to be. For example, Fensham, Gunstone & White (1994) assert that constructivist teaching 'does not give students licence to claim that their meaning is as good as scientists' meaning, no matter what its form'. Moreover, they continue, constructivism 'does not mean "anything goes"; some meanings are better than others. Means for determining what is better are then significant'. They then endorse criteria for explaining a natural phenomenon that are very familiar in the rhetoric of Western science, namely, that an explanation should be 'elegant and parsimonious and connected with other phenomena, as well as having... intelligibility, plausibility and fruitfulness... and be testable' (p. 6). They present these criteria as unquestioned assertions. But why *should* an aesthetic criterion like elegance apply to scientific explanations? Why *should* an arbitrary criterion such as parsimony be applied? Like all of the other criteria that Fensham et al. (1994) recommend, their meanings are embedded in the historically specific practices of interpretation and testimony that characterize the narrative traditions of Western science.

Rather than trying to determine that 'some meanings are *better* than others', Hayles (1993) suggests that 'within the representations we construct, some are ruled out by constraints, others are not' (p. 33). In Hayles's (1993) terms, 'by ruling out some possibilities... constraints enable scientific inquiry to tell us something about reality and not only about ourselves':

> Consider how conceptions of gravity have changed over the last three hundred years. In the Newtonian paradigm, gravity is conceived very differently than in the general theory of relativity. For Newton, gravity resulted from the mutual attraction between masses; for Einstein, from the curvature of space. One might imagine still other kinds of explanations, for example a Native American belief that objects fall to earth because the spirit of Mother Earth calls out to kindred spirits in other bodies. No matter how gravity is conceived, no viable model could predict that when someone steps off a cliff on earth, she will remain suspended in midair. This possibility is ruled out by the nature of physical reality. Although the constraints that lead to this result are interpreted differently in different paradigms, they operate universally to eliminate certain configurations from the range of possible answers (pp. 32–3).

Hayles (1993) emphasises that constraints do not – indeed cannot – tell us what reality *is* but, rather, that constraints enable us to distinguish which representations are consistent with reality and which are not. For example, the limit on how fast information can be transmitted with today's silicon technology is usually explained as a function of how fast electrons move through a semiconductor. 'Electron' and 'semiconductor' are social constructions, but the limit is observed no matter what representation is used. If atomic theories had been formulated around the concept of waves rather than particles, then we might now explain the limit in terms of indices of resistance and patterns of refraction rather than electrons and semiconductors. Hayles notes that for any given phenomenon, there will always be other representations, unknown or unimaginable, that are consistent with reality: 'The representations we present for falsification are limited by what we can imagine, which is to say, by the prevailing modes of representation within our culture, history, and species' (p. 33).[8] Hayles (1993) calls this position 'constrained constructivism':

> Neither cut free from reality nor existing independent of human perception, the world as constrained constructivism sees it is the result of active and complex engagements between reality and human beings. Constrained constructivism invites – indeed cries out for – cultural readings of science, since the representations presented for disconfirmation have everything to do with prevailing cultural and disciplinary assumptions (pp. 33–4).

Hayles articulates very clearly a philosophical position that should commend itself to science educators – a position that problematises the non-discursive 'reality' of nature without collapsing into antirealist language games. Constrained constructivism is not 'anything goes' but neither does it disallow representations that fail to meet criteria that disguise their Eurocentric and androcentric biases behind claims for universality. But science educators – including those who espouse constructivism – often seem to do the precise opposite of what Hayles suggests by requiring learners to *confirm* representations that conform to 'cultural and disciplinary assumptions' that no longer prevail even in the West.[9]

This brings me to Gunstone's example 3, which has almost nothing to do with what it claims to be exemplifying – students' interpretations of natural phenomena – but typifies a rhetorical strategy that is common in school textbooks. The strategy is to reject students' understanding of an everyday word (in this case 'force') and to replace this word's meaning with a formula (here, $F = ma$). A textbook recently

---

[8] It should be noted that an analysis of the consistency between reality and a representation is different from applying Karl Popper's (1965) doctrine of falsification, since Popper maintained that congruence is a conceptual possibility. But as Hayles (1993) explains, the most we can say is that a representation is 'consistent with reality as it is experienced by someone with our sensory equipment and previous contextual experience. Congruence cannot be achieved because it implies perception without a perceiver' (p. 35).

[9] Lorraine Code (2000) similarly argues that 'responsible global thinking *requires*... a *mitigated epistemological relativism* conjoined with a "healthy skepticism"' (p. 69, emphases in original). She continues:

> I am working with a deflated conception of relativism remote from the 'anything goes' refrain which anti-relativists inveigh against it. It is 'mitigated' in its recognition that knowledge-construction is always constrained by the resistance of material and human-social realities to just any old fashioning or making. Yet, borrowing Peter Novick's words, it is relativist in acknowledging 'the plurality of criteria of knowledge... and deny[ing] the possibility of knowing absolute, objective, universal truth' (1988, 167). Its 'healthy skepticism' in this context manifests itself in response to excessive and irresponsible global pretensions, whose excesses have to be communally debated and negotiated with due regard to local specificities and global implications. (p. 69)

used in Australian schools illustrates the extraordinary lengths to which science educators will go to ensure that learners are, to repeat Driver et al. (1994) words, 'initiated into the ideas and practices of the scientific community' and to insist that learners find 'these ideas meaningful at an individual level', even if these ideas no longer constitute 'contemporary scientific ways of knowing' (p. 6). In the textbook to which I refer, Malcolm Parsons (1996) introduces the topic of 'Work and energy' with a half-page freehand illustration of a girl pushing hard against a brick wall. She is grimacing with the effort and beads of sweat are bursting from her brow. She is watched from their perch on overhead wires by two puzzled birds (both are wide-eyed and one has a question mark over its head) with the characteristic colours and features of galahs (this is a nice local touch: among Australians of European descent the galah is an emblem of extreme foolishness – the village idiot of birdland). Its caption indicates that this drawing is no mere decoration but a substantial component of the text: 'FIGURE 8.1: Considerable force is being applied here. How much work is being done?' (p. 150)

Occupying a narrow but very prominent column on the left hand side of the page (bold black print over a bright yellow box) is a so-called 'Fact File' (a regular feature of this particular text) which reads, in part:

> A scientist considers that no work has been done on an object if the object has not moved through a distance. For example, if you spend all day pushing hard against a wall, but the wall does not move, then no work has been done on it! (p. 150)

Consider the cumulative effect of the exclamation mark, the positioning of the above sentences in a 'Fact File', the illustration I have described (a girl, two galahs) and its caption. These textual strategies appeal to commonsense understandings of an everyday word, reject this understanding, and then replace the meaning of the word with a formula by insinuating that work is 'really' the product of force and distance. All of these graphic and semantic ploys are directed towards establishing the textbook's claims to being the repository of authoritative knowledge of what 'work' means. It is claiming that any other meanings for 'work' are deficient, unscientific, intuitive, even foolish (clinging to your commonsense understandings makes you a bit of a galah).

Such stipulative definitions are not, and cannot ever be, 'scientific' truth claims. The assertion that 'no work has been done' if we try but fail to move an object does not belong in a 'Fact File'. There is not, and cannot be, one privileged 'fact' informing what 'work' means. Words *in fact* mean whatever they are used to mean, and 'work' is used to mean 'force multiplied by distance' only in very restricted circumstances. AsDavid Chapman (1992) writes:

> The intellectually honest way to present this concept would be to invent a new word for it, say 'woozle'. Woozle is the product of force and distance. Actually, we are going to need new words for those too, so woozle is the product of frizzle and drizzle. We could go through a physics book and systematically substitute these new words in and we'd get a new book that wouldn't be making claims to ownership of any ordinary-language words. I believe that students

> would have a much easier time with such a book; it would be much easier to learn the new words than to deal with the cognitive dissonance involved in abandoning old ones (n.p).[10]

Many science educators might say that this is not done because relating a physics concept, such as woozle, to an everyday concept, such as work, allows learners to use their commonsense understanding of this phenomenon as a stepping-stone to understanding the 'correct' scientific concept. However, as the proposed rewritten text entry above would make clear, work has very little to do with woozle, and saying that woozle is 'work' is confusing. If this was no more than a recycling of the word, students could understand that 'work' has two meanings, which would present them with little or no difficulty; but the claim that is imposed on the students by the textbook is that woozle is the *true* meaning of 'work' and that they must abandon other meanings. Chapman concludes:

> I believe the actual reason physics continues to claim 'work' for its own can be seen if we imagine the fully-renamed physics book about woozle and frizzle. The problem with this book is that it never makes contact with reality. It's a nice consistent mathematical system that isn't about anything. If it is going to describe the world, it either has to have some ordinary words in it to ground it, or else we need to have instruments that measure woozle and frizzle rather than work and force.

But, as most physicists will acknowledge if the point is pressed, this cannot be done. The real world is not programmed to run according to the rules of Newtonian mechanics or any of the other representations that Western scientists, in their astonishing arrogance, have come to call 'laws'.

At this stage I must emphasise that I am neither seeking to diminish the significance of Newtonian mechanics in the history of Western science nor suggesting that Newton's work should be ignored in science education. The point at issue here is that if students perceive an incoherence between their commonsense understandings of reality and a scientific representation of it, science educators should not assume that the 'fault' lies with students or that it is their sacred duty to coerce students towards an orthodox belief. On the contrary, science educators should be helping students to *understand the incoherence* rather than to fudge it – to *demonstrate* that gaps between reality and representation are inevitable rather than to deploy rhetorical tricks in an effort to persuade students that it all makes sense.

For example, another textbook currently used in Australian schools (Cooper et al., 1988)[11] asserts that: '*All masses attract each other.* No one has yet discovered why' (p. 160). Yet the very next paragraphs state that:

---

[10] Quoted from an email message posted to the list-server <postech@weber.ucsd.edu> on Friday 13 March 1992. The subject of Chapman's message is: 'Science is stupid, part nineteen'.

[11] I did not select the textbooks by Cooper et al., (1988) and Parsons (1996) because they provide examples that suit my own rhetorical purposes particularly well. Rather, I chose them because they are, respectively, the textbooks prescribed for use in my daughter's high school science classes in 1996 (grade 9) and 1997 (grade 10).

> The force of attraction between masses is known as *gravitational force*. The larger and closer the masses, the greater is the pull between them.
> The gravitational force between the earth and the sun holds the earth in orbit around the sun. The gravitational force between the moon and the earth holds the moon in orbit around the earth (p. 160; emphasis in original).

These statements appear to be contradictory: if 'no one has yet discovered why' all masses attract each other, how can the attraction be attributed to 'gravitational force'? Moreover, if students do not recognise the contradiction, then science educators should draw attention to it by distinguishing between phenomena and their representations and elucidating the local and historical determinants of privileged representations. If this were done, it might become much more obvious to students (and teachers?) that the representations that constitute Newtonian mechanics are culturally determined, socially constructed, context dependent – and certainly not the only, let alone the 'best', interpretations of natural phenomena that are consistent with reality.

Examples such as these do not support Cobern's (1996) contention that constructivist thought supplies a transcultural view of learning. Rather, they suggest that the strategic rhetoric of constructivist science education is compounding the problem of scientific illiteracy by continuing to reflect and reproduce monocultural models of inquiry, representation, interpretation and explanation as if these were 'natural'.

I have no desire to reach closure on the issues discussed in this chapter, and so I have no 'conclusions' to offer. Much of this essay has been concerned with identifying what Jon Wagner (1993) calls the 'blind spots and blank spots' (p. 16) that configure the 'collective ignorance' of science education researchers at the present time. In Wagner's schema, what we 'know enough to question but not answer' are our blank spots; what we 'don't know well enough to even ask about or care about' are our blind spots: 'areas in which existing theories, methods, and perceptions actually keep us from seeing phenomena as clearly as we might'. Science educators are beginning to fill in blank spots in their emerging understandings of the extent to which science education is a cross-cultural activity and the implications of seeing it as such. My principal concern here has been with the blind spots that may remain in the vision of science educators who are particularly attentive to the cultural contexts of science and science education. Although we may not be able to speak from outside our own Eurocentrism, continuing to ask questions about the globalisation of the cultural practices we call science education will, I hope, help to make both the limits *and strengths* of the knowledge tradition we call Western science increasingly visible.

3. Acknowledgments

My thanks to Hank Bromley, Susan Edgerton, and David Shutkin for their constructive comments on an earlier draft of this paper. I also acknowledge prior publication of an earlier version of this chapter in *Educational Policy* (Gough, 1998).

4. References

Australian Education Council (1991). A National Statement on Science for Australian Schools: Complete Consultative Draft #1 (Carlton, Victoria: Australian Education Council).

Bowers, C. A. (2005). *The false promise of constructivist theories of learning: A global and ecological critique*. New York: Peter Lang.

Cobb, P. (1994). Constructivism in mathematics and science education. *Educational Researcher, 23*(7), 4.

Cobern, W. W. (1996). Constructivism and non-western science education research. *International Journal of Science Education, 18*(3), 295–310.

Code, L. (2000). How to think globally: stretching the limits of imagination. In U. Narayan & S. Harding (Eds.), *Decentering the center: Philosophy for a multicultural, postcolonial, and feminist world* (pp. 67–79). Bloomington and Indianapolis: Indiana University Press.

Cooper, V., Pople, S., Ray, B., Seidel, P., & Williams, M. (1988). *Science to sixteen 1*. (Revised Australian ed.). Melbourne: Oxford University Press.

Driver, R., Asoko, H., Leach, J., Mortimer, E., & Scott, P. (1994). Constructing scientific knowledge in the classroom. *Educational Researcher, 23*(7), 5–12.

Fensham, P. J., Gunstone, R. F., & White, R. T. (Eds.). (1994). *The content of science: A constructivist approach to its teaching and learning*. London: The Falmer Press.

Gough, N. (1998). All around the world: science education, constructivism, and globalization. *Educational Policy, 12*(5), 507–524.

Gross, P. R., & Levitt, N. (1994). *Higher superstition: The academic left and its quarrels with science*. Baltimore and London: The Johns Hopkins University Press.

Gunstone, R. F. (1988). Learners in science education. In P. Fensham (Ed.), *Developments and dilemmas in science education* (pp. 73–95). London: The Falmer Press.

Harding, S. (Ed.). (1993). *The 'Racial' economy of science: Toward a democratic future*. Bloomington and Indianapolis: Indiana University Press.

Harding, S. (1994). Is science multicultural? Challenges, resources, opportunities, uncertainties. *Configurations: A Journal of Literature, Science, and Technology, 2*(2), 301–330.

Harré, R. (1986). *Varieties of realism*. Oxford: Blackwell.

Hayles, N. K. (1993). Constrained constructivism: locating scientific inquiry in the theater of representation. In G. Levine (Ed.), *Realism and representation: Essays on the problem of realism in relation to science, literature and culture* (pp. 27–43). Madison WI: University of Wisconsin Press.

Jansen, S. C. (1990). Is science a man? New feminist epistemologies and reconstructions of knowledge. *Theory and Society, 19*, 235–246.

Knudtson, P., & Suzuki, D. (1992). *Wisdom of the elders*. Sydney: Allen & Unwin.

Maybury-Lewis, D. (1991). *Millennium: Tribal wisdom of the modern world*. London: Viking.

Parsons, M. (1996). *Science 4*. Melbourne: Heinemann.

Peat, F. D. (1997). Blackfoot physics and European minds. *Futures, 29*(6), 563–573.

Petitjean, P., Jami, C., & Moulin, A. M. (Eds.). (1992). *Science and empires: Historical studies about scientific development and european expansion*. Dordrecht and Boston: Kluwer.

Popper, K. (1965). *Conjectures and refutations: The growth of scientific knowledge*. (2nd ed.). New York: Basic Books.

Rorty, R. (1979). *Philosophy and the mirror of nature*. Princeton NJ: Princeton University Press.

Rorty, R. (1989). *Contingency, irony, and solidarity*. Cambridge MA: Cambridge University Press.

Sardar, Z. (Ed.). (1988). *The revenge of athena: Science, exploitation and the third world*. London: Mansell.

Sardar, Z. (1989). *Explorations in islamic science*. London: Mansell.

Serres, M. (1982). *Hermes: Literature, science, philosophy* (J. V. Harari and D. F. Bell, eds). Baltimore: The Johns Hopkins University Press.

Simon, P. (1986). *All around the world; or, the myth of fingerprints* [Song]. Burbank CA: Warner Bros. Records Inc.

Slattery, L. (1995, July 15–16). The big questions. *The Australian Magazine*, 12–19.

Sutton, C. (1992). *Words, science and learning*. Buckingham and Philadelphia: Open University Press.
Third World Network. (1988). *Modern science in crisis: A third world response*. Penang, Malaysia: Third World Network.
Tobin, K. (1990). Social constructivist perspectives on the reform of science education. *Australian Science Teachers Journal, 36*(4), 29–35.
Turnbull, D. (1991). *Mapping the world in the mind: An investigation of the unwritten knowledge of the micronesian navigators*. Geelong: Deakin University Press.
Wagner, J. (1993). Ignorance in educational research: or, how can you *not* know that? *Educational Researcher, 22*(5), 15–23.
Wynne, B. (1994). Scientific knowledge and the global environment. In M. Redclift & T. Benton (Eds.), *Social theory and the global environment* (pp. 169–189). London: Routledge.

4

GEOPHILOSOPHY, RHIZOMES AND MOSQUITOES: BECOMING NOMADIC IN GLOBAL SCIENCE EDUCATION RESEARCH

Noel Gough
La Trobe University, Australia

Abstract: This chapter enacts an approach to global science education research inspired by Gilles Deleuze and Félix Guattari's figurations of rhizomatic and nomadic thought. It imagines rhizomes 'shaking the tree' of modern Western science and science education by destabilising arborescent conceptions of knowledge as hierarchically articulated branches of a central stem or trunk rooted in firm foundations, and explores how becoming nomadic might liberate science educators from the sedentary judgmental positions that serve as the nodal points of Western academic science education theorising. This is demonstrated by commencing a rhizomatic textual assemblage that makes multiple, hybrid connections among the parasites, mosquitoes, humans, technologies and socio-technical relations signified by malaria in order to generate questions, provocations and challenges to dominant discourses and assumptions of contemporary science education.

Keywords: rhizomatic, nomadic, hybrid connections

1. Methodology and Mess

Figure 1 is my attempt to represent a mess. Of course, it's not 'really' a mess – it's too sparse and contrived to be that – but my purposes will be served if you can *imagine* the mess I am trying to represent here. When I think about issues of internationalisation and globalisation in relation to mathematics and science education – especially as I have experienced these in various nations/regions – including Australia, China, Europe, Iran, New Zealand, and southern Africa[1] – I imagine

[1] The experiences to which I refer are both direct (such as teaching or conducting research in these nations/regions) and vicarious (such as supervising or examining research conducted by doctoral students in these nations/regions).

B. Atweh et al. (eds.), Internationalisation and Globalisation in Mathematics and Science Education, 57–77.

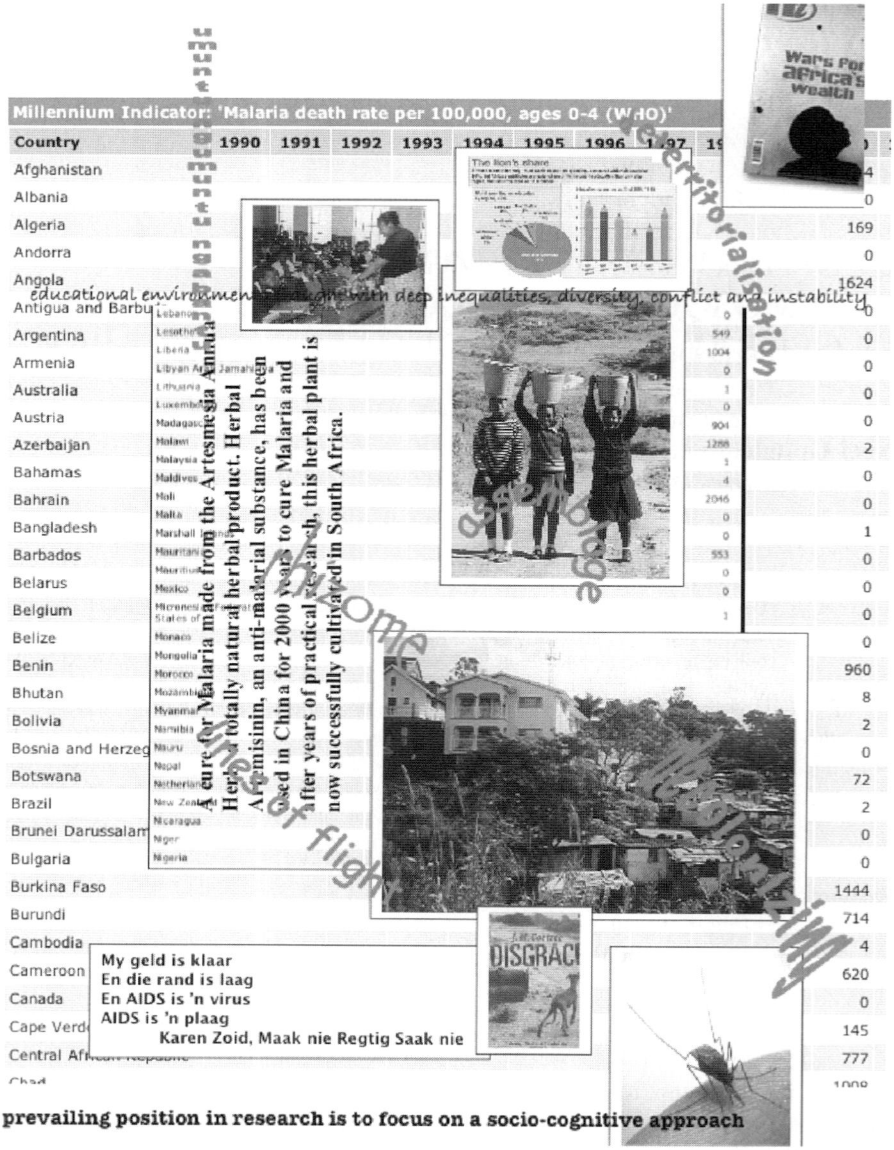

Figure 1. 'If this is an awful mess... then would something less messy make a mess of describing it?' (Illustration inspired by – and caption quoted from – John Law, 2003, pp. 2–3)

a mess. For example, southern Africa presents science education researchers with rapidly changing educational environments that are fraught with deep inequalities, diversity, conflict and instability, and if research is to be responsive and relevant to these circumstances we need to develop methodologies for knowing mess that help us to understand the politics of mess and messiness. My mess, therefore, is made from samples of texts (in the broadest sense of the term) that represent some of my understandings of the inequalities, diversities, conflicts and instabilities that constitute science education in many southern African nations.

I should say immediately that I do not think that so-called 'developing' nations are necessarily any messier than those we characterise as 'developed'. Deep inequalities, diversity, conflict and instability are certainly not unknown in the educational environments of Australia and other late capitalist nations, although they may be less pervasively and persistently visible (in colloquial terms they are less 'in your face') than in many 'developing' nations/regions. Moreover, the messiness that presents itself to educational researchers in Australia has some qualitatively different characteristics from that which confronts our colleagues in, say, South Africa. For example, we live with the cultural residues of the White Australia policy rather than of apartheid, but it is relatively easy for Australian researchers to ignore the historical traces of institutionalised racism as a component of our mess because they manifest themselves in more subtle ways than they do to South Africans.

However, the particularities of the image of mess that I have assembled in figure 1 are not important. Imagine your own picture of the messiness of mathematics and/or education in any 'developed' or 'developing' nation/region with which you are familiar, and then reflect on the caption I have borrowed from John Law (2003): 'If this is an awful mess... then would something less messy make a mess of describing it?' (p. 3). This is a rhetorical question. Law wants you to agree that *simplification does not help us to understand mess.*

Law (2003) asserts (and I concur) that: 'the world is largely messy' and that 'contemporary social science methods are hopelessly bad at knowing that mess'; furthermore, 'dominant approaches to method work with some success to repress the very possibility of mess' (p. 3). He invites us to imagine method more imaginatively, to imagine what method – and its politics – might be 'if it were not caught in an obsession with clarity, with specificity, and with the definite' (p. 3). Law (2003) argues that social science inquiry is mostly 'a form of hygiene':

> Do your methods properly. Eat your epistemological greens. Wash your hands after mixing with the real world. Then you will lead the good research life. Your data will be clean. Your findings warrantable. The product you will produce will be pure. Guaranteed to have a long shelf-life. So there are lots of books about intellectual hygiene. Methodological cleanliness. Books which offer access to the methodological uplands of social science research...
>
> In practice research needs to be messy and heterogeneous. It needs to be messy and heterogeneous, because that is the way it, research, actually is. And also, and more importantly, it needs to be messy because that is the way

> the largest part of the world is. Messy, unknowable in a regular and routinised way. Unknowable, therefore, in ways that are definite or coherent. That is the point of the figure. Clarity doesn't help. Disciplined lack of clarity, that may be what we need (p. 3).

In *After Method: Mess in Social Science Research*, Law (2004) elaborates upon this argument at much greater length. He does so in his own way, drawing on his immersion in the discourses of actor-network theory (ANT) and its successor projects. I also find ANT to be very generative in thinking about methodology but my current preference is to engage messy and heterogeneous objects of inquiry through the frames and figurations provided by Deleuze and Guattari's 'geophilosophy', especially their concepts of *rhizome* and *nomad*.[2]

2. Deleuze and Guattari's Geophilosophy

> Nietzsche founded geophilosophy by seeking to determine the national characteristics of French, English and German philosophy. But why were only three countries collectively able to produce philosophy in the capitalist world?
>
> —Gilles Deleuze & Félix Guattari, *What is Philosophy?* (1994, p. 102)

Deleuze and Guattari (1994) map the 'geography of reason' from pre-Socratic times to the present, a geophilosophy describing relations between particular spatial configurations and locations and the philosophical formations that arise therein. They characterise philosophy as the *creation* of concepts through which knowledge can be generated,[3] and create a new critical language for analysing thinking as flows or movements across space. Concepts such as *assemblage, deterritorialisation, lines of flight, nomadology,* and *rhizome/rhizomatics* clearly refer to spatial relationships and to ways of conceiving ourselves and other objects moving in space. For example, Deleuze and Guattari (1987) distinguish 'rhizomatic' thinking from 'arborescent' conceptions of knowledge as hierarchically articulated branches of a central stem or trunk rooted in firm foundations. As Umberto Eco (1984) explains, 'the rhizome is so constructed that every path can be connected with every other one. It has no center, no periphery, no exit, because it is potentially infinite. The space of conjecture is a rhizome space' (p. 57; see figure 2).

---

[2] Law (1997) recognises the convergences between ANT and Deleuze and Guattari's approach to the 'naming of parts' (p. 2) of complex systems. Elsewhere (see Gough, 2004c) I have coined the term 'rhizomANTic' to name a methodological disposition that connects rhizomatics with ANT and with Haraway's (1991) feminist, socialist, materialist technoscience.

[3] As Michael Peters (2004) points out, this is very different from the approaches taken by many analytic and linguistic philosophers who are more concerned with the *clarification* of concepts – Deleuze and Guattari complicate the question of philosophy: 'by tying it to a geography and a history, a kind of historical and spatial specificity, philosophy cannot escape its relationship to the City and the State. In its modern and postmodern forms it cannot escape its form under industrial and knowledge capitalism' (p. 218).

Figure 2. A tangle of rhizomes (drawing: Warren Sellers)

The space of educational research can also be understood as a 'rhizome space'. Rhizome is to a tree as the Internet is to a letter – networking that echoes the hyper-connectivity of the Internet. The material and informational structure of a tree and a letter is relatively simple: a trunk connecting two points through or over a mapped surface. But rhizomes and the Internet[4] are infinitely and continually *complicating*. They are irreducibly messy.

3. Commencing a Rhizome (shaking the tree)

Souma yergon, sou nou yergon, we are shaking the tree[5]

—Peter Gabriel & Youssou N'Dour, 'Shaking the Tree' (1989)

[R]hizomes are anomalous becomings produced by the formation of transversal alliances between different and coexisting terms within an open system.

—Gilles Deleuze & Félix Guattari, *A Thousand Plateaus* (1987, p. 10)

---

[4] See, for example, the Burch/Cheswick map of the Internet as at 28 June 1999 at http://research.lumeta.com/ches/map/gallery/isp-ss.gif accessed 3 August 2006.

[5] 'If we had known, if we only had known, we are shaking the tree'.

I interpret Deleuze and Guattari's (1987) figurations of *rhizome* and *nomad* as tools for 'shaking the tree' of modern Western science,[6] science education curricula and science education research. Thinking rhizomatically and nomadically destabilises arborescent and sedentary conceptions of knowledge as hierarchically articulated branches of a central stem or trunk rooted in firm and fixed foundations. The materials from which we can commence making such rhizomes are readily to hand, and can be found among the many and various arts, artefacts, disciplines, technologies, projects, practices, theories and social strategies that question and challenge the monocultural understandings of science reproduced by many science education programs and professors.[7] These materials include not only academic discourses/practices – such as feminist, queer, multicultural, sociological, antiracist, and postcolonialist cultural studies and/or science studies – but also the products and effects of popular arts and arts criticism.[8]

Peter Gabriel and Youssou N'Dour's song, 'Shaking the Tree', is emblematic of my project because it is a call to change and enhance lives that complements Deleuze and Guattari's geophilosophy. Both Gabriel and N'Dour compose and perform songs about taking action to do something about particular problems in the world, and Deleuze (1994) similarly argues that concepts 'should intervene to resolve local situations' (p. xx). 'Shaking the Tree' is a North-South collaboration (Gabriel is British, N'Dour is Senegalese) between two men who celebrate and affirm the women's movement in Africa, where patriarchal traditions and gender discrimination remain pervasive. Thus, as a popular song, 'Shaking the Tree' represents marginalised knowledges[9] in form as well as content – both popular media/culture and non-Western knowings tend to be ignored or devalued within many forms of Western science education, including those that have been exported to 'developing' nations. These exclusions contribute to the Eurocentric and androcentric character

---

[6] Elsewhere in this volume (see Gough, 'All around the world') I acknowledge and address the problematic 'West versus the rest' dualism that this formulation – 'modern *Western* science' – appears to introduce.

[7] This characterisation might appear to create a straw target but I firmly believe that it is defensible, For example, I have demonstrated elsewhere (Gough, 1998a, 2003, 2004a) that a number of science and environmental educators who argue cogently (and I believe sincerely) for more multiculturalist approaches – including William Cobern (1996), Victor Mayer (2002), and David Yencken, John Fien and Helen Sykes (2000) – nevertheless reproduce monocultural (and even culturally imperialistic) representations of science through their use of rhetorical strategies that implicitly privilege Western science and take for granted its capacity to produce universal knowledge.

[8] I therefore intend the word 'studies' to convey not only the conventional academic sense of pursuing some 'branch' of knowledge but also to suggest the various meanings of the noun 'study' in the arts, such as a sketch made as a preliminary experiment for a picture or part of it, or a musical composition designed to develop skill in a particular instrumental technique.

[9] I recognise that 'marginalised knowledges' might be interpreted to be an arborescent concept but would argue that representing some knowledges as 'marginalised' or 'subjugated' serves a useful referential function within an arborescent sign system. As Claire Colebrook (2002) points out, Deleuze and Guattari use binary oppositions – such as the distinction between rhizome and tree – to create pluralisms: 'You begin with the distinction between rhizomatic and arborescent only to see that all distinctions and hierarchies are active creations, which are in turn capable of further distinctions and articulations' (p. xxviii).

of much science education that can result from concentrating students' attention on two main ways of representing science, namely, documentary media (especially the science textbook and its equivalents in other media) and the 'theatre' of school laboratory work (see Gough, 1993b, 1998b). Representations of science in the arts and popular media – and the wide variety of contributions that they make (or might make) to the public understanding of science as a social, cultural, and historically contingent process tend to be given much less attention.[10] But I argue that a contemporary media genre such as SF – an acronym that designates much more than 'science fiction'[11] – provides many of the most convincing and publicly accessible demonstrations of how 'nature', as an object of knowledge, is culturally determined. Andrew Ross (1996) alludes to this capacity of SF to illuminate the social and cultural meanings and consequences of scientific research when he writes: 'Outside of *Jurassic Park*, I have yet to see a critique of Chaos Theory that fully exposes its own kinship with New Right biologism, underpinned by the flexible economic regimes of post-Fordist economics' (p. 114).[12]

---

[10] This is not to say that science educators and science education researchers ignore the effects of the arts and popular media on public understandings of science but, rather, that their attention to the range and variety of these effects is somewhat limited. For example, Norris, Phillips and Korpan (2003) point out that although research on *reading* science media reports extends back at least four decades, science educators have only recently begun to recognise their value in teaching and assessing scientific literacy; these authors also draw attention to 'the relatively small corpus of work that has been completed in this area' (p. 124). As I have demonstrated elsewhere (Gough, 1993b, 2001), many of the texts that purport to 'teach science fact through science fiction' (e.g., DeSalle & Lindley, 1997; Dubeck, Moshier, & Boss, 1988, 1994) portray popular media as sites of fantasy or of scientific 'misconceptions'. Thus, a common use of popular media in science education is to encourage students to identify these 'misconceptions'. For example, Dubeck et al (1988, 1994) devote two whole books to exposing 'pseudoscience' in more than fifty movies. More recently, in a special issue of *ENC Focus* on the theme of 'Becoming literate in mathematics and science', Frank Baker (2001) writes: 'Stereotypes and misconceptions are frequently generated by television and movie producers. Classroom teachers can take advantage of students' interest in popular movies to help them analyze the misconceptions' (n.p). Such readings constitute very narrow interpretations of popular media and implicitly devalue their educative potential by suggesting that their representations of science are in some way deficient unless they illustrate conventional textbook science 'correctly'. This obscures the ways in which particular works of art and popular media function as critical and creative probes of issues in science, technology and society that their creators and consumers see as problematic.

[11] As Haraway (1989) explains, since the late 1960s the signifier SF has designated 'a complex emerging narrative field in which the boundaries between science fiction (conventionally, sf) and fantasy became highly permeable in confusing ways, commercially and linguistically'; thus, SF refers to 'an increasingly heterodox array of writing, reading, and marketing practices indicated by a proliferation of "sf" phrases: speculative fiction, science fiction, science fantasy, speculative futures, speculative fabulation' (p. 5). Electronic games and web-based media and activities have added to the complexity and heterodoxity of this array. In addition, many of the interrogations of technoscience produced by visual, installation and performance artists that once might have been localised in a small number of galleries or exhibition spaces now reach a much broader audience via the websites that almost invariably accompany such exhibitions. See, for example, *Gene(sis): Contemporary Art Explores Human Genomics* at www.gene-sis.net <12 September 2004>

[12] 'New Right biologism' (like social Darwinism before it) is the selective and strategic deployment of biological ideas in the pursuit of conservative political and economic goals. Thus, free-market economists privilege selected interpretations of chaos and complexity theories in order to 'naturalise' the desirability

In his Preface to *Difference and Repetition* Deleuze (1994) asserts that a text in philosophy 'should be in part... a kind of science fiction' (p. xx) in the sense of writing 'at the frontiers of our knowledge, at the border which separates our knowledge from our ignorance and transforms the one into the other' (p. xxi). Much of my previous research and writing on science education has been concerned with demonstrating the possibilities of 'synthetically growing a post-human curriculum' – to quote John Weaver's (1999) interpretation of my work – by expanding and diversifying the cultural materials and tools that science educators deploy in their curriculum practices.[13] Here I will explore some of the immanent but hitherto unexplicated implications of this work and demonstrate how Deleuzean concepts might generate transformative possibilities for theorising science education and performing science education research. Like Laurel Richardson (2001), 'I write in order to learn something that I did not know before I wrote it' (p. 35), and so this text too is 'a kind of science fiction' that rewrites my philosophy of science education as geophilosophy.

For example, I have previously argued for adapting to the natural sciences a proposal that Richard Rorty (1979) makes in respect of the social sciences: 'If we get rid of traditional notions of "objectivity" and "scientific method" we shall be able to see the social sciences as continuous with literature – as interpreting other people to us, and thus enlarging and deepening our sense of community' (p. 203). I argued that seeing the natural sciences also as 'continuous with literature' means, to paraphrase Rorty, seeing both science and literature as interpreting the earth to us and thus 'enlarging and deepening our sense of community' with the earth. The consequences for science education research would then best be understood in terms of storytelling – of abandoning what Harding (1986) calls 'the longing for "one true story" that has been the psychic motor for Western science' (p. 193). Rather, we should deliberately treat our stories of science education research as metafictions – self-conscious artefacts that invite deconstruction and scepticism (Gough, 1993a, p. 622).

I initially drew my support for this argument from scholars who work at the intersections of literary criticism and science studies. For example, David Porush (1991) argues persuasively that in the world of complex systems revealed to us by postmodernist science – protein folding in cell nuclei, task switching in ant colonies, the nonlinear dynamics of the earth's atmosphere, far-from-equilibrium chemical reactions, etc. – 'reality exists at a level of human experience that literary tools are best, and historically most practiced, at describing' and that 'by science's own terms, literary discourse must be understood as a superior form

---

of global economic deregulation and oppose national development models that seek to internally regulate and articulate the agricultural and industrial sectors of a nation's economy. Instead of building national economies from networks of interlocking primary and secondary industrial assembly lines, post-Fordism promotes a global market in which efficiencies are achieved through, for example, farm concentration and specialisation.

[13] I have drawn particular attention to the significance for science education of contemporary cultural trajectories in popular media and global (eco)politics (see, e.g., Gough, 1993a, 1998a, 2001, 2002, 2004b).

of describing what we know' (p. 77). Such arguments are consistent with those offered by scholars whose work is identified with cultural studies of science (e.g., Haraway, 1994; Harding, 1994; Rouse, 1993), the discursive production of science (e.g., Bazerman, 1988, 1999), and sociological studies of scientific knowledge (e.g., Collins & Pinch, 1998; Latour, 1987, 1988, 1993, 1999; Latour & Woolgar, 1979; Shapin, 1994; Woolgar, 1988). Science education researchers who have drawn on these areas of inquiry include Jay Lemke (1992), who queries the defensibility of the 'simulations and simulacra of science' that students encounter in conventional school science courses (see also Lemke, 1990, 2001). Researchers who have paid particular attention to representations of science in popular media/culture include Peter Appelbaum (1995, 2000; see also Appelbaum & Clark, 2001) and Matthew Weinstein (1998a; 1998b; 2001a; 2001b).

Although I agree that literary and artistic modes of representation might be more defensible for many purposes in science education than the supposedly more 'objective' accounts of professional scientists and textbook authors, I now want to go beyond debating the merits and demerits of competing representationalist philosophies. In this previous work I moved away from the fixity and centeredness of a conventional scientist's (or mainstream literary scholar's) point of view, but I still worked within the limits of grounded positions, albeit positions found as I moved (in Rorty's terms) from one temporary standpoint to another along various continua between literature and science.[14]

Thus, in this chapter I explore what it means to be becoming nomadic in theorising science education, to liberate science education research from the sedentary points of view and judgmental positions that function as the nodal points of Western academic science education discourse. What happens when we encourage random, proliferating and decentred connections to produce rhizomatic 'lines of flight' that mesh, transform and overlay one another? Imagining knowledge production in a rhizomatic space is particularly generative in postcolonialist educational inquiry because, as Patricia O'Riley (2003) writes, 'Rhizomes affirm what is excluded from western thought and reintroduce reality as dynamic, heterogeneous, and nondichotomous' (p. 27), and this is precisely what I will attempt to demonstrate here by commencing a rhizome, a textual assemblage that I hope will generate questions, provocations and challenges to some of the dominant discourses and assumptions of contemporary science education research.

The approach I adopt here is inspired by the method Donna Haraway (1989) uses in the final chapter of *Primate Visions*, wherein she 'reads' primatology as science fiction, and vice versa:

> Placing the narratives of scientific fact within the heterogeneous space of SF produces a transformed field. The transformed field sets up resonances among all of its regions and components. No region or component is 'reduced' to

[14] As journals such as *Configurations* (published by Johns Hopkins University Press for the Society of Literature, Science, and the Arts) demonstrate, these are increasingly well-trodden paths.

> any other, but reading and writing practices respond to each other across a structured space. Speculative fiction has different tensions when its field also contains the inscription practices that constitute scientific fact. The sciences have complex histories in the constitution of imaginative worlds and of actual bodies in modern and postmodern 'first world' cultures (p. 5).

I depart from Haraway by imagining the 'transformed field' I produce as a rhizomatic and nomadic space rather than a 'structured space'. However, the final sentence in the quotation above opens a connection to the rhizome that I commence here, because it also resonates with 'Shaking the Tree'. The sciences not only have complex histories in the constitution of imagined worlds and actual bodies in modern and postmodern 'first world' cultures, but also in 'third world' cultures – and the silences about such histories *and geographies* in 'first world' science education textbooks and science journalism are aspects of scientific illiteracy that are doubly troubling when they are exported to (or imported by) 'developing' nations under the guise of 'best' practice (where 'best' is usually a euphemism for 'West').

4. Mosquito Rhizomatics

> We are writing this book as a rhizome. It is composed of plateaus... Each morning we would wake up, and each of us would ask himself what plateau he was going to tackle, writing five lines here, ten lines there. We had hallucinatory experiences, we watched lines leave one plateau and proceed to another like columns of tiny ants.
>
> —Gilles Deleuze & Félix Guattari, *A Thousand Plateaus* (1987, p. 22)

> [Rhizomes] implicate rather than replicate; they propagate, displace, join, circle back, fold. Emphasizing the materiality of desire, rhizomes like crabgrass, ants, wolf packs, and children, de- and reterritorialize space.
>
> —Patricia O'Riley, Technology, Culture, and Socioeconomics (2003, p. 27)

Ants have already inspired me to make a rhizome (see Gough, 2004c) so now I will allow mosquitoes to suggest connections, lines of flight and opportunities for deterritorialisation. I commenced these 'mosquito rhizomatics' in July 2004 when a number of initially separate threads of meaning – a research article in *Public Understanding of Science*, a *Time* magazine cover story, recollections of an SF novel and of various studies in the sociology of scientific knowledge read over the past decade and more – coincided, coalesced, and eventually began to take shape as an object of inquiry.

The *Public Understanding of Science* article that caught my attention was Stephen Norris, Linda Phillips and Connie Korpan's (2003) empirical-analytic study of university students' interpretations of scientific media reports, which followed an earlier (and similarly designed) study of high school science students by two of these authors (Norris & Phillips, 1994). Both studies sought to measure certain aspects of the students' interpretations of the meanings of five media reports, with particular reference to the degree of certainty with which various statements were expressed, the scientific status of statements (e.g., cause, observation, method) and the role of statements in each report's chain of reasoning (e.g., justifications for what ought to be done, evidence for other statements made in the report). According to the measurement instruments devised by these researchers, both high school and university students had difficulties with all aspects of the task, displaying a certainty bias in their responses to questions regarding truth status, confusing cause and correlation, and also confusing statements reporting evidence with statements reporting justifications. Although Norris, Phillips and Korpan (2003) admit that they designed their research to *assess* the interpretive abilities of high school students (rather than to 'explain' them), they nevertheless speculated that: 'The performance of these students suggests that the science curriculum had not prepared them well to interpret media reports of scientific research' (p. 125). The authors attempted to address this limitation in their study of university students by obtaining additional data, such as participants' self-assessments of their background knowledge and interests and the reading difficulty they ascribed to each report. Within the analytic framework of a correlational study, these self-assessments 'explained' virtually none of the variance in the interpretive performances of the university students who, 'in general, had an inflated view of their ability to understand the five media reports' (p. 123). Norris, Phillips and Korpan (2003) conclude:

> Generally, science textbooks used in high school and early university do not provide information on why researchers do their research, on the histories of research endeavors, on the motivation underlying particular studies, on how scientific questions arise out of the literature or anomalous events, or on the texture and structure of the language used in science. By contrast, scientific media reports often include the history and background to studies, information on the motivation for the reported research, and a variety of textured and structured language. In short, there is a mismatch. If media reports of science are to serve as an effective source of life-long scientific learning and support for public deliberation on science-related social issues, then much change is needed in high school and university science instruction to make this dream a reality. Otherwise, highly educated individuals having, as most of them do, little education specifically in science will help to run the major systems of our society without the benefit of being able to interpret and evaluate simple media reports of the latest scientific developments upon which those systems crucially depend (p. 141).

Norris, Phillips and Korpan's (2003) characterisations of the differences between science textbooks and scientific media reports converge with my own and others' qualitative studies of these types of textual materials (see, for example, Dunwoody, 1993; Gough, 1993b). But asserting that these differences constitute a 'mismatch' begs the question of why these distinct genres of science/education text should 'match' at all, given that science textbooks and scientific media reports serve different purposes. I suspect that what Norris, Phillips and Korpan might be suggesting here is that science instruction in high schools and universities equips students to interpret science textbooks but not scientific media reports, which therefore diminishes the potential effectiveness of the latter as resources for life-long scientific learning and for more increased public understanding of science and science-related social issues. However, positioning scientific media reports as some sort of desirable Other to science textbooks leaves unanswered the question of whether either type of text counters the Eurocentrism and/or androcentrism that tends to pervade science education wherever it is performed.

Shortly after reading Norris, Phillips and Korpan's (2003) article, I also read 'Death by Mosquito', a cover story in *Time* magazine by Christine Gorman (2004). The story begins by pointing out that malaria sickened 300 million people in 2003 and killed 3 million, most of them under age 5. In the same period AIDS killed just over 3 million people. 'What makes the malaria deaths particularly tragic', Gorman writes, 'is that malaria, unlike AIDS, can be cured'. She asks: 'Why isn't that happening?' Some selected excerpts from her story follow:

> Countries in sub-Saharan Africa have suffered the brunt of this renewed assault, but nations in temperate zones, including the U.S., are not immune...
>
> Doctors have long suspected that the malaria problem was getting worse, but the most searing proof has come to light in just the past year. Researchers believe the average number of cases of malaria per year in Africa has quadrupled since the 1980s. A study in the journal *Lancet* last June reported that the death rate due to malaria has at least doubled among children in eastern and southern Africa; some rural areas have seen a heartbreaking 11-fold jump in mortality...
>
> Recognition of malaria's toll on the global economy is growing. Economist Jeffrey Sachs, director of Columbia University's Earth Institute, estimates that countries hit hardest by the most severe form of malaria have annual economic growth rates 1.3 percentage points lower than those in which malaria is not a serious problem. Sachs points out that the economies of Greece, Portugal and Spain expanded rapidly only after malaria was eradicated in those countries in the 1950s. In other words, fighting malaria is good for business – as many companies with overseas operations have long understood...
>
> To better understand why malaria has become such a threat and what can be done to stop the disease, it helps to know a little biology.

I compared the 'little biology' provided by the *Time* article with the single page on malaria in a very popular Australian senior secondary school biology textbook,

Nature of Biology Book 2 (Kinnear & Martin, 2000)[15], which devotes one of its sixteen chapters to 'Disease-Causing Organisms'. The two texts exemplify the generic differences to which Norris, Phillips and Korpan (2003) draw attention, with the *Time* article providing more information on the history of research on malaria and on the social, political and economic motivations for current research efforts. In most other respects, both sources tell a very familiar story: malaria is caused by protozoan parasites and is spread among humans by mosquitoes. The *Nature of Biology* account provides little more than a brief description of the parasite's lifecycle and brief biological explanations for the symptoms of mosquito bites (itchy swellings) and malaria (chills and fevers) and concludes with the following paragraph:

> Although drugs are available for the treatment of malaria, a complete cure is difficult. This is because the parasite can remain dormant for many years in the liver before becoming active again. Different drugs are used against the different stages of the malarial parasite. Malaria is still one of the most serious infections in the world and is particularly common in some tropical and sub-tropical areas. The *Anopheles* mosquito, the main carrier of malaria, is common in these areas. Australians travelling to such areas are advised to take anti-malarial drugs both before, during and after visiting the area although this does not guarantee prevention of infection (Kinnear & Martin, 2000, p. 209).

Other than advising Australian travellers to take precautions against malaria, Kinnear and Martin say nothing about what makes it 'one of the most serious infections in the world' – nowhere do they mention that malaria can be fatal, or that it bears comparison with AIDS, tuberculosis and dysentery in currently being among the world's deadliest diseases. Not only do they tacitly diminish malaria's effects but there is also very little in Kinnear and Martin's account that might enable readers to understand at least some of malaria's social, cultural and historical determinants. The *Time* article provides explanations for the increased vulnerability of pregnant women and young children, and for variations in human resistance to the disease. *Time* also addresses issues on which *Nature of Biology* is completely silent, such as how malaria parasites have become increasingly drug-resistant, why controlling anopheles mosquitoes in tropical regions is so difficult, and why the most effective pharmaceutical responses currently available are beyond the financial resources of the poorest nations of the world, particularly those in Africa. Admittedly, *Time's* treatment of malaria is more than twice as long as that found in *Nature of Biology*, but Kinnear and Martin could still have made different choices about what to include in (and exclude from) their account. For example, is alerting Australians to the risks of malarial infection if they travel to certain regions really more important than alerting them to the massive human tragedy of millions dying of malaria in

[15] Kinnear and Martin's textbook was written specifically for the Victorian Certificate of Education biology study design for Units 3 and 4 (year 12) and was widely prescribed by teachers of the subject.

the West's tourist destinations and, moreover, that this is a tragedy that Western nations have the resources to ameliorate?

However, neither *Time* nor *Nature of Biology* provide readers with any alternatives to understanding malaria within what David Turnbull (2000) calls 'the knowledge space of Western laboratory science' wherein 'malaria is made to appear as a natural entity in the world, ... embedded in a conceptual framework which portrays the discovery and elucidation of its causes as occurring within the gradual unfolding of an emanent scientific logic, which will culminate in a physicochemical solution to the malaria problem' (p. 162). To step outside this logic requires, in Deleuze's terms, an act of *deterritorialisation* which, as Kaustuv Roy (2003) describes it, is 'a movement by which we *leave the territory*, or move away from spaces regulated by dominant systems of signification that keep us confined to old patterns, in order to make new connections' (p. 21; italics in original). Roy (2003) continues:

> To proceed in this manner of deterritorializing, we make small ruptures in our everyday habits of thought and start minor dissident flows and not grand 'signifying breaks,' for grand gestures start their own totalising movement, and are easily captured. Instead, small ruptures are often imperceptible, and allow flows that are not easily detected or captured by majoritarian discourses (p. 31).

I am disposed to produce such 'small ruptures' and 'minor dissident flows' by reading questions for inquiry in science education within an intertextual field that includes SF. In this instance, as I read the *Time* and *Nature of Biology* texts, I also recalled reading *The Calcutta Chromosome: A Novel of Fevers, Delirium, and Discovery*, a mystery thriller in the SF sub-genre of alternative history by Amitav Ghosh (1997). Like most works of SF, Ghosh's novel offers a semiotic space that is *not*, to recall Roy's words, 'regulated by dominant systems of signification', and that therefore invites readers to think beyond the sign regimes of Western laboratory science. Ghosh's novel becomes another filament in my mosquito-led rhizome by offering a speculative counterscience of malaria that connects with (but does not replicate) the 'real' history of Western medicine's explorations of the disease. For example, one of the key protagonists in Ghosh's alternative history is Ronald Ross, whose work on the lifecycle of the plasmodium parasite for the British Army's Indian Medical Service in late nineteenth century Calcutta eventually brought him a Nobel prize (although, as fictionalised by Ghosh, Ross is a much less heroic character than the one that official histories provide). *The Calcutta Chromosome's* action takes place in three temporal frames (the late 1890s, the mid-1990s and a very near future) and in locations that range from India to the Americas. Characters, places and events are connected by the mosquito, a vector that easily crosses boundaries between human and animal, rich and poor, here and there, and (with the parasite it transmits) also blurs the border between then and now.

Whereas *Time* and *Nature of Biology* occlude malaria's complex heterogeneity, Ghosh's novel dramatically foregrounds the ways in which outbreaks of malaria in particular places and times are manifestations of numerous complex interactions among parasites, mosquitoes, humans and various social, political (often military), administrative, economic, agricultural, ecological and technological processes. Although some of the interactions Ghosh depicts might be figments of his imagination, malaria's irreducible multiplicity is substantiated not only by malariologists (see Turnbull 2000, pp. 162–165) but also by scholars in other disciplines. For example, political historian Timothy Mitchell (2002) demonstrates that a terrible outbreak of malaria in Egypt during the 1940s cannot be understood as a predictable unitary event but as the effect of a series of complicated interconnections involving war, disease and agriculture:

> War in the Mediterranean diverted attention from an epidemic arriving from the south, brought by mosquitoes, that took advantage of wartime traffic. The insect also moved with the aid of the prewar irrigation projects and the ecological transformations those brought about. The irrigation works made water available for industrial crops, but left agriculture dependent upon artificial fertilizers. The ammonium nitrate used on the soil was diverted for the needs of war. Deprived of fertilizer, the fields produced less food, so the parasite carried by the mosquito found its human hosts malnourished and killed them at a rate of hundreds a day... The connections between a war, an epidemic and a famine depended upon connections between rivers, dams, fertilizers, [and] food webs... What seems remarkable is the way the properties of these various elements interacted. They were not just separate historical events affecting one another at the social level. The linkages among them were hydraulic, chemical, military, political, etiological, and mechanical (p. 27).

In the light of such examples, is it easy to see why sociologists of science such as Turnbull (1989) see malaria as a political disease 'resulting from the dominance of the Third World by the colonial and mercantile interests of the West' (p. 287). Indeed, the development of tropical medicine as a specialisation within Western medical science can itself be understood as a response by colonial administrators to the devastating effects of malaria and other tropical diseases on imperial demands for resources and labour. For example, Latour (1988) quotes a French colonial official who complained in 1908: 'Fever and dysentery are the "generals" that defend hot countries against our incursions and prevent us from replacing the aborigines that we have to make use of' (p. 141).

Diane Nelson (2003) has recently drawn on Ghosh's novel – and other works in social and postcolonialist studies of science, technology and medicine – to explore Paul Rabinow's (1989) proposition that Europe's colonies were 'laboratories of modernity' (p. 289). Although Latour's (1987, 1988) work is an obvious and acknowledged influence on Nelson's analysis, the introduction to her essay evokes rhizomatic connectivity as much as actor-networks:

> Ghosh imagines the science of malaria, a disease dependent on multiple connections, enmeshed in the logics of a colonial counterscience. In turn, I argue that the hybrid form of social science fiction may be the most adequate way to think about the delirious products and unlikely networks of these colonial laboratories. Malaria as a disease figures largely there, an emblem of the simultaneously faithful and fickle nature of postcolonial connectivity (p. 246).

Similarly, I argue that deliberately seeking and/or making multiple, hybrid connections between the texts of science education, scientific media reports, social studies and histories of science, SF and SF criticism (as I have demonstrated here with respect to the assemblage of parasites, mosquitoes, humans, technologies and socio-technical relations that malaria signifies) enables generative lines of flight from the defined territories of Western science and science education. In South Africa the need for science educators and researchers to move beyond the arborescent knowledge space of Western laboratory science is given further urgency by the increasing complexity of the linkages between traditional cultural practices (such as the production of herbal medicines by traditional gatherers and healers) and the activities of transnational corporations (such as large pharmaceutical companies). In this respect the *Time* article points out that during the past few years research on treating drug-resistant malaria has demonstrated the efficacy of combining several compounds – the most powerful of which is artemisinin, a 2000-year old Chinese herbal remedy derived from *Artesmesia annua* (sweet wormwood) that cures 90% of patients in three days. This plant is now being successfully cultivated in South Africa where some of the sources of traditional herbal malaria remedies have been over-harvested to near extinction. Given that around 80% of the black population in South Africa consult a traditional healer either before, after, or in preference to consulting a Western doctor, the interest of large pharmaceutical companies in traditional medicines is unsurprising, as is the potential for economic exploitation and environmental degradation (see, for example, Bbenkele, 1998). Europe's former colonies are still laboratories, not of colonialist modernity but of a neocolonialist postmodernity driven by the new imperialism of global corporate hypercapitalism. They are still laboratories for producing resources from Other people's labour in which the colonisers perform the experiments and the colonised are the guinea pigs. For example, as Sonia Shah (2002) reports, many non-Western countries have a thriving and largely unregulated industry in providing subjects for drug trials to multinational pharmaceutical companies.

I suggest that science educators and science education researchers in Europe's former colonies have a moral obligation to find the fissures in the arborescent and sedentary knowledge space of Western laboratory science and science education and begin to experiment with what Helen Verran (2001) calls 'postcolonial moments':

> Postcolonialism is not a break with colonialism, a history begun when a particular 'us,' who are not 'them,' suddenly coalesces as opposition to colonizer... Postcolonialism is the ambiguous struggling through and with

> colonial pasts in making different futures. All times and places nurture postcolonial moments. They emerge not only in those places invaded by European (and non-European) traders, soldiers, and administrators. Postcolonial moments grow too in those places from whence the invading hordes set off and to where the sometimes dangerous fruits of colonial enterprise return to roost (p. 38).

I have tried to demonstrate here that Deleuze and Guattari's (1987) figurations of rhizomatic thought and nomadic inquiry might nurture such postcolonial moments within the messy contexts of science education and research in regions such as southern Africa by disrupting and transgressing the epistemological and methodological colonialism (and even racism[16]) that results from the vast majority of epistemologies and methodologies currently legitimated in education having arisen almost exclusively from the social histories of the dominant Western cultures (and White races).

5. Losing *the* way: Becoming Nomadic in Science Education Research

> History is always written from the sedentary point of view and in the name of a unitary State apparatus, at least a possible one, even when the topic is nomads. What is lacking is a Nomadology, the opposite of a history.
> . . . nomads have no history; they only have a geography.
>
> —Gilles Deleuze & Félix Guattari, *A Thousand Plateaus* (1987, pp. 23, 393)

Ronald Bogue (2004) argues that Deleuze and Guattari's binary opposition of nomadic and sedentary is a '*de jure* distinction of pure differences in nature', that is, the nomadic and the sedentary are 'pure tendencies that are real, yet that are experienced only in various mixed states. They are qualitatively different tendencies co-present across diverse social and cultural formations' (p. 173). Bogue adds:

> Deleuze and Guattari's nomadic thought is inherently unstable in that its use of binary oppositions is intended to be generative and mutative, but it is not therefore to be pursued in a haphazard and careless fashion. . . yet their effort is not to fix categories and demarcate permanent essences, but to make something pass *between* the terms of a binary opposition, and thereby to foster a thought that brings into existence something new. In this regard, the categories of pure differences in nature are themselves generative forces of differentiation, which through their mutual opposition function to displace and transform one another (p. 178; italics in original).

---

[16] See James Scheurich and Michelle and Young (1997) for a more detailed discussion of epistemological racism.

Western science and science education also tend to be written from a sedentary point of view and I have thus tried to demonstrate the generativity of performing a nomadic subjectivity. Like Rosi Braidotti (2002), I understand that 'the nomadic subject is a myth, or a political fiction, that allows me to think through and move across established categories and levels of experience' and that choosing to 'become nomad' is 'a move against the settled and conventional nature of theoretical and especially philosophical thinking' (n.p.). However, I emphasise that my disposition to 'wander' away from the conventional semiotic spaces of science education textbooks and scientific media reports, and to experiment with making passages to hitherto disconnected systems of signification, is neither 'haphazard' nor 'careless' but a deliberate effort to unsettle boundary distinctions and presuppositions.

John Zilcosky (2004) notes that some poststructuralist and postcolonialist theorists see merit not only in wandering but also in getting lost: 'losing one's way – literally and philosophically – leads to a deterritorialization of knowledge: literary wandering subverts... and resists the systematisation of the world' (p. 229). However, I agree with Zilcosky that such claims might be disingenuous, that 'lostness presupposes a state of being found' (p. 240). I prefer to imagine nomadic wandering in the discursive fields of science education research not as 'losing *one's* way' but as losing *the* way – as losing any sense that just one 'way' could ever be prefixed and privileged by the definite article. Like rhizomes, nomads have no desire to follow one path.

6. Acknowledgement

I acknowledge prior publication of parts of this chapter in a special issue of *Educational Philosophy and Theory* on the philosophy of science education (Gough, 2006).

7. References

Appelbaum, P. (1995). *Popular culture, educational discourse, and mathematics*. Albany NY: State University of New York Press.

Appelbaum, P. (2000). Cyborg selves: Saturday morning magic and magical morality. In T. Daspit & J. A. Weaver (Eds.), *Popular culture and critical pedagogy: Reading, constructing, connecting* (pp. 83–115). New York and London: Garland.

Appelbaum, P., & Clark, S. (2001). Science! Fun? A critical analysis of design/content/evaluation. *Journal of Curriculum Studies, 33*(5), 583–600.

Baker, F. (2001). Media literacy: yes, it fits in math and science classrooms. *ENC [Eisenhower National Clearinghouse for Mathematics and Science Education] Focus, 8*(3), 48–49.

Bazerman, C. (1988). *Shaping written knowledge: The genre and activity of the experimental article in science*. Madison: University of Wisconsin Press.

Bazerman, C. (1999). *The languages of edison's light*. Cambridge, MA: MIT Press.

Bbenkele, E. C. (1998). *Enhancing economic development by fostering business linkages between the pharmaceutical companies and the traditional medicines sector*. Retrieved 17 January, 2005, from http://www.microfinancegateway.org/files/13974_Six_Studies_of_SMMEs_in_KwaZulu_Natal_and_Northern_Province_.htm#N_2_

Bogue, R. (2004). Apology for nomadology. *Interventions: International Journal of Postcolonial Studies, 6*(2), 169–179.

Braidotti, R. (2002). Difference, diversity and nomadic subjectivity. *Labrys. Études féministes*, (1–2), n.p.

Cobern, W. W. (1996). Constructivism and non-western science education research. *International Journal of Science Education, 18*(3), 295–310.

Colebrook, C. (2002). *Understanding deleuze*. Crows Nest NSW: Allen & Unwin.

Collins, H., & Pinch, T. (1998). *The golem: What you should know about science* (2nd ed.). Cambridge: Cambridge University Press.

Deleuze, G. (1994). *Difference and repetition* (P. Patton, Trans.). New York: Columbia University Press.

Deleuze, G., & Guattari, F. (1987). *A thousand plateaus: Capitalism and schizophrenia* (B. Massumi, Trans.). Minneapolis: University of Minnesota Press.

Deleuze, G., & Guattari, F. (1994). *What is philosophy?* (G. Burchell & H. Tomlinson, Trans.). London: Verso.

DeSalle, R., & Lindley, D. (1997). *The science of jurassic park and the lost world or, how to build a dinosaur*. London: HarperCollins.

Dubeck, L. W., Moshier, S. E., & Boss, J. E. (1988). *Science in cinema: Teaching science fact through science fiction films*. New York: Teachers College Press.

Dubeck, L. W., Moshier, S. E., & Boss, J. E. (1994). *Fantastic voyages: Learning science through science fiction films*. Woodbury NY: AIP Press.

Dunwoody, S. (1993). *Reconstructing science for public consumption: Journalism as science education*. Geelong: Deakin University.

Eco, U. (1984). *Postscript to the name of the rose* (W. Weaver, Trans.). New York: Harcourt, Brace and Jovanovich.

Gabriel, P., & N'Dour, Y. (1989). *Shaking the tree [Song]*. London/Paris: Peter Gabriel Ltd/Editions Virgin Musique.

Ghosh, A. (1997). *The calcutta chromosome: A novel of fevers, delirium and discovery*. New York: Avon Books.

Gorman, C. (2004, 26 July). Death by mosquito. *Time, 164*(4), 50–52.

Gough, N. (1993a). Environmental education, narrative complexity and postmodern science/fiction. *International Journal of Science Education, 15*(5), 607–625.

Gough, N. (1993b). *Laboratories in fiction: Science education and popular media*. Geelong: Deakin University.

Gough, N. (1998a). All around the world: science education, constructivism, and globalization. *Educational Policy, 12*(5), 507–524.

Gough, N. (1998b). 'If this were played upon a stage': School laboratory work as a theatre of representation. In J. Wellington (Ed.), *Practical work in school science: Which way now?* (pp. 69–89). London: Routledge.

Gough, N. (2001). Teaching in the *(Crash)* zone: Manifesting cultural studies in science education. In J. A. Weaver, M. Morris & P. Appelbaum (Eds.), *(Post) Modern science (Education): Propositions and alternative paths* (pp. 249–273). New York: Peter Lang.

Gough, N. (2002). Thinking/acting locally/globally: Western science and environmental education in a global knowledge economy. *International Journal of Science Education, 24*(11), 1217–1237.

Gough, N. (2003). Thinking globally in environmental education: implications for internationalizing curriculum inquiry. In W. F. Pinar (Ed.), *International handbook of curriculum research* (pp. 53–72). Mahwah NJ: Lawrence Erlbaum Associates.

Gough, N. (2004a). Global science literacy [book review]. *Science Education, 88*(1), 146–148.

Gough, N. (2004b). Living in a material world. In W. A. H. Scott & S. R. Gough (Eds.), *Key issues in sustainable development and learning: A critical review* (pp. 236–240). London and New York: RoutledgeFalmer.

Gough, N. (2004c). Rhizomantically becoming-cyborg: Performing posthuman pedagogies. *Educational Philosophy and Theory, 36*(3), 253–265.

Gough, N. (2006). Shaking the tree, making a rhizome: Towards a nomadic philosophy of science education. *Educational Philosophy and Theory, 38*(5), 625–645.

Haraway, D. J. (1989). *Primate visions: Gender, race, and nature in the world of modern science*. New York: Routledge.

Haraway, D. J. (1991). *Simians, Cyborgs, and Women: The Reinvention of Nature*. New York: Routledge.

Haraway, D. J. (1994). A game of cat's cradle: Science studies, feminist theory, cultural studies. *Configurations: A Journal of Literature, Science, and Technology, 2*(1), 59–71.

Harding, S. (1986). *The science question in feminism*. Ithaca NY: Cornell University Press.

Harding, S. (1994). Is science multicultural? Challenges, resources, opportunities, uncertainties. *Configurations: A Journal of Literature, Science, and Technology, 2*(2), 301–330.

Kinnear, J., & Martin, M. (2000). *Nature of biology book 2* (2nd ed.). Milton, Queensland: John Wiley & Sons.

Latour, B. (1987). *Science in action: How to follow scientists and engineers through society* (C. Porter, Trans.). Milton Keynes: Open University Press.

Latour, B. (1988). *The pasteurization of france* (J. Law, & A. Sheridan, Trans.). Cambridge, MA: Harvard University Press.

Latour, B. (1993). *We have never been modern* (C. Porter, Trans.). Cambridge, MA: Harvard University Press.

Latour, B. (1999). *Pandora's hope: Essays on the reality of science studies* (C. Porter, Trans.). Cambridge, MA: Harvard University Press.

Latour, B., & Woolgar, S. (1979). *Laboratory life: The social construction of scientific facts*. Beverly Hills: Sage Publications.

Law, J. (1997). *Topology and the naming of complexity*. Retrieved 17 January, 2005, from http://www.comp.lancs.ac.uk/sociology/papers/Law-Topology-and-Complexity.pdf.

Law, J. (2003). *Making a mess with method*. Retrieved 25 January, 2005, from http://www.comp.lancs.ac.uk/sociology/papers/Law-Making-a-Mess-with-Method.pdf.

Law, J. (2004). *After method: Mess in social science research*. London: Routledge.

Lemke, J. L. (1990). *Talking science: Language, learning, and values*. Norwood NJ: Ablex Publishing Corporation.

Lemke, J. L. (1992). *The missing context in science education: Science*. Paper presented at the Annual Meeting of the American Educational Research Association, Atlanta, GA.

Lemke, J. L. (2001). Articulating communities: Sociocultural perspectives on science education. *Journal of Research in Science Teaching, 38*(3), 296–316.

Mayer, V. J. (Ed.). (2002). *Global Science Literacy*. Dordrecht: Kluwer.

Mitchell, T. (2002). *The rule of experts: Egypt, techno-politics, modernity*. Berkeley CA: University of California Press.

Nelson, D. (2003). A social science fiction of fevers, delirium, and discovery: *The Calcutta Chromosome*, the colonial laboratory, and the postcolonial new human. *Science Fiction Studies, 30*(2), 246–266.

Norris, S. P., & Phillips, L. M. (1994). Interpreting pragmatic meaning when reading popular reports of science. *Journal of Research in Science Teaching, 31*(9), 947–967.

Norris, S. P., Phillips, L. M., & Korpan, C. A. (2003). University students' interpretation of media reports of science and its relationship to background knowledge, interest, and reading difficulty. *Public Understanding of Science, 12*(2), 123–145.

O'Riley, P. A. (2003). *Technology, culture, and socioeconomics: A rhizoanalysis of educational discourses*. New York: Peter Lang.

Peters, M. (2004). Geophilosophy, education and the pedagogy of the concept. *Educational Philosophy and Theory, 36*(3), 217–231.

Porush, D. (1991). Prigogine, chaos and contemporary SF. *Science Fiction Studies, 18*(3), 367–386.

Rabinow, P. (1989). *French modern: Norms and forms of the social environment*. Chicago: University of Chicago Press.

Richardson, L. (2001). Getting personal: Writing-stories. *International Journal of Qualitative Studies in Education, 14*(1), 33–38.

Rorty, R. (1979). *Philosophy and the mirror of nature*. Princeton NJ: Princeton University Press.

Ross, A. (1996). Earth to gore, Earth to gore. In S. Aronowitz, B. Martinsons, M. Menser & J. Rich (Eds.), *Technoscience and cyberculture* (pp. 111–121). New York and London: Routledge.

Rouse, J. (1993). What are cultural studies of scientific knowledge? *Configurations: A Journal of Literature, Science, and Technology, 1*(1), 1–22.

Roy, K. (2003). *Teachers in nomadic spaces: Deleuze and curriculum*. New York: Peter Lang.

Scheurich, J. J., & Young, M. D. (1997). Coloring epistemologies: are our research epistemologies racially biased? *Educational Researcher, 26*(4), 4–16.

Shah, S. (2002, 1 July). Globalizing clinical research: Big pharma tries out first world drugs on unsuspecting third world patients. *The Nation, 275*, 23–28.

Shapin, S. (1994). *A social history of truth: Civility and science in 17th century England*. Chicago: University of Chicago Press.

Turnbull, D. (1989). The push for a malaria vaccine. *Social Studies of Science, 19*(2), 283–300.

Turnbull, D. (2000). *Masons, tricksters and cartographers: Comparative studies in the sociology of scientific and indigenous knowledge*. Amsterdam: Harwood Academic Publishers.

Verran, H. (2001). *Science and an African logic*. Chicago and London: University of Chicago Press.

Weaver, J. A. (1999). Synthetically growing a post-human curriculum: Noel Gough's curriculum as a popular cultural text. *Journal of Curriculum Theorizing, 15*(4), 161–169.

Weinstein, M. (1998a). Playing the paramecium: science education from the stance of the cultural studies of science. *Educational Policy, 12*(5), 484–506.

Weinstein, M. (1998b). *Robot world: Education, popular culture, and science*. New York: Peter Lang.

Weinstein, M. (2001a). Guinea pig pedagogy: critiquing and re-embodying science/education from other standpoints. In A. C. Barton & M. D. Osborne (Eds.), *Teaching Science in Diverse Settings: Marginalized Discourses and Classroom Practice* (pp. 229–250). New York: Peter Lang.

Weinstein, M. (2001b). A public culture for guinea pigs: US human research subjects after the Tuskegee Study. *Science as Culture, 10*(2), 195–223.

Woolgar, S. (1988). *Science: The very idea*. London: Tavistock.

Yencken, D., Fien, J., & Sykes, H. (Eds.). (2000). *Environment, education and society in the Asia-Pacific: Local traditions and global discourses*. London and New York: Routledge.

Zilcosky, J. (2004). The writer as nomad? The art of getting lost. *Interventions: International Journal of Postcolonial Studies, 6*(2), 229–241.

5

SCIENCE EDUCATION AND CONTEMPORARY TIMES: FINDING OUR WAY THROUGH THE CHALLENGES

Lyn Carter
Trescowthick School of Education, Australian Catholic University, Melbourne, Australia

Abstract: This chapter argues for science education's engagement with contemporaneity, and for a repositioning of its research directions to better address the theoretical and methodological challenges raised. To this end, this chapter utilises the more usual discourses of globalisation (for example, Delanty, 2000; Harvey, 2000; Jameson, 1998), as well as Lash's (2002) formulation of global information culture. It begins by briefly recounting the impact of globalisation on education, and consequently, science education (Carter, 2005), before describing some of Lash's (2002) perspectives on global information culture relevant to contemporary science education. It sketches out some possible research directions for science education, as well as identifies some of the crucial issues of contemporaneity with which we as science educators can only begin to grapple

Keywords: Globalism, information culture, policy, reform, cultural globalisation

1. The Contemporary Realities of Science Education

We are condemned to live in interesting times; complex, challenging and rapidly changing times where economic, political and social transformations have profoundly reorganised the ways we interpret our world and its increasing interconnectivity (Carnoy & Rhoten, 2002; Delanty, 2000). While there is little disagreement over the existence of such change, the best ways in which it may be characterised are somewhat more contentious. Scholars emphasise different trends within a multiplicity of constructs. For instance, discourses that focus on changing economic conditions can include those on post Fordism (Sennett, 1998), the rise of global communications (Castells, 1996), the digital flows of international capital exchange (Lash & Urry, 1994), and the emergence of global urban environments or 'citistats' as the new organisational units (Dear & Flusty ,1999). Other theorists

B. Atweh et al. (eds.), Internationalisation and Globalisation in Mathematics and Science Education, 79–94.

investigate the consequences of some of these changes highlighting, for example, global ecological degradation (Mander & Goldsmith, 1996), and the growth of manufactured technological risk (Beck, 1992; O'Mahony, 1999). Still others focus on changing cultural life-forms including the bricolaged nature of identity (Coombes & Brah, 2000), the dominance of the image (Baudrillard, 1996), the rise of the global information order (Lash, 2002), cosmopolitanism (Friedman, 2000), and the move towards hybridity and transculturality (Welsch, 1999). These and other characterisations act as inscriptions upon contemporaneity, implying a familiarity and order to a complexity that otherwise threatens to overwhelm. They encode the desire to stake out the terrain of investigation that as if in their naming, they will render the contemporary world more tangible and explicable.

The discourses of *globalisation* referring to the recent transformations of capital, labour, markets, communications, technological innovations, distributions and ideas stretching out across the globe, are one such set of discourses that have become fundamental for constructing our understandings of the contemporary world. As the most macro of all of the discourses, globalisation draws from many disciplines and perspectives in an attempt to make sense of major social, cultural, technological and economic changes (Delanty, 2000; Harvey, 2000). The everyday consciousness is now one of a global imaginary, making us feel connected to far-flung places and events. Delanty (2000) is amongst the many theorists who group various characterisations of globalisation into the political economic transformations of globalism and sociocultural changes (see also Beck, 2000; Jameson, 1998; Scholte, 2000; Tomlinson, 1999). Within the former, the processes of convergence foster an increasingly hegemonic homogenisation embodied in the growth of neoliberal ideologies and of supra national regulation, the decline of the nation state, the extension of the enterprise form to scientific and technological innovation, and the expansion of Western capitalism and culture. Sociocultural characterisations on the other hand, emphasise the divergence in local adaptations of larger global forces so that diversity, identity and fragmentation become the leitmotifs of the global age (Paolini, 1999). Globalisation thus, can be thought of as a complex dialectic of both political-economic and sociocultural transformations that are still to be fully configured even as they work themselves into the materiality of the everyday (Jameson, 1998).

Yet not all scholars believe that theories of globalisation best capture the complexities of contemporaneity. For example, Lash (2002; 1999) prefers to focus on global information culture rather than globalisation because it emphasises the unifying principle of contemporaneity's architecture, that is, information itself. By information Lash (2002) means not only the informational or knowledge-enriched goods and processes of post-industrial society, but also the more recent form of information as message. Small, message-sized bites of information like the latest stock market figures, the most recent sport score, the hottest trend, the most topical political story and so on, incessantly circulate the globe constantly being updated or superseded. In contrast to the long-tested wisdom base held in the narrative and discursive structures of industrial society, informational knowledge possesses a limited currency that constantly defers to the latest. For Lash (2002), within a time that promotes the rationality of knowledge production as a consequence of neoliberal market ideologies, there has become an irrationality of information overload, misinformation, disinformation and out of control information. Hence in global information culture, the symbolic power resides with information and intellectual property that gets compressed to the immediacy of the particular, and that is quickly outdated and replaced and leaves almost no time for reflection. Clearly Lash's (2002) views hold profound implications for contemporary education heavily invested in knowledge as it is, including science education, which McLaren and Fischman (1998) see as still rooted in the same categories of educational debate as it has been for the last two to three decades. This apparent reticence to explore the changing global landscape limits our understanding of education's, and for our purposes here, science education's contemporary conceptual and material realities. Consequently, there is a need for science education to inquire into the complexities of contemporaneity, be they expressed as global information culture, globalisation, or any one of a number of other discourses, so as it can engage in dialogues about key issues that are practically and intellectually urgent, and which must be addressed if it is to remain relevant for our current times.

In this chapter then, I argue for science education's urgent and critical engagement with contemporaneity, and for a repositioning of its research directions to better address the theoretical and methodological challenges raised. To this end, I utilise the discourses of globalisation as well as Lash's (2002) formulation of global information culture. I have already argued elsewhere that despite science education's preference for the traditional categories of analysis, globalisation is clearly at work in science education's more recent policy and practical transformations (see Carter, 2005). The latest *Science for All* policy reform movement and its development of scientific literacy as the universalised goal of science education is a case in point. Here, I want to extend the discussion by considering some of the challenges science education faces if it takes the theorisations of globalisation and global information culture seriously. I begin then, by briefly recounting the impact of globalisation on education, and consequently, science education (Carter, 2005), before describing some of Lash's (2002) perspectives on global information culture relevant to contemporany science education. I attempt to sketch out some possible research directions for science education, as well as identify some of the crucial issues of contemporaneity with which we as science educators can only begin to grapple.

2. Educational Reform for the New Global Order

While most globalisation theorists would acknowledge education is, or should be, implicated in accounts of globalisation, their apparent preoccupation with elaborating its political, economic, legal, civic and other material and cultural dimensions unfortunately seems to have marginalised education as a key field within these categories. This is somewhat surprising given that knowledge and information is globalisation's fundamental resource, and education is a major player in its production, rationalisation and allocation (Delanty, 1998). It is left then, to the education literatures to explore the way globalisation constructs contemporary education, and education represents and circulates globalisation. Following Delanty (2000), these literatures can be thought of as drawing together around globalisation's two main characterisations as economic-political globalism, and sociocultural diversity. The educational policy research for example, have investigated the knowledge/power implications of global economic-political restructuring (globalism) manifest in various national educational reforms (see Apple, 2001; Astiz, Wiseman & Baker 2002; Carnoy, 2000; Daun, 2002; Levin, 1998; Li, 2003; Lingard & Rizvi 1998; Morrow & Torres 2000; Stromquist & Monkman, 2000; Torres, 2002; Wells, Slayton & Scott, 2002), while globalised cultural flows and diversity have begun to be explored within comparative and multicultural education discourses (for example, McCarthy & Dimitriades, 2000; McCarthy, Giardina, Harewood & Park, 2003; Stoer & Cortesao 2000).

Educational policy scholars have argued that the discourses of neoliberalism and neoconservatism have dominated the agenda of educational reform precipitated by globalism (see for instance Apple, 2001; Morrow & Torres, 2000; Wells et al., 2002). Neoliberalism is an economic and political fundamentalism that generalises the economic form to all human conduct including education (Burchell, 1993). Neoconservatism aims to reassert Eurocentric cultural control, protecting the 'Canon' from the contamination of competing narratives and practices newly available in the globalising world. Neoliberal and neoconservative forces work in tandem to marketise and reform, and as reform proceeds, to (re)distribute power back to traditional elites. While neoliberalism and globalisation are distinct phenomena, their intimate intertwining that sees neoliberalism open economic and political entities to globalisation, and globalisation foster neoliberalism, ensures that neoliberalism is generally regarded as the ideology of globalisation (Monkman & Baird, 2002). Pertinent for education is the dynamic relationship between the nation state, neoliberalism and globalism, as it is usually at the level of the nation state that educational reform policies and procedures are produced and enacted. Neoliberalism's imperatives of reduced governance and the rule of the markets has meant for the nation state, restructuring around the twin tendencies of centralisation and decentralisation. Decentralisation is achieved through devolution of administrative and other structures to the local site, while centralisation reconstitutes selected areas of strategic control with procedures for increased surveillance and accountability. In effect, this has generally meant there are fewer restrictions on educational institutional infrastructures with fiscal and other responsibilities being assumed at the school level. At the same time, control over areas like teacher autonomy and professionalism, and the curriculum, which were once at the discretion of school communities, have been tightened and centralised. Control is now exerted through standardised curricula, testing and auditing procedures across a range of student and teacher performance indicators, constituting schools as performative spaces providing increasing amounts of feedback upwards. Drawing on a combination of new institutional economics, new managerialism and performativity (Ball, 1998, 2000), both tendencies involve incentives for institutional change, the adoption of business practices such as privatisation and strategic planning and quality assurance. These practices both discursively and structurally determine the ways in which education can exist.

2.1 Science Education and Globalism

In common with most other areas of education, science education in many parts of the world has recently undergone an era of major reform consistent with these decentralising and centralising tendencies. A powerful influence on this latest phase of reform has been the American reports, Project 2061: *Science for All Americans* (American Association for the Advancement of Science, 1989), and the National Academy of Science's National Science Education Standards (National Science Council, 1996). These documents were produced in response to the perceived crisis in science education, and its implicated role in international challenges to the technoscientific supremacy, and the subsequent declining economic fortunes, of the United States identified in *A Nation at Risk* (1983). Together with other similar reports, Project 2061 and the National Science Education Standards reiterated the prevailing orthodoxy in place since the Second World War in national policies of all sorts, that of 'science, and by extension science education, for economic development' (see Drori, 2000). This model established the causal link between the amount and type of science taught, the objectives of national economic development, and international competitiveness. It took a utilitarian view of science, and assumed that a systematic programme for the development of a scientifically and technologically skilled workforce would lead to greater economic progress. Despite the dominance of this developmental model, Drori (2000) has shown that its policy assumptions have been rarely tested, and any evidence provided by the small number of studies investigating the connection between science education and economic development are at best, inconclusive. Nonetheless, Project 2061 and the National Science Education Standards have been highly influential within this conceptual model, and through their international dissemination have in effect, crystallised the directions for the curricula and teaching reform agendas for science education globally.

Influenced by these and comparable British reports into science, technology, economic development and education, Australia has developed national and state-level standards in science education (Dekkers & de Laeter, 2001) that promote the mastery of scientific knowledge, and changes in teaching and learning practices. For example, in the Australian state of Victoria from which I write, the official school curriculum now comprises standards-based, planning documents known as the *Curriculum and Standards Framework (CSF)* (Board of Studies 1995; 2000; 2005), organised into eight key learning areas, two of which are science and technology. They are the basis for curriculum planning and implementation, and student assessment, for the compulsory years of schooling (Preparatory-Year 10). Such standards are usually couched within a rhetoric of access, equity and diversity, but more likely conceptualised in precise and predictive terms benchmarked against international 'best practice' utilising national and international testing. In this vein, Australia alongside with other countries, participated in the Third International Maths and Science Study (TIMSS), and will in 2006, participate in the OECD's Programme for International Student Assessment (OECD/PISA) evaluation of scientific literacy in the 15-year-old cohort. Goodrum, Hackling & Rennie, 2001 suggest that the OECD/PISA assessments represent strong international agreement about the purposes of science education, as well as a new commitment by OECD countries to monitor outcomes of education systems in terms of the functional higher-order knowledge and skills important for national economic development and competiveness in the globalising world (also Drori, 2000).

Project 2061, the National Science Education Standards, the Victorian CSF and other similar, usually state-based, science education reform documents and standards aim to achieve their purposes through the development of scientific literacy as the main goal of science education. Embodied within the slogan of *Science for All* by which these reforms have become known, scientific literacy is regarded as an essential characteristic for living in a world increasingly shaped by science and technoscience. DeBoar (2000) has reviewed significant government position papers, policies, reports and scholarship to conclude there are up top nine meanings of scientific literacy. He argues that the version of scientific literacy adopted within Project 2061 and the National Science Education Standards was particularly narrow. Based upon the achievement of knowledge standards, usually expressed as scientific concepts and processes, scientifically literate students become those able to meet the specified standards. The standards are drawn from a narrow interpretation of what knowledge can constitute science, legitimating only canonical science. Such science for Harding (1998) and others (see Gough, 2003) is that endeavour produced in Europe during a particular historical period, and whose cultural characteristics have endured to dominate and regulate the boundaries of global understandings of science. Hence, in these and similar documents, a contracted meaning of scientific literacy has come to prevail, conflated with the mastery of readily-implementable, content-based standards and habits of mind.

While an overt acknowledgement of the relationship between globalisation and science education is mostly absent from the science education literatures (exceptions include Drori, 2000; Gough, 1999, 2003; McKinley, Scantlebury & Jesson, 2001), the clearest manifestation of globalisation within science education is in the pervasive reach of these science education reforms, that is, the widespread adoption of the hegemonic and homogenising educational model favouring curriculum and teaching standards coupled to sophisticated regimes of surveillance. This model has been comprehensively described in the educational policy literatures as a consequence of the globalisation's extension of the enterprise form to education. As knowledge is globalisation's fundamental resource and education is essential to its production and distribution, the imperatives for education reform including science education reform, have been largely generated beyond national borders, ideologically conceived with neoliberal and neoconservative market principles, discursively structured, and ultimately regulated by supra national institutions with little consultation with the broader educational research community. Although economic imperatives for science education are not new, what is new is the unique combination of neoliberalism and neoconservatism ideologies in which they are now embedded. Neoliberalism 'marketise' everything, even notions of subjectivity, desire, success, democracy and citizenship, in economic terms at the same time neoconservatism works to preserve traditional forms of privilege and marginalise authentic democratic and social justice agendas.

In general terms then, we can see in science standards neoconservative desire for 'real canonical knowledge' where legitimated agents, in this case academic scientists and selected science education professionals, work to reassert control, alongside the neoliberal desires for increased regulation, accountability and surveillance. Fensham's (1992) critique of Project 2061 for instance, makes it apparent that neoconservative forces have envisaged school science, yet again, as a steady induction into a particularised canonical version of science, despite new views emerging from fields like science studies and multiculturalism that have broadened our understandings of science. They recapitulate the 1960s curriculum projects into contemporary standards-based science curricula that Hurd (2002) observes is simply "updating the traditional principles and generalizations of science disciplines and labelling them standards" (p. 5). The science reforms also perform the neoliberal desire for increased regulation, surveillance and accountability apparent in the increasing acceptance of standardised content and student tests. In a climate where the need to develop measurable definitions for OECD/PISA testing has conflated scientific literacy within a narrow range of indicators, students, classes, schools or systems must show quantifiable results, so as improvements can be claimed and deficiencies blamed. Good test results, including the OECD/PISA scientific literacy indicators, are constructed as the value-added productivity, reiterating Carnoy and Rhoten's (2002) view that within a global economy, there is a need to measure national knowledge production and hold education workers (usually teachers) accountable.

In summary, it is clear that neoliberal and neoconservative education reform agendas of globalisation permeate a broad range of science education. Science education works somewhere in the spaces between globally influenced nation state policy production formulated as a consequence of globalism, and local sites of practice, strongly influenced by the continuing trajectories of normative science education. Thus, there is a naturalisation of globalisation's shaping forces, influencing and changing science education in ways that remain largely underacknowledged and opaque. These relationships have remained unexplored because I suggest, science education tends to ignore theorisations and analyses of contemporaneity prominent in the broader social sciences of which it is a part, to focus self-referentially on aspects of practice with which it has been traditionally concerned. In much the same way as the reforms have been problematised within the educational policy literatures, science education needs to problematise its reforms so as their connections to globalising contemporaneity and its ideologies can be fully investigated and elaborated. Moreover, science education needs to adopt more insightful and multifaceted conceptualisations of contemporaneity to truly rise to the overwhelming complex challenges that are before us. Hence, in the next section I describe some of Lash's (2002, 1999) relevant perspectives on global information culture as a way of extending science education's view of contemporaneity.

3. Global Information Culture

The complexity of our times is theorised by Lash (1999), who describes the world of global information culture as "a swirling vortex of microbes, genes, desire, death, onco-mice, semiconductors, holograms, semen, digitized images, electronic money and hyperspaces in a general economy of indifference" (p. 344) that increasingly fragments all before it as it separates us from our moorings. We are shifting, Lash (1999) argues, from a first epistemological modernity where knowing subjects constructed the objects of knowledge, to a second or reflexive modernity of ontology where objects themselves have become possessed of being. The rise of global information culture shifts us again, to somewhere else yet to be configured, but somewhere that sees human singularity decline as these objects begin to think. It is the age "of the inhuman, the post-human and the non-human, of biotechnology and nanotechnology" (p. 12), of an object material culture in which technologies, objects of consumption, lifestyles and so forth come to dominate the cultural landscape and take on the power to constitute, track, and judge. Like Benjamin's melancholic we have no recourse but to find our way, Lash (1999) laments, among the objects of this new time so resolutely upon us and irreversible.

At the heart of Lash's (2002) global information order is the shift from a national industrial society with its accumulation of goods and capital, and its social and civil institutions and norms, to a global information order where informational processes dominate, and individualisation is the new norm of sociation. Even material goods are informationalised with their knowledge-intensive designs, regulated content, global reach, inbuilt obsolescence, and their branding and trade-marking that can confer an instant recognition and worth often beyond the utility of the object. The intellectual value of relentless innovation disembeds objects from real value, producing a knowledge intensive rather than work intensive society with design studios and R&D laboratories replacing the factories that have moved offshore. For Lash (2002), the principal actors are more and more key global cities and amorphous groups in forms of 'disorganisation', rather than countries with hierarchical organisations. These new global networks of communication and exchange are non-linear, socio-cultural-technical assemblages that produce flexible, mobile, value and issue oriented groups as need dictates, and are highly differentiated in terms of access and control. The consequences include a networked global elite that identifies more with itself than with others, and an underclass excluded from the informational structures and flow. As low skill labour becomes more and more irrelevant to knowledge accumulation, power works less from unequal distribution and exploitation as it did in manufacturing order to a new form of exclusion and inclusion that Lash (2002) argues is inherently more violent and devastating.

Integral to Lash's (2002) global information order is the theory of unintended consequences. The paradoxical effect of knowledge intensive production argues Lash (2002), is a ubiquitous overload of information that flows and circulates, overwhelms and consumes as it spins out of control. This is disinformation, that is, information compressed to the immediacy of the present, message-sized, fact-based rather than abstract, instantaneously relevant, and whose speed and ephemerality leaves almost no time for reflection. Informational knowledge is quickly outdated and replaced in an environment of overload that can prevent our engagement because there is just too much to which to pay attention. In the swirl of these information flows, brand names, trademarks, platforms, regulations and standards become fixed reference points that help ameliorate the otherwise anarchy of overload. Without the time to develop narrative and discursive structures, deep meaning disappears leaving only contingent, everyday meaning with its application of algorithms. For Lash (2002), the rationality of knowledge intense production has resulted in out of control information, causing a chronic dialectic of disordering, reordering and disordering again that threatens to dumb us all down as we swirl around within its flow.

Like other globalisation and cultural theorists, Lash (2002) does not discuss education at length. However, his formulation of global information culture has profound implications for science education, and indeed education more generally, particularly as science itself has also shifted from fundamental inquiry towards tangible products as a consequence of informationalisation. Lash (2002) argues that knowledge intensive production requires an education that reflects the highly rational and analytical nature of that knowledge. Such knowledge is discursive, based upon abstraction, selection, and complexity reduction, and emphasises highly codified mathematical, verbal and computing skills. It problematises and tests out concepts, applies systematic rules, subsumes particulars, looks for connections, and attempts to be reflexively aware of all possibilities. It perpetuates a Cartesian attitude of subject/object binary in contrast to the hands-on, practical knowledge of manufacturing society that produced its mass commodity outcomes in conjunction with machines. This education emphasises the production of abstract outcomes like rationally-argued essays or research papers as proof of knowledge gained. The focus is on the knowledge production as intellectual property rather than the labour process that has been invisibly located elsewhere.

Yet, there is a rub. Reich (1991) has identified the 'symbolic-analytic', 'routine-production services' and 'in-person services' as three emergent categories of work in this new knowledge order. Symbolic-analytical workers are relatively small in number, stable in identity and proportion, and are involved in knowledge intensive production and services. They are the networked global elites. But with the material demands of an embodied life still with us, and the consequent outsourcing of most aspects of living from cleaning to child care, Reich (1991) argues that the greatest job expansion is really in lower knowledge and skill categories of routine-production services and in-person services. Reich's (1991) categories when connected to Lash's

(2002) view of a discursive education raise important questions about real purposes and distribution of a discursive education to which I will return below.

4. Some Possible Research Directions for Science Education

Not only have these complexities of contemporaneity been largely underacknowledged within science education, albeit as formulations of globalisation or Lash's (2002) global information order, they are also under researched and theorised. Without such insights, science education risks limiting the conceptual and analytical frameworks of much of its present and future scholarship, as well as misjudging the forces that have a direct impact on science classrooms. Elsewhere I have begun to sketch out some areas of research that will help elaborate science education's relationship to globalisation (see Carter, 2005). I reiterate these directions here and move on to extend the discussion to include the implications for science education of Lash's (2002) view of the knowledge intensive culture and its unintended consequence of information overload.

Firstly, within the policy arena there is potential for close analyses of policy documents, curriculum projects, research studies and a range of other science education policy texts using key concepts from globalisation theory and education policy. In Australia, Goodrum et al.'s (2001) influential report promoting 'scientific literacy for all' for example, could be deconstructively read to examine and judge the adequacy of the authors' theoretical discussion against the global imperatives for change. Moreover, there is scope to investigate decentralising tendencies and related policy issues like Drori's (2000) work on science education and global policy. Studies like these and others still to be developed may go someway towards explaining the inherent difficulties and range of issues involved when centralised reform agendas are devolved to decentralised agents responsible for their implementation. Secondly, beyond the policy arena case study research on the relationship between globalisation and specific local sites of science education like some of the research included in this volume needs to be extended. Such scholarship focuses on the nature of the interactions between the global and the local, and how their interpenetration becomes a mediating influence to what constitutes science education at any given site (after Monkman & Baird, 2002). This information would provide a fuller picture of science education important for both local stakeholders and the broader science education community to make better decisions about ways they wish to proceed.

Thirdly, researching the relationship between science education and globalisation's neoliberal and neoconservative reform agendas gives us another perspective on some of science education's current tensions, ambiguities and paradoxes. For example, one paradox lies in the conceptualisation of science standards as neoliberal demands for flexible practices that can be in tension with neoconservative desires for canonical concepts and processes. Hurd (2002) is one of the few science education scholars to recognise that globalisation's massive changes to science itself "has created the demand for a reinvention of school science" (p.7). Hurd's (2002) view is supported by Duggan and Gott's (2002) investigation of the science competencies required by current 'symbolic-analytic workers' in science-based industries. They found that whilst procedural understanding was vital, conceptual understanding was so specific that it was acquired only on a need-to-know basis. Notwithstanding these developments, those like Goldsmith & Pasquale (2002) continue to call for more rigorous conceptual understanding as part of science education reform. Clearly, there is a tension between the traditional canonical knowledge well favoured by neoconservatives and the demands of the knowledge-intensive workplace. The Victorian CSF provides a further particular example of this tension. The CSF science curriculum has been developed as largely 19th century, canonical knowledge with its few applications presented in the post-war linear model of 'pure' research and 'applied' technology. By contrast, the CSF technology curricula has been constructed as a type of post-Fordist vocationalism that promotes generic design and problem solving skills, intertwined research and application, just-in-time learning, and flexible specialisation, rather than the transmission of non-transferable knowledge and skills.

These examples represent various aspects of neoliberalism and of neoconservatism that when considered from within a frame that understands globalisation, while they are clearly contradictory, at the same time they are also complimentary. This view embodies the very nature of globalisation itself as simultaneously able to manoeuvre between/within/around, colonising all contexts, and consummate at creating the conditions for its own success. In other words, as the most macro of all discourses, globalisation is large enough to tolerate, accommodate and even encourage, competing and opposing tendencies, so that all bases are covered in order to maximise its success. Apple (1999) regards these tensions and contradictions as compatible with information supplied through increased surveillance enabling markets to make choices between options and so work better as markets. These competing tendencies in science education consequently represent different aspects of the larger discourse, and are integral to the reform processes themselves. It becomes a moot point as to whether they ever could be resolved.

5. The Implications of Global Information Culture for Science Education

While researching the impact of globalisation on science education along the lines suggested here is an obvious priority, Lash's (2002) prescient theorisations of global information culture provides us with further insights relevant to reviewing contemporary science education. In particular, his view of knowledge intensive culture and its unintended consequence of information overload raise crucial challenges to which science education needs to respond. For example, Lash's (2002) description of the highly abstract knowledge and discursive education required by the new knowledge intensive society has profound implications not only for the type of science education deemed appropriate but also for whom it is made available. This discursive approach continues to privilege, and radically intensifies, the narrow form of knowledge prevailing in science education since the massive reform efforts of the 1960s. As we have already seen, Fensham (1992) argues these reforms emphasised highly abstract scientific conceptual knowledge and processes explicitly geared to train future scientists and engineers. They were, and remain, in tension with a more general education required by the diverse learners staying on longer at school. The broadening of science education that occurred in response to the failure of these reforms (see amongst many other scholars Gardiner, 1999 for a discussion of broadening learning styles, and Aikenhead & Jegede, 1999; Cajete, 1999; Lee & Fradd, 1998; Michie, 2003 for more diverse cultural approaches to science education), is under increasing threat. A reinvigorated narrowing of science education to highly codified and abstract forms of knowledge in the interests of successful knowledge production brings with it an emerging constellation of power and inequality issues that we can only just begin to grasp. When coupled to Reich's (1991) description of the relatively small number and proportion of 'symbolic-analytic' workers, or global elite, supported by the global information economy, it clear that such an education is suitable for only a very few. Those that succeed with discursive knowledge have not only have access to the more secure, higher paid specialized jobs, but are also able to better negotiate their way through the complexities of global information society.

Yet difficult as these issues are, perhaps the most challenging aspects of contemporaneity to which all education, including science education, must respond is Lash's (2002) description of disinformation society, with its speed and ephemerality, new forms of connectivity and sociation, and its chaos of information overload. Flows of money, images, utterances, people, objects, communications, technologies, ideas and so on, solidify sometimes into standards and platforms like brands names and trademarks, sound bites and mantra-like rhetoric, open systems, and non-places. They can only be temporary and exist for ever decreasing intervals of time as the relentless speed of innovation makes them consistently redundant. If nothing else, Lash's (2002) view of disinformation gives us a perspective on the development of educational standards as part of the existence of other types of standards, regulations and platforms that attempt to fix reference points and impose some order over informationalised flows. It is only when points can be set despite their inbuilt obsolescence, that they can be utilised within the neoliberal market that drives contemporaneity. The rise of neoconservatism attempting to influence such standards with what were already know and value, is eminently understandable as we try to grapple with the overwhelming speed and nature of disinformation and its flows.

In what ways then, can science education enframe disinformation to help make sense out of the anarchy of such flows? In the sea of information, to what should science education pay attention, and for what purpose? How do we help our students develop skills for this complex new world? Such questions are only just becoming apparent as crucial to formulating a 21st century approach. Generating possible answers is another matter, and the danger is that our strategies will have too much of a past flavour as we are likely to start from the restricted social and cultural forms McLaren and Fischman (1998) see as still gripping educational debate. This danger is apparent in much of the science education literature with its under-theorised view of contemporaneity and under-examined assumptions, polemic even, about increases in student interest, motivation and learning destined to flow from current innovations in pedagogy and curricula. Ogawa's (2001) comments about the decreasing desire of Japanese youth to be involved with science and technology despite a very high receptivity to technoscientific products and services are extremely relevant here. Ogawa (2001) quotes the Japanese sociologist Kobayashi in arguing the inevitability of such disinterest in advanced techno-scientific informational society, and sees many implications for a contemporary science education seeking to reengage youth. Ogawa's (2001) observations speak to the relentless global circulation and superficial consumption of informationalised, technologised and taken-for-granted products and processes in which today's youth are embedded. At the same time, Lather (1998) asks what happens in terms of student knowledge and interest if the sociocultural and political contexts of science are taken into account. Such questions are important because they scrutinise assumptions about students' motivation to engage in science education of whatever persuasion, particularly now as young people well versed in dis/informational flow and technoscientific artefacts may need to be convinced that any science education is worthy of their time and attention.

6. Conclusion

Elaborating the relationship between science education and contemporaneity, be it as either theorisations of globalisation or global information culture, forces us to ask some hard questions about the relevance and future directions of science education. To what extent must science education reflect national responses to neoliberal and neoconservative global economic restructuring and the imperatives of the supra national institutions, or can spaces be forged for other types of reforms? In the sea of information, to what should science education pay attention, and for what purpose? In what ways can science education enframe disinformation to help make sense out of the anarchy of such flows, and what strategies should we help students develop? These questions challenge us to confront the type of science education we wish to work towards. This remains an intensely difficult and enduring dilemma. Personally, I want to work towards developing science education that values noncommodified forms of knowledge, relationships, activities and aspects of life, and that includes cultural recognition and social redistribution and inclusion within its agenda. While the form this may take is yet to be configured, an important part of its development is elaborating the relationship between globalisation, global information and science education. As science educators, we need to roll up our sleeves because there is a lot of hard thinking to do!

References

Aikenhead, G. S., & Jegede, O. J. (1999). Cross-cultural science education: a cognitive explanation for a cultural phenomenon. *Journal of Research in Science Teaching, 36*(3), 269–287.

American Association for the Advancement of Science, (1989). *Project 2061: Science for all Americans: Benchmarks for scientific literacy*. Washington, D. C.

Apple, M. W. (1999). *Power, meaning and identity. Essays in critical educational studies*. New York: Peter Lang Publishing, Inc.

Apple, M. W. (2001). *Educating the "Right" way*. New York: RoutledgeFalmer.

Astiz, M. F., Wiseman, A. W., & Baker, D. P. (2002). Slouching towards decentralization: Consequences of globalization for curricular control in national education systems. *Comparative Education Review, 46*(1), 66–91.

Ball, S. (1998). Big policies/small worlds: An introduction to international perspectives in education policy. *Comparative Education, 34*(2), 119–130.

Ball, S. J. (2000). Performativities and fabrications in the educational economy: Towards the performative society? *Australian Educational Researcher, 27*(2), 1–24.

Beck, U. (2000). *What is globalization?* Cambridge, UK: Polity Press.

Baudrillard, J. (1996). *The perfect crime* (C. Turner, Trans.). London: Verso.

Beck, U. (1992). *Risk society: Towards a new modernity* (M. Ritter, Trans.). London: Sage Publications Ltd.

Board of Studies, (2000). *Curriculum and standards frameworks II*. Carlton, Victoria: Board of Studies

Burchell, G. (1993). Liberal government and techniques of self. *Economy and Society, 22*(3), 267–282.

Cajete, G. A. (1999). Igniting the sparkle: An indigenous science education model. Skyand, NC: Kivaki Press.

Carnoy, M. (2000). Globalisation and educational reform. In N. P. Stromquist & K. Monkman (Eds.), *Globalization and education: Integration and contestation across cultures* (pp. 43–62). Lanham, Maryland: Rowman & Littlefield Publishers, Inc.

Carnoy, M., & Rhoten, D. (2002). What does globalization mean for educational change? A comparative approach. *Comparative Education Review, 48*(1), 1–9.

Carter, L. (2005). Globalisation and science education: rethinking science education reforms. *Journal of Research in Science Teaching. 42*(5), 561–580.

Castells, M. (1996). *The information age: Economy, society and culture: Vol. 1. The rise of the network society*. Oxford: Blackwell Publishers Ltd.

Coombes, A. E., & Brah, A. (2000). *Hybridity and its discontents. Politics, science and culture*. London: Routledge.

Daun, H. (Ed.). (2002). *Educational restructuring in the context of globalization and national policy*. New York and London: RoutledgeFalmer.

Dear, M., & Flusty, S. (1999). The postmodern urban condition. In M. Featherstone & S. Lash (Eds.), *Spaces of culture: City-Nation-World* (pp. 64–85). London: Sage Publications Ltd.

De Boar, G. E. (2000). Scientific literacy: Another look at its historical and contemporary meanings and its relationship to science education reform. *Journal of Research in Science Teaching, 37*(6), 582–601.

Dekkers, J., & de Laeter, J. (2001). Enrolment trends in school science education in Australia. *International Journal of Science Education, 23*(5), 487–500.

Delanty, G. (2000). *Citizenship in a global age: Society, culture, politics*. Buckingham: UK: Open University Press.

Delanty, G. (1998). The idea of the university in the global era: From knowledge as an end to the end of knowledge? *Social Epistemology, 12*(1), 3–25.

Drori, G. S. (2000). Science education and economic development: Trends, relationships, and research agenda. *Studies in Science Education, 35*, 27–58.

Duggan, S., & Gott, R. (2002). What sort of science education do we really need? *International Journal of Science Education, 24*(7), 661–672.

Fensham, P. J. (1992). Science and technology. In P. W. Jackson (Ed.), *Handbook of research on curriculum* (pp. 789–829). New York: Macmillan Education Pty Ltd.

Friedman, J. (2000). Americans again, or the new age of imperial reason? Global elite formation, its identity and ideological discourses. *Theory, Culture & Society, 17*(1), 139–146.

Gardiner, H. (1999). *Intelligence reframed: Multiple intelligences for the 21st century* New York, Basic Books.

Goldsmith, L. T., & Pasquale, M. M. (2002). Providing school and district-level support for science education reform. *Science Educator, 11*(1), 24–32.

Goodrum, D., Hackling, M., & Rennie, L. (2001). *The status and quality of teaching and learning of science in Australia schools*. Canberra, ACT: Department of Education, Training and Youth Affairs.

Gough, N. (1999). Globalization and school curriculum change: Locating a transnational imaginary. *Journal of Education Policy, 14*(1), 73–84.

Gough, N. (2003). Thinking globally in environmental education: Some implications for internationalizing curriculum inquiry. In W. F. Pinar (Ed.), *Handbook of international curriculum research*. New York: Lawrence Erlbaum Associates.

Harvey, D. (2000). *Spaces of hope*. Edinburgh, Scotland: Edinburgh University Press.

Harding, S. (1998). *Is science multicultural? Postcolonialisms, feminisms and epistemologies*. Bloomington, IN: Indiana University Press.

Hurd, P. D. (2002). Modernizing science education. *Journal of Research in Science Teaching, 39*(1), 3–9.

Jameson, F. (1998). Notes on globalization as a philosophical issue. In F. A. Jameson & M. Miyoshi (Eds.), *The cultures of globalization* (pp. 33–54). Durham, NC: Duke University Press.

Lee, O., & Fradd, S. H. (1998). Science for all, including students from Non-English-Language backgrounds. *Educational Researcher, 27*(4), 12–21.

Lash, S. (2002). *The critique of information*. London: Sage Publications Ltd.

Lash, S. (1999). *Another modernity a different rationality*. Oxford: Blackwell Publishers Ltd.

Lash, S., & Urry, J. (1994). *Economies of sign and space*. London: Sage.

Levin, H. M. (1998). Educational performance standards and the economy. *Educational Researcher, 27*(4), 4–10.

Li, H. -L. (2003). Bioregionalism and global education: A reexamination. *Educational Theory, 53*(1), 55–70.

Lingard, B., & Rizvi, F. (1998). Globalization and the fear of homogenization in education. *Change Transformations in Education, 1*(1), 62–71.

Mander, J., & Goldsmith, E. (Eds.). (1996). *The case against the global economy*. San Francisco, CA: Sierra Club Books.

McCarthy, C., & Dimitriades, G. (2000). Globalizing pedagogies: power, resentment, and the re-narration of difference. In N. C. Burbules & C. A. Torres (Eds.), *Globalization and education: Critical perspectives* (pp. 187–204). New York: Routledge.

McCarthy, C., Giardina, M. D., Harewood, S. J., & Park, J. -K. (2003). Contesting culture: Identity and curriculum dilemmas in an age of globalization, postcolonialism, and multiplicity. *Harvard Educational Review, 73*(3), 449–460.

McKinley, E., Scantlebury, K., & Jesson, J. (2001). *Mixing metaphors: Science, culture and globalisation*. Paper presented at the American Education Research Association, New Orleans: USA.

McLaren, P., & Fischman, G. (1998). Reclaiming hope: Teacher education and social justice in the age of globalization. *Teacher Education Quarterly, Fall*, 125–133.

Michie, M. (2003). The role of cultural brokers in intercultural science education: A research proposal. Paper presented at the Australasian Science Education Research Association, Melbourne.

Monkman, K., & Baird, M. (2002). Educational change in the context of globalization. *Comparative Education Review, 46*(4), 497–505.

Morrow, R. A., & Torres, C. A. (2000). The state, globalization, and educational policy. In N. C. Burbules & C. A. Torres (Eds.), *Globalization and Education: Critical Perspectives* (pp. 27–56). New York: Routledge.

National Science Council, (1996). *National science education standards*. Washington, D.C.: National Academy Press.

Ogawa, M. (2001). Reform Japanese style: Voyage into an unknown and chaotic future. *Science Education, 85*, 586–606.

O'Mahony, P. (Ed.). (1999). *Nature, risk, and responsibility. Discourses of biotechnology*. Houndmills, UK: Macmillan Press Ltd.

Paolini, A., ed Elliot, A., & Moran, A. (1999). *Navigating modernity*. Boulder, CO: Lynne Reiner Publishers.

Reich, R. B. (1991). *The work of nations : Preparing ourselves for 21st century capitalism*. London: Simon & Schuster.

Scholte, J. A. (2000). *Globalization: A critical introduction*. New York: St. Martins Press.

Sennett, R. (1998). *The corrosion of character: The personal consequences of work in the new capitalism*. New York: W.W. Norton & Co.

Stoer, S. R., & Cortesao, L. (2000). Multicultural and educational policy in a global context (European perspectives). In N. C. Burbules & C. A. Torres (Eds.), *Globalization and education: Critical perspectives* (pp. 253–274). New York: NY: Routledge.

Stromquist, N. P., & Monkman, K. (Eds.). (2000). *Globalization and education: integration and contestation across cultures*. Lanham, Maryland: Rowman & Littlefield Publishers, Inc.

Tomlinson, J. (1999). *Globalization and culture*. Cambridge, UK: Polity Press.

Torres, C. A. (2002). Globalization, education, and citizenship: Solidarity verses markets? *American Educational Research Journal, 39*(2), 2–14.

Wells, A. S., Slayton, J., & Scott, J. (2002). Defining democracy in a neoliberal age: Charter school reform and educational consumption. 39, *2*(337–357).

Welsch, W. (1999). Transculturality: The puzzling form of cultures today. In M. Featherstone & S. Lash (Eds.), *Spaces of culture: City-nation-world*. London: Sage Publications Ltd.

6

SOCIAL (IN)JUSTICE AND INTERNATIONAL COLLABORATIONS IN MATHEMATICS EDUCATION

Bill Atweh[1] **and Christine Keitel**[2]

[1] *Curtin University of Technology, Australia*

[2] *Free University of Berlin, Germany*

Abstract: The literature mathematics education contains several references to issues related to social justice, including gender, racial and multicultural aspects, and perhaps to a lesser degree, socioeconomic factors. More commonly, this literature discusses social justice in terms of "equity" and "equal opportunity". However, very rarely the term social justice is theorised. This chapter aims to: (a) present a theoretical discussion of the construct "social justice" from a variety of perspectives, and (b) apply the theoretical discussion to raise issues of social justice behind several types of international contacts and collaborations between educators in the discipline

Keywords: Globalisation; International contacts; Social justice; International collaboration

Social justice concerns are no longer seen at the margins of mathematics education research and practice. Issues relating to gender, multiculturalism, ethnomathematics, and the effects of ethnicity, indigeneity, socio-economic and cultural backgrounds of students are regularly discussed in the literature and many of these have found their way into education policies in many countries around the world. More recent concerns about access to appropriate mathematics education by students with learning difficulties and special needs, the gifted and talented, and the so called "what about the boys" agendas are increasingly being constructed as social justice issues. Undoubtedly, different writers have different understandings of social justice – at times leading to contradictory conclusions and demands.

We note that in the early literature in mathematics education, claims for social justice have often presented their case in isolation from each other. Gender in mathematics is perhaps the first to be established as a strong social justice movement

B. Atweh et al. (eds.), Internationalisation and Globalisation in Mathematics and Science Education, 95–111.

within mathematics education. In 1987, the affiliation of the International Organization of Women in Mathematics Education was the result of decades of persistent research and lobbying by women educators from many countries. Following the history of this movement, we note a pattern of widening attention to other claims of social justice to include ethnicity and culture, more usually referred to in the USA as "race", social class and cultural differences. The yearbook of the National Council of Teachers in Mathematics on "Multicultural and Gender Equity in the Mathematics Classroom" in 1997 illustrates this diversification of the social justice agenda.

In this current chapter, we deal with issues of social justice as they relate to one area of mathematics education practice that is not usually discussed in the literature; namely, international collaborations and contacts among mathematics educators. This issue is already raised in Atweh, Clarkson, and Nebres (2003, p. 224,) where the authors argue that

> international contacts and exchanges in mathematics and mathematics education have … increased in the new age of globalization and will continue to exponentially increase in the future with further developments in technology, ease of travel and population movements. While we do not construct such contacts as necessarily either good or bad, the outcomes of these processes should be carefully scrutinized world wide as to the benefits and losses that might arise from them. This can only be achieved through deliberate and targeted research, reflection and debate.

This chapter contributes to this debate by raising questions about social justice issues behind such collaborations. We commence by problematising the social justice agenda in mathematics education, pointing out its lack of explicit conceptualisation and relating it to more prevalent concepts of equity and diversity. This is followed by a short discussion on the different theoretical models presented in general education literature. Based on these theoretical considerations, the third section discusses major issues of social (in)justice in international collaboration. The Chapter concludes with a discussion of a case study of an international project and how it dealt with these issues in its design and its policies.

1. Problematising the Social Justice Agenda in Mathematic Education

Although social justice represents a strong area of research in mathematics education, the term itself remains under-theorised (Gewirtz, 1998)[1]. Social justice remains a "contested area of investigation" (Burton, 2003, p. xv). Our contention here is that working *for social justice* necessitates *working for theorising* its

[1] Since the writing of this chapter, the Montana Mathematic Enthusiast published a monograph on International perspectives on social justice in Mathematics Education.

meaning(s) – and vice versa. To use the terminology of Mamokgethi Setati (2005), we need *working with the concept* as well as *working on the concept*. Often, debates on social justice are debates between different understandings of the concept as much as lack of commitment to its ideals. Rather than attempt to present a universally accepted definition of social justice – a task, we argue, is neither possible nor desirable – we will attempt to unpack a multidimensional concept and develop a language in which we can discuss different issues in social justice and overcome the fragmentation of its agenda by identifying the limitations of its various understandings. We will begin our discussion of social justice by relating it to the terms "equity" and "diversity".

Firstly, at the risk of essentialising the difference between the USA's and Europe's writings on social justice, there seems to be some difference between its conceptualisation in the two contexts – at least in mathematics education. The dominant view from the USA associates social justice with equity. Hart (2003) asserts that "Because the terms *equity*, *equality* and *justice* have been used in different ways in the literature, it is important to briefly consider some of the meanings of these terms" (p. 29). Using Secada's (1989) conceptualisation of equity as "our judgement about whether or not a given state of affairs is just" (p. 29), implies that equity is the measure to know if social justice has been done. Hart (2003) uses a multidimensional definition of equity as equal opportunity, as equal treatment and as equal outcome and concludes by saying that "I will use equity, as Secada … did, to mean justice" (p. 23). In the same volume, Secada, Cueto, and Andrade (2003) note that "the viewing of group-based inequality as an issue of equity has a long tradition within policy-relevant social science research … and in different forms of educational research in particular" (p. 108).

In an attempt to differentiate between equity and social justice, Burton (2003) from the UK, in her introduction to her book "Which Way Social Justice in Mathematics Education", argues that there is a "shift from equity to a more inclusive perspective that embraces social justice" (p. xv). She goes on to say "the concept of social justice seems to me to include equity and not to need it as an addition. Apart from taking a highly legalistic stance, how could one consider something as inequitable as socially just?" (p. xvii)

Secondly, the social justice agenda in mathematics education is at times discussed in relation to *diversity* – also a term that has its origin in the USA literature (Loden & Rosener, 1991). While the concept of *equity* arose from, and is often associated with – though not exclusively – gender concerns, the concept of *diversity* arose from, and is often associated with – though not exclusively – concerns about cultural and linguistic diversity (Sepheri & Wagner, 2000; Thomas, 1996). Plummer (2003), however,presents an overview of what he calls the "big 8" dimensions of diversity: *race, gender, ethnicity/nationality, organisational role, age, sexual orientation, mental/physical ability and religion*. In this context, social justice is constructed as "managing diversity" (Cox, 1991; Krell, 2004). For our purposes here, we note that, undoubtedly, the increasing diversity of students in most mathematics classrooms, and the persistent research evidence that some groups of students are not achieving or participating in mathematics as much as other students raise serious social justice issues. However, the diversity discourse might lead to essentialising the differences between the different groups and it may fail to take into consideration the changing constructions of these labels and their contextual understanding in time and place. Similarly, the diversity discourse fails to take into consideration one of the biggest threats to social inequality, namely socio-economic background and poverty that are difficult to construct as a diversity issue in the same meaning as cultural values and practices.

2. Conceptualisations of Social Justice

One of the few writers in mathematics education to have attempted to theoretically define social justice is Cotton (2001), from the UK. In his chapter entitled Mathematics Teaching in the Real World, the author presents, perhaps more implicitly, what can be taken as alternative definitions of social justice. For the author, "social justice is linked directly to issues of *power and control*" (p. 24; our italics). He explains that "injustice has been done when someone takes a decision that affects us personally or emotionally and with which we disagree but against which we have no power to argue" (p. 25). The author goes further to argue that social justice is also linked with "*individual rights*: rights to education, rights to individual life choices, without being denied access to certain chances through discriminatory practices, and the rights to fight against practices we perceive as unjust" (p. 25; our italics). Further in his chapter, he presents a definition of social justice as "*fairness*" based on the writing of John Rawls and the metaphor of halving an apple: "If two people are sharing an apple, one person cuts the apple and the other has the choice which half they want" (p. 26). Lastly, Cotton goes on to acknowledge "the approaches to social justice explored so far are, in the main products of the deliberations of groups of men" (p. 27). Using some feminist theorisations he argues: "the push towards autonomy within society leads to a detached view of an individual, living within a hierarchically ordered society, whereas the values of care and attachment create a network or *relationships*" (p. 27; our italics). He concludes: "In fact somewhat paradoxically, for me that the concept of social justice represents a shift in thinking *away from* equality in classrooms. 'Equality' can suggest a norm towards which we should strive. It does not easily accept and value difference" (p. 28; our italics).

In the following section we will examine three main models of theorising social justice. We will present the main tenets of each model and discuss some of their limitations. Due to the limited theorisation of the concept from within mathematics education, we will rely on works from outside the discipline itself.

2.1 Market Model

Arguably, the model of social justice based on the market logic of free competition and de-regulation is not prominent in mathematics educational discourse; however, this rationality is highly influential in the neo-liberal thinking which characterises many governments educational funding and policies. According to this model, social justice is a "desert" depending on one's labour and what one deserves rather than what one possesses or acquires. It is the icing on the cake to which people who struggle and achieve are entitled. Based on Plato's conception of justice, i.e., giving to each what is rightfully theirs, Nozic's (1981) construction of inequality is a matter of bad luck or lack of effort, rather than injustice. Wealth is the right reward for effort. Society is simply to act as a protector of the ability for personal justice – "its function is to pass laws that protect individual ability to pursue good" and "a just society can then be called one in which the wrong are punished, and the rest let prosper". In educational practice, McInerney (2004) identifies the rise of individual rights to education and the thinking of common sets of learning exemplified by the National Curriculum in the UK as reflections of this model of social justice. Similarly, we can understand trends towards privatisation, deregulation and the demise of public education in many countries in the light of such constructions of justice.

By focusing on individual effort, this model of social justice fails to take into consideration the role of social structures and relations in determining individual or group success and achievement. It does not question issues such as cultural capital and the unlevel playing field that students start from based on factors such as their gender, socioeconomic background, and ethnicity. Similarly, it neglects the role of exam regimes, the prescribed curriculum, and the language of instruction that privilege certain students over others.

2.2 Distributive Model

Distributive models of social justice focus more on unequal opportunities in society rather than mere outcomes. McInerney (2004) argues that a society cannot be called just unless "it is characterized by a fair distribution of material and non material resources" (p. 50). Rawls (1973) claims "the primary subject of social justice must be the basic structure of society, or, more precisely, 'the way in which the major social institutions distribute fundamental rights and responsibilities and determine the division of advantages from social cooperation'" (in McInerney, 2004, p. 50). At the same time as he is affirming the individual rights to pursue goods, he is insisting that distribution of wealth, income, power and authority are justifiable if they work to maximize the benefit of the least advantaged in society. Gewirtz (1998) identifies two forms of distributive justice: a weak form, equality of opportunity, and a strong form, equality of outcome.

In education, distributive models of social justice are reflected in compensatory programs allocating designated resources for the disadvantaged. However, this model does not question the curriculum itself, the pedagogy or the regimes of testing used in the classroom and their role in creating educational inequality. Further, it constructs the disadvantaged as individuals and not as parts of a collective. Finally, it does not take into account the reasons for the inequality that have historical roots and are socially and politically determined. Here, we note that the majority of compensatory programs to increase the achievement of target groups in education follow this construction of social justice.

2.3 Recognition Model

Several poststructuralist feminist writers have critiqued distributive models. Gewirtz (1998) argues that *relational* understandings of social justice are needed in order to "theorize about issues of power and how we treat each other, both in the micro face-to-face interactions and in the sense of macro social and economic relations which are mediated by institutions such as the state and the market" (p. 471). Relational models of social justice deal with "the *nature* and *ordering* of social relations" (p. 471, italics in original). Gewirtz goes on to indicate that "the relational dimension is holistic and non-atomistic, being essentially concerned with the nature of inter-connections between individuals in society, rather than with how much individuals get" (p. 471).

Similarly, Young (1990) argues that traditional conceptions of social justice are based on "having" rather than "doing". Grounding social justice in individual solutions that allow little room for the consideration of multiple social groups is inadequate. Furthermore, extending models developed on the distribution of material goods to other goods such as self-respect, honour opportunity, and power, is problematic. To understand the struggles for social justice experienced by a variety of social groups, a new model of social justice is needed based on the principle of *recognition*. Nancy Fraser (1995) expounds:

> Demands for "recognition of difference" fuel struggles of groups mobilised under the banners of nationality, ethnicity, 'race', gender and sexuality. … And cultural recognition replaces socio-economic redistribution as the remedy of social injustice and the goal of political struggle.
>
> (p. 68)

In response to the critique that giving attention to cultural recognition might have devalued economic inequality that is best alleviated through a distribution model, Fraser (2001) argues that social justice today requires *both* redistribution and recognition measures. She presents a model of "parity of participation" as a guiding principle that incorporates both models.

Whereas it might be problematic to define social justice in concrete terms, feminist authors often resort to identifying practices that are unjust. In particular, Young (1990) presents five signs of injustice: *exploitation, marginalisation, powerlessness, cultural imperialism,* and *violence*. In the following section, we will use these markers of social injustice to raise some issues of social justice in international collaboration in mathematics education.

3. Can International Collaboration be Unjust?

3.1 Can International Collaborations be Exploitative?

Traditionally, this Marxist concept is used to refer to conditions of, and returns to, the different social groups from work carried out by themselves or by others. It "does not consist only in the distributive fact that some people have great wealth while most people have little" (Young, 1990, p. 49), but "enacts a structural relation between social groups" (pp. 49–50). For the discussion here, we are interested in questions of exploitation with respect to symbolic goods such as knowledge, and in research as knowledge production, rather than with the traditional understanding of labour.

The first example relates to the phenomenon of "brain drain" from less industrialised to industrialised countries. One aspect of the globalised world we live in is the increase of movement of people between different countries facilitated by factors such as a decline in costs of travel and an increased awareness of different cultures. Some such movement, such as that by international students, has been undertaken in order to gain knowledge, but international movement has also been a means to access new research sites. Undoubtedly great benefits for individuals and countries may result from such exchanges among educational personnel. Less industrialised countries can benefit from research and curriculum development and knowledge generated in more affluent contexts that, in principle, can lead into a *distribution* of knowledge from the *haves* to the *have-nots*. However, such benefits may or may not be mutual or equitable; in some cases it has led to catastrophic results in less industrialised countries. To start with, such interactions are often not based on models of *recognition* that acknowledge the equal contribution to global knowledge of educators from less industrialised countries. Often, theories and research results are taken uncritically from industrialised countries into contexts that are quite different in values and resources leading to predictable failure.

Further, such interactions often lead to brain drain from the less industrialised countries. In a previous publication, Atweh (2003) discusses concerns of mathematics educators from the Philippines relating to a loss of some of their best teachers and academics from schools and teacher training institutes for overseas destinations. While the phenomenon of transition of university educators to overseas destinations is perhaps not new (UNESCO, 1998), the Philippines is experiencing an escalating brain drain to countries such as the United States. Although there are no concrete statistics on this phenomenon, one informant talked about at least twenty members of one cohort of her students requesting early transcripts of their results because they wanted to move overseas to fill a shortage of teachers mainly in the United States. One should note, however, that migrant academics often find themselves in lower positions than in their home countries – that is, academics teaching in schools and school teachers becoming childcare workers. Even so, considering the low socio-economic conditions in the country, such movement is very attractive to the individual teachers. However, in an ironic sense, a country like the Philippines is providing teacher education and preparation to one of the richest countries in the world.

Secondly, questions of exploitation can be raised when consideration is given to who benefits from international research activities and whose views are expressed in them. To illustrate this concern, we will take the specific example of research into ethnomathematics. By taking this example, we do not imply that questions of exploitation are specific or intrinsic to this area of research. On the contrary, researchers from this perspective have in the main been guided by concerns about social justice aimed at recognition and emancipation of groups of disenfranchised learners of mathematics. Yet, Dowling (1998) makes the observation that nearly all research and writing in mathematics education comes from researchers from within cultural groups who have identified with the dominant "Western" mathematics tradition. These researchers, "external" to the cultures they have studied, have looked at the practices of other cultural groups. Further, Vithal and Skovsmose (1997) maintain that while ethnomathematics researchers have been able to study the development of mathematics as interactions of power "between" different cultural groups, they have not done the same with power interactions "within" the different cultural groups. Questions need to be raised as to the effect of seeing the mathematics by outsiders on changing the lived reality of the people from the inside. In particular, how can this ethnomathematics be used by the insiders to challenge their subordination from within and from outside their particular culture? Hence, even with the best intentions of studying the knowledge of the voiceless in international educational debates, there still remains the concern about whose knowledge is bring represented and who is benefiting from such studies.

3.2 Can International Contacts Lead to Marginalisation?

Young (1990) argues that marginalisation is "perhaps the most dangerous form of oppression" (p. 53). It occurs when a "whole category of people is expelled from useful participation is social life and thus potentially subjected to severe material deprivation and even extermination" (p. 53). In line with the focus of this chapter on knowledge rather than material goods we will raise a few questions about marginalisation of educational interests, needs and voices from less industrialised nations in international contacts in mathematics education.

In mathematics education, international conferences play a key role in internationalisation of the discipline. On one hand, for many educators from less industrialised and affluent nations, they are the primary, and in some instances the sole face-to-face contact that they have with the international scene in mathematics education. Undoubtedly, such contact might have led to further collaboration between educators outside the boundaries of the organisation itself. On the other hand, academics from less industrialised countries are often unable to participate because of high costs of travel and accommodation overseas. Very rarely are these international gatherings held in less expensive locations. For example, the last two International Congress of Mathematic Education (ICME) conferences were held in Japan and Denmark – two highly expensive locations even for educators from industrialised countries. Atweh and Clarkson (2002b) reported on interviews conducted with educators from Colombia where mathematics educators expressed a great feeling of isolation from international debates in education due in the main to their lack of participation in international gatherings. This lack of participation implied, among other things, that the great achievements in the education systems of their country, such as the Escuela Neuva (New School) (Constanza, 2000), an innovation internationally recognised for its excellence,[2] remain virtually unknown in international educational publications and theory. While the internet has contributed to diminishing the feeling of isolation for some of the educators in the country, it has failed to make the dialogue with the international community genuinely reciprocal.

Further, questions about marginalisation can be raised regarding the choice of research questions and methodologies used in research. Bishop (1992) comments on the divergence of research questions and methodologies adopted by educational researchers around the world. Here, we note two aspects that can be construed as a marginalisation of mathematics education in less industrialised countries. First, research questions investigated in less industrialised countries often mirror research questions in international publications from more industrialised countries. Questions that are of direct concern to industrialised countries, such as teaching in large classes and resource poor contexts, have not achieved the same prominence in mathematics education literature as other topics. Atweh (2003) reports on educators from the Philippines describing researchers in the country as being "very much influenced by what they see in [international] journals". At times, the research questions are not judged by their contribution to improving the practice of teaching in the local context. Some, indeed, were seen as researching "trivial topics". Similarly, these academics argued how some methodologies developed and stressed in less industrialised countries, for example action research, remain at the margin of methodologies in the Anglo-European research.

3.3 *How is Powerlessness Constructed in International Collaboration?*

Young (1990) claims that the powerless are "those who lack authority or power ... those over whom power is exercised without exercising it; the powerless are situated so that they must take orders and rarely have the right to give them" (p. 56). We acknowledge the problematisation of the concept of power in terms of poststructuralist writing. However, the lack of reciprocity in sharing knowledge between countries raises serious questions for the mathematics education community about the power of educators and policy makers in developing countries to make decisions about their systems based on their locally produced knowledge.

This concept of powerlessness may be of assistance in understanding a phenomenon referred to by Atweh et al. (2003) of calls from certain educators from less industrialised countries for a global curriculum. At the ICME regional conference in Australia in 1995, the president of the African Mathematical Union (Kuku, 1995) warned against the over-emphasis on culturally oriented curricula for developing countries that act against their ability to progress and compete in an increasingly globalised world. He called for "a global minimum curriculum below

[2] In 1988, it was chosen by the World Bank as one of three most outstanding projects undertaken in developing countries, with UNESCO hailing it as one of the most important educational innovations in recent years.

which no continent should be allowed to drift, however under developed" (p. 407). Also, at the same conference, a similar call was given by Sawiran (1995), a mathematics educator from Malaysia. Sawiran based his comments on the belief that "our experience shows that mathematics is an important ingredient of technology and therefore is a key element to 'progress'" (p. 603) (quotes in original). He concluded his address by saying that "[t]he main thrust in enhancing better quality of education is through 'globalisation' of education. In this respect, it is proper to consider globalisation in mathematics education" (p. 608) (quotes in original). He added that the most important step in globalisation is through "collaborative efforts" (p. 608).

Many educators in Western countries are concerned about standardisation of the curriculum for its lack of sensitivity to differences due to the cultural and social backgrounds of students (Apple, 1993) and their effect on demoralisation and deprofessionalisation of teachers (Hargreaves, 1994). While the two authors Kuku and Sawiran referred to here are not representative of all voices in the industrialised and less industrialised countries, such divergent views can be partially explained by the contexts from which they speak. Undoubtedly, marginalisation of less industrialised countries as discussed above leads to the feeling of disempowerment by educators from many developing countries. However, the economic situation in many countries is seen by many educators as a major limitation for them to develop their own local research and curriculum development programs. Hence, it may not be surprising to see how international standards and curriculum might be constructed as viable alternatives.

Atweh et al. (2003) discuss how Colombian educators have expressed a great sense of disempowerment when it comes to international collaborations. As mentioned above, Colombian mathematics educators operating in a globalised world sense of a lack of reciprocity and a limited ability to "exchange" with overseas countries on equal terms. One academic made the distinction between "copying" and "appropriating" ideas from outside the country. Due to limited resources, the former means of international exchange was seen as more dominant in their situation and as a form of colonialism. According to one educator, "we feel we are in a diminished situation, so minimal, that we are only a small piece in the big board".

3.4 Can International Contacts Lead to Cultural Imperialism?

Young (1990) defines cultural imperialism as "how the dominant meanings of a society render the particular perspective of one's own group invisible at the same time as they stereotype one's group and mark it as the Other" (p. 59).

The dominance of Anglo-European views of mathematics and mathematics education has often been contested in the literature on mathematics education. Questions can be raised about the proliferation of curricula around the world that were developed by educators from and based on research conducted in industrialised countries. In a publication on ethnomathematics, the editors, Powell and

Frankenstein (1997), have chosen the subtitle: Challenging Eurocentrism in Mathematics Education. Research on the history of mathematics has demonstrated that the contribution of non-Mediterranean cultures to the development of mathematics is often marginalised. Commenting on the ICME7 conference, Rogers (1992) laments that "all our theories about learning [of mathematics] are founded in a model of the European Rational Man, and that this starting point might well be inappropriate when applied to other cultures" (p. 22). He goes on further to assert that "the assumptions that mathematics is a universal language, and is therefore universally the same in all cultures cannot be justified. Likewise, the assumptions that our solutions to local problems ... will have universal applications is even further from the truth" (p. 23).

The issues discussed above under the sections of marginalisation and powerlessness contribute to the dominance of Anglo-European knowledge on the international scene. Here, we will discuss the potential of international comparative studies in promoting cultural imperialism in mathematics education. The publication of results from the recent Third International Mathematics and Science Study (TIMSS) has ignited interest in a type of research that is based on cross-country comparisons in curriculum and student achievement. This type of study has generated a considerable amount of controversy within the mathematics education literature. Robitaille and Travers (1992) give the case for international studies on achievement while others identify concerns about their validity, usefulness, misuses and abuses. Keitel and Kilpatrick (1999) raise several political questions about comparative studies. They argue that the outcomes of these studies are perceived as biased towards the host country; that is, of those who do the data collection, the analysis and the funding. These authors question whether this is to the detriment of other countries and their concerns about improving education systems. The authors add that "no allowance is made for different aims, issues, history and contexts across the mathematics curricula of the systems being studied" (p. 243). They conclude that comparative testing is not really useful as an educational tool, as it does not produce a clear view of what is really happening in the classroom and why.

In a comprehensive discussion of international studies, Clarke (2003) summarises the potential dangers of the misuse of such activities as follows:

> (i) Through the *imposition* on participating countries of a global curriculum against which their performance will be judged; (ii) Through the *appropriation* of the research agenda by those countries most responsible for the conduct of the study, the design of the instruments, and the dissemination of the findings; and (iii) through the *exploitation* of the results of such studies to disfranchise communities, school systems, or the teaching profession through the implicit denigration of curricula or teaching practices that were never designed to achieve the goals of the global curriculum in which such studies appear predicated.
>
> (p. 178)

3.5 Can International Collaboration Contain an Element of Violence?

It is true that many educators in mathematics education live under constant threat of violence from within and from without their immediate society. If violence is taken as use of force to cause physical damage, then this criterion of injustice may be less relevant to studying international contacts in mathematics education. However, if violence is taken to mean the use of coercion to perform a certain action, then the means of imposing certain forms on developing countries should be questioned as they relate to symbolic violence. Here, we will discuss mathematics educators' concerns about certain educational reforms tied to funding projects from the World Bank and their effect on mathematics research and teaching in less industrialised countries.

Atweh et al. (2003) discuss the role of the World Bank in several less industrialised countries. The authors argue that to understand the role of the World Bank in education, it is essential to understand that it is primarily a financial banking institution governed by the logic of sound investment. Accountability to its lenders is a paramount concern behind its decision-making. It is not an organisation for policy and theory development. While its impact on policy in education in many less industrialised countries is significant, it is not to be seen as having the same role as UNESCO, for example, in its role to generate new ideas and broad educational vision. Nor is it the usual aid or social welfare agency. The Bank's programs are based on commercially sound investments and not necessarily on the aspirations of the recipients. In discussing the World Bank from this angle it is not to be taken that all of its activities are evil and harmful. Undoubtedly, it has been highly influential in developing mathematics education programs in many developing countries (Jacobsen, 1996).

Atweh and Clarkson (2002a) reports on a focus group on globalisation with some Brazilian mathematics educators in which some participants discussed the role of the World Bank and its equivalent international funding organisations on the education systems in their country. It should be recalled here that Brazil is a country that suffers from massive foreign debt. A large portion of the country's budget goes towards paying the many loans that the country has taken in the past 40 years. It was portrayed as a continuation of the process of colonialisation and described as "perverse globalisation". A similarity has been drawn between paying taxes to the colonial powers of the past and paying taxes to the new financial colonials of our age: "Now … when the United States revolted against the taxes paid to England … they were against taxes paid to the [English] crown. [In the same way, the] independence of Latin America was about revolt [against] the taxes paid to the [Spanish] crown. Now we are paying taxes to another crown that is the international financial system. … This is the way they just keep getting taxes and they keep getting richer and richer".

Another country that was affected by the priorities of the World Bank is Colombia. Higher education, which had been expanding throughout the 1990s, has seen a reversal in its growth. Beginning in 1998 and continuing until the present, the number of new entrants to tertiary education has been declining. The coverage rate currently stands at just 15% which compares unfavourably to other countries in the region and to the OECD country average of 54%. At the insistence of the World Bank, public institutions have increasingly shifted their revenue base towards cost-recovery where 49% of revenues came from students as of 2000. As a consequence, the number of entrants into tertiary education declined by 19%. Private providers enrol more than two-thirds of students. This makes the higher education sector in Colombia far less accessible, and hence, far more inequitable than ever before. Only 192 students were enrolled in Doctoral level studies in the country in 2003.

4. A Case Study: The project "The Learner's Perspective Study" (LPS)

The Learners Perspective Study is a collaborative project between an expanding network of researchers from several countries including Australia, China (Mainland, Hong Kong and Macao), Czech Republic, Germany, Great Britain, Israel, Japan, Korea, Norway, the Philippines, Singapore, South Africa, Sweden, Portugal, and the United States. The project commenced with a shared concern about methodological as well as social justice issues about international comparative studies such as TIMSS (cf., Clarke, 2001; Keitel & Kilpatrick , 1999). In one sense, the LPS is an international collaborative research project, designed to use powerful and innovative data collection methods, which aims at integrating complementary analyses of the substantial international data set generated through the combined efforts of the participating researchers. However, it is also a study that gives social justice considerations a high priority. The project has recently compiled two volumes reporting on the first analyses and interpretations, which provide an insider's view, as well as comparative accounts under specific themes that had been considered of mutual interest and worthwhile for in-depth collaboration (Clarke, Emanuelson, Jablonka, & Mok, 2006; Clarke, Keitel, & Shimizu, 2006).

4.1 Social Justice in Methodology

Unlike previous studies, the LPS does not look at classroom practice in isolation. It constructs the classroom within a wider social context. It is based on the assumption that classroom culture cannot be understood as a result of a single lesson observation. Further, rather than merely concentrating on teachers' actions alone, the study focuses the attention on the student as well as the teacher in order to obtain a more holistic understanding of what is happening in the classroom. This is not only good methodological consideration but also a social justice principle giving a voice to all participants in the classroom interactions, especially the students. Likewise, the study is based on the belief that the substance of a social practice, such as that found in a mathematics classroom, cannot be fully understood without trying to reconstruct the meanings that the participants attribute to their actions.

The full methodology is found in Clarke, Keitel, et al. (2006, Chap.2). Here, we will merely identify its main features. The LPS:

- consisted of videotaping a sequence of at least ten consecutive lessons;
- used three cameras at once to record teachers' talk and student activities; and
- conducted post class interviews with individuals or groups of students to explain their understandings and actions in the classroom.

All lessons and interviews were transcribed and translated into English as the principal language of the project. However, the analysis and project deliberations included chances to discuss differences in the languages represented within the project, often giving rise to challenging discussions about meaning and understanding.

The data in the project was subjected to three types of analysis. First, project-wide analysis was conducted using themes agreed to by the whole group aimed at identifying ways in which role-related asymmetries and culturally sanctioned ways of interaction serve as an orientation for the participants in mathematical classrooms. Second, subgroups of the project that shared specific interests conducted analysis on subsets of the data. Last, provisions were made for individual researchers to conduct their own analysis of the data as they saw fit.

4.2 Social Justice in Group Organisation

Research in LPS is based and deeply dependent on an equal collaboration of the members of the research teams from each participating country. All research decisions were negotiated amongst the researchers. Alternative interpretations of the data were shared and adopted or contested. The LPS is guided by a belief that people need to collaborate and learn from each other.

LPS is atypical insofar as it is a completely non-hierarchal project. Each partner in the group has equal rights and support within the project, and decisions are only taken unanimously. The complete set of data is accessible to each and every partner country. However, negotiation needs to be conducted to access data from another country. Because the teachers were equal partners, they had access to the data of their classroom videos – excluding the video-stimulated-recall interviews with their individual students – and they also could use some videos themselves to discuss with their students and colleagues in their school.

It is worth noting that the countries involved in the project were not deliberately chosen based on research design principles. Involvement grew out of personal contacts and the interests of individual researchers. Questions about who may join the project, what are the members' obligations, how to make sure that decisions are made in a just and fair way, how to increase the competencies to collaborate effectively and with equal rights, were constant concerns and subject to debates at each meeting of partners.

A couple of regular meetings are scheduled each year alongside international conferences that are attended by the majority of the group of the partners. In addition, one regular meeting is usually held once a year in Melbourne, Australia, where the International Centre for Classroom Research has been established as an appropriate place to house technological installations and data sets and to facilitate mutual exchange and collaboration: The Centre is also open to partners researchers, who share their expertise and use the most up-to-date equipment.

Perhaps the participation by the Philippines is particularly interesting for our discussion. Although the Philippines' educators wanted to join the international team, they were concerned about the lack of Philippine funds available to conduct such a study, as well as their ability to participate at the group's international meetings. To encourage participation, other project participants elected to subsidise the Philippines by sending them equipment previously used in the Australian study as well as contributing to expenses to join the group meetings.

Collaboration with South African participants followed a different approach. One member of the original founding group spent a period of several months between 1999 and 2001 assisting the local partners in an application for funding from the South Africa National Research Foundation and participating in the first round of data collection and analyses. This collaboration assured the full ownership of the project by local researchers, as well as providing intellectual support and discussion and referring to experiences of other data from other countries. These comparisons were inserted in the project from its first stages and have proved invaluable for the local colleagues to reflect on their newly developed and ambitious curricula. In terms of social justice, this collaboration has supported local researchers in providing some additional financial support without taking away their sense of ownership of the project. It also allowed for the provision of useful comparisons with other countries consistent with a major aim of the project.

5. Concluding Remarks

This chapter has employed a theoretical discussion of the issue of injustice developed by the feminist writer Young (1990) to analyse some findings from a study of the globalisation and internationalisation of mathematics education. It was demonstrated that each criterion presented by Young matches some concerns discussed in the literature in mathematics education. However, one striking feature of the discussion above is the complexity of issues when it relates to making social justice decisions on international collaborations. None of the issues discussed above lends itself to simple classifications of being socially just or unjust. International contacts in education may be said to be exploitative if the knowledge of one social group is advanced at the expense of another group. While research into marginalised social and cultural groups may give voice to the voiceless, questions of whose point of view and who is benefiting should remain at the forefront of critical evaluation of all academic action. Similarly, international contacts can lead to marginalisation of some participants if their participation is limited on economic and language grounds. Further, if the research questions and methodologies of some countries dominate international research at the expense of issues of concern of other nations, then the latter can be said to be marginalised. In addition to exploitation and marginalisation, economic situations in many less industrialised nations limit the capacity of educators from those countries to take an active and equal role in international academic activities and hence can lead to a sense of powerlessness. Further, the non-critical transfer of curricula and research results from one country with a certain perceived higher status to another can be said to be a form of cultural imperialism. In particular, the assumed direct correlation of Western mathematics to economic development and the assumption of the universality of mathematics can lead to imposing certain forms of mathematics that may not be appropriate or relevant to many students around the world. Finally, the tying of international aid and development monies to the impositions of agendas, policies and priorities developed in Western countries can be regarded as a form of violence on less affluent nations.

References

Apple, M. (1993). The politics of official knowledge: Does a national curriculum make sense? *Teachers College Record*, 95(2), 222–241.

Atweh, B. (2003). *International aid activities in mathematics education in developing countries: A call for further research*. Auckland, New Zealand: AARE.

Atweh, B., & Clarkson, P. (2002a). Globalisation and mathematics education: From above and below. *Proceedings of the Annual Conference of the Australian Association of Research in Education*. Brisbane: University of Queensland, AARE.

Atweh, B., & Clarkson, P. (2002b). Some problematics in international collaboration in mathematic education. In B. Barton, K. Irwin, M. Pfannkuch, & M. Thomas (Eds.), *Mathematics education is the South Pacific: Proceedings of the 25th annual conference of the Mathematics Education Research Group of Australasia*. MERGA: University of Auckland.

Atweh, B., Clarkson, P., & Nebres, B. (2003). Mathematics education in international and global context. In A. Bishop, M. A. Clements, C. Keitel, J. Kilpatrick, & F. Leung (Eds.), *The second international handbook of mathematics education* (pp. 185–229). Dordrecht: Kluwer Academic Publishers.

Bishop, A. J. (1992). International perspectives on research in mathematics education. In D. Grouws (Ed.), *Handbook of research on mathematics teaching and learning* (pp. 710–723). New York: Macmillan.

Burton, L. (Ed.). (2003). *Which way social justice in mathematics education?* London: Praeger.

Clarke, D., Emanuelson, J., Jablonka, E., & Mok, I. (Eds.). (2006). *Mathematics classrooms in 12 countries: Bridging the gap* (LPS Series Volume 2). Rotterdam, NL: SENSE Publishers.

Clarke, D., Keitel, C., & Shimizu, Y. (Eds.). (2006). *Mathematics classrooms in 12 countries: The insider's perspective* (LPS Series Volume 1). Rotterdam, NL: SENSE Publishers.

Clarke, D. J. (Ed.). (2001). *Perspectives on meaning in mathematics and science classrooms*. Dordrecht: Kluwer Academic Publishers.

Clarke, D. J. (2003). International comparative studies in mathematics education. In A. J. Bishop, M. A. Clements, C. Keitel, J. Kilpatrick, & F. K. S. Leung (Eds.), *Second international handbook of mathematics education* (pp. 145–186). Dordrecht: Kluwer Academic Publishers.

Constanza, A. (2000). Escuela Nueva in Colombia goes urban. In *World education forum*. Retrieved May 25, 2002, from http://www2.unesco.org/wef/en-news/colombia.shtm

Cotton, T. (2001). Mathematics teaching in the real world. In P. Gates (Ed.), *Issues in mathematics teaching* (pp. 23–37). London: Roudledge Falmer.

Cox, T. H. (1991). The multicultural organization. *Academy of Management Executive*, 5(2), 34–47.

Dowling, P. (1998). *The sociology of mathematics education: Mathematical myths/pedagogic texts*. London: The Falmer Press.

Fraser, N. (1995). From redistribution to recognition: Dilemmas of justice in a post-socialist society. *New Left Review*, July–August, 68–93.

Fraser, N. (2001). Social justice in the knowledge society. *Invited keynote lecture at conference on the "Knowledge Society,"* Heinrich Böll Stiftung: Berlin. Retrieved July, 2, 2005 from http://www.wissensgesellschaft.org/themen/orientierung/socialjustice.pdf

Gewirtz, S. (1998). Conceptualizing social justice in education: Mapping the territory. *Journal of Educational Policy, 13*(4), 469–484.

Hargreaves, A. (1994). *Changing teachers, changing times: Teachers' work and culture in the postmodern age*. London: Cassell.

Hart, L. (2003). Some directions for research on equity and justice in mathematics education. In L. Burton (Ed.), *Which way social justice in mathematics education?* (pp. 25–49). London: Praeger.

Jacobsen, E. (1996). International co-operation in mathematics education. In A. Bishop, et al. (Eds.), *International handbook of mathematics education* (pp. 1235–1256). Dordrecht: Kluwer Academic Publishers.

Keitel, C., & Kilpatrick, J. (1999). Rationality and irrationality of international comparative studies. In G. Kaiser, I. Huntley, & E. Luna (Eds.), *International comparative studies in mathematics education* (pp. 241257). London: Falmer Press.

Krell, G. (2004). Managing diversity: Chancengleichheit als Wettbewerbsfaktor. (Managing diversity: Equity of chances as a factor of competitiveness). In G. Krell (Ed.), *Chancengleichheit durch Personalpolitik* (pp. 41–56). Wiesbaden: Gabler.

Kuku, A. (1995). Mathematics education in Africa in relation to other countries. In R. Hunting, G. Fitzsimons, P. Clarkson, & A. Bishop (Eds.), *Regional collaboration in mathematics education* (pp. 403–423). Melbourne: Monash University.

Loden, M., & Rosener, J. (1991). *Workforce America: Managing employee diversity as a vital resource*. Homewood, IL: Irvin Inc.

McInerney, P. (2004). *Making hope practical: School reform for social justice*. Queensland: Post Pressed.

Nozic, R. (1981). *Philosophical explanations*. Cambridge: Harvard University Press.

Plummer, D. L. (Ed.). (2003). *Handboook of Diversity Management*. Lanham, MD: University Press of America.

Powell, A., & Frankenstein, M. (Eds.). (1997). *Ethnomathematics: Challenging eurocentrism in mathematics education*. Albany: Sunny Press.

Rawls, J. (1973). *A theory of justice*. Oxford: Oxford University Press.

Robitaille, D. F., & Travers, K. J. (1992). International studies of achievement in mathematics. In D. Grouws (Ed.), *Handbook of research on mathematics education* (pp. 687–709). New York: Macmillan.

Rogers, L. (1992). Then and now. *For the Learning of Mathematics, 12*(3), 22–23.

Sawiran, M. (1995). Collaborative efforts in enhancing globalisation in mathematics education. In R. Hunting, G. FitzSimons, P. Clarkson, & A. Bishop (Eds.), *Regional collaboration in mathematics education* (pp. 603–609). Melbourne: Monash University.

Secada, W. (1989). *Equity in education*. Philadelphia: Falmer.

Secada, W., Cueto, S., & Andrade, F. (2003). Opportunity to learn mathematics among Aymara-, Quechua-, and Spanish-speaking rural and urban fourth- and fifth-graders in Puno, Peru. In L. Burton (Ed.), *Which way social justice in mathematics education?* (pp. 103–132). London: Praeger.

Sepheri, P., & Wagner, D. (2000). Managing diversity – Wahrnehmung und Verständnis im Internationalen management. *Personal, Zeitschrift für Human Resource Management*, 52(9), 456–462.

Setati, M. (2005). Researching teaching and learning in school from "with" or "on" to "with" and "on" teachers. *Perspectives in Education, 23*(1), 91–101.

Thomas, R. R. (1996). *Redefining diversity*. New York: Amacom.

UNESCO. (1998). World declaration on higher education for the twenty-first century: Vision and action. Retrieved www.unesco.org/education/educprog/wche/index.html

Vithal, R., & Skovsmose, O. (1997). The end of innocence: A critique of "ethnomathematics". *Educational Studies in Mathematics, 34*, 131–157.

Young, I. M. (1990). *Justice and the politics of difference*. Princeton, NJ: Princeton University Press.

7

GLOBALISATION, ETHICS AND MATHEMATICS EDUCATION

Jim Neyland
Victoria University, Wellington, jim.neyland@vuw.ac.nz

Abstract: Mathematics education has been a tool of cultural imperialism, and continues to acquiesce to its forces. The current mode of cultural imperialism includes a strengthening globalisation. The momentum of modern globalisation is towards a new and growing social stratification resulting in increasing bureaucratic domination of the poor. Accordingly, for the poor, the newly emerging globalised mathematics education is likely to result in a transaction that is mathematically trivialised, and educationally degraded

Keywords: globalisation, social stratification, justice, the literary curriculum

1. Mathematics Education and Imperialism

Mathematics, Bishop pointed out, is mostly considered to be "universal and, therefore, culture-free." For "most people", it has "the status of a culturally neutral phenomenon . . . free from the influences of any culture." But mathematics is not culture-free. Mathematical ideas are "humanly constructed" and have a "cultural history" (1995, p. 71). In addition, mathematics education has a history of collaboration with the forces of cultural imperialism. It has been a means by which the West has imposed its culture on others. This has involved, in the first instance, a contrived blindness to the non-European roots of mathematics. Joseph wrote that "the standard treatment of the history of non-European mathematics exhibits a deep-rooted historiographical bias in the selection and interpretation of facts, and . . . mathematical activity outside Europe has as a consequence been ignored, devalued or distorted" (1991, p. 3). This disinterested neglect has been accompanied by a colonial agenda. In Bishop's view, "it is thoroughly appropriate to identify 'western mathematics'" as having "played . . . a powerful role in achieving the goals of [western imperialism]." Three "mediating agents in the process of cultural invasion

B. Atweh et al. (eds.), Internationalisation and Globalisation in Mathematics and Science Education, 113–128.

of colonised countries by western mathematics [were]: trade, administration and education." Education played

> a critical role in promoting western mathematical ideas, and, thereby, western culture . . . At worst, the mathematics curriculum was abstract, irrelevant, selective and elitist . . . governed by structures like the Cambridge Overseas Certificate, and culturally laden to a very high degree. It was part of a deliberate strategy of acculturation—international in its efforts to instruct in 'the best of the West', and convinced of its superiority to any indigenous mathematical systems and culture.
>
> (1995, p. 73)

Mathematics carried with it certain values that extend beyond the discipline itself. These, according to Bishop, were of "far more importance, particularly in cultural terms." Indeed, western conceptions now have a taken-for-granted status that may be impossible to overturn. "From colonialism through to neo-colonialism, the cultural imperialism of western mathematics has yet to be fully realised and understood. . . . one must wonder," he concluded, "whether its all-pervading influence is now out of control" (1995, p. 75).

Willinsky and Bauman also linked education with an imperialist agenda. Willinsky argued that education has had a major role in serving the interests of imperialism. The "will to know" and the "will to display" were at the root of the colonial enterprise. There was "a desire to take hold of the world" in "an unrelenting enthusiasm for learning". This, of course, is not a problem. What is significant, though, is the degree to which this enthusiasm was "dedicated to defining and extending the privileges of the West." Following Foucault, he argued that the sciences instigated a programme aimed at ordering the world. Thus occurred a major shift in European thought from a focus on semblance to one of difference. Increasingly fine calculations identified a world of difference (1998, p. 26). Particular care is needed in interpreting this newly emerging focus on difference. This was not a celebration of diversity. Bauman, in his analysis of the emergence of difference, points out that what happened here was not the sudden recognition of the reality of diversity, but the birth of the perception that this diversity, to a considerable degree, results from human educational activity. It is not that no one noticed differences before this time. What changed is that human nature came to be seen as the product of "cultivation", and therefore amenable to human control via education. This led to differences being seen as things that could be changed, and, crucially, should be changed. Difference became marked off for normalisation, or extinction (Bauman, 1992). Education aimed for homogenisation. Willinsky showed that, behind all this, there is some sense of a blueprint; an ideal mapped out in advance. I will return to the notions of "blueprint" and "mapping" in a later part of this chapter when I discuss the "artificially designed" mathematics curriculum.

2. Imperialism and Scientific Management

The "scientific management of education" – and its natural offspring, the "outcomes-based curriculum" – is now largely taken for granted by mathematics educators in a growing number of countries (Levin, 1997; Smyth, 1995; Smyth & Dow, 1998; Wise, 1979). Importantly, in each case the country concerned has drawn on the same basic ideas (Levin, 1997). It is a movement with a single source: a set of ideas born of the assembly-line-efficiency sector in the United States and applied there to education. It now represents the orthodox consensus in education theory throughout a growing portion of the world. Where it is found, it is impossible to avoid. This is because the central feature is its mandatory nature. It is made compulsory through governmental legislation. When education is managed scientifically through legislation, four things are required: (i) an unambiguous statement of what the legislation requires; (ii) a theory of control that ensures compliance; (iii) agencies that will monitor the degree of compliance; and (iv) instrumentally oriented research to aid rationalistic decision making (Neyland, 2004). The scientific management of mathematics education was not chosen by mathematics teachers. It was driven mainly by management theorists and economists (Wise, 1979). It was, in effect, a blueprint designed in one influential country and imposed upon teachers there. Subsequently, it was imposed on teachers elsewhere. It is a component of modern-day cultural imperialism with many of the hallmarks of its precursors.

Resistance in mathematics education circles has been somewhat muted. This is because what we have now in mathematics education is growing evidence of what Whyte called "organisation man": organisationally acculturated people increasingly living in a cage; a gilded cage. They work, more and more, for others – for the faceless voices of authority enshrined in legislation and outcomes statements – and less and less for themselves and their students in direct (ethical) response to individual learning situations (Neyland, 2004). There has been a marked drop in the level of adventurousness teachers are prepared to display (Neyland, 2002), and teachers are subject to what Whyte called a "benign tyranny" (Whyte, 1957). The scientific management of education and the accompanying rise in the organisationally acculturated mathematics teacher has led to the mathematical equivalent to what Sennett identified as "the fall of public man" (1977). We have now "the fall of the public mathematics curriculum," and an accompanying privatisation. The modern "privatised" mathematics curriculum is pyramidal. It is a blueprint designed by so-called experts, with little or no consultation with teachers,[1] and made mandatory via legislation. Public curriculum space, within which in former times teachers could debate about and co-construct a mathematics curriculum, has mostly disappeared. The mathematics curriculum is less and less fashioned by education

[1] According to some of the theories that underlie the scientific management of education it is important *not* to consult teachers because consultation will lead to an unwanted "provider capture" of the process of education.

communities in the public space of dialogue and debate. Instead, in a new phase of imperialism, the tendency now is for a privatised curriculum to be imposed. The globalising scientific management of education has an ally: the "international comparisons" movement. These two consort to bring about the same homogenisation in education and culture – the same eradication of diversity – that was first evident at the dawn of the eras of cultural imperialism and globalisation.

3. Intensifying Globalisation

Globalisation is old news. But it cannot, for this reason, be dismissed as unremarkable. This is because the effects of globalisation are intensifying. There are two reasons. First, we are currently witnessing a steady increase in (i) the further shifting of power and control from the public to a tiny and un-elected private sector, (ii) the impact of transnational pseudo-capitalism, (iii) the efficiency and affordability of communication technologies, and (iv) related to all of the above, the embedding of cultural-imperialism and neo-colonialism based largely around the promotion of consumerism. Second, we are increasingly experiencing an out-of-control disorder. This experience has three causes. (1) Market pseudo-capitalism craves certainty. This is achieved by subjecting the majority to the opposite; through "flexibility" mechanisms and the like. (2) We are now living with an unexampled (i) global interconnectedness that is transforming out experience of time and location, (ii) pluralism of authority and centrality of personal choice in the constitution of self-identity, and (iii) disembodiment and de-contextualisation of know-how and value. (3) We are witnessing the failure of the project of universalisation. I referred, a moment ago, to pseudo-capitalism. This is because transnational corporations pay only lipservice to free-market ideals. They prefer, instead, "monopolies, cartels, and government contracts" (Wright, 2004, p. 129). Passat calculated that "purely speculative inter-currency financial transactions reach a volume of $1,300 billion a day – 50 times greater than the volume of commercial exchanges and almost equal to the total of . . . all the reserves of all the 'national banks' of the world" (Bauman, 1998, p. 66). Chomsky argued that the free-market is actually under threat from global capital (1996).

Let me turn to the first group of reasons mentioned earlier. Wallerstein identified the emergence of capitalism, which "never allowed its aspirations to be determined by national boundaries," as being responsible for the decline in influence of nation-states and accordingly in aiding the process of globalization (Giddens, 1990, pp. 68–69). For Tariq Ali, transnational capitalism is at the heart of globalisation. But he also argued that this goes hand-in-hand with American imperialism, and that the involvement of the United States military in South America, the Middle East, Europe and the Far East, over the last century, was mainly at the service of United States capitalism (2002). The analysis of globalisation as a new colonialism, or a transmutation of European imperialism features prominently in the literature. Aijaz Ahmad, for instance, argued that the period since World War II has seen the emergence of a "Super Imperialism". Advanced capitalism "has now reached a level of global self-organisation [that has] given the imperialist countries a kind of unity that was inconceivable even fifty years ago" (1992, p. 313). Hobsbawm views colonialism as a precursor of the market society and of eventual globalisation of capitalism (1987).

The United States is the new recipient of the colonial legacy, and puts it into operation through the programmatic normalisation of mass production, communication, and consumption, and the growing dominance of transnational capitalism. Suzuki recognised a new wave of colonialism: "The first wave of colonialism forced everyone to become Christians, farmers or producers of wealth. Now we want everyone to watch TV, drink Coke, wear Nikes and consume the goods we offer" (Suzuki & Dressel, 1999, p. 231). Globalisation, in its dominant manifestation, erodes the ability of governments and the people to freely make decisions about their country's welfare.[2] One of the main reasons for this is the fact that the public sector – the civic part of society, the bond that keeps societies together – has been ground down. In 1997, $167 billion worth of public assets around the world – including education – transferred to private hands (Suzuki & Dressel, 1999). At the dawn of the 21st century the world's three richest people had a combined wealth greater than that of the poorest 48 nations (Wright, 2004). Of the 100 biggest economies in the world, 51 are private. The number of countries in this group is steadily declining. Mitsubishi is bigger than Denmark, Thailand and Indonesia. Royal Dutch Shell is bigger than Norway. Exxon is bigger than Finland. Wal-Mart is bigger than Poland (Suzuki & Dressel, 1999). In 1998, 29 transnational corporations had larger economies than New Zealand (Olssen, Codd & O'Neill, 2004). The true economic power in the world is no longer the United States, but 200 large companies whose sales are equivalent to over 25% of global economic activity. These companies basically control 25% of the world's wealth. Do they employ an equivalent workforce? No. These 200 largest companies employ less than 0.003% of the global workforce.

Nation-states are weakened, but not dismantled. They are left with their "powers of repression" and their "means of violence" (Giddens, 1990, p. 58). Their independence has been "annulled"; but not their relevance. They have become "a simple security service for the megacompanies" (Castoriadis cited in Bauman, 1998, p. 66). This weakening of the nation-state is potentially catastrophic. This is because, as Olssen et al. pointed out, global survival needs global governance, but the latter is dependent on strong democratic nation states (2004). Bauman reminded us that large corporations do not want global governance because global legislative and policing powers would be "detrimental [to their] interests" (1998, p. 69). Thus, economic globalisation, weak states and political fragmentation can be seen to be mutually complementary. Olssen et al. argued that there is another "key force affecting (and undermining) nation-states today." It is the "imposed policies of

[2] The well-publicised global opposition to the activities of the World Trade Organisation and other similar institutions is evidence of an enhancement of participatory democracy in a globalised world. But this remains inconsequential when compared with the more pervasive dismantling of democracy that has accompanied globalisation.

neoliberal governmentality" (2004, p. 13). These are a threat to democracy. Importantly, for the present discussion, these policies are the same policies that gave rise to the scientific management of mathematics education discussed earlier.

Now, the second group of reasons: those that lead to the experience of disorder. What is the nature of this experience? Bauman's description is this: "The deepest meaning conveyed by the idea of globalisation is that of the indeterminate, unruly and self-propelling character of world affairs; the absence of a centre, of a controlling desk" (1998, p. 59). For Giddens, living in the modern global world is, he summarised, "more like being aboard a careering juggernaut [than] in a carefully controlled and well-driven motor car" (1990, p. 53). People now experience ambiguity and confusion. The consequences of the "intensification of worldwide social relations" that accompany globalisation is, he said, "not necessarily, or even usually, a generalised set of changes acting in a uniform direction, but consists in mutually opposed tendencies" (1990, p. 64). Parts of this experience are not directly contributable to particular causes; they are emergent in a complex world. This increases their capacity to foster the feeling that things are out of control. Other experiences of being out of control, such as those that accompany the mechanisms of modern "flexible capitalism", are orchestrated. Markets required stability and predictability. In order to achieve this, capital visits havoc on the lives of many. Such orchestrated uncertainty is a recognised mode of monopolistic control. Those who pull the strings maintain maximum autonomy and regularity for themselves while ensuring that the subjugated are kept in a constant state of uncertainty.

The experience of disorder is also attributable to the combined effect of what Giddens refers to as three "dynamisms" in operation (1990, p. 53). The first, is a "stretching process" in which "modes of connection between different social contexts or regions become networked across the earth's surface as a whole" (1990, p. 64). What this boils down to is an unprecedented separation of time and space. I no longer need to allow a significant amount of time to pass in order to change location either physically or electronically. Time and location now can be changed independently of each other. Global interconnectedness has thus transformed my experience of both time and space. In the electronic age, shifting information does not need to involve the movement of the physical bodies within which information is embodied. This leads to meaning either being dissipated, or becoming established in a new distinctively late modern way: by association with other disembodied pieces of information.

The second dynamism is "reflexivity". This refers to the fact that social practices are "constantly examined and reformed in the light of incoming information about those very practices, thus constitutively altering their character" (1990, p. 38). When this second feature works in consort with the first, reflexivity extends to incorporate massive spans of time-space (1990).Reflexivity is not a new phenomenon. It is characteristic of all social life (Barnes, 1995). But modern global reflexivity is a magnified version of earlier modes. The reflexive reshaping of knowledge has a crucial role in a world characterised by a pluralism of authority, and the centrality of choice in shaping modern self-identity. On the positive side, it has created the conditions that allow forms of global resistance, and it works against any tendency towards globalisation shaping a single integrated culture. One the negative side, it is necessarily unstable (Giddens, 1990).

The third dynamism is "disembedding". This refers to the "lifting out of social relations from local contexts of interaction and their restructuring across indefinite spans of time-space" (Giddens, 1990, p. 21). In this way aspects of life are deprived of their "situatedness" in specific locales (1990, p. 53). Know-how becomes disembodied, and value becomes de-contextualised. Giddens identified two principal modes of disembedding: "symbolic tokens" and "expert systems" (1990, p. 21). We are familiar with one instance of a symbolic token: money. It is pure commodity. Education credentials are another. Credentialing, especially in the form it takes under the scientific management of mathematics education, results in mathematics losing its real value and assuming only an exchange value. Increasingly, especially through the influence of the "international comparison" movement, credentials are becoming universal and homogeneous. Through symbolic tokens the medium of exchange effectively negates the content of goods, services, and individual human characteristics. In their place is substituted impersonal and tradable standards.

We are familiar with expert systems. Bauman argues that the dubious notion of "expertise" is based on the assumptions that: (i) doing things properly requires particular knowledge held by only a few, (ii) these few are therefore responsible for how things are done, and ought to be in charge, and (iii) for the rest, responsible action involves following the advice of these experts (1992). Again, we have here a disembedding. Individual responsibility becomes free-floating, and actions drained of ethical significance. Eroded responsibility is replaced by technical accountability. Thus, wrote Bauman, it is the "technology of action, not its substance, which is subject to assessment as good or bad, proper or improper, right or wrong" (1992, p. 160). How is this relevant to our present discussion? Through the scientific management of mathematics education, together with the "international comparisons" movement, we have an expert system: otherwise called a top-down research-development-dissemination (RDD) model of curriculum. This produces, not globalised togetherness, but globalised homogeneity. Disembedding results in the corrosion of interpersonal trust. Symbolic tokens and expert systems demand a form of trust. But it is not trust in people. It is trust in an abstract capacity, or in a technical-expert position. To summarise: the joint influence of (i) the separation of time and space, (ii) reflexivity, and (iii) disembedding results in the experience of being torn from the connections that provide meaning and sustain interpersonal relationships; the same phenomena are found in current mathematics education.

The final reason why globalisation is intensifying is the failure of the project of universalisation. When this project broke down what we were left with was globalisation. Mander, Head of both the International Forum on Globalisation and the Foundation for Deep Ecology, said this: "Globalisation tends to be described as something that evolved, as though it were some sort of force of nature. But [it's] not an accident, and it's not just a natural evolutionary process. It is a designed system, set up for the specific purpose of enabling corporations to make the rules of economic activity globally. The underlying belief in that designed system is that they, and not any kind of government, are the logical, rational system to do that" (Suzuki & Dressel, 1999, p. 189). He is talking here about the deliberate design of a programme; a blueprint for economic global activity to be operational universally. This is the project of universalisation, and resembles earlier blueprints in its determination to eradicate diversity and manufacture regularity. It was designed by a particular group of economists, corporate CEOs and government leaders in the 1920s and 1930s. It came to fruition in the Bretton Woods Agreements (1944). Bretton Woods gave the United States military and corporate institutions unlimited access to minerals, oil, markets and cheap labour (Pilger, 2002).

The project of universalisation sought to eradicate unpredictability and create order. But it failed. It failed for three reasons. First, Nixon allowed the deregulation of financial markets, thus working against some of the thrust of Bretton Woods. This led to the "haemorrhage of capital from the real economy (investment and trade) to financial manipulations that now constitute 95% of foreign exchange transaction (as compared with 10 percent of a far smaller total in 1970)" (Chomsky, 1996, p. 130).[3] Second, a growing chasm between rich and poor eventually led to major social disruption. Third, because human beings and social worlds are complex, systems of rules aimed at eradicating unpredictability were never going to succeed. This is for two reasons. (1) Rules can never be numerous enough or sufficiently flexible to govern behaviour. There is always the need for reference to an ethical horizon; one that cannot be reduced to a code (Barnes, 1995; Taylor, 1991). (2) Such systems are based on a false assumption: society causes goodness. This is what Bauman calls the aetiological myth (1989). The truth is that social institutions do not cause goodness, they manipulate our capacity for it, sometimes for good, sometimes for ill. Systems of rules aimed at manufacturing regularity end up "eroding" the social bond and causing a breakdown in interpersonal responsibility (Levinas, 1998). When, at the same time, social relationships are deliberately undermined, as occurs in modern flexible capitalism, people's sense of goodness and identity becomes "corroded" (Sennett, 1998). Modern globalisation, then, is radically different from the universalisation from which it evolved. The programme of universalisation aimed at imposing a regularity upon events. Globalisation is the effect of the failure of this programme. It no longer refers to an order-making initiative. It is not about what we might "hope to do", wrote Bauman, it is about "what is happening to us all" (1998, p. 60). Thus, globalisation is a manufactured disorder. It is what Giddens refers to as a "manufactured jungle" (Bauman, 1998, p. 60).

---

[3] It is notable that a simple tax on foreign exchange transactions – the proposed Tobin Tax – would have considerably slowed this haemorrhaging. Such a tax was never instigated.

4. A Growing Social Stratification

Will modern global capital reduce the number living in poverty? Seabrook's answer is simple and unflinching: "Poverty cannot be cured [by global capitalism]," he wrote. "It is not a disease of capitalism. It is a sign of robust good health" (Bauman, 1998, p. 79). Globalisation, then, requires a class of unfortunates. Rees and Wackernagel, in their analysis, also cut to the chase (1996). They developed a way of measuring how much of the planet's resources each of us uses: our individual "ecological footprint". They asked: How much land and sea does it take to support the physical needs of a human being? What did they find? Human beings are skilfully capable of making up for areas of scarcity. However, unless we colonise a few more planets, we are unable to invent our way around one recalcitrant fact: we each require a certain productive area on the surface of the (finite) Earth to maintain our physical needs and absorb our wastes. The "ecological footprint" of average Canadians – each individual's food, wood, paper, assimilation of carbon dioxide, and so on – is 7 hectares per person, including 0.7 hectares of marine component. To put this in context, the equivalent figure for many in Bangladesh is 0.5 hectares. This begs the question, What would happen if we shared the Earth's resources equally? We are approximately 6 billion people, and have globally 8.9 billion hectares of agricultural land. We would each be allocated approximately 1.5 hectares. Adding the productive part of the ocean, our fair share is 2 hectares per person. So, if the Earth's resources were to be shared, those in the developed world would need to reduce individual consumption from 7 to 2 hectares. Put differently, to raise the standard of all to that of the average in developed countries would require four or five more Earths.[4] Can we all keep consuming more and more? Clearly not. Further, if some consume more, others must necessarily survive on less. Behind this is greed. "Capital," Wright observed, "lures us onward like the mechanical hare before the greyhounds, insisting that the economy is infinite and sharing therefore irrelevant" (Wright, 2004, p. 124). Shareholders require corporations to maximise profits; in fact, to increase profits from one year to the next. The corporate system demands greed, certainly of its shareholders, and, to a considerable degree, of the consumer.

There has always been social stratification. What is alarming is its rate of growth. In the United States, in the late 1970s, the ratio between the salary of a CEO and that of a shop-floor worker was 39:1. Today it is 1,000:1 (Wright, 2004). In the "Rhine" economies[5] the rate of unemployment has risen significantly. This is especially so in Germany, France and Italy where there was a six per cent increase in the rate between 1980 and 1995. In America, the average weekly wage (adjusted for inflation) of the bottom 20% fell by 18% from 1973 to 1995. Over the same period, the pay of the corporate elite rose by 19% before taxes, and by 66% once

[4] The "ecological footprint" of the world population is 35% larger than the ecological capacity of the planet. How? We are using up the Earth's natural capital through deforestation, depletion of water tables, and so on.

[5] The Netherlands, Germany, France, Italy, Japan, Scandinavia and Israel.

the tax accountants had manipulated the tax laws to their advantage. In Britain, the top 20% of the working population earn seven times that of the bottom 20%. Twenty years ago the figure was four times (Sennett, 1998). What is the situation more globally? A UNESCO study divided the world's population into fifths and compared the richest fifth with the poorest fifth. In 1960 the richest group received 30 times more than the poorest. In 1990 the richest group received 59 times more than the poorest. The ratio doubled. In 1960 the richest 20% controlled 70.2% of global GNP. By 1990 it was 82%. In 1960 the poorest 20% controlled 2.3% of global GNP. By 1990 that figure had decreased to 1.4% (Sennett, 1998).

Is all this inevitable? We all need to consume; and fair trade is desirable.[6] But consumerism and greed are optional and chosen. Kung, writing in 1991, observed that every minute US$1.8 million is spent on armaments, while every hour 1,500 children die from hunger, and every month the global economy adds US$7.5 billion to the debt burden of developing countries already weighed down by US$1,500 billion (Kung, 1991). During the 20th century, while the population increased by a factor of 4, the economy increased by a factor of 40. If the gap between rich and poor had stayed the same, all people would have been ten times better off. Yet the number in abject poverty today is the same as the world population at the turn of the 20th century (Wright, 2004). In 1998 the United Nations estimated that US$40 billion could provide clean water, sanitation, and other basic needs for the poorest on earth. Wright noted that the richest person in the world – who owns more than poorest 100 million Americans combined – could alone provide that and still have US$11 billion left for himself (Wright, 2004).

5. Consumerism, Social Control and Domination

Let me summarise the main points of the argument so far. (1) Past trends suggest that mathematics education will fall in with the momentum of cultural imperialism. (2) Cultural imperialism is leading towards an intensifying globalisation. (3) One consequence of globalisation is an increasing gap between rich and poor. (4) A second consequence is a dramatic increase in ambiguity and uncertainty, and, associated with this, in differentiated capacities for overcoming the time-space coupling. (5) A third consequence is the "disembedding" of knowledge and value, and, the consequential erosion of the public sector. In this section I will remind the reader that consumerism is more than an impulsive acquisitiveness. It is a mode of social control. But, importantly, it is inadequate as such.

Consumerism is, according to Giddens, "one of the prime driving forces behind modern institutions." It is "essentially a novel phenomenon." Through

[6] Care has to be taken when interpreting trade figures. Increased trade in food, Vandana Shiva showed, has not always meant more food security. For instance, in India, during 1998, there was increased trade in food, but a decline in food consumption domestically. Grain was grown and exported, leaving the poor with insufficient food. So grain was imported. Thus figures (falsely) showed a large volume of trade (Suzuki & Dressel, 1999).

consumerism, "appearance replaces essence as the viable sign of successful consumption [outweighing] the use-value of goods and services." More insidiously, the "project of the self as such" may also be becoming "heavily commodified" (1991, pp. 198–199). Consumerism is also tied up with modern notions of freedom. In modern society, Bauman wrote, "individual freedom is constituted as, first and foremost, freedom of the consumer" (1988, pp. 7–8). The fully mature consumer must feel the impossibility of living otherwise. The will-to-consume must reveal itself in the guise of personal free choice (Bauman, 1998). Markets and consumers collaborate. The consumer market seduces its customers by arousing desire through the production of temptation. Customers desire to be seduced. The secret of present-day society, Seabrook argued, lies in "the development of an artificially created and subjective sense of insufficiency" since "nothing could be more menacing" to its foundational principles "than that the people should declare themselves satisfied with what they have" (Bauman, 1998, p. 94). The result has been the embedding of a new mode of social control. Earlier, more repressive and costly methods of control can be set aside and the soliciting of conduct entrusted to the market. This is what Tocqueville referred to as "soft despotism" (Taylor, 1991), and Arendt, as the "silk bonds of necessity" (Bauman, 1988). It is Bourdieu's modern mode of domination, in which seduction substitutes for repression, public relations for policing, advertising for authority, and needs-creation for norm-imposition (Bauman, 1987).

But not everyone can be a consumer. So consumerism is insufficient as a mode of social control. The modern globalised market, we have seen, necessarily stratifies society. Rich and poor alike can feel a desire to consume. But the poor cannot play the consumer game; they cannot afford it. In a consumer society levels of consumption are measures of worth. "Insiders are wholesome persons," Bauman observed, "because they exercise their market freedom. Outsiders are nothing else but flawed consumers." But the barriers that divide true from "flawed" consumers seldom appear as such. Instead, "they are thought of as commodity prices, profit margins, capital exports, taxations levels" (1988, p. 93). The desire to consume, then, is necessary but not sufficient. To be genuinely amenable to the soliciting of conduct, hope is also needed. But, for the poor, hope is futile (Bauman, 1998). So alternative means of control are needed, and in a consumer society, the only alternative to consumer freedom is "bureaucratically administered oppression" (Bauman, 1988, p. 86). In fact, according to Bauman, the effectiveness of a consumer-based system is determined by its "success in denigrating, marginalising or rendering invisible all alternatives to itself except blatant bureaucratic dominance" (1988, p. 93).

6. The Emerging Globalised Mathematics Education

Where is mathematics education headed? With globalisation has come consumerism. The flip side of "soft", consumerist, social control, is firm bureaucratic domination. The principal modes are: surveillance, routinisation, and confinement. The barriers that separate the mobile from the confined in education are rarely seen as such. They are thought of as bulk funding, voucher systems, equality of access to credentials, and pure (without any normative dimension) standards-based assessment systems. There are two other global phenomena affecting education: the scientific management of education through legislation described earlier, and the "international comparisons" movement. Through the first, the tools of scientific management theory,[7] backed by government legislation, have been applied to education. Mathematics educators have flirted with scientific management since its inception. What distinguishes contemporary scientific management from earlier forms are two critical developments: it has become compulsory, and it has gone global. I noted earlier that the first of four requirements for the scientific management of education through legislation is an unambiguous and detailed statement of the curriculum outcomes teachers must produce. This requirement has led to the now familiar outcomes-based mathematics curriculum. The "international comparisons" movement has resulted in a number of studies – for instance, TIMSS, PIRLS and PISA[8] – that aimed to compare national education systems and practices. Together, these two global phenomena have led to increasing levels of homogenisation, routinisation, privatisation, and disembedding in mathematics education globally. The net effect has been that the mathematics curriculum has become a literary curriculum, an artificially designed curriculum, an elite-ruled curriculum, and a privatised curriculum.

A literary mathematics curriculum is one that is fully legible and logical. It can be recounted in writing in every minute detail. A literary curriculum contains – or is presented as containing – nothing ineffable or illegible; nothing evades clear representation. Seeking this sort of transparency is not new. What is new is the systematic pursuing of transparency. It is now systematically pursued because the transparency of curriculum space is a major stake in the battle for sovereignty over that space. Curriculum "experts" – researchers who technologise pedagogy, or who standardise in order to make comparisons – play a major role in making curriculum space hospitable and readable. The opposite of the legible curriculum is not an illegible one, it is one that is locally emergent and complex. Those who recount the literary curriculum need to override the communal practices of localised territories. This is another manifestation of the "disembedding dynamism" of globalisation, discussed earlier. As a result the locals are deprived of their established means of orientation. Local teachers are left disoriented, confused, and with the feeling that things are somehow out-of-control. Contrary to the well-meaning intentions of the curriculum expert, the imposed order rebounds in the disintegration of localised orienting practices, spontaneous ethical know-how, and locally fashioned protective nets. Thus, paradoxically, the literary mathematics curriculum leaves teachers illiterate. The (globalised) literary mathematics curriculum produces a manufactured disorder.

---

[7] Developed by Frederick Taylor.

[8] Trends in International Mathematics and Science Study; Progress in International Reading Literacy Study; and Programme for International Student Achievement.

The literary mathematics curriculum is not a map of curriculum space, although it is mistakenly, or duplicitously, presented as such. It is a blueprint. It is an artificially designed curriculum. The distinction between a map and a blueprint is important. In earlier times, when the practice of mapping was in its first phase, maps were designed to conform to the territories they represented. Later, for the purposes of taxation, the assignment of levies, and so on, physical and social spaces were reorganised and standardised. The territories were made to conform to the maps which now functioned as blueprints. The same has now occurred in the mapping of curriculum space. For the purposes of teacher accountability, comparison-making, the provision of resources, and so on, curriculum space has been made to conform to a curriculum map, or blueprint. The blueprint – the sworn enemy of diversity and spontaneity – makes curriculum space legible and administrable. Thus, in order to make it comprehensible, local variety is neutralised, and the curriculum is standardise and routinised. There are two steps in the process of producing an artificially designed curriculum. First, the practice of cartography has to be normalised. This is the crucial step and the hardest to take. Second, once the practice of mapping is accepted, control is taken of the office of cartography. In this way, without much resistance, territory can be reshaped in the likeness of the cartographer's design. In the mathematics curriculum a straightforward illustration of this process is this: first, normalise the practice of planning and specifying outcomes of learning; second, take control of the plans and specify the outcomes centrally. In these two steps curriculum control is wrested from teachers and communities in a bloodless revolution – in fact, in a revolution that no one notices. Who takes control of the office of cartography? The experts and those in authority. This is how the curriculum becomes elite-ruled; and how its opposite, the communally designed and administered curriculum, eradicated. Outcomes, standards, and learning sequences come to be specified at the top, and, increasingly, globally. But, just in case the locals do notice, the sovereignty of supra-communal administration of the curriculum is maintained by ensuring that mathematics teachers are immobilised through requirements to test frequently, to record continuously, and to subject their practice to continual external- and self-surveillance through measurements against "best evidence syntheses" and the like. Is this confinement of teachers deliberate? You decide. The person who seized the office of cartography in New Zealand said this in an address to an Educational Council conference in Australia: "Implement reform by quantum leaps. Moving step by step lets vested interests mobilise. Big packages neutralise them. Speed is essential. It is impossible to move too fast. . . . Once you start the momentum rolling never let it stop" (Lauder, 1991, p. 8).

The confinement of teachers effectively leads to the curriculum becoming privatised. That is, what was formerly public curriculum space – the teacher-class environment, local mathematics associations, local communities, and so on – is demolished and then privatised or government subsidised. Public curriculum space is where norms are created, ideals articulated, social bonds forged, and shared criteria for evaluation originated. Within public space concrete situations are the point of departure for professional development. Concrete situations are described and compared to similar ones encountered elsewhere. Evaluative frameworks and a common understanding emerge from these complex interactions. They are not caused, as if by a deterministic process, directed by experts. Privatised curriculum space is characterised by the presence of protocols, procedures, norms, and evaluative verdicts handed down from on high. These extra-territorial artefacts enter locally-bound life as caricatures; as mutants.[9] The experts design the rules and set them in stone. State sponsored curriculum developers – bearers of the message from on high – disseminate these directives to teachers who are viewed as passive, and somewhat inept, receivers of the message. Along with this an ersatz professional mode becomes authorised: technical accountability. Put differently, the mathematics teacher as public intellectual, becomes the mathematics teacher as technocrat.

In summary, the work of mathematics teachers is increasingly programmed,[10] and they are becoming immobilised. Such routinisation, standardisation and dehumanisation is leading to an erosion of professional ethical autonomy. Teachers are becoming Whytes's "organisation" people in "gilded cages", avoiding adventureness, and being subject to "benign tyranny" (1957). Curriculum space is treated as if it were a complicated clockwork mechanism. It is managed, made legible, and operated as a "disembedded expert system". But, like the project of universalisation, it is failing, and leading to disorder. When all this happens to teachers, the same inevitably is visited upon students. The notion of "education" thus becomes degraded. It becomes a mechanism for exchanging symbolic tokens of doubtful value for the demonstrated performance of specified and mostly inconsequential operations. The real value of mathematics becomes the exchange value of credentials, and the control value that ensures confinement.

7. None of This is Inevitable

I have sketched, in broad brushstrokes, a sobering picture. I set out to outline the problem we face. There is, I feel sure, a solution. But problems must be stated first. No one is interested in fixing something they don't believe to be broken. Globalisation is a present reality. But things need not unfold as I have predicted. Mathematics education should be different. So, in a final few words, I will attempt to indicate where we should turn.

What, in the light of the globalised world analysed above, ought the characteristics of mathematics education be? Four things seem important. (1) Mathematics education must be conceived as a public good in the service of a world community oriented by wanting to live together in just institutions. (2) Mathematics education must take a pivotal role in the construction of a participatory democracy – this

[9] Freudenthal (1978) makes this argument.

[10] Brown and Lauder argue in a recent paper (2003) that the work of the professional strata in general is increasingly becoming routinised and programmed. They show, for instance, that localised and contextualised decision making by bank managers is being taken over by computer applications that require little personal insight, experience or personal judgment.

being the best available form of social organisation – aimed at world governance. (3) The organisation of mathematics teaching at the local level must be based on genuine trust, and on personal responsibility for others expressed as benevolent spontaneity. (4) Mathematics must be presented to students in programmes of work that emphasise its humanistic qualities and its basis in human ideas. Other legitimate goals for mathematics education can be aspired to, but these must always be secondary to the primary horizon of significance partially articulated in the above four characteristics.

References

Ahmad, A. (1992). *In theory: Classes, nations, literatures*. London: Verso.

Ali, T. (2002). *The clash of fundamentalisms: Crusades, jihads and modernity*. London: Verso.

Barnes, B. (1995). *The elements of social theory*. London: University College London Press.

Bauman, Z. (1987). *Legislators and interpreters: On modernity, post-modernity and intellectuals*. New York: Cornell University Press.

Bauman, Z. (1988). *Freedom*. Minneapolis: University of Minnesota Press.

Bauman, Z. (1989). *Modernity and the holocaust*. New York: Cornell University Press.

Bauman, Z. (1992). *Intimations of postmodernity*. London: Routledge.

Bauman, Z. (1998). *Globalization: The human consequences*. New York: Columbia University Press.

Bishop, A. (1995). Western mathematics: The secret weapon of cultural imperialism. In B. Ashcroft, G. Griffiths, & H. Tiffin. (Eds.), *The post-colonial reader* (pp. 71–76). London: Routledge.

Brown, P., & Lauder, H. (2003). *Globalisation and the knowledge economy: Some observations on recent trends in employment, education and the labour market*. Working paper series: Paper 43. Cardiff: School of Social Sciences, Cardiff University.

Chomsky, N. (1996). *Power and prospects: Reflections on human nature and the social order*. New South Wales, St Leonards: Allen & Unwin.

Freudenthal, H. (1978). *Weeding and sowing*. Dordrecht: Reidel Publishing Company.

Giddens, A. (1990). *The consequences of modernity*. Stanford: Stanford University Press.

Giddens, A. (1991). *Modernity and self-identity: Self and society in the late modernage*. Oxford: Polity Press.

Hobsbawm, E. (1987). *The age of imperialism*. London: Weidenfeld & Nicolson.

Joseph, G. (1991). *The crest of the peacock: Non-European roots of mathematics*. London: Penguin Books.

Kung, H. (1991). *Global responsibility: In search of a new world ethic*. New York: Crossroad.

Lauder, H. (1991). *Tomorrow's education, tomorrow's economy*. Wellington: New Zealand Council of Trade Unions.

Levin, B. (1997). The lessons of international education reform. *Journal of Education Policy, 12*(4), 253–266.

Levinas, E. (1998). *Otherwise than Being: Or beyond essence* (A. Lingis, Trans.). Pittsburgh: Duquesne University Press.

Neyland, J. (2002). Rethinking curriculum: An ethical perspective. In B. Barton, K. Irwin, M. Pfannkuch, & O. Thomas (Eds.), *Mathematics education in the South Pacific: Proceedings of the 25th Annual Conference of the Mathematics Education Research Group of Australasia Incorporated* (pp. 512–513). Auckland: University of Auckland.

Neyland, J. (2004). Towards a postmodern ethics of mathematics education. In M. Walshaw (Ed.). *Mathematics education within the postmodern, Greenwich* (pp. 55–73) Conneticut: Information Age Publishing.

Olssen, M., Codd, J., & O'Neill, A. -M. (2004). *Education policy: Globalisation, citizenship and democracy*. London: Sage.

Pilger, J. (2002). *The new rulers of the world*. London: Verso.
Rees, W., & Wackernagel, M. (1996). *Our ecological footprint: Reducing human impact on the Earth*. Gabriola Island, British Columbia: New Society Publishers.
Sennett, R. (1977). *The fall of public man*. New York: Knopf.
Sennett, R. (1998). *The corrosion of character: The personal consequences of work in the new capitalism*. New York: W. W. Norton & Co.
Smyth, J. (1995). What's happening to teachers' work in Australia? *Education Review, 47*, 189–198.
Smyth, J., & Dow, A. (1998). What's wrong with outcomes? Spotter planes, action plans, and steerage of the educational workplace. *British Journal of Sociology of Education, 19*(3), 291–303.
Suzuki, D. & Dressel, H. (1999). *Naked ape to superspecies: A personal perspective on humanity and the global ecocrisis*. New South Wales, St Leonards: Allen and Unwin.
Taylor, C. (1991). *The ethics of authenticity*. Cambridge: Harvard University Press.
Whyte, W. H. (1957). *Organisational man*. London: Jonathan Cape.
Willinsky, J. (1998). *Learning to divide the world*. Minneapolis: University of Minnesota Press.
Wise, A. (1979). *Legislating learning: The bureaucratization of the American classroom*. Los Angeles: University of California Press.
Wright, R. (2004). *A short history of progress*. Melbourne: Text Publishing.

8

THE POLITICS AND PRACTICES OF EQUITY, (E)QUALITY AND GLOBALISATION IN SCIENCE EDUCATION: EXPERIENCES FROM BOTH SIDES OF THE INDIAN OCEAN

Annette Gough
RMIT University, Australia

Abstract: Gender, equity, equality, quality and globalisation are political issues which are interwoven into the discourses and practices of science education and education writ large. In this chapter I firstly review the status of the gender agenda in education, particularly science education, within a global context, and then explore the complicated curriculum conversation that constitutes gender in South African (science) education and the ways in which gender is/is not an educational issue. I then discuss the tensions between equality and quality in educational discourses in South Africa and Australia, and how the resolution of social justice issues such as gender equality are so tightly interwoven into issues of democratic education

Keywords: gender, policy, social justice, science education, quality education, Australia, South Africa

1. Introduction

Gender has been on the agenda in science education, and in education writ large, for over three decades now, at least in the Western world. During this time feminist researchers have experienced a game of snakes and ladders as each advance (up a ladder) has, sooner or later, been followed by a diminution of status. That gender issues are often seen as a Western project has many implications for the work of feminist researchers and curriculum developers in science education in a globalised, world, particularly when Western feminists are working with colleagues from different cultural contexts.

In order to consider internationalisation and globalisation in science education from a feminist perspective, I first chart research and curriculum transformations

B. Atweh et al. (eds.), Internationalisation and Globalisation in Mathematics and Science Education, 129–147.

around girls and science education in the Western world (with a focus on Australia). I then reflect on the "complicated conversation" (Pinar, Reynolds, Slattery, & Taubman, 1995, p. 848) constituted by a confluence of international project experiences, local perspectives and research literature, as I discuss tensions between equality and quality in South African educational discourses and how the difficulties of resolving social justice issues such as gender equality are so tightly interwoven with issues of quality in education, including science education.

Science education from a feminist perspective in a global context shares an agenda with moves towards a more democratic science education that "may enable less partial and distorted descriptions and explanations" (Harding, 1991, p. 301), and is concerned with "pointing out how better understandings of nature result when scientific projects are linked with and incorporate projects of advancing democracy; [and that] politically regressive societies are likely to produce partial and distorted accounts of the natural and social world" (Harding, 1993, p. ix). Such democratic moves are consistent with the South African Bill of Rights (Mangena, 2006) and other declarations discussed in this chapter.

1.1 Research on Girls and Science

Research on gender and science education has been a focus of two issues of the journal *Studies in Science Education,* in 1982 and in 1998. In the first of these, both Kaminski (1982) and Manthorpe (1982) observed that the (then) current research into girls and science education approached the issue from three distinct, though not unrelated, perspectives:

- the intellectual potential of girls was seen as a significant but untapped labour source for science and technology, or
- a concern with equity and a desire to identify and reform those factors which were seen to impede girls' achievement in science, or
- a concern with the under-representation of women in science, combined with an argument that the male nature of the practice of science was oppressive for women, hence their "science avoidance".

Overall, at this time, the emphasis was on research which would result in getting more girls to study science and follow scientific careers.

During the 1980s these related concerns associated with gender and science developed into and alongside a more broadly focused education discourse which sought to enhance girls' post-school options by altering their relationship to school subjects not traditionally associated with girls. When given a choice, girls were to be encouraged both to select such subjects and to make "non-traditional" choices within such subject groupings, such as physical rather than biological sciences, or higher level rather than lower level mathematics. Teachers and others were to develop educational means by which girls would achieve greater success in and/or a stronger identification with such subjects when they are part of the compulsory school curriculum. Further, girls were to become "empowered" through the reconstruction of the processes and contents of the curricula in these areas.

Kenway and Gough (1998) analysed and critiqued the gender and science discourse in the late 1990s, highlighting the difficulties and dilemmas which confronted the advancement of the discourse. They concluded that there were four main themes in the discourse:

- documenting and reporting differences between girls' and boys' participation, achievement, attitudes and types of engagement with learning strategies in science education;
- arguing for greater participation of girls and women in science education and careers;
- explaining girls' patterns of participation in terms of a lack of the appropriate aptitudes, attitudes, experience and knowledge, i.e. a deficit of girlhood, or in terms of the masculinist nature of science curriculum materials and science classrooms;
- changing girls' choices and enhancing their participation and success through changing the girls, changing the curriculum, changing the pedagogy and/or changing the learning environment.

In the context of this chapter it is significant that Kenway and Gough (1998, p. 4) noted an emergent recognition that "gender equity research ought to transcend the boundaries of race, ethnicity, class and socio-economic identities" (Krockover & Shephardson, 1995, p. 223), and such research is now emerging through work such as that of Barton, Rivet, Tan, and Groome (2006). The shift from a concern with girls to a focus on gender equity is a recurring theme in the area of gender and (science) education on both sides of the Indian Ocean, and elsewhere, as discussed below. Later in this chapter I also examine current gender and science education discourses in Australia and South Africa in the light of the themes identified by Kenway and Gough (1998), above, and others that are emerging.

2. A Complicated Curriculum Conversation Across the Indian Ocean

In the late 1990s I participated in an AusAID funded institutional links project entitled *Educating for Socio-Ecological Change: Capacity Building in Environmental Education*. The project brought together academics from Deakin and Griffith universities in Australia and from Rhodes, Stellenbosch and Venda universities and Shingwedzi, Thlabane and Tshisimani Colleges of Education in South Africa.

This project was timely for post-apartheid South Africa because, as Lesley Le Grange (2003, p. 497) wrote in reference to the project, "a changing socio-political climate created new spaces for social engagement and knowledge production in South Africa after apartheid". These new spaces potentially included ones created by curriculum documents such as Curriculum 2005 (which was first introduced in Grade 1 in 1998) and reports such as that of the Department of Education's Gender Equity Task Team (1997). From an Australian perspective, where we too had recently seen the work of a comparable group result in a national Gender Equity Framework for Schools (MCEETYA, 1997), we were surprised by the relative silence on women's issues and feminist research methodologies in environmental and science education in South African schools and universities.

Attempts by the Australian partners to raise these issues within the project were very largely ignored or marginalised, although I eventually published an article on feminist research in environmental education in the *Southern African Journal of Environmental Education* (Gough, 1999). Subsequently, as discussed below, researchers such as Enslin and Pendlebury (2000) and Gouws (2004) have helped to explain some of tensions around gender equity in South Africa. In more recent times there has been an increase in public awareness of the need to get girls more engaged in science studies and science careers in South Africa and the government has introduced a number of programs to promote women and science (Mangena, 2006). As Deputy Minister Brigitte Mabandla (2002) noted:

> South Africa also understands the need to develop human capital in the maths and science field to fully exploit the role of science and technology in the development of this country. Strategies have been devised and implemented to promote maths and science to girls within the school education system, for example, by having girls' camps and achievement awards for girls in maths and science.

However, sustaining gender on the education agenda will not be an easy task. According to the Council of Education Ministers (2004, as quoted in Pandor, 2005, p. 23) gender and race have complex intersections in South African schools with education being "both a producer and a product of gender discrimination." This assertion is supported by Carolyn McKinney's (2005) research into the relationship between language, identity and conditions for learning in four urban, racially desegregated schools in Johannesburg. In addition, as Naledi Pandor, the Minister of Education, points out, "gender is not so much hidden as absent" (2005, p. 22) in documents such as a review of human resource development in South Africa (Kraak & Perold, 2003) where age and race are discussed, but there is nothing about gender.

In the dominant educational discourses the focus is still on the tension between equality writ large and quality in education. Also, the introduction of gender studies in South Africa has been viewed as a Western project (Gouws, 2004) and professional development for South African teachers is seen as a vexed issue (Reddy, 2004) with gender and rights enmeshed in other issues around democracy and education. As Unterhalter and Samson (1998, p. 1) point out:

> The South African transition to democracy and the initiatives in educational transformation that have accompanied this have both been the outcomes of

> global processes, as much as local aspirations… [and] living with globalization is … simultaneously a local and a global cultural product.

2.1 Gender Policy in Australia

Concerns about girls and schooling have a long history in Australia, as in much of the Western world. For example, in 1975 the Schools Commission published a report on *Girls, School & Society*, with a focus on girls and gender equality rather than gender equity:

> Equality of opportunity cannot be achieved where the forces acting on all girls, irrespective of social class or ability, cause them to have lower self esteem and less self confidence than boys, and where those forces go unquestioned and unchallenged. Nor can young people of either sex be equipped to cope in a considered way with the changes taking place in social roles of the sexes unless their frame of reference includes relevant information and some appreciation of the processes of social change.
>
> (McKinnon, in Schools Commission, 1975, p. iii)

This report was followed in 1987 by the *National Policy for the Education of Girls in Australian Schools* (Australian Education Council, 1987). The key aspects of this policy were that:

- Gender is not a determinant of capacity to learn.
- Girls and boys should be valued equally in all aspects of schooling.
- Equality of opportunity and outcomes in education for girls and boys may require differential provisions, at least for a period of time.
- Schools should educate girls and boys for satisfying, responsible and productive living, including work inside and outside the home.
- Schools should provide a challenging learning environment which is socially and culturally supportive and physically comfortable for girls and boys.
- Schools and systems should be organised and resources provided and allocated to ensure that the capacities of girls and boys are fully and equally realised.

The National Policy was augmented in 1993 by the *National Action Plan for the Education of Girls in Australian Schools 1993–1997*, but both of these documents were superseded in 1996 with Ministerial endorsement of *Gender Equity: A Framework for Australian Schools* (MCEETYA, 1997).

This *Gender Equity* document was strongly influenced by the "What about the boys" backlash movement of the late 1990s in Australia (Kenway & Willis, 1998, see also House of Representatives Standing Committee on Education and Training,

2002), but in many ways its position differed only slightly from earlier work. The principles for action included:

- Equitable access to an effective and rewarding education, which is enhanced rather than limited by definitions of what it means to be female and male, should be provided to all girls and boys.
- Girls and boys should be equipped to participate actively in a contemporary society which is characterised by changing patterns of working, civic and domestic life.
- Schools should be places in which girls and boys feel safe, are safe, and where they are respected and valued.
- Schools should acknowledge their active role in the construction of gender, and their responsibility to ensure that all organisational and management practices reflect commitment to gender equity.
- Understandings of gender construction should include knowledge about the relationship of gender to other factors, including socioeconomic status, cultural background, rural/urban location, disability and sexuality.
- Understanding and accepting that there are many ways of being masculine and feminine will assist all students to reach their full potential.
- Effective partnerships between schools, education and training systems, parents, the community, and a range of other agencies and organisations will contribute to improvement and change in educational outcomes for girls and boys.
- Intervention programs and processes should be targeted towards increasing options, levels of participation and outcomes of schooling for girls and boys.
- Anti-discrimination and other relevant legislation at state, territory, federal and international levels should inform educational programs and services.
- Continuous monitoring of educational outcomes and program review should inform and enhance decisions on the development, resourcing and delivery of effective and rewarding education for girls and boys.

In recent times gender, from a girls' perspective in particular, has mainly been a silence on the educational agenda in Australia, to the extent that the Australian Science Teachers Association's (1994) policy on *Girls and Women in Science* – "ASTA must encourage women to pursue science for its career opportunities" – seems to be no longer mentioned. Indeed, in recent years the focus has been on curriculum reform and achievement testing – particularly that associated with international programs such as TIMSS and PISA. For example, Masters (2005, p. 10) argues that

> If Australia is to lift its performance in mathematics and science over the next decade, then greater attention will need to be given to the teaching of basic factual and procedural knowledge and the development of teachers' confidence and competence in teaching primary school mathematics and science.

> The focus of the past decade on *what* is taught (the curriculum) needs to be accompanied by a greater focus on *how* subject matter is taught (research-based pedagogy).

There is little or no mention equity issues, yet there is much research to support the contention that gender, race and class as well as educational opportunity all have an effect on achievement (see, for example, Teese & Polesel, 2003).

2.2 Gender on the Agenda in (South) Africa

In the apartheid years, "education was marked not only by segregation, under resourcing for the majority of children and punitive regimes of control for pupils and teachers, but also by curricula infused with racism, sexism and approaches to knowledge and culture which failed all but a tiny minority" (Unterhalter & Samson, 1998, p. 2). The curriculum of this period was used to promote social inequality, racial segregation and differentiation on the basis of perceptions of ability. It also was examination driven, gave a central role to teachers, emphasised rote learning by passive students, and had little public accountability and exposure to public debate.

In the 1980s globalisation was experienced in South Africa through non-government organisations "which saw themselves as articulating global alternative visions for new curricula linked to emancipatory pedagogy" (Unterhalter & Samson, 1998, p. 4). But, in the 1990s, as the apartheid forces were dismantled, globalisation came to South Africa through conventional international organisations such as the World Bank and UNICEF, and in funded visits for South Africans to "successful" curriculum reform projects in the UK and USA.

Feminism was not a popular mobilising force in South Africa in the early 1990s, partly because of the enormous divisions between academic feminists (mainly whites) and activist feminists (mainly blacks). However, from the mid 1990s academic feminist education policy makers opened up a space for feminism in discussions about educational change and made links with a transnational community of feminists working on feminist initiatives in education. These initiatives accorded well with those of UNICEF, UNESCO and the World Bank who were similarly concerned with analysing and implementing the gender dimensions of educational policy (because of "human capital theorists 'discovery' of apparently significant correlations between the education of women and girls, growth of GDP, and reduction of fertility" (Unterhalter & Samson, 1998, p. 5). It was within this framework that the previously mentioned Gender Equity Task Team (1997) produced its report.

The global agenda continues to highlight the importance of the education of girls in international strategies such as the

- United Nations' Millennium Development Goal process which provides targets for international actions to bring such visions into reality by: overcoming poverty; improving child, maternal and sexual health; expanding educational provision and redressing gender inequalities in education; and developing national strategies for sustainable development.

- World Education Forum Education for All Dakar Framework of Action has six goals concerned with extending the reach of basic education to every child and adult (and thus overlaps with the MDGs on the provision of primary education and gender equality in education).
- United Nations Literacy Decade which is situated within the Education for All movement with literacy as a thread through the six EFA goals.
- United Nations Decade on Education for Sustainable Development (2005–2014) which is closely linked to the above three efforts. Here gender equality and basic education are also emphasised: "The broader goal of gender equality is a societal goal to which education, along with other social institutions, must contribute." (UNESCO, 2004, p. 17)

There is also the UN Girls' Education Initiative (UNGEI) which is working towards the fulfilment of the EFA goals of gender parity in education (by 2005) and gender equality (by 2015).

Within the African Union gender has gained increasing prominence in recent times. A succession of meetings and declarations has foregrounded the importance of women's rights and the need for the mainstream participation of women in African society. These include

- Durban Declaration on Mainstreaming Gender and Women's Effective Participation in the African Union (30 June 2002)
- Dakar Strategy on Mainstreaming Gender and Women's Effective Participation in the African Union (26 April 2003)
- Maputo Declaration on Gender Mainstreaming and the Effective Participation of Women in the African Union (24 June 2003)

Subsequently, on July 11 2003, the African Union summit in Maputo, Mozambique adopted the Protocol to the African Charter on Human and Peoples' Rights on the Rights of Women in Africa. The protocol entered into force in early 2004 after it was ratified by fifteen states. An important part of these declarations and the Protocol is education of women so they can participate effectively in society.

With this quantity of declarations, protocols, movements and decades focussing on gender and education issues it would seem reasonable to expect that gender would be part of the education discourse in South Africa, but instead there remains tension. As Naledi Pandor (2005, p. 19) commented to a 2004 conference on gender equity in South African education: there has been no lack of seminars and conferences on gender inequality in the South African education system yet "(t)here are clear indications that South African educators and policy-makers hold the view that there is no gender equity challenge confronting girls and women in the education sector".

2.3 *(In)Equality, Democracy and Quality as Contours of Educational Discourse*

An indicator of an increasing awareness of the tension between quality and equality as contours of the educational discourse in South Africa was the declaration of "(In)equality, democracy and quality" as the theme for the 2005 Kenton Conference for South African educational researchers. As part of their rationale for this theme the organisers wrote:

> In 2004, a "Decade of Democracy" has passed and educationalists looked back and reflected on what had been achieved. This review of educational progress revealed that, counter to the vision of the policy, inequalities had increased, exacerbated by the deepening of poverty and its impact on education ... There remained serious questions with regard to access to quality education; for example, the issue of "language" remains problematic and represents one of the greatest challenges to equality of access, as does access to ICTs ... An educational debate on access via school fees raised its head in public last year, raising old questions about class, access and race within a framework of emerging elites and ongoing inequalities...
>
> In retrospect, it seems that the attempt to improve the quality of education, to bring about equality and support democracy in a globalising, market-oriented society, has met with mixed results. Policies seem to have been idealised and are remote from contextual realities. Democracy in education appears to exist in name only and falls short in its actualisation.

However, this rationale statement also continues the silence around gender equity as an educational issue while recognizing class and race as significant issues. Perhaps not unexpectedly, the 2005 Kenton conference papers that touched on inequality issues were generally silent on gender (one exception was McKinney, 2005) and reflected a concern with outcomes-based education (Mason, 2005), maintaining standards and providing quality education to all learners in terms of their right to education (Joubert & Bray, 2005) and inclusive education in lower socio-economic schools (Geldenhuys & Pieterse, 2005).

Yet gender is considered a significant issue in wider society, as is evidenced by the reports of the Commission on Gender Equality, in speeches by government ministers (for example, Mabandla, 2002; Mangena, 2006; Pandor, 2005), and in South African newspapers (see, for example, Meintjes (2003) in *The Sunday Independent* and Win (2005) in the *Mail & Guardian*).

Even though Gouws (2004) argues that gender studies in South Africa has been viewed as a Western project, in an age of globalisation and democracy, and given the goal of social transformation in a post-apartheid era, the silence around women's rights and gender issues in educational discourses remains mysterious to outsiders. It is also a mystery to some insiders: for example, Claudia Mitchell (2005, p. 104), in her review of *Changing Class: Education and Social Change in Post-apartheid South Africa* (Chisholm, 2004), notes the lack of analysis on gender equity within education:

> Where are the women? Where is gender? Given the elaborate gender machinery in the country since 1994 – from the Gender Commission to the Office of the Status of Women, and the key challenges posed by the Gender Equity Task Team Report (1997) and the Human Rights Watch Report *Scared at school* (2001), this is an omission that one hopes might be addressed in a volume two of Changing class, should such a collection materialize.

Other South African writers such as Enslin and Pendlebury (2000) and Gouws (2004) have helped to explain some of tensions around gender equity in South Africa. For example, Enslin and Pendlebury (2000, p. 432,) note that "the inconsistent treatment of gender and rights in Curriculum 2005 risks perpetuating the oppression of South African girls and women", and Gouws (2004, p. 69) argues that "education for nation building has been one of the ideals of the post-apartheid government… [but] the nation is being built through male values and symbols to the exclusion of women… [and] the nation state also suffers from the increasing pressure of expanding globalisation". However, in the dominant educational discourses the focus is still on the tension between equality writ large and quality in education.

2.4 Quality and Equality in South African Education Discourses

Noel Gough (2006) points out that "quality" can be understood as an example of what Deleuze and Guattari (1987, p. 79,) call "order-words" (*mots d'ordre*), words that presuppose or create a socio-political order or performs an ordering function, and thus produce different effects in different locations. In post-apartheid South Africa, "quality" in education and schooling presupposes political imperatives toward social transformation. Johann Steyn (2004) characterises transformation in South African education as:

- the transformation from a fragmented educational system to a unified system;
- the efforts to remover inequalities and the move towards equal education;
- the shift away from a monocultural educational system;
- the intention to shift from a content based education to Outcomes Based Education;
- the repealing of anti-democratic policies;
- the transformation from a closed society to a more open society;
- the "catching up" with leaders in the field of education, and
- the intention to create a just system that provides for access to quality education (pp. 101–102).

A distinctive characteristic of South African discourses of educational transformation is that the enunciation of "quality" orders conversations around "equality", and *vice versa*. For example, Willem Du Plessis (2000, p. 65) argues that during the apartheid years "all good quality education was the sole property of schools for Whites, in White residential areas, beyond the reach of non-White students". Similarly, writing in the run-up to the first democratic elections in 1994, Pam Christie (1993, p. 11) asserts that the pursuit of quality education "has become a catch cry limiting the influence of Black students on the existing practices of historically privileged schools". Five years after those elections, Ken Hartshorne (1999, p. 7) insists that little has changed: "quality education is only a strategy to slam doors in the faces of Black learners". More recently, Steyn (2004, p. 106) characterises contemporary perceptions of quality and equality in South African education as opposing positions in a "debate", with some protagonists arguing that "the quest for quality education is an attempt to maintain standards in White schools and universities and to exclude Black learners", and others arguing that "the eradication of gross inequalities is not a viable option in the light of the hard [economic] realities".

Thus, in South Africa, the enunciation of "quality" not only orders conversations around "equality" but also orders these concepts into an inverse or adversarial relationship. For example, Steyn (2004, p. 97) describes "balancing quality and equality" as a "dilemma" and as "a kind of juggling act", which implies that increasing one's commitment to quality necessarily reduces one's commitment to equality (and *vice versa*). This is not the case in a number of other nations from which South Africa has made policy borrowings, where equality (or equity) is understood to be a necessary condition of quality.

South Africa's discourses of social transformation produce a socio-political ordering of education and schooling that emphasises economic and racial equity, which in turn leads to a positioning of equality as being in tension with quality. This contrasts with "equality" in nations such as Australia and the UK, which produces "orders" (such as policy directives) on equity issues that extend beyond race and class to include gender, sexuality, disability, etc. In Australia it is relatively easy to demonstrate that gender equity is an *achievable condition* of quality education, rather than something that is economically or socially "beyond the reach" of the majority of learners.

2.5 Globalisation as a Tension in Local/Global Knowledge Production

Just as with gender and (e)quality in South African educational discourses there is tension around discussions of globalisation and its impact on curriculum transformation in South Africa. For some South Africans, such as Ramphele and Kraak (in Unterhalter & Samson, 1998, p.7), globalisation is "an unstoppable force, which carries considerable promise for South Africa" and "their position entails either an eclipse of the significance of local knowledge or a diminution of the significance of the local in relation to the global". In addition, "all new knowledge is generated, codified and disseminated in a global context", and "local knowledge is reduced to a passive resource mined in order to add value to the global processes of knowledge creation" (pp. 7–8). However, according to Untherhalter and Samson (1998, p. 9), Kraak and Ramphele fail "to acknowledge the inherently gendered and racialised nature of the privileged spheres which are positively affected by globalization" and this "prevents them from identifying that rather than promoting transformation and elimination of racial and gender equalities, globalization entrenches these".

Most South African critics of globalisation view it in economic terms, and none except the Gender Equity Task Team (GETT) give it a gendered dimension. According to GETT (in Unterhalter & Samson, 1998, p. 10), "while the preparation of students for the labour market is usually seen as the direct outcome of education", labour markets are gender biased and "globalization has led to the feminisation of the labour market, although this is often linked to harsh conditions of work for women who are seen as more 'disposable' workers than their male counterparts". For critics of globalisation, local knowledge is seen as "the foundation for an education system designed to assist in the strategic engagement with the global sphere" (Unterhalter & Samson, 1998, p. 11). Although they view globalisation as ungendered, critics such as Nzimande and Chisholm (in Unterhalter & Samson, 1998, p. 12), see unequal gendered power as an issue in South Africa.

In the Curriculum 2005 documents globalisation is completely absent, although there is reference to "good global citizenship", but there is strong emphasis given to the local, generally uncoupled from the global. However, "the critical cross-field outcomes of Curriculum 2005, which are intended to act as the basis for the entire curriculum, make no reference to the existence of gender inequality, the need to overcome gender stereotypes, or the need for the curriculum to be anti-sexist. Similarly, none of the specific outcomes for any of the eight learning areas include any reference to gender issues" (Unterhalter & Samson, 1998, p. 16), although there is reference "to the role which curriculum plays in building a democratic, non-racist and equitable society" (Unterhalter & Samson, 1998, p. 11).

Where gender is mentioned in Curriculum 2005 it is either

- a lofty policy goal with no specific enabling mechanisms,
- a disabling characteristic of female students, where the role of the new curriculum is to help girls overcome this disability,
- a marker of a homogenised female community whose different experiences, ideas and histories need to be considered, or (most commonly)
- an ahistorical variable for discrimination, without consideration of present structural factors.

There is a blind spot that the actual curriculum content could be gender biased, "that femininity might shape a powerful identity for some (for example when linked with race and class privileges) and a disempowering identity for others; [and] there is no consideration of identities shifting in relation to changing political and economic conditions" (Unterhalter & Samson, 1998, pp. 16–17). In addition, although "as a result of Curriculum 2005, 'critical thinking, rational thought and deeper understanding – central principles of the new education system – will soon begin to break down class, race and gender stereotypes'... the material and structural basis of inequality is therefore dismissed and the potential for economic globalisation to entrench and perpetuate these inequalities remains unexplored" (Unterhalter & Samson, 1998, p. 17).

2.6 *(E)quality and Science Education*

In this section I draw on the work of Kreinberg and Lewis (1996) and their analysis of the politics and practices of equity across the Pacific, and Kenway and Gough (1998) review of research into gender and science, and discuss the relationship between globalisation and gender issues across the Indian Ocean.

Kreinberg and Lewis (1996) developed a model for analysing gender reform and curriculum change in science education (Figure 1), adapted from the work of Schuster and Van Dyne (1984) on stages of curriculum transformation with respect to women in the liberal arts. Although I have some concerns with their model (as noted below), I draw on it in the following discussion of the current status of gender reforms and science curriculum change and the posing of some challenges for the future.

The first two stages of Kreinberg and Lewis' model provide a useful reminder of the starting point for much of the "girls and science" discourse in schools. With respect to Stage 1, the absence of women not being noticed, despite "broad social recognition that women are missing from science and that social justice requires action to change this", and all the activities relating to "girls and science" of past decades, Kreinberg and Lewis (1996, p. 194) note that "there is an ongoing need for more bridges to be built with 'non-converted' teachers who are currently not concerned with these equity issues". All South African gender equity literature cited in this chapter – and my experience – supports this on-going need.

Getting more girls interested in studying science also requires changes in teacher education. Kreinberg and Lewis (1996, p. 194) assert that "there is an ongoing need for more bridges to be built with 'non-converted' teachers who are currently not concerned with these equity issues", but, as previously noted, in a South African context, professional development for teachers is a vexed issue (Geldenhuys & Pieterse, 2005; Reddy, 2004) as gender and rights become enmeshed in other issues around democracy and education.

Stage 2, the search for the missing women, was a compensatory exercise to identify women scientists. It resulted in a number of publications and activities in the 1980s which did little more than add to the existing data within conventional

Stage 1: Absence of women in science not noticed
Stage 2: The search for the missing women in science
Stage 3: Why are there so few women in science?
Stage 4: Studying women's experience in science
Stage 5: Challenging the paradigm of what science is
Stage 6: The transformed, gender balanced (gender-free) science curriculum

Figure 1. Kreinberg and Lewis' Model for Looking at Gender Reform and Curriculum Change in Science Education

paradigms (see the critique of this search for role models and strategies for encouraging more girls to pursue science careers in Kenway & Gough, 1998). There is some evidence that South African science education is currently working at this level.

As Kreinberg and Lewis (1996, p. 195) describe it, the emphasis in Stage 3 was that "women were disadvantaged because individual females missed out on being part of male achievement. It was a protest rather than a direct challenge to the conventional paradigm of science and science education". South Africa's Minister of Science and Technology, Mosibudi Mangena (2006), recently voiced a similar sentiment:

> Recognising the existing gender disparity across the broad spectrum of cultural, institutional, and organisational spheres of society, South Africa decided to entrench gender equality in our Bill of Rights as a fundamental principle for all, irrespective of race, class, age or disability. In so doing, we have acknowledged that gender discrimination can, and will, be 'un-learned' and changed in this country.
>
> At a broader level, the discrimination confronting South African women in science is a little different from that faced by women in other countries. Women have been seeking access to science and technology education and careers for well over a century, but their efforts in this regard have been met by opposition – often subtle and sometimes blatant. As a result, there is a general lack of support for women within the science system.
>
> ... in order to address the challenges women face in entering and pursuing careers in SET, a qualitative five-year longitudinal study was instituted in 2005. We expect the study to give us better insight into the complexity of the issues that typically hinder women's advancement in the system. Amongst other things, the study is also intended to give us the opportunity to address concerns at school-going age, and to identify and introduce preferential funding mechanisms for women at post graduate level.

However, Kreinberg, and Lewis also noted that there was growing recognition that the curriculum itself needed to be challenged – and that science, not girls, had to change. Both South Africa and Australia can be seen as being stuck at this stage with the foci on changing girls' choices, beliefs, perceptions, attitudes and aptitudes.

Several aspects of the notion of "changing girls" are evident in Kreinberg and Lewis' Stage 4, studying women's experience in science. Here, curriculum change activities focused on developing a gender-inclusive curriculum in science that emphasised teaching strategies which (Kreinberg & Lewis, 1996, p. 196):

- provided active learning contexts for students (e.g. constructing with *Lego*),
- described alternative ways of organising the classroom (e.g. cooperative groups), or
- reorganised the curriculum (e.g. starting from and valuing students' experiences).

Their goal was a different science in the classroom, with "increasing emphasis on the science contexts of girls' lives and on the ways in which girls prefer to learn" (Kreinberg & Lewis, 1996, p. 196).

Kreinberg and Lewis (1996, p. 197) describe Stage 5 in terms of challenging the dominant science education paradigm through constructing the curriculum through students studying science "issues" in their lives:

> Constructing curriculum in this way can challenge the dominant paradigm in science through the inclusion of all the issues that surround students' lives ... Science classrooms can be places where the practices of science are discussed, different views expressed, alternative information considered and local, national or international action initiated.

However, Kreinberg, and Lewis' Stage 5 seems to neglect the gendered, classed and racist nature of the scientific knowledge that constitutes the science curriculum.

The first five stages of Lewis's model can be described as a gender-inclusive curriculum movement "focussed on the inclusion of female experiences and perspectives to change curricula and teaching" (Kreinberg & Lewis, 1996, p. 197). Stage 6, the transformed, reconstructed gender-free curriculum, "goes beyond the restrictions of male and female ... examines critically the assumptions behind the culture and practice of science and ... the social construction of masculinity and femininity" (Kreinberg & Lewis, 1996, p. 197). In this curriculum the feminist critiques of science (see, for example, Harding, 1993, 2004; Keller, 1985) would be taken into account and the science itself would be challenged. I should note that I have some reservations about Lewis's use of the term "gender-free" in this context in that the traditional objectivist science curriculum claimed to be gender-free, but this has been found to be a universalist masculine field where "man" ruled supreme. However, with the qualifiers of "transformed" and "reconstructed" it is probable that a different notion of "gender-free" is intended.

Kenway and Willis (1993) make a sharp distinction between the approaches of changing choices/changing girls – where the responsibility for change rests with girls themselves – and of changing the curriculum/changing the learning environment, where "the implication is that while girls must accept some responsibilities, they don't have to shoulder all the burden and can be encouraged by the sense that other people are changing for and with them" (Kenway & Willis 1993, p. 83). This is a significant difference as it ideally involves girls, boys and teachers working together at the centre of educational change.

Such a focus brings together concerns of quality and equality, not as alternatives but as mutually beneficial goals in "new spaces for social engagement and knowledge production" (Le Grange, 2003, p. 497). At the moment, on both sides of the Indian Ocean, there are rhetoric-reality gaps. The policy frameworks exist at government level but evidence of implementation in science curricula is generally absent. Gender as an education issue is seen as a task for the Department of Education, and gender as a science education issue is seen as a task for the Department of Science and Technology. The challenge is for equity and quality in science education to be seen as a task for schools.

3. Conclusion

Catherine Odora-Hoppers (2005, pp. 55–56) writes of gender as an heuristic tool:

> Speaking gender means trying to understand how society made you what you have become, how it shaped your behavior, your aspirations, and your attitude towards yourself as well as towards society at large … [Gender sensitivity] should be part of a vision of development that both redresses gender inequalities, and which constructs a new ethical basis for continuing development.

Within science education discourses, in a globalisation context, gender equity is an issue for society and the curriculum in schools. On both sides of the Indian Ocean there are issues around girls and science. Both Australia and South Africa recognise that there are different levels of participation in science by boys and girls, and both countries have policies or strategies to encourage more girls to study science and embark upon science careers. Also, in both countries, there is still a need for gender equity research and curriculum transformation "to transcend the boundaries of race, ethnicity, class and socio-economic identities" (Krockover & Shephardson, 1995, p. 223).

Science educators need to be engaging the question: "How much of the nature of science is bound up with the idea of masculinity, and what would it mean for science if it were otherwise?" (Keller, 1985, p. 3). This is not just a question for women but a question for all. There is a need for some fundamental changes in the way science is represented in schools – moving away from science as objective and dispassionate, reassessing the nature of evidence and explanation (and their relationship to each other), and reviewing the status of scientific knowledge, especially the philosophical and psychosocial aspects of learning environments. Evans (1996, p. 74) proposes that we need to develop "science education courses which are gender-critical and gender-inclusive and which regender the field accordingly; this is men's work too and this must be confronted". To this I would add Kreinberg and Lewis's notion of a transformed, reconstructed, gender-free curriculum which goes beyond gender inclusivity. By engaging in developing and studying such courses, unconcerned men and women in teacher education and teaching (such as those who have not made it to Kreinberg and Lewis's Stage 1 or beyond) would be confronted with their own prejudices and practices and challenged to change.

Achieving gender equity across race, class, ethnicity and socio-economic boundaries is an issue in all societies, and science education – as part of formal education – has a role in this transformation. Science education has made some advances on both sides of the Indian Ocean. The absence of women in science is now noticed, questions are asked about why there are so few women in science, and women's experiences in science are now being studied. The challenging of the paradigm of what science is, and transforming the science curriculum is largely still to happen – especially taking into account race, class, ethnicity and socio-economic boundaries – but it is a necessary part of equity, equality and globalisation in science education.

There is a need for a rethinking of "girls and science" – recast as a more democratic science education – as a globally relevant issue in education. Such a democratic science education examines "Western science's complicity with racist, imperialist, [gendered] and Eurocentric projects [and] enables us to gain a more critical, more scientific perspective on an important part of that Western 'unconscious'" (Harding, 1993, p. 19). It also requires the provision of an adequate science education for *all* citizens. This is an issue of equity, quality and equality: a Science for All.

Acknowledgements

An earlier version of this chapter was presented as a paper co-authored with Noel Gough at the 2005 British Educational Research Association Conference. His assistance in the preparation of this chapter is gratefully acknowledged, as are a number of conversations with South African colleagues Heila Lotz-Sisitka, Lesley le Grange and Chris Reddy. I also thank Bill Atweh for his boundless patience.

References

Australian Education Council. (1987). *National policy for the education of girls in Australia*. Canberra: Australian Government Publishing Service.

Australian Science Teachers' Association (ASTA). (1994). *Policy on girls and women in science*. Canberra: ASTA.

Barton, A. C., Rivet, A., Tan, E., & Groome, M. (2006, April). *Urban girls' merging science practices* (pp. 7–12). Paper set presented at the annual meeting of the American Educational Research Association, San Francisco.

Christie, P. (1993). Equality in curriculum in post-apartheid South Africa. *Journal of Education, 18*(1), 5–18.

Chisholm, L. (Ed.). (2004). Changing class: Education and social change in post-apartheid South Africa. Cape Town: HSRC Press and London: Zed Books.

Deleuze, G., & Guattari, F. (1987). *A thousand plateaus: Capitalism and schizophrenia* (B. Massumi, Trans.). Minneapolis: University of Minnesota Press.

Department of Education, Republic of South Africa. (1997). *Gender equity in education: Report of the gender equity task team*. Pretoria: Department of Education.

Du Plessis, W. (2000, 27–28 September). *Official policy related to quality and equality in education: A documentary study*. Paper presented at the Democratic Transformation of Education in South Africa, Stellenbosch, South Africa.

Enslin, P., & Pendlebury, S. (2000). Looking others in the eye: Rights and gender in South African education policy. *Journal of Education Policy* 15(4), 431–440.

Evans, T. (1996). Under cover of night: (Re)gendering mathematics and science education. In L. H. Parker, L. J. Rennie, & B. J. Fraser, (Eds.), *Gender, science and mathematics: Shortening the shadow* (pp. 67–76). Dordrecht, The Netherlands: Kluwer.

Geldenhuys, J. L., & Pieterse, G. (2005). *Preparedness of eastern cape educators in lower socio-economic schools for inclusive education*. Paper presented at the 2005 Kenton at Mpekweni Conference, South Africa.
Gough, A. (1999). The power and the promise of feminist research in environmental education. *Southern African Journal of Environmental Education, 19*, 28–39.
Gough, N. (2006). Quality imperialism in higher education: A global empire of the mind? *ACCESS: Critical Perspectives on Communication, Cultural & Policy Studies*, *25*(2), 1–15.
Gouws, A. (2004). Democracy and gendered citizenship: Educational institutions as sites of struggle. In Y. Waghid & L. le Grange (Eds.), *Imaginaries on democratic education and change* (pp. 69–81). Pretoria/Matieland: South African Association for Research and Development in Higher Education/Stellenbosch University.
Harding, S. (1991). *Whose science? Whose knowledge? Thinking from women's lives*. Ithaca, NY: Cornell University Press.
Harding, S. (Ed.). (1993). *The "Racial" economy of science: Toward a democratic future*. Bloomington: Indiana University Press.
Harding, S. (Ed.). (2004). *The feminist standpoint theory reader*. New York: Routledge.
Hartshorne, K. (1999). *The making of education policy in South Africa*. Cape Town, Southern Africa: Oxford University Press.
House of Representatives Standing Committee on Education and Training. (2002). *Boys: Getting it right*. Report on the inquiry into the education of boys. Canberra: The Parliament of the Commonwealth of Australia.
Joubert, R., & Bray, E. (2005). *Practical manifestation of equality, democracy and quality in education: A legal perspective*. South Africa, Paper presented at the 2005 Kenton at Mpekweni Conference.
Kaminski, D. M. (1982). Girls and mathematics and science. An annotated bibliography of British work (1970–1981). *Studies in Science Education, 9*(2), 81–108.
Keller, E. F. (1985). *Reflections on gender and science*. New Haven: Yale University Press.
Kenway, J., & Gough, A. (1998) Gender and science education in schools: A review with 'attitude'. *Studies in Science Education*, *31*, 1–30.
Kenway, J., & Willis, S. (1993). *Telling tales: Girls and schools changing their ways*. Canberra: Department of Employment, Education and Training.
Kenway, J., & Willis, S. (with Blackmore, J., & Rennie, L.) (1998). *Answering back: Girls, boys and feminism in schools*. Sydney: Allen & Unwin; London/New York: Routledge.
Kraak, A., & Perold, H. (Eds.). (2003). *Human resources development review 2003: Education, employment and skills in South Africa*. Cape Town: HSRC Press.
Kreinberg, N., & Lewis, S. (1996). The politics and practice of equity: Experiences from both sides of the Pacific. In L. H. Parker, L. J. Rennie, & B. J. Fraser, (Eds.), *Gender, science and mathematics: shortening the shadow* (pp. 177–202). Dordrecht, The Netherlands: Kluwer.
Krockover, G. H., & Shephardson, D. P. (1995). Editorial: The missing links in gender equity research. *Journal of Research in Science Teaching*, *32*(3), 223–224.
Le Grange, L. (2003). The role of (dis)trust in a (trans)national higher education development project. *Higher Education, 46*, 491–505.
Mabandla, B. (2002). *Third World Organisation for Women in Science (TWOWS) WSSD science forum event – Panel discussion: Women's perceptions of science and technology for sustainable development*. Retrieved September 5, 2006, from http://www.dst.gov.za/media/speeches.php?id=112&print=1
Mangena, M. (2006). *Keynote address by the minister of science and technology, at the 2006 South African women in science awards, Friday 4 August*. Retrieved September 2, 2006, from http://www.dst.gov.za/media/speeches.php?id=204&print=1
Manthorpe, C. A. (1982). Men's science, women's science or science? Some issues related to the study of girls' science education. *Studies in Science Education, 9*(2), 65–80.
Mason, M. (2005). *The quality of learning and its contribution to the development of equality and democracy: What really works to enhance learning*. Paper presented at the 2005 Kenton at Mpekweni Conference, South Africa.

Masters, G. (2005). International achievement studies: Lessons from PISA and TIMMS. *Research Developments 13*, 6–10.

McKinney, C. (2005). *Language, identity and (subverting) assimilation in a South African desegregated suburban school*. Paper presented at the 2005 Kenton at Mpekweni Conference, South Africa.

Meintjes, S. (2003, August 24). Rating the gender equality scorecard. *The Sunday Independent*, p. 7.

Ministerial Council on Education, Employment, Training and Youth Affairs. (MCEETYA). (1997). *Gender equity: A framework for Australian schools*. Canberra: Australian Capital Territory for MCEETYA.

Mitchell, C. (2005). Review of Changing class: Education and social change in post-apartheid South Africa. *Compare, 35*(1), 101–104.

Odora-Hoppers, C. (2005). Between 'mainstreaming' and 'transformation': Lessons and challenges for institutional change. In L. Chisholm & J. September (Eds.), *Gender equity in South African education 1994–2004* (pp. 55–73). *Perspectives from Research, Government and Unions. Conference Proceedings*. Cape Town: HSRC Press.

Pandor, N. (2005). The hidden face of gender inequality in South African education. In L. Chisholm & J. September (Eds.), *Gender Equity in South African education 1994–2004* (pp. 19–24). *Perspectives from Research, Government and Unions. Conference Proceedings*. Cape Town: HSRC Press.

Pinar, W. F., Reynolds, W. M., Slattery, P., & Taubman, P. (1995). *Understanding curriculum: An introduction to the study of historical and contemporary curriculum discourses*. New York: Peter Lang.

Reddy, C. (2004). Democracy and in-service processes for teachers: A debate about professional teacher development programmes. In Y. Waghid & L. le Grange (Eds.), *Imaginaries on democratic education and change* (pp. 137–146). Pretoria/Matieland: South African Association for Research and Development in Higher Education/Stellenbosch University.

Schools Commission. (1975). *Girls, school and society*. Report by a study group to the schools commission. Canberra: Australian Government Publishing Service.

Schuster, M., & Van Dyne, S. (1984). Placing women in the liberal arts: Stages of curriculum transformation. *Harvard Educational Review, 54*(4): 413–430.

Steyn, J. (2004). Balancing the commitment to quality education and equal education in South Africa: Perceptions and reflections. In Y. Waghid & L. Le Grange (Eds.), *Imaginaries on democratic education and change* (pp. 97–110). Pretoria/Matieland: South African Association for Research and Development in Higher Education/Stellenbosch University.

Teese, R., & Polesel, J. (2003). *Undemocratic schooling: Equity and quality in mass secondary education in Australia*. Carlton: Melbourne University Press.

UNESCO. (2004). *United Nations Decade of Education for Sustainable Development 2005–2014*. Draft implementation scheme. October 2004. Retrieved August 1, 2005, from http://portal.unesco.org/education/en/file_download.php/03f375b07798a2a55dcdc39db7aa8211Final+IIS.pdf

Unterhalter, E., & Samson, M. (1998, April). Unpacking the gender of global curriculum in South Africa. Paper presented at the AERA Annual Meeting, San Diego.

Win, E. (2005, 4–10 March) Plus ça change ... *Mail & Guardian*, p. 23.

SECTION 2

ISSUES IN GLOBALISATION AND INTERNATIONALISATION

9

CONTEXT OR CULTURE: CAN TIMSS AND PISA TEACH US ABOUT WHAT DETERMINES EDUCATIONAL ACHIEVEMENT IN SCIENCE?

Peter J Fensham
QUT, Brisbane

Abstract: Most mainstream researchers in science education are weak in their inclusion of the wider educational, personal and social contexts in which their studies have been conducted. The TIMSS and PISA projects, on the other hand, have both had the status and resources to include a great deal of data about these wider contexts, nationally and cross-nationally. The success and failure of these projects in relation to elucidating strong relations between contextual constructs and science achievement is considered. The methodological choices of these cross national studies and the theoretical perspective they have adopted for these interactions are critically appraised. An alternative approach is then explored

Keywords: Influence of context, science achievement, culture of science education, international comparative studies

1. Introduction

In the 1960s the National Science Foundation in USA and the Nuffield Foundation in United Kingdom funded a number of projects to reform the curriculum of school science education. Science education authorities in many other countries, also concerned about the inadequate state of school science for the scientific and technological challenges facing societies at that time, likewise expressed interest. School science education had begun to be, not just a national issue, but one with global implications.

The new materials (texts, teachers' guides, laboratory manuals, etc.) for teaching science were of an extent and quality of production that far exceeded what had been hitherto available. By the later 1970s these new materials had been available long enough to assess their impact – directly in the countries of origin, and indirectly in the other countries that adopted, adapted, or used them as sources of new ideas.

B. Atweh et al. (eds.), Internationalisation and Globalisation in Mathematics and Science Education, 151–172.

The findings fell far short of the high expectations of the funding bodies and of those enthusiastic scientists and science teachers involved for several years to bring each project's materials to fruition. At the national level, Stake and Easley (1978) in case studies of US school districts illustrated the problems of suddenly innovating in science education, when schools or districts had other urgent agenda to contend with.

Layton (1973), Young (1971), and Waring (1979) in Britain and Gintis (1972) and Apple (1979) in USA, from the wider perspectives of social and institutional change, concluded that these attempts to reform school science education had been undertaken, as if this curriculum area existed in a social and political vacuum.

Forty years later, the papers in the international journals for science education research, often tempt me to paraphrase this conclusion as "*science education research is conducted as if science education exists in a social or political vacuum*".

Jenkins (2004, p. 117), expressed the same concern: "*I become worried that so many persons seem to think that complex educational questions can be answered by some fairly straightforward empirical test. I am bothered that the researcher so readily moves from a particular context in Israel or New Zealand to claim some strong generalization.*"

It is very rare to see in such articles a cautionary acknowledgement of the possibility that things happen in the way the researcher found, because of the political, social, economic or institutional constraints determining the situation being studied, albeit that these lie beyond the researcher's scope and capacity to explore.

It would not be easy for these individual researchers to pursue these wider contextual matters, requiring as they would considerable additional resources and unfamiliar issues of design and methodology, and great problems of accessing the sources of these constraints. In comparison, international projects that are supported by, and have the authority of governments, are able to consider both a particular aspect of education and its wider social determinates. This paper uses two ongoing international projects in the area of science education to consider the issues of contextual design and methodology that then arise. More than forty countries have participated in each of these projects that have involved comparative international testing of students' achievements in science – major examples of the globalization of school science.

The International Association for Evaluation in Education (I.E.A.) began in the early 1990s an ongoing project with several component studies, the Third International Mathematics and Science Study (TIMSS, now re-titled Trends in Mathematics and Science Study). The Organization for Economic and Co-operative Development (O.E.C.D.) in the later 1990s launched the Program for International Student Achievement (PISA) for the study of reading, mathematical and scientific literacies. The designs of both these projects involve the collection of a deal of wider contextual data, from a number of countries.

These projects clearly lie within comparative education as defined by Noah and Eckstein (1996, p. 127), namely, "*an intersection of the social sciences, education and cross national study [which] attempts to use cross national data to test propositions about the relationships between education and society and between teaching practices and learning outcomes.*" Furthermore, the projects' emphases on context are consistent with Bereday's (1964) classic model of comparative educational research, and with Le Métais' (2001, p. 197) hopeful observation that "*context……goes a long way to explain the success or failure of specific teaching and learning approaches*"

Sjøberg (2004) and Ogawa (2001,2005) have raised serious concern about the unwarranted educo-political influences such global and apparently authoritative projects and their findings can have on national priorities for educational reform. In this chapter, however, my interest is rather on these projects as research studies that set out to illuminate the links between wider contextual features and school science education. I examine the extent to which this expectation is fulfilled, and the two issues of methodology and research perspective, that relate to the shortfall in these expectations.

1.1 Paradigms for Comparative Research

Research perspective and methodology are, as they should be, quite interwoven and determining of each other. Both TIMSS and PISA have adopted a positivistic perspective in which student science achievement and a number of chosen contextual constructs of possible influence are conceived as having cross-national relevance. Furthermore, the contextual constructs are assumed to be measurable as variables so that their degree of influence on student achievement can be statistically analysed, independently or in conjunction with other constructs. The primary methodology is survey research, with centrally developed and trialed questionnaires that are administered intra-nationally by local agents. The data that ensue from the closed items in these questionnaires lead to measures of the contextual constructs and of the student achievement, that are then analysed statistically. The findings largely consist of tables and figures derived from these statistical analyses.

This research approach contrasts with another perspective in comparative education that conceives more holistically of educational contexts and performance within them. Individual features of such a context can be considered and discussed, but are likely to be so interdependent with others that their influences cannot be separately discerned. Another way of putting this is to say that constructs, that relate to collectable data, may only be manifestations of more fundamental values and complex mores, that are not readily accessible with such an external and commonly assumed methodology.

In this second approach, the paramount methodologies for exploring the complexity of the national contexts and their constructs' interdependence would involve open ended interviewing of local experts and stake holders and "*careful observation and listening*" of the issue in question (Hayhoe, 2004 , p. 77).

To the extent that some commonalities seem to emerge from the case studies, these may then be further explored by data collected by survey methods. However, these survey instruments may still only be useful intra-nationally, because their items would need to be couched in local terms that reflect the specific meanings of the constructs intra-nationally. Narrative descriptions are often the most effective way to report this second paradigm's findings.

The methodologies used in most of the TIMSS studies and in PISA locate them very definitely within the first perspective of comparative education. Besides the main Achievement Study, two supplementary studies in TIMSS did use data gathering methodologies other than questionnaires but, in each case, the shape of the data was predetermined by an externally designed format that was to have common applicability. The Performance Achievement Study in TIMSS in 1994/1995, involved students in a number of practical tasks; and its Curriculum Study collected and content analysed documentary curriculum materials, to complement the categories in a commonly administered survey questionnaire. TIMSS did conduct a Television Study of a representative sample of grade 8 mathematics classrooms in three countries, that got somewhat closer to the second paradigm's perspective. In this study there was direct recording (*observing*) of the classes in action by means of video-recording. Many of the observed differences needed local interpretations (*listening*) to enable sense to be made of the analyses of the data into findings.

To foreshadow the argument that will be central to my discussion of the projects, I shall associate **"Context"** with the first paradigm and **"Culture"** with the second one.

2. Context in TIMSS and PISA

In the case of the TIMSS, an explicit Conceptual Framework was developed (see Figure 1) that sets out the contextual constructs and their relations with science achievement that the project managers recognised as possibly influencing science achievement. This Framework, with some modifications, has been used in TIMMS Repeat in 1999 and in TIMSS as Trends in Mathematics and Science Study in 2003.

In a rather similar approach to its study design, the PISA project set out to find contextual factors from school, family background, and student characteristics (gender and educational), that could be associated with students' scientific literacy, within and across the participating countries.

2.1 Descriptions of Contexts in TIMSS

The contextual data collected by TIMSS for its initial testings in 1994/1995 provided a great deal of information about each country's educational system and the role of science within it. Indeed, Robitaille (1997) used these data to edit a volume, **National Contexts for Mathematics and Science Education: An encyclopaedia of the education systems participating in TIMSS.** 1997 Pacific Educational Press: Vancouver, Canada. Each country is described in 7–10 pages under the common headings: *Country Profile; The Education System; TIMSS Populations; Mathematics*

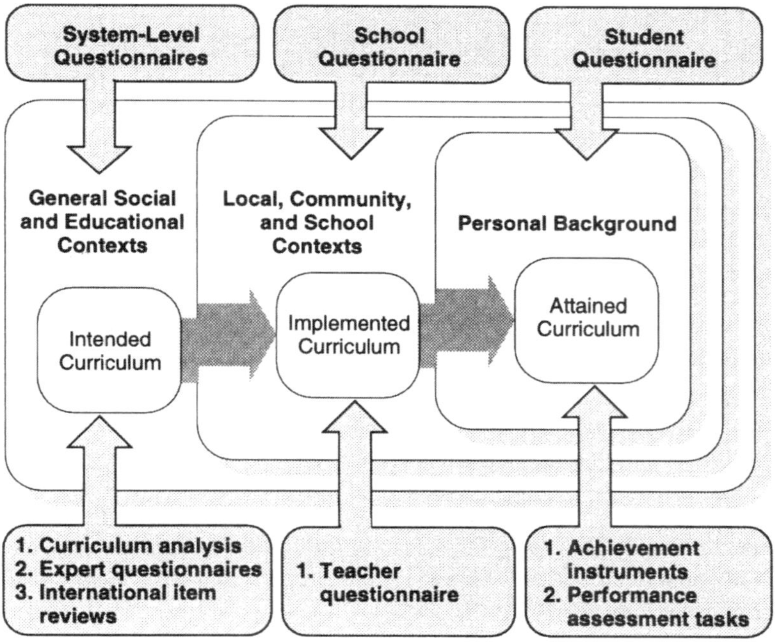

Figure 1. Conceptual framework for the contextual and achievement data collection in TIMSS.

Source: Robitaille, D. F., & Garden, R. A. (1996). Design of the study. In D. F. Robitaille & R. A. Garden (Eds.). TIMSS Monograph No.2: Research questions and study design. Vancouver: Pacific Educational Press.

Curriculum and Pedagogy; Science Curriculum and Pedagogy; Evaluation Policies and Practices; References and Sources for Further Reading (including references for quoted statistics).

This volume is a mine of detailed information, indeed an encyclopaedia. However, each country's account provides little or no insight into how these separate details operate together (and with others not listed that spring to mind) to form the distinctive ways science education occurs in its school classrooms. Yet it is to these distinctive ways of science education that teachers would identify, and to which the words, "culture of science education" would apply, for it is within these educational cultures that the science teachers in each country teach and their students endeavour to learn.

3. Context: Simplistic Responses

The initial responses to the national relativities among the students' mean science scores in TIMSS (and to the similar rankings in PISA) has, not surprisingly, focused on how they might be explained by factors that may be closely related to science teaching and learning.

After the international and national TIMSS reports in 1996, a number of US authors rushed into print. Some simply criticised the design of both the achievement test and of the measures of the contextual constructs in order to downplay the significance of findings that US students performed relatively poorly compared with a number of other countries (e.g. Bracey (1997a&b, 1998, 2000, 2002a&b); Gibbs & Fox (1999); Holliday & Holliday (2003); Wang (1998)). The most prolific of these critics was Bracey who, in a stream of short articles, questioned the sampling of students, the items in the tests, the comparative validity of some contextual variables, and even the significance of the middle school results as indicators of national economic well-being (2002a).

Other authors highlighted one or two contextual variables of concern to them to explain the relative differences in achievement. For McKnight and Schmidt (1998) and Callahan, Kaplan, Reis and Tomlinson (2000), it was lack of science curricular focus and low degree of rigour in US middle school science. Zach (1997) picked up the phrase "splintered vision" used by Schmidt McKnight and Raizen (1997), to refer to both the relatively diffuse nature of the US curriculum (a TIMSS measure) and to the complexities of American educo-political reform (not a TIMSS measure).

Another group of authors ignored the contextual data altogether, and sought to explain the relative differences between scores in USA and other countries in terms of their own sense of the key influences (e.g. Blank & Wilson, 2001 ; Jones, 1998; McCallister, 2002, chose *modes of instruction*; while Rakow, 2000; Riley, 1997, chose *teachers' expectations*).

Some voices called for very different forms of response. Baker (1997) expressly warned against the emerging temptation in the USA to interpret TIMSS with naïve explanations of performance in terms of isolated contextual variables. He urged for a much more measured approach, that would see TIMSS and its multitude of data, not as answers but as an opportunity to ask interesting questions, and to reflect on them in relation to one's own familiar practices.

Atkin and Black (1997) also raised doubts about simplistic analyses of the TIMSS findings. For example, they pointed out that it is naïve to argue that the higher *degree of focus* in the curriculum in Japanese schools compared with those in USA schools is a reason for Japan's higher performances, when the same relative difference in *curricular focus* occurs between the USA and other countries that have achieved lower scores than the USA. Comparisons, from a study involving forty plus countries, should not be selected on the basis of educo-political interest in just a pair of countries. Such piecemeal comparisons, and worse still, subsequent attempt to imitate in one country the measure of the chosen context variable in another country, will not be solutions to improved science performance. These cautionary comments were repeated by Keitel and Kilpatrick (1999) in relation to the mathematics component of TIMSS.

Atkin and Black had good grounds to counsel caution. Earlier in the 1990s for the OECD they had been involved in a study of curricular innovations in mathematics, science and technology (Black, Atkin and Pevsner, 1995; Atkin, 1998). Among its thirteen countries there was widespread dissatisfaction with the state of science education. There was, however, no observable link between the contextual goals and criteria these countries had set for improvement and their scores in TIMSS.

Simplistic approaches to context were not confined to US scholars. Puk (1999) compared teaching practices in Alberta and Ontario in Canada and in Singapore, and found an association between what he describes as an '*implicit formula for success*' and high scores among the middle school students. House (2000) found a relationship among students in Ireland between *student self-efficacy* and *achievement*, but it was not sustained across the other countries he investigated. In Germany, Möller and Köller (1998) discussed the relativities in terms of a construct, *social relativities*, for which they had data from other studies.

The publication of the PISA Science results in 2001 have been followed by another spate of simplistic comments, in a number of countries (e.g. Germany and Norway, and Japan after the publications at the end of 2004 of TIMSS 2002 and PISA 2003).

Fuchs (2003) in Germany, echoing Baker (1997) above, has tried to shift the educational and public focus in that country from the PISA rankings and suggesting short term remedies, to the aims of PISA Science and their underpinning educational concepts, as worthy of serious consideration for Germany's current emphases for schooling. Similarly, Messner (2003) tried to direct attention to the socially grounded demands in PISA's three literacies and linked them to a rethinking of basic education.

4. Contextual Influences: More Extended Responses

A special issue of **Educational Research and Evaluation** reported attempts by international authors to carry out analyses, using several of the TIMSS measures of context, to explore their influence on student performance. Kuiper and Plomp (1999), the editors, discussed how the TIMSS Framework (see Figure 1) ought to have allowed a scale or path analysis to be used to identify the contextual factors of influence in different countries. In practice, they concluded that "*the instruments used in the project do not contain sufficiently well-tested scales to operationalise all the important constructs*" (p. 177).

Keys (1999) discussed the unusual finding in England among the European countries that 9 and 13 year olds both performed better in science than in mathematics. She pointed out that the former group's performance could perhaps be associated with the increased attention on science in primary schooling in England since 1991, but that this would not explain the latter group's findings. She then tried to find a comparative explanation through four likely contextual constructs for which TIMSS did collect data – *intended curriculum, lesson time, homework, and practical activity in class*. So few consistent relationships emerged between England and other European countries, that she concluded that TIMSS could not provide any simple answer, and that many other factors must be involved, only some of which are measured in TIMSS. "*TIMSS provides a broad-brush picture of what happens in science classrooms. More in depth research, based on observation and interviews, is required to build up a detailed understanding*" (p. 199).

In the Netherlands the testing occurred during a period of change in the curriculum for science, and so Knuver (1999) pointed out that it was encouraging when the Dutch students performed quite highly, comparable with students in England at the lower age population. However, the extent of this changing curriculum was somewhat undermined, when Dutch teachers, who were closer to the "experienced science curriculum" of students inside and outside of schooling, estimated *the relevance of the science items in the TIMSS trials* much higher than did the curriculum experts.

Cheng and Cheung (1999) reviewed these European articles from Taiwan's position in the East Asian group of very high achieving countries in TIMSS. They gently point out that a number of variables, such as *the item relevance* from the Test Curriculum Matching Analysis (TCMA), that European authors used to explain their students' low performance are also applicable in East Asia. More fundamentally however, they agreed with Kuiper and Plomp's assessment that the quality and extent of the measures of contextual constructs in TIMSS are not good enough to make valid international comparisons. "*No clear theory or knowledge has advanced from TIMSS that can explain how and why students' achievements in mathematics or science can be enhanced through manipulation of the factors at individual, class, school or system level.*" (p. 233).

These authors blame the TIMSS project for the quality of its measures of the contextual constructs in which they were interested. Given the range of constructs TIMSS (and PISA) set out to measure in the short time available for collecting data from students, it was inevitable that a number of their constructs would be dependent on student responses to just one or two items, that may also be open to different understanding within and across countries. *Organisation of schools, instructional approaches, hours of television watching, hours of homework in grade 4,morale of teachers*, etc. are examples of constructs with high inference measures.

Nevertheless, there are constructs in both TIMSS and PISA that do meet the criteria of best practice in educational studies of social contexts and students' cognitive learning. It may be that Young, Webster and Fisher (1999), psychometric researchers with a keen interest in science education, had such higher quality measures in mind, when they claimed:

> *the usefulness of this research for enhancing the scientific and technological skills of a country is established, both in terms of the quality and uniqueness of the data, the untapped potential of the data bases, advanced statistical techniques, previous research experience, availability of expert advice and resources, the identification of gender and socioeconomic issues and the problem of lack of equity in mathematics and science achievement.*

However, from the reports discussed so far in this chapter, it would seem that this statement may result from confusing the quantity and statistical reliability of the project's data, with their capacity to deliver strong comparative relationships for improving science teaching and learning. Nevertheless, the analyses of the findings involving these higher quality variables needs now to be considered.

4.1 Gender Analyses: A Quality Measure

Perhaps the construct with the lowest inference (highest quality of measure) is the gender of the students, and both projects (with science learning differently defined) provide confident relationships of gender with science achievement across their participating countries. TIMSS, as a study of science curriculum learning that has now been repeated with three cohorts of students across almost a decade, shows some countries have gender equality, while a number have a gender bias in favour of boys at both the primary and middle school levels. Countries, that have taken deliberate steps to encourage the girls in relation to school science (since recognition of this issue in the early 1980s), can now assess the effectiveness and sustainability of their efforts, while others can assess and re-assess the persistence of gender as a bias in schooling.

The PISA Science test in 2000 was based on the scientific literacies involved in critically appraising media reports involving science – an unfamiliar form of test for 15 year olds in most, if not all of the 32 countries. Given that the same students showed very strong bias to girls in the main Reading test, the absence of significant gender differences in the Science testing in 26 countries is all the more remarkable, since the format of the science test required an unusual amount of reading. It can only be concluded that both boys and girls found the journalistic stories introducing each block of items to be engaging – a strong indication for a curriculum direction worth exploring in the face of the current lack of interest in sciences among most of the countries in PISA.

4.2 Other Quality Measures

TIMSS 2002 gave considerable attention to constructs associated with the science curriculum, the school and its teachers. Separate questionnaires ask about *the structure and content of the intended curriculum for science, the preparation, experience and attitudes of teachers, the actual taught content, the instructional approaches used*, and *the organization and resources of schools*. For the students, a questionnaire asked about aspects of students' home and school lives, including *classroom experiences, self-perception* and *attitudes about science, homework* and *out-of-school supports*.

The PISA project chose to give emphasis to the students' family background and the students' awareness of themselves as learners. For the constructs related to these aspects of context, sophisticated measures were developed. An index of *economic, social and cultural status (esc index)* was created to capture more aspects of a student's family background than those commonly used in research studies of family influence on educational achievement, namely, parents' occupation in conjunction with parents' education and income (*the socioeconomic status index).*

The more elaborated *esc* index was derived from the *highest international socio-economic of occupational status* for father or mother, *the highest level of education of father or mother converted into years of schooling, the number of books in the home,* as well as *the access to home educational and cultural resources*. The student's *access to home educational and cultural resources* was compounded from responses to a wide list of possible resources in the questionnaire. These *access* variables are an alternative (and a more salient) indicator to include than the simple *parental income*.

Other items on the student questionnaire relating to the family enabled indices for *immigration background, language used at home, home educational resources,* and *possessions relating to classical culture in the family home* to be calculated.

In response to emphases in the recent research literature on the importance of *self-regulated cognition* in students, PISA included a large number of items in the student questionnaire in 2000 for Reading, and in 2003 for Mathematics, (and in 2006 for Science) that lead to scaled measures of constructs relating to the students' cognition and metacognition in the form of indices for *learning strategies of elaboration, memorisation/rehearsal, control strategies, competitive learning,* and *co-operative learning* (OECD, 2001).

With such elaborated and sophisticated measures of a student's family and personal educational backgrounds, there was great scope in PISA for statistical exploration of the relationships between them and the student's performance on the corresponding achievement tests.

Some examples of their use in analyzing contextual influence are now considered.

5. Findings of Influence

The TIMSS testing in 2000 and the PISA testings in 2000 and 2003 have led to international and national reports that include a large number of Figures and Tables the analyses of their data (Martin et al., 2004; OECD, 2001, 2004). The findings that emerge from single variate and multivariate analyses involving the high quality measures are exemplified by those for the socio-economic and cultural indices in PISA.

For each country and indeed for the total international data set, the very considerable scatter of student achievement score against student *economic, social and cultural status* was reduced to provide a line of best fit. The slope of this line is a country's *socio-economic gradient*. These gradients differed quite considerably with a few countries showing quite shallow gradients (less than seven percent of variance in student performance) through increasing slopes to some with very marked gradients (more than 20% of variance). These gradients are gross indicators of how well socio-economic equity is achieved in the school system nationally. Some of the shallower gradient countries (e.g. Finland, Korea, Japan, Hong Kong) had very high mean performance scores, but others like Sweden and Norway did not score so highly.

Some countries with significantly steeper gradients also scored highly (e.g. Australia), indicative of highly differentiated (on socio-economic terms) school systems, in which students from families with a higher *esc* index are performing very highly indeed, compensating for the lower performances among lower *esc* families.

The OECD concluded that ***quality and equity need not be seen as competing policy objectives.*** The same relations are, however, found for the Reading and Mathematics scores, so there is no specific message for improving Science performance, and optimizing both objectives is unlikely to be a quick and easy fix for many countries.

When the analysis is repeated using schools as the socio-economic unit rather than student, the gradients are generally higher, reflecting deliberate school recruitment policies or a marked geography of socio-economic and residential differentiation.

The OECD concluded ***there is advantage (and in a number of countries, substantial advantage) in attending higher esc index schools***, and ***provided this involves only small numbers from lower index families, these students are likely to benefit in performance terms from being part of a high index school community.*** Countries that extended such a placement policy would, however, disadvantage further their lower *esc* schools.

The socio-economic and cultural contributions to the variance in student performance are, in almost all countries, large for the between-schools variance, and smaller, but still a major contribution, for the within-schools variance. That there are strong socio-economic and cultural influences on student learning is hardly new information to the participating countries, although their strength for students' literacies in reading, mathematics and science may not have been so well quantified before PISA reported them. The international PISA Report speculates about the educational differences that lie behind these strong socio-economic and cultural influences, acknowledging that ***they are likely to be a combination of in-school aspects and aspects of parental support and motivation that, together, create positive learning environments at school and at home.*** But these combinations are beyond the capacity of the projects' data to reveal. Nor were any data collected by the project about the participating countries' recent policies that may bear on their socio-economic gradients, and would certainly add some reality to the two conclusions reported above. Australia, for example, has for a decade now been widening these differences across its school system through government funding policies that promote the movement of students from the public school system to the many types of private schools that now enjoy substantial funding support from government.

Almost every other contextual variable in each of the projects shows a similarly wide scatter in relation to student achievement internationally that reduces somewhat at a national level. Many do have a statistically significant relation with student achievement, but the correlations are too low for them to be educationally significant and the basis for targeted innovation. The PISA project constructed indications of the magnitude of the influence from the distribution of each of these variables in schools across the OECD countries. For example, a plus standard deviation of difference in school contextual features, like *students' use of school resources, level of school autonomy, teacher-student relations,* and *disciplinary climate* was associated with between 5 and 18 points of the higher achievement score that has 500 for its international mean. The insignificance of these constructs educationally is, however, underscored when the same difference in the average *economic, social and cultural index* is associated with 67 points.

5.1 Multiple Influences

When multivariate analysis, rather than single variate, is applied to the measures it is possible at a national level to decompose the variance in the students scores in terms of the contributions made by the constructs as factor variables. These analyses are now illustrated for Australia in the PISA 2000 testing in Table 1 where the contributions of significant contextual constructs to the between-schools and within-schools variances for the science scores are given (ACER/OECD, 2001). Overall, the Difference in the Students' Scores in Scientific Literacy is made up of ~ 20% between-schools and ~80% within schools. This suggests that schools and science teachers have considerable scope to change what happens in their science classsrooms, whether or not the education system as a whole tackles the issues of demographic differentiation, in which family variables (*student esc, family wealth, cultural possessions*, etc.) contributing almost two thirds of the smaller between-schools variance.

The large set of contextual constructs that were chosen as likely influences account for only 17.8% of the much larger within-schools variance. PISA has provided a

Table 1. Variances decomposition for scientific literacy in Australia (Table 8.4 in ACER/OECD, 2001)

Factor	Percentage of between school variance accounted for	Percentage of within school variance accounted for
Socioeconomic status	25.0	6.9
Family wealth	9.8	0.7
Home educational resources	0.8	2.1
Cultural possessions	6.2	0.3
Parents' education	3.0	0.2
Living with a guardian	0.0	0.2
Immigrant status	0.4	0.7
Number of siblings	0.3	0.2
Time spent on homework	7.1	2.1
Comfort with computers	0.0	1.0
Attitude to computers	1.1	0.3
Control strategies in learning	0.0	0.9
Social communication	0.6	0.4
Student determination to do well	0.0	0.6
Elaboration strategies in learning	0.0	0.7
Mean school SES*	6.1	0.3
Disciplinary climate*	4.5	0.0
Teacher support*	4.9	0.0
Achievement press*	1.7	0.0
Instruction time*	4.3	0.1
Total Variance accounted for	75.9	17.8

*School level variables

disappointingly small amount of insight into the more directly influenced part of the total variation in student scores. Furthermore, within this small identified amount of the within-schools variance, the family *sec* measures are again the major contributors.

Accordingly, PISA's advice is rather lame, limited to *instructional time,*and *time on homework*, as conditions schools might attend to in order to improve performance. The high quality constructs, *student strategies in learning* and *social communication*, that schools do have capacity to vary, turn out to play quite minor roles, as do the school's *disciplinary climate* and *teacher support.*

The unidentified 82.2% of the within-schools variance is frustrating after so much data has been collected. It suggests that important features of a school's science classes and their interactions with students have been missed by this intensive and extensive study, and that these interact very differentially with the students within the same classes. More science-specific constructs about the science classroom have been added for PISA 2006.

In the Australian report of TIMSS 2002, Table 2 indicates more of the Australian variance has been exposed – 84% and 53% respectively, of the smaller between-schools and the larger within-schools (ACER/IEA, 2004). The *ethnic and language backgrounds* of students at Year 8 are prominent contributors, as well as *gender, educational aspiration* and *self-confidence in science.* Although data for 14

Table 2. Estimates of influences on science achievements in Australian schools, year 8 (Table 5.18 in ACER/IEA, 2004)

	Coefficient (standard error)
Intercept	557.2 (6.4)
Student Level Variables	
Indigenous	–33.9 (5.7)
Language background	18.5 (4.7)
Aspirations	14.1 (1.4)
Books	13.2 (1.3)
Gender	–13.0 (2.3)
Self-confidence in science	11.8 (1.5)
Computer usage	6.6 (1.7)
Parents' education	5.4 (1.4)
Valuing science	4.3 (1.5)
Age	–3.1 (1.4)
Classroom level variables	
Disadvantage	–12.2 (3.0)
Variance	
Explained by the model	37%
Unexplained school level (between-schools)	16%
Unexplained student level (within-schools)	47%

classroom variables were tried in the TIMSS interaction model, only two,*principal's perception of a good school and class attendance,* and *proportion of students from disadvantaged (family) backgrounds,* were significant contributors. None of the more direct science teaching variables were significant.

TIMSS also collected data on *time spent on homework* and found it varied considerably across the countries, with little association with average achievements. The report concludes rather weakly that ***a certain level of homework may be needed in a system, before positive effects become visible*** What constitutes science homework in different countries, and its relation to other out-of-school education are obvious comparative questions to ask, but they lie beyond the project's capacity to collect data.

6. Towards Culture as an Explanation

Ramesier (2001) delved more deeply into the TIMSS achievement scores to look for patterns that are common across groups of countries. By analyzing these data in terms of different cognitive demands among the test items, he sought to relate them to emphases in the science curriculum in Switzerland. He compared performance on items with high demand and low terminological difficulty, with items with the reverse features. This rather unusual view of successful science learning led him to patterns of performance in eight other European countries that were similar to Switzerland, and to a reverse pattern in ten countries, largely from northern and western Europe. Rather than seeking parallel patterns in the contextual measures from TIMSS, he suggested that these patterns may result from commonalities within each group's *educational history and culture.* The presence in each of his achievement groupings of countries with clear historical differences was, he acknowledged, a difficulty for his proposal.

Kelly (2002), in a similar manner, looked for consistencies among the science items that distinguished the higher performing American students in both the fourth and eighth grades' tests. She used these more difficult items to establish some "benchmarks of performance", namely, *students' use of science terminology and content, their use of principles compared with facts*, and *their ability to communicate.* Expert panels were asked to assess the weight of these benchmarks in the science curricula of 13 countries as a means of relating the achievements of the fourth and eighth grade students.

Turmo (2004) took advantage of both the sophisticated measure of family backgrounds and the clear difference that existed in the PISA Science test between items for conceptual understanding and for scientific processes. He grouped the ten sub-constructs of family background to create measures of three forms of *family capital – cultural, social* and *economic. Cultural capital in* Norway and Denmark are close to the OECD mean value, while Sweden, Iceland and Finland are below this mean. Within the relatively weak relationship between *esc* and students' scientific literacy in these Nordic countries, he found *cultural capital* made a surprisingly strong contribution to the variance, with *social* and *economic capital* contributing much less. The correlations of the student responses to the seven items making up the measure of *cultural capital* with the students' sub-scores on the test items for *conceptual understanding* and for *scientific processes* in the PISA Science units were only slightly different, but were mostly higher for the process ones. From this he concluded that a cultural approach to science education (e.g. Aikenhead, 1996) may be helpful to children in Scandinavian families with lower *cultural capital.*

6.1 Context Gives Way to Culture

Neither TIMSS nor PISA refer to their contextual data as "cultural data", and the only mention of the term, *culture*, in TIMSS was by the AERA Think Tank, an outsider group, who referred to the *role of culture* as being significant in its brief commentary on the project's analysis of the curriculum study and on the development of the achievement study (AERA, 1994, 1995).

It is, thus, of interest to follow an evolution of thinking by Schmidt, the leader of the curriculum project for science in TIMSS, reported in **Many Visions, Many Aims Volume 2**, (Schmidt, Raizen, Britten, Bianchi & Wolfe, 1997). In a number of subsequent published journal articles Schmidt has further analysed the extensive data that were collected.

Valverde and Schmidt (1998) combined data from the curriculum analysis and the achievement study to claim a relationship between achievement and the level of expected science learning in the intended curriculum across the 22 higher achieving countries. This finding across a sub-set of countries was similar to the curricular – oriented ones mentioned above by Puk (1999) and Kelly (2002).

In a second paper Valverde and Schmidt (2000) related the decline in relative achievement by students in USA from grade 4 to grade 8 to contextual variables such as *the degree of focus in the curriculum, the dispersion of educational control, a fixed conception of basic learning* and the *absence of an integrated approach to reform.*

In these and in other papers published with McKnight in 1995 and 1998, Schmidt reported on the pervasive variations that existed at almost every level of their cross-national analyses and spoke of a *climate* or *context of variation* in which differences among countries in some variables are similarities for others. Nevertheless, Schmidt continued to maintain that these variables, albeit confusing in their variation, were distinct enough to be discussed as possible determinants of the learning outcomes of the students.

Finally, in yet another account of the curriculum study, Cogan, Wang and Schmidt (2001) describe the variations in content standards, textbooks and teachers' instruction, as indications of distinctly different *cultural approaches*. They relate the content of what is taught in school science to *cultural contexts of curriculum policy* and acknowledge *'the folly of adopting in a wholesale fashion the curricular patterns observed in an alien country'* (p. 130). These authors go on to speak of the need "*to learn from other cultures, but these lessons must be thoughtfully analysed and creatively translated into our own unique cultural context for education*" (p. 130) – an echo of the call by Baker (1997) four years earlier. It would seem that continued reflection on the bewildering, but fascinating variations in the curriculum data of TIMSS has moved Schmidt from a contextual perspective to a cultural perspective.

7. Culture as an Alternative Paradigm

The comparable performances of Japanese and American students in fourth grade gave way in the grade 8 study to the Japanese students performing at almost a grade level higher than the Americans. This decline led Linn, Lewis, Tsuchida & Songer (2000), an American team, to observe, record and analyse a number of science lessons in Japan itself in order to gain insight into contextual differences in science education and in Japanese education more generally. These authors described teacher and learner activities they regularly observed in the Japanese classes in terms of what in USA would be a model or ideal science program. They also found synergies between these regular classroom activities and features of the wider educational system that were not generally present in the US educational scene. For example, the exchanges of information between students they saw so frequently in the science lessons were supported by long term social and ethical emphases in Japanese education that nurture respectful discussion in small groups from preschool onwards.

The direct observational and interrogation methods used by Linn et al. to obtain their rich descriptions of these science classes and to make these comparisons about contextual connections were very different from those used by TIMSS to gather contextual data. Indeed they belong to the *Culture paradigm* for comparative education. It is thus strange that the team referred in their conclusion to that project's contextual data as "an unprecedented resource" for identifying significant features of other countries' systems. They seem to have made almost no use of this resource themselves!

7.1 The TIMSS Television Study

The complementary study in TIMSS of actual classroom teaching and learning involved the video recording of a large random sample of Grade 8 mathematics classes in USA, Germany and Japan. This large study of classrooms in action threw open a further window on the connections referred to by Linn et al. between classroom activities and other features of Japanese education.

The video-ed classes were direct observations of classroom practices. They were then externally analysed in many different ways, but one outstanding finding was the surprisingly distinctive pattern of classroom activity that occurred in each of these three countries. In Japan, this pattern extended to how lessons involving particular mathematical content were prepared and presented by the teachers (*the lesson study*). Because the grade 8 Japanese students scored higher than those in most other countries, much attention has been given subsequently to the role of this *lesson study* in Japanese schools.

Stigler and Hiebert (1999), the directors of study in TIMSS, were clearly impressed by the practice of the lesson study which results in teachers in a school, or across schools, working together to plan and improve the teaching of a particular topic. Their direct discussions about it with Japanese colleagues convinced them that teaching itself is a cultural activity. By this they meant that it is an activity that evolves in a manner that is consistent with a web of beliefs and other behaviours that are part of a much wider culture. Nevertheless, they still believe that something akin to *lesson study* would be advantageous to mathematics teaching in USA, and outline some principles and initiatives that would be needed if such a cultural change was ever to be achieved among American teachers.

During an extended visit to Japan in 2003 I made enquiries about how the lesson study was seen by the Japanese themselves. The web of beliefs and behaviours in Japan that support teaching as a shared wisdom was quite astonishing. Each was alien to the very individualistic sense of teaching with which I am familiar in Australia and a number of other western countries in which I have worked.

My informants affirmed that there was, indeed, a great deal of sharing among teachers about the teaching of topics within subjects, although it was recognised that individual teachers, as they become more experienced, would initiate their own variations in the pedagogy. Ogawa (2004) has reported on the "cultural climate" among Japanese teachers that allows easy communication between peers to happen in a single school and across schools.

Schools have "Open Days", not for parents and outsiders as in Australia, but for teachers from other schools to attend to observe and discuss how their teachers are teaching the various topics and subjects (Bishop, 2005).

Textooks in Japan are not encyclopaedic volumes as they are in western countries. They are more like what teachers in western countries might call a set of bound worksheets. That is, they are learning materials, rather than repositories of what is to be learned (and so often of what is transmissively taught). The "insignificance" of the Japanese student texts leads to a much more significant role for the teaching suggestions in the Japanese national teachers' guides, than such a volume has for teachers in western countries.

Student teachers in Japan (at the university of my visit) spent only a very short time (about 4–5 weeks) sharing a teaching practicum at a school with a large number (10–15) of other students. Each student may have been personally responsible for just a couple of lessons, but these were attended by most of the peer group of student teachers. After each such lesson, the group with their mentors discussed that lesson as an example of teaching its particular topic – *how it might be changed, and improved.* In their first year of appointment, a new teacher in Japan has 60 hours of mentored sharing in their schools and 30 hours (with other new teachers) in a regional education centre. The focus of this sharing is once again on how particular lessons can be improved.

These experiences of Japanese teachers, their pre-service teacher education, their first year of teaching, the importance of teachers' guides, and the Open Days, converge to form an homogeneous culture that makes the practice of the *lesson study* a natural and an obvious way to tap the collective wisdom of one's colleagues in the school and beyond. These features of the Japanese teaching culture do not resonate with those in western countries with which I am familiar. Hence to imagine that one piece of this Japanese classroom culture, namely, the *lesson study*, could be extracted, as if it is an independent variable, and imported to a culturally different country like USA or Australia, is to fail to understand the complexity of the processes of teaching/learning, and how deeply embedded they are in cultures of education, defined in quite unique ways by national histories.

8. Conclusion

Both TIMSS and PISA, as large scale multi-country projects, have been extensive and very expensive studies of contextual influences on student science achievement. It seems that the Context approach they adopted has achieved very little in providing insights to educational authorities, schools, and teachers about the factors and conditions that foster better quality science learning (TIMSS) or scientific literacy (PISA). The reports of these two projects give very little sense of what the students are experiencing day by day with their teachers in the science classrooms, and how this can be improved.

From large scale studies of primary schooling in England and four other countries, Alexander (2001) argued that "the central educational activity, pedagogy,… is a window on the culture of which it is a part, and on that culture's underlying tensions and contradictions as well as its publicly declared educational policies and purposes." (p. 4). He goes on to emphasise the importance of building up rich data sets from classrooms in action, using methodologies that seek to balance atomistic and holistic data.

His, and other evidence from the Culture approach, suggests a richer landscape picture of comparative science education might be achieved, if it could be pursued comprehensively. The set of pictures that would emerge may well encourage the questioning and comparative reflection that Baker (1997) suggested was the beginning of improvement.

Eckstein and Noah's (1991) comparative study of senior secondary examinations is an example of the Culture comparative paradigm. They used national documents to see how each country framed its discourse about these examinations, visited each country, and worked with local scholars and the range of participants in the examining procedures. On site, they could observe and conduct openly-structured interviews to elucidate what they saw and heard. In these ways they gathered the web of data in each country that led to their strikingly vivid set of national narratives, that describe the examinations and their systemic settings through two students undertaking these examinations from very different family backgrounds.

The narratives set the scene for the cross-country comparisons of features of these examinations that follow in the later chapters.

It is challenging to consider what the cost, in personnel and money, would be, if an approach in the Culture paradigm approach were to be employed in the continuing comparative studies of science achievement. It would involve (as Eckstein and Noah did) rotating teams of expert comparativists (as in OECD national audits) who would spend, say one month or so in each country, working with local consultants at all levels and their interpreters, to construct a coherent picture of ***how science education occurs*** and ***why it is like it is***.

If each expert covered five countries and if a team consisted of three members, then 24 experts would (with support staff) be able, over a 6–12 month period to cover forty countries. The quantitative data that would emerge may be generally less precise than in TIMSS or PISA, but it would be complemented by qualitative accounts (quite absent in the Context approach) that should provide much richer insights into the strengths and weaknesses of each country's culturally embedded science education. The cost/benefit equations of such a paradigm shift, if the benefit is *how science education might be improved*, would be most interesting to compare.

References

ACER/IEA. (2004). *Examining the evidence: Science achievement in Australia's schools in TIMSS 2002.* Camberwell Victoria: ACER.

ACER/OECD. (2001). *How literate are Australia's students?* Victoria: Camberwell.

AERA Think Tank. (1994). *Report on TIMSS achievement project.* Washington DC: AERA Grants Program Advisory Board.

AERA Think Tank. (1995). *Report on TIMSS curriculum analysis project.* Washington DC: AERA Grants Program Advisory Board.

Aikenhead, G. (1996). Science education: Border crossing into the sub-culture of science. *Studies in Science Education, 27*, 1–52.

Alexander, R. (2001). Pedagogy and culture: A perspective in search of a method. In J. Soler, A. Craft, & H. Burgess (Eds.), *Teacher development: Exploring our own practice* (pp. 4–25). London: Paul Chapman Open University.

Apple, M. W. (1979). *Ideology and curriculum.* London: Routledge and Kegan Paul.

Atkin, J. M. (1998). The OECD study of innovations in science, mathematics and technology education. *Journal of Curriculum Studies, 30*(6), 647–660.

Atkin, J. M., & Black, P. (1997). Policy perils of international comparisons: The TIMSS case. *Phi Delta Kappan, 79*(1), 22–28.

Baker, D. P. (1997). Surviving TIMSS, or everything you have forgotten about international comparisons. *Phi Delta Kappan, 79*(4), 295–300.

Bereday, G. Z. F. (1964). *Comparative method in education.* New York: Holt, Rinehart and Winston

Bishop, A. (2005). Private communication from Japan, March.

Black, P., Atkin, M., & Pevsner, D. (1995). *Changing the subject: Innovation and change in science mathematics and technology education.* New York: Routledge.

Blank, R. K., & Wilson, L. D. (2001). Understanding NAEP and TIMSS results. *ERS Spectrum, 30*(1), 23–33.

Bracey, G. W. (1997a). Accuracy as a frill. *Phi Delta Kappan, 78*(10), 801–802.

Bracey, G. W. (1997b). More on TIMSS. *Phi Delta Kappan*, 78(8), 656–657.

Bracey, G. W. (1998). Rhymes with dims: As in "Witted". *Phi Delta Kappan, 79*(9), 686–687.

Bracey, G. W. (2000). "Diverging" American and Japanese science scores. *Phi Delta Kappan, 81*(10), 791–792.

Bracey, G. W. (2002). Facing the consequences. Using TIMSS for a closer look at US mathematics and science education. *Science Education, 86*(5), 730–733.

Bracey, G. W. (2002). Test scores, creativity and global competitiveness. *Phi Delta Kappan, 83*(10), 738–739.

Callahan, C. M., Kaplan, S. N., Reis, S. N., & Tomlinson, C. A. (2000). TIMSS and high ability students: Measures of doom or opportunities for reflection, *Phi Delta Kappan, 81*(10), 787–790.

Cheng, Y. C., & Cheung, W. M. (1999). Lessons from TIMSS in Europe. *Studies in Educational Evaluation, 5*(2), 227–236.

Cogan, L. S., Wang, H., & Schmidt, W. H. (2001). Culturally specific patterns in the conceptualisation of the school science curriculum: Insights from TIMSS. *Studies in Science Education, 36*, 105–133.

Eckstein, M. A., & Noah, H. J. (1991). *Secondary school examinations: International perspectives on policies and practice*. New Haven and London: Yale University Press.

Fuchs, H. -W. (2003). Towards a world curriculum: The concept of basic literacy underlying PISA and the tasks allocated to schooling. *Zeitschrift für Pädagogik, 49*(2), 161–179.

Gibbs, W. W., & Fox, D. (1999). The false crisis in science education. *Scientific American, 281*(4), 86–93.

Gintis, H. (1972). 'Towards a political economy of education'. *Harvard Educational Review, 42*(2), 70–96.

Hayhoe, R. (2004). *Full circle: A life with Hong Kong and China* (p. 77). Hong Kong: Comparative Education Research Centre, University of Hong Kong.

Holliday, W. G., & Holliday, B. W. (2003). Why using international comparative mathematics and science achievement data is not helpful. *Educational Forum, 67*(3), 250–257.

House, J. D. (2000a). Students' self-belief and science achievement. *International Journal of Instructional Media, 27*(1), 107–115.

Jenkins, E. W. (2004) quotation in P. J. Fensham (2004). *Defining an identity: The evolution of science education as a field of research* (p. 117). Dordrecht, The Netherlands: Kluwer Academic Publishers.

Jones, R. (1998). Solving problems in mathematics and science education, *American School Board Journal, 185*(7), 16–21.

Keitel, C., & Kilpatrick, J. (1999). The rationality and irrationality of international comparative studies. In G. Keiser, L. Eduardo, & I. Huntley (Eds.), *International comparisons in mathematics education* (pp. 241–256). London: Falmer.

Kelly, D. L. (2002). The TIMSS 1995 international benchmarks of mathematics and science achievement: Profiles of world class performance at fourth and eighth grades. *Educational Research and Evaluation, 8*(10), 41–54.

Keys, W. (1999). What can mathematics educators in England learn from TIMSS? *Studies in Educational Evaluation, 5*(2), 195–213.

Knuver, A. (1999). National and cross-national perspectives on the Population 2 and 3 findings. *Studies in Educational Evaluation, 5*(2), 214–226.

Kuiper, W., & Plomp, T. (1999). Modelling TIMSS data in a European comparative perspective: Explaining influencing factors on achievement in mathematics in grade 8. *Educational Research and Evaluation, 5*(2) 157–179.

Layton, D. (1973). *Science for the people*. London: Allen and Unwin.

Le Métais, J. (2001). Approaches to comparing educational systems. In K. Watson (Ed.), *Doing comparative education research: Issues and problems* (pp. 197–209). Oxford: Symposium Books.

Linn, M., Lewis, C., Tsuchida, I., & Songer, N. B. (2000). Beyond fourth-grade science: Why do U.S. students and Japanese students diverge? *Educational Researcher, 29*(3), 4–14.

Martin, M. O., Mullis, I. U. S., Gonzalez, E. J & Chrostawski, S. J. (2004). *TIMSS 2003: International Report*. Chestnut Hill MA: Boston College.

McCallister, G. (2002). A proposal to improve science education in the public schools, *American Biology Teacher, 64*(4), 247–249.

McKnight, C. C., & Schmidt, W. H. (1998). Facing facts in US science and mathematics education: Where we stand, where we want to go. *Journal of Science Education and Technology, 7*(1), 57–76.
Messner, R. (2003). PISA and general education. *Zeitschrift für Pädagogik, 48*(3), 400–412.
Möller, J., & Köller, O. (1998). Dimensional and social comparisons regarding school results. *Zeitschrift für Entwicklungspsychologie und Pädagogische Psychologie, 30*, 118–127.
Noah, H. J., & Eckstein, M. A. (1969). *Towards a science of comparative education*. New York: Macmillan.
OECD. (2001). *Knowledge and skills for life: First results from PISA 2000.* Paris: OECD
OECD. (2004). *Learning for tomorrow's world: First results from PISA 2003*. Paris: OECD.
Ogawa, M. (2001). Reform Japanese style: Voyage into an unknown and Chaotic future. *Science Education, 85*(5), 586–606.
Ogawa, M. (2004). How is the novice getting to be the expert?: A preliminary case study of Japanese science teachers. *Journal of Korean Association for Research in Science Education. 22*(5), 1082–1102.
Ogawa, M. (2005). *Recent Affairs in Japanese Science Education*. Keynote Lecture at Korean Association for Research in Science education Annual Conference, Seoul, February.
Puk, T. (1999). Formula for success according to TIMSS or the subliminal decay of jurisdictional educultural integrity? Canada's participation in TIMSS. *Alberta Journal of Education Research, 45*(3), 225–238.
Rakow, S. J. (2000). NSTA's response to TIMSS, *AWIS Magazine*, *67*(1), 61.
Ramseier, S. J. (2001). Scientific literacy of upper secondary students: A Swiss perspective. *Studies in Educational Evaluation, 27*(1), 47–64.
Riley, R. W. (1997). From the desk of the secretary of education: TIMSS Benchmarks. *Teaching Pre-K.8, 27*(4), 6.
Robitaille, D. F. (Ed.). (1997). *National contexts for mathematics and science education: An encyclopedia of education systems participating in TIMSS.* Vancouver, BC: Pacific Educational Press.
Robitaille, D. F., & Garden, R. A. (Eds.). (1996). *Reasearch questions & study design. TIMSS MONOGRAPH NO. 2*. (p. 50). Vancouver, BC: Pacific Educational Press.
Schmidt, W. H., & McKnight, C. C. (1995). Surveying educational opportunity in mathematics and science: An international perspective. *Educational Analysis and Policy Evaluation, 17*(3), 337–353.
Schmidt, W. H., & McKnight, C. C. (1998). What can we really learn from TIMSS. *Science, 282*(5395), 1830–1831.
Schmidt, W. H., McKnight, C. C., & Raizen, S. (1997). *A splintered vision: An investigation of US science and mathematics education.* Dordrecht, The Netherlands: Kluwer Academic Publishers.
Schmidt, W. H., Raizen, S. A., Britten, E. D., Bianchi, L. J., & Wolfe, R. G. (1997). *Many visions, many aims volume 2: A cross national investigation of curricular intentions in school science*. Dordrecht, The Netherlands: Kluwer Academic Publishers.
Sjøberg, S. (2004, July 25–30). Science and technology in the new millennium – Friend or foe? In *Proceedings of the 11th IOSTE Symposium*, 1–2, Lublin, Poland.
Stake, R., & Easley, J. (1978). *Case studies in science education.* Urbana-Champaign: CIRCE and CCC.
Stigler, J. W. and Hiebert, J. (1999). *The teaching gap: Best ideas from the world's teachers for improving education in the classroom.* Los Angeles CA: The Free Press.
Turmo, A. (2004). Scientific literacy and socio-economic background among 15 year olds: A Nordic perspective. *Scandinavian Journal of Educational Research, 48*(3), 287–306.
Valverde, G. A., & Schmidt, W. H. (1998). Refocusing US mathematics and science education. *Issues in Science and Technology, 14*(2), 60–66.
Valverde, G. A., & Schmidt, W. H. (2000). Greater expectations: Learning from other nations in the quest for "World Class Standards" in US school mathematics and science. *Journal of Curriculum Studies, 32*(5), 651–687.
Wang, J. (1998). International achievement comparison, *School Science and Mathematics, 98*(7), 376–382.
Waring, M. (1979). *Social pressures on curriculum innovation: A study of the Nuffield Foundation science teaching project.* London: Methuen.

Young, M. F. D. (1971). An approach to the study of curricula as socially organised knowledge. In M. F. D. Young (Ed.), *Knowledge and control*. London: Collier McMillan.

Young, D., Webster, B., & Fisher, D. (1999). Gender and socioeconomic equity in mathematics and science education: A comparative study. Paper presented at AREA Conference, Montreal, Canada, April.

Zach, K. (1997). US mathematics and science education in an international context, *AWIS Magazine, 26*(3), 21–33.

10

QUIXOTE'S SCIENCE: PUBLIC HERESY/PRIVATE APOSTASY

Paul Dowling
Institute of Education, University of London

Abstract: This chapter is concerned with modes of authority and interaction in educational discourses and technologies. In particular, it explores, through an illustrative analysis of some of the assessment items of the Trends in International Mathematics and Science Studies, the construction of what may be referred to as *mathematicoscience*, a technology that may be associated with what may be publicly recognised as legitimate forms of relation to the empirical and legitimate forms of argument; it regulates, in other words, public forms of rationality. The globalising of this legitimating discourse through such mechanisms as international comparative studies of schooling performances, effectively privatises real concerns and seduces social criticism with its offer of an appearance on the global stage. The chapter also introduces two analytic frames (from Dowling's broader organisational language) that enable the organisation and constructive description of educational technology and discourse

Keywords: Authority, interaction, technology, mathematicoscience, private discourse, public discourse, social activity theory

> At this point, they come in sight of thirty forty windmills that there are on plain, and as soon as Don Quixote saw them he said to his squire, "Fortune is arranging matters for us better than we could have shaped our desires ourselves, for look there, friend Sancho Panza, where thirty or more monstrous giants present themselves, all of whom I mean to engage in battle and slay, and with whose spoils we shall begin to make our fortunes; for this is righteous warfare, and it is God's good service to sweep so evil a breed from off the face of the earth"
>
> (Miguel de Cervantes)

El Don Quixote was right, of course; windmills in Cervantes' Europe were monstrous giants, though wrong (as he eventually discovered) in his chivalrous crusade. If the enhanced performance of this new technology over hand milling

B. Atweh et al. (eds.), Internationalisation and Globalisation in Mathematics and Science Education, 173–198.

didn't persuade the locals to pay the miller's fee, then the destruction of their querns by or on behalf of the wealthy mill owners – local lords or the church – would chivvy them into the new era.[1] Did the introduction of windmills change people's lives? Even this brief account points in the direction of a division of, labour.[2] There are entrepreneurs, shall we say (the owners of the mill), there are millrights (employed by the entrepreneur), there is the miller, and there are producers of grain, there are the henchmen who take a hammer to household handmills in a kind of Luddism in reverse. The millright's skills had been developing for half a millennium before Quixote took exception to them, but, essentially, all of these positions were in place, *mutatis mutandis*, before the building of the first mill. The appearance of the giant on the landscape signalled an enhancement in the organization of this division of labour that effected a movement in the demarcation of the public and the private; the deterritorialisation of domestic flour production and its reterritorialisation as a publicly available (at a cost) service.[3] So, people's lives changed, but the change constituted and was constituted by a developing sophistication in the division of labour of which the windmill stood as a material sedimentation. Quixote's error was in mistaking a signifier for the social organization that it signaled, though his lance would never have been a match for either.

This, essentially, was the line of argument that I offered in Dowling (1991a), although in that essay I was concerned not with "the windmill", but with "the computer" and, more than a decade later, I might want to replace the latter by "the internet" which, of course, I can access via my mobile phone or my TV as well as my MacBook Pro and which can be imagined as a very visible sedimentation of the globalised division of labour. That is to say, I am conceiving of *technology* as a regularity of practice; the kind of regularity, indeed, that enables us to recognize the internet as such. This regularity is emergent upon the formation of diverse oppositions and alliances that we can think of as social action and that carries on at all levels of analysis from state activity down to the strategies and tactics of individual players (see Dowling, 2004a).

A curriculum is a technology. It exists in at least two forms, an official or general form and its realization in local instances (cf Bernstein, 1996/2000). A technological determinist kind of argument might conceive of the local curriculum, in its enactments in classrooms and lecture theatres, as only relatively autonomous

---

[1] See "The history of flour milling" at http://www.cyberspaceag.com/kansascrops/wheat/flourmilling-history.htm. The extract is from the opening of Chapter VIII of John Ormsby's translation of *Don Quixote*, *http://www.online-literature.com/cervantes/don-quixote*

[2] Perhaps the term "division of labour" is somewhat unfashionable in educational studies, these days. I retain it both to acknowledge a residual debt to Marx – a debt of the same character, perhaps, as that acknowledged by Foucault (I forget where) – and because it is now sufficiently anachronistic to stand out and thus allow me to avoid a neologism for that which brings together definable (and, of course, hierarchically organised) social groups with specific regularities in practice the articulation of which activities is constitutive of the sociocultural order.

[3] The terms, "deterritorialisation" and "reterritorialisation" are from Lacan via Deleuze and Guattari (1984) (see also Holland (1999)), whose position is not entirely inconsistent with my own in this essay.

with respect to the official form. In this conception, emphasis would be placed on the effects on local practices of changes in the official form as well as, perhaps, the nature of and limitations upon the autonomy of the classroom. Consider, though, the push for modern or new mathematics in many parts of the world in the 1960s (see Cooper, 1983, 1985; Dowling, 1990; Moon, 1986). Here, the crucial bourbakiist message was ultimately dissipated as the central organizing language of set theory was recontextualised as a pedagogic resource in the primary classroom (hoops and chalk circles for organizing objects) and as merely another topic on the secondary curriculum. The strong classification in the division of labour between mathematicians and school mathematics teachers survived quite intact the intervention of the former in the activities of the latter.

Similarly in Higher Education, being required (by quality assurance scrutineers) to provide explicit lists of intended learning outcomes for postgraduate seminars results merely in the production of an official, local curriculum and has little impact on the local, local curriculum in which the professor is still established as author rather than relayer of knowledge, albeit within a tradition of discourse, a discipline, perhaps. Here, the division of labour closely associates the person of the professor with the institutionalised practice of the discipline so that they may claim what I refer to (after Weber (1964), *mutatis mutandis*) as *traditional* authority. This mode of authority action is most likely to be effective under conditions of relative stability. Thus, back in school, in a period of healthy supply of mathematics graduates, those appointing mathematics teachers are in a position to stipulate that a degree in mathematics is a requirement for a successful application. Such a stipulation brings together a particular category of person and a particular technology (the mathematics curriculum) in authorizing its appointee who may, of course, teach mathematics, but not science, which is the exclusive territory of graduates in that field. But, as an "expert", the qualified mathematics teacher may claim a degree of authority over the mathematics curriculum giving rise to the dominance of the local over the official, the private over the public.[4]

In 1970s London the supply of mathematics graduates wanting to enter teaching had fallen below demand to such an extent that the possession of a mathematics degree was more of a rarity than a requirement for a mathematics teacher. Indeed, I was appointed as a teacher of mathematics despite having only a degree in physics and no professional or academic teacher education. I was appointed head of department less than three years later. The crisis continued throughout that and much of the next decade and teachers from all sorts of academic backgrounds found themselves teaching mathematics. As head of department I found myself working with physical education specialists, language teachers and geographers as well as a fair number of fellow natural scientists. Clearly, authorizing strategies had reined back on the specificity of the author – the teacher. However, many schools in London

[4] Those teaching in England in the 1970s and 1980s may remember the "mode 3" public examination syllabuses which were under the control of teachers and could even be established at the level of an individual school.

began adopting a student-centred scheme of school mathematics called SMILE.[5] This was a workcard-based scheme that had been designed specifically in response to the shortage of specialist mathematics teachers. That which was principally demanded of the teacher was skill in classroom management and administration. In addition, local meetings at which workcards would be revised and new cards produced would also function as in-service training for the teachers. The effect was the constitution of an official curriculum over which individual teachers may be disinclined to claim individual authority. Rather, their role would be, to a substantial extent, defined by the curricular technology so that the authority would reside in the role or practice rather than in the person. I refer to this as *bureaucratic* authority (again recontextualising Weber). Naturally, with the weakening of the autonomy of the teacher, this mode of authority action is likely to be associated with an assertion (or reassertion) of the dominance of the official over the local, the public over the private.

Now in a more recent paper (Dowling, 2001a) I offered some examples of current trends in the development in the division of labour that entail the production of disembodied analogues of competence in what I am referring to as technologies. The unification and codification of school curricula in England and Wales (see Dowling & Noss, 1990; Flude & Hammer, 1990) and the development of national qualifications frameworks here and elsewhere are examples as are spellcheckers and other software developments such as Adobe Creative Suite which (amongst a great deal more) allows me – a sociologist, not a photographer – to produce quite acceptable digital images from the rather amateur RAW files captured on my Canon 10D (a technology already obsolescent less than four years after its unveiling and which I've now replaced with the latest 5D model). These bureaucratising technologies are emergent upon the weakening of the esoteric control of the traditional expert over the form of institutionalisation of the practices to which they relate. The digital codification of these practices operates rather like the mass media, which, as Becker and Wehner (2001) point out, serve as "reduction mechanisms", rendering their messages accessible to the public.

What appears to have happened is not that technologies have been invented that are able to achieve this – the technologies still have to be acceptable to their audiences – but that changes in the division of labour have effected a shift in the mode of relationship between (certain) categories of traditional "expert" and their audiences. With the "expert" exercising traditional authority, this relationship is what I refer to as *pedagogic* (Dowling, 2001a). This means that the author in an interaction retains, or seeks to retain, control over the principles of evaluation of their utterance. The kind of change that I am describing here gestates as this mode of authority becomes increasingly non-viable and the "expert" is increasingly held to account for their actions. The relationship takes on more of the character of an

[5] Secondary Mathematics Learning Experiment – later, "experiment" was replaced by "experience" in the title. This was a teacher-led response to the changing situation, particularly in London; the state response was somewhat slower.

exchange mode (ibid.) whereby the principles of evaluation are devolved to the audience. The bureaucratic technology that facilitates this, through its "reduction mechanisms," signifies the presence in the division of labour of a mediating or competing authority: the state, in the case of curricula and qualifications frameworks; software houses etc in the case of spellcheckers. The significance of such developments is that to some extent (perhaps to an increasing extent) the voice of the expert may be heard only in terms of the public forms of their practice that are codified in and by the technology; I will return to this in the closing of this essay.

In the UK, the change in the field of education was signalled when, in 1962, the then Minister of Education referred to the school curriculum as a "secret garden" (see Kogan, 1978). The invasion of this garden by politicians and capital over the ensuing forty years established the curriculum as a national park. The mathematical region of this park has been discussed in Dowling and Noss (1990).[6] However, with corresponding public spaces opening up in other national systems and being freely available on the internet, the impact of each national government's policies becomes comparable in terms of a further "reduced", international curriculum. A key representative of this technology is to be found in the series of comparative *Trends in International Mathematics and Science Study* (TIMSS) carried out under the auspices of the International Association for the Evaluation of Educational Achievement (IEA) (see http://www.iea.nl/iea/hq/, also http://timss.bc.edu/ and http://nces.ed.gov/timss/). The results of this study and diverse reflections on the performances of participating nations[7] are available globally for recruitment in struggles relating to the bureaucratising of education at national level. This is how it is put on the National Center for Educational Statistics (NCES) website:

> With the emergence and growth of the global economy, policymakers and educators have turned to international comparisons to assess how well national systems of education are performing. These comparisons shed light on a host of policy issues, from access to education and equity of resources to the quality of school outputs. They provide policymakers with benchmarks to assess their systems' performances, and to identify potential strategies to improve student achievement and system outputs. (http://nces.ed.gov/surveys/international/IntlIndicators/)

Given the trend towards the globalising of English (see Crystal, 2003), what we have in this technology is a globally visible public educational discourse; the secret garden has blossomed into a world heritage site.

The first point to note about this discourse is that its subject focus establishes mathematics and science as the global public face of schooling, relegating most other areas to a relatively private sphere. It is easy to see why this is bound to

---

[6] Though this was published at a time when we had to rely on paper publication of the National Curriculum.

[7] See, for example, Symonds (2004) on the US and Wolf (2002) on Chile, both referring to poor performances on TMSS.

be the case. As the exponents of ethnomathematics and ethnoscience have been energetic in pointing out, mathematical and scientific knowledge has long been appropriated by the dominant and self-styled "developed" nations as their own. At the same time, most other areas of school knowledge – such as history and art – are closely and enthusiastically allied with individual national identities. A study entitled, Trends in International Poetry and Painting would present engaging methodological as well as political problems and Trends in International History would certainly provoke belligerent uproar.[8] Comparative literacy rates are clearly of political interest (see, for example, the Progress in International Reading Literacy Study (PIRLS), http://www.iea.nl/iea/hq/, also an IEA study), but they do not (and, at the moment could not) specify the language (what with English, Spanish, Arabic and Chinese all legitimately vying for global hegemony). Perhaps sport comes closest to exhibiting the global status of (western) mathematics and science, but really only at the level of elite performance, which is clearly not the primary concern of formal schooling.

This observation is consistent with, at the global level, a *public* curricular sphere consisting of mathematics and science in which context other curricular areas are relegated to a national, which is to say comparatively *private* sphere; there is an important exception to this division to which I will return later. Stanley Fish localises in time and place the hegemony of science:

> ... in our culture science is usually thought to have the job of describing reality as it really is; but its possession of that franchise, which it wrested away from religion, is a historical achievement not a natural right. (Fish, 1995, p. 72)

Now I do not, in any case, subscribe to a theory of natural rights – here, at least, I am a happy (perhaps unhappy) positivist[9] – and so I will certainly go along with Fish in understanding western science as a cultural arbitrary.[10] This particular cultural arbitrary, however, is now constituted as one key element in a global hegemony. Furthermore, the contrast in modes of authority that are deployed by religious and scientific practices, respectively, is also consistent with the public ownership of the latter at the expense of the relative privatising of the former. Specifically, religious practices commonly involve the development of a traditional priesthood in one form or another. The developments in science and mathematics curricula that I am referring to here, on the other hand, facilitate bureaucratic authority which tend to render individuals interchangeable: we can all be scientists to the extent that we can have public access to the principles of evaluation of scientific texts; but only a Catholic priest may hear a confession.[11]

---

[8] See, for example, the furore in South Korea and China over a Japanese school history textbook that, it is claimed, downplays Japanese militarism and war crimes committed by Japanese troops http://news.bbc.co.uk/2/hi/asia-pacific/4678009.stm.

[9] See Crotty (1998) for a discussion of naturalist and positivist philosophies in the fields of research and law.

[10] "Arbitrary" in the sense of Bourdieu & Passeron (1977).

Rather than tilt at my windmill, I want to explore it further to determine just what kinds of relationships (between author and audience) and practices it privileges. As my empirical object I shall take the US government TIMSS website at http://nces.ed.gov/timss/ (see Figure 1).[12] I have no space for a detailed analysis of this site. Rather, I shall use aspects of it to illustrate the points that I want to make. Firstly, concerning the form of the technology, this is fairly conventional hypertext

Figure 1. TIMSS(USA) Home Page

[11] There is a corresponding contrast between the modes of authority deployed as, in Western culture, science replaces literature as the apogee of erudition. The origins of the humanities in British universities was predicated upon a sense of embodied literature and other artistic faculties as the necessary prerequisite of a cultivated English gentleman.

[12] All screenshots were made in September 2004.

site, so that each page consists of a set of common elements – a standard header, a menu to the left (including links to the parent NCES site), page-specific text (which may or may not contain links) to the right, below all of this are plain text links to the NCES site, and above are links to a site map, the US Department of Education site, the NCES site, and a search engine. The righthand section of the home page contains a graphic link (a cartoon frog) to some of the questions used in TIMSS, "For Students!" Below this are two windows, one showing "What's New" and the other "International Fast Facts", the content of which changes when the page is refreshed, apparently on the basis of a random selection from a file of "facts".

This design presents, on each page, the key claims to bureaucratic authority – established by the links to other government sites in the page header and footer[13] – and the structure of the site – principally in the menu – which consistently frames the page-specific content. On this site the page-specific content is generally linear, discursive text. In addition, this page-specific content is, in most cases, marked, which is to say that it carries one or more links. These links are generally to other pages in the same site or the parent NCES site.[14] The design conforms to what Michael Joyce (1995) has described as an "exploratory" rather than a "constructive" hypertext. James Sosnoski succinctly describes the difference as follows:

> The exploratory (or expository) hypertext is a 'delivery or presentational technology' that provides ready access to information. By contrast, constructive hypertexts are 'analytic tools' that allow writers to invent and/or map relations among bits of information to suit their own needs.
>
> (Sosnoski, 1999, p. 163)

In my terms, the site establishes pedagogic relations between its author and audience; this is unsurprising, of course, in a government publication. It is, however, worth pointing out that even were the site to include multiple links to other, non-governmental sites, this would itself remain a pedagogic action insofar as it is a privileging of marked over unmarked text; the TIMSS site asserts a stronger pedagogic claim by additionally retaining control over the targets of links to marked text. Unmarked text is, of course, open to interrogation – any term or terms may be copied into a non-governmental search engine. However, such alternative readings are privatised by the TIMSS site. Similarly, the reader may formulate alternative structures for the site – this is essentially what I am doing here. Again, though, such strategies are privatised by the pedagogic site, which deploys bureaucratic authority strategies and essentially privileges an explicit taxomony and marked text over contingent organisation and unmarked text. So, the educational technology that I have been discussing signals (which is to say, is arguably emergent upon – see

[13] The authority action is bureaucratic because government *per se* is bureaucratic insofar as its authority is taken to reside in the office (practices) rather than in individuals. Of course, other modes of authority may be deployed in establishing the legitimacy of government.

[14] Although it is possible to exit the NCES site by following some of the links as I will illustrate below.

Dowling (2004a)) the establishment of a public/private partitioning of educational discourse that locates mathematics and science and strongly institutionalised modes of reading within the public sphere and other areas of knowledge and alternative modes of reading in the private.

The next question to be considered relates to the nature of the public mathematical and scientific knowledge. In order to address this I will click the frog link on the TIMMS homepage (Figure 1). This takes me to a page on another site parented by NCES, the "Students" Classroom' (http://nces.ed.gov/nceskids/index). The particular page is titled "Explore Your Knowledge" (http://nces.ed.gov/nceskids/eyk/index and see Figure 2). The page gives access to assessment items from the TIMSS study and also from the Civic Education Study (CivEd) to which I shall return later. From the page in Figure 2 I select my subject, grade and the number of questions (5, 10, 15 or 20) and am presented with the required number of test items; examples of these are shown in Figures 3–12. After making my selections from the multichoice radio buttons I can click "show me the answers" and my page is replaced with an answers page including a score given as a percentage – Figure 13 shows part of an answer page. Clicking on the globe

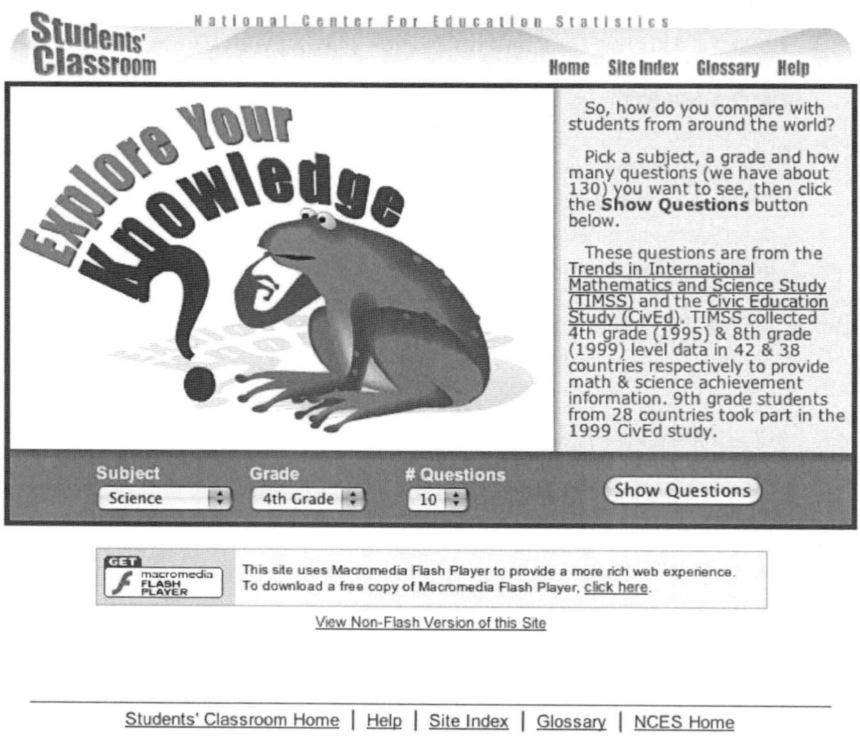

Figure 2. "Explore Your Knowledge" Page, NCES Site

2. Some children were trying to find out which of three light bulbs was brightest. Which one of these gives the best START toward finding the answer?

- ○ "One bulbs looks brightest to me, so I already know the answer."
- ○ "All the bulbs look bright to me, so there cannot be an answer."
- ○ "It would help if we had a way to measure the brightness of a light bulb."
- ○ "We can take votes and each person will vote for the bulb he or she thinks is the brightest."

Figure 3. TIMSS Test Item for Grade 4 Science

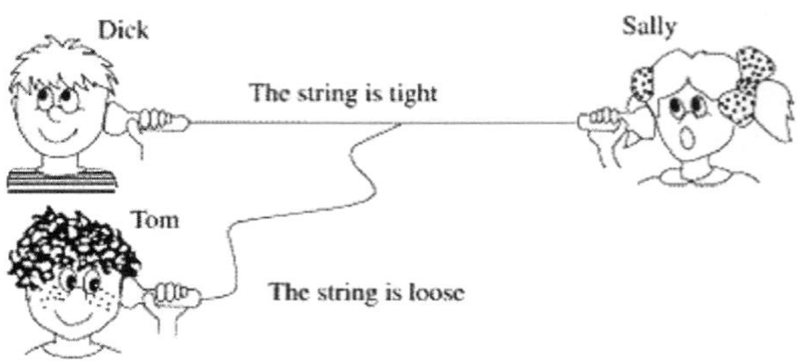

14. The picture shows Dick and his friends playing with a string-telephone. Sally is speaking. Dick and Tom are trying to listen. Which of them can hear her speak?

- ○ Both of them can hear equally clear.
- ○ Neither of them can hear.
- ○ Only Tom can hear clearly.
- ○ Only Dick can hear clearly.
- ○ Both of them can hear equally faintly.

Figure 4. TIMSS Test Item for Grade 4 Science

16. Four children can feel and smell an object inside a bag, but they cannot see it. Which of the following is NOT an observation about the object?

- ○ "It is flat at one end and round at the other."
- ○ "It smells like peppermint."
- ○ "It has a bump on it."
- ○ "I hope it is candy."

Figure 5. TIMSS Test Item for Grade 4 Science

button – one is given for each item – opens a pop-up window (Figure 14) showing the US national performance and the international average for the item; buttons in

17. Alexander Fleming noticed that bacteria growing on a plate of agar did not grow next to a mold that was growing on the same plate. He wrote in his laboratory report: "The mold may be producing a substance that kills bacteria." This statement is described as:

- ○ an observation
- ○ a hypothesis
- ○ a generalization
- ○ a conclusion

Figure 6. TIMSS Test Item for Grade 8 Science

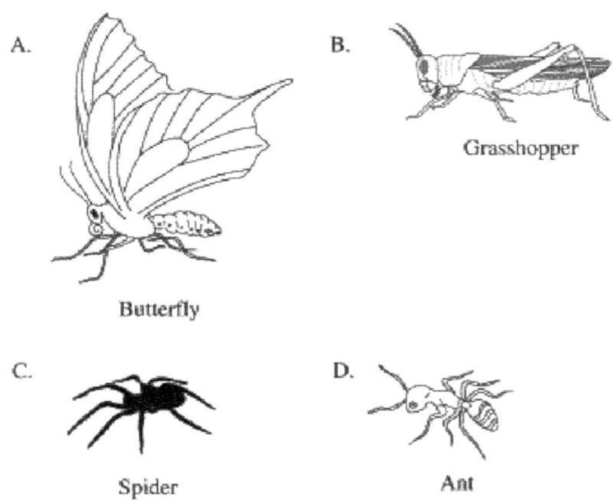

1. Which picture does NOT show an insect?

- ○ A
- ○ B
- ○ C
- ○ D

Figure 7. TIMSS Test Item for Grade 8 Science

13. Two open bottles, one filled with vinegar and the other with olive oil, were left on a window sill in the Sun. Several days later it was observed that the bottles were no longer full. What can be concluded from this observation?

- ○ Vinegar evaporates faster than olive oil.
- ○ Olive oil evaporates faster than vinegar.
- ○ Both vinegar and olive oil evaporate.
- ○ Only liquids containing water evaporate.
- ○ Direct sunlight is needed for evaporation.

Figure 8. TIMSS Test Item for Grade 8 Science

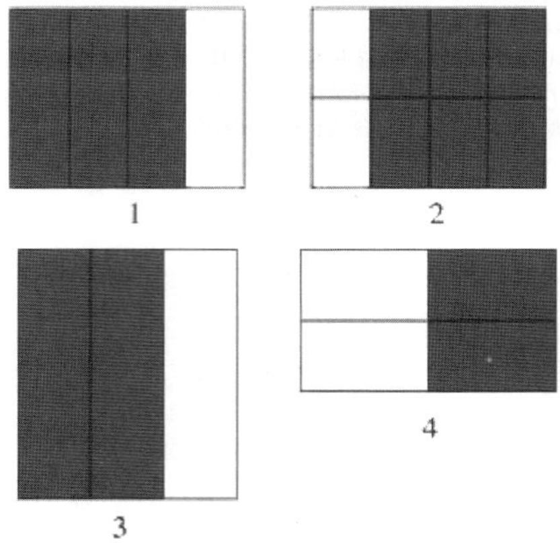

9. **Each figure represents a fraction. Which two figures represent the same fraction?**

○ 1 and 2
○ 1 and 4
○ 2 and 3
○ 3 and 4

Figure 9. TIMSS Test Item for Grade 4 Mathematics

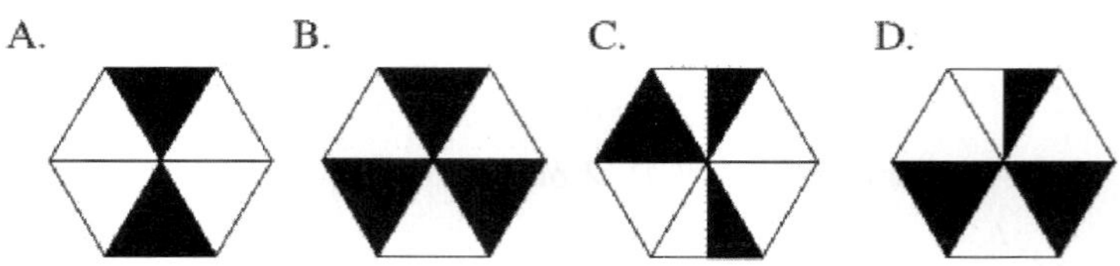

4. **Samantha drops a stone onto each of these targets. The stone has the best chance of landing on a shaded space in which target?**

○ A
○ B
○ C
○ D

Figure 10. TIMSS Test Item for Grade 4 Mathematics

other country locations on a world map[15] will replace the US flag and performance with that of the relevant country.

Before proceeding to look at some items, I will briefly make two preliminary observations based on the description thus far. Firstly, the provision of the world

[15] The full list of TIMSS participating countries is given at http://nces.ed.gov/timss/countries.asp. Each information map shows only a small selection, though the US is always included (it being a US site).

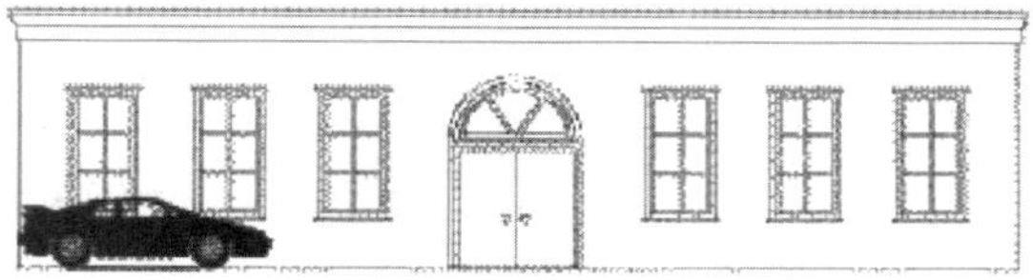

14. The car is 3.5 m long. About how long is the building?

- ○ 18 m
- ○ 14 m
- ○ 10 m
- ○ 4 m

Figure 11. TIMSS Test Item for Grade 8 Mathematics

x	y
2	5
3	7
4	?
7	15

18. The table represents a relation between x and y. What is the missing number in the table?

- ○ 9
- ○ 10
- ○ 11
- ○ 12
- ○ 13

Figure 12. TIMSS Test Item for Grade 8 Mathematics

map and clickable international comparisons is a good illustration of my point that we are talking about global public discourse here, even if only in its larval stage. Secondly, the combination of multichoice radio buttons and definitive "correct" answers is a particularly effective privatising of alternatives by a strongly pedagogic technology. The multichoice test item (and the precoded questionnaire and countless other digitisings) is a technology that is emergent upon a drive to render all commensurable, all accountable to a public discourse via the exclusion of the private.

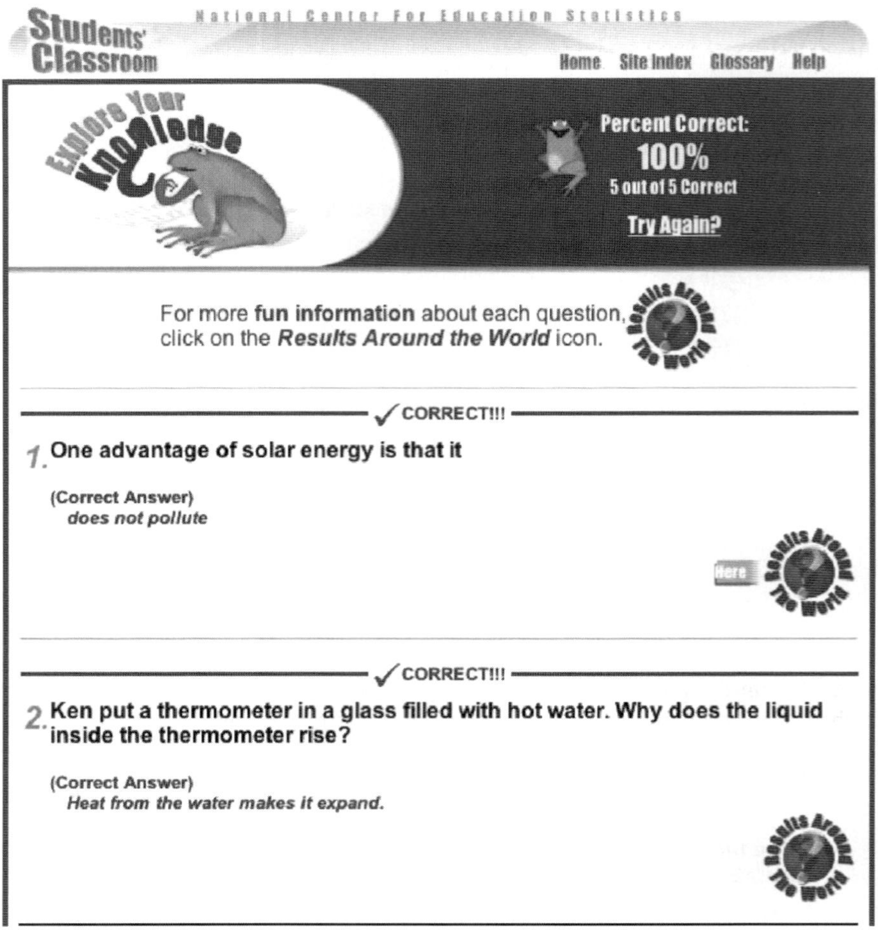

Figure 13. Answers Page

The TIMSS test items construct scientific and mathematical knowledge in a familiar way, perhaps. Firstly, they constitute formal modes of expression (see Figure 6) and content (see Figure 7, which invokes a taxonomy) that represent what I refer to as the *esoteric domain* (Dowling, 1998) of mathematical or, in these cases, scientific knowledge. The esoteric domain consists of discourse, which is strongly marked out from other areas of practice and contrasts with the *public domain,* which is weakly marked out.[16] Thus, contrasting with Figures 6 and 7, the item in Figure 4 refers to a children's game using a tin can phone – a public

[16] I have been referring, throughout this essay, to public/private divisions; this use does not correspond to the esoteric/public domain distinction that I am making here although there is clearly some relation

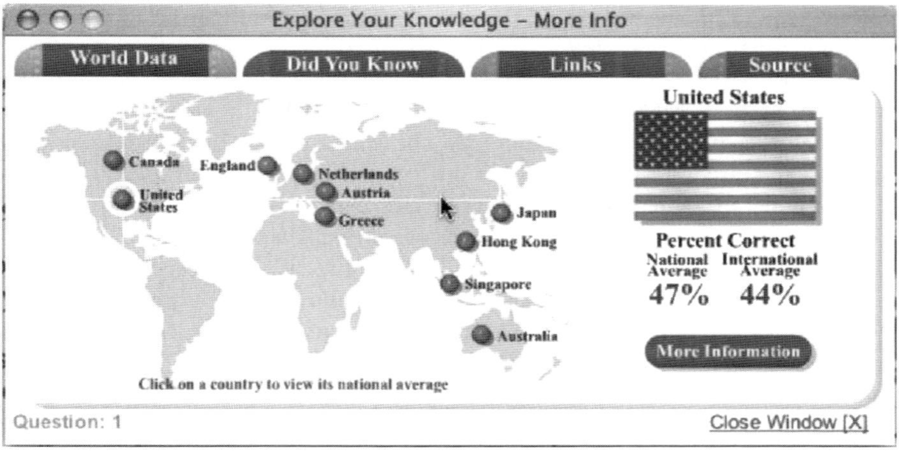

Figure 14. Information about International Performances on Selected TIMSS Test Item

domain setting. The item in Figure 10 also employs a public domain setting and it is significant to note that the term, "probability" is substituted by "chance". This is consistent with my findings in my analysis of a major British textbook scheme that the theme of probability was (at least at that time and in that place) very substantially taught within the public domain (Dowling, 1998). School science and, especially, mathematics constitute esoteric domains that are strongly institutionalised. This is to say that scientific and mathematical language are deployed with a high degree of regulation – far more so than in most other areas of the curriculum. If I may gloss mathematics, as such, as the study of formal systems, then it is clear why its esoteric domain must be strongly institutionalised. Science, then, might be thought of as the study of partially- or to-be-formalised systems and its esoteric domain language emerges out of (induction) and is projected onto (deduction) the systems that are to be formalised. Science too, then, is predicated upon a strongly institutionalised esoteric domain. However, public domain text renders invisible the esoteric domain structuring that makes a task mathematical or scientific rather than something else. In the item in Figure 5, the response, "I hope it's candy" is indeed an observation about the object in the bag,[17] but not in the scientific sense which must exclude the subjective. "Intensity" has been replaced by "brightness" in the item in Figure 3; which bulb is "brightest" may well relate to colour (frequency) as well as to intensity and so call for a subjective response; again, subjectivity must be excluded from formal school science. The item in Figure 8 is particularly interesting in that the most likely public domain response – someone has been making salad – is not

between them. For the sake of clarity here it is best to think of "public domain" as a single term rather than an adjective-noun pair.

[17] The statement may be reformulated as, "the object in the bag is something that I hope is candy", thus making the object in the bag the subject of the principal clause.

offered as an option; there is a sense in which this item might be thought of as teaching rather than assessing.

Some of the mathematics test items (Figures 9–12) may be interpreted as tending to undermine esoteric domain mathematics and science. The Figure 9 item represents a standard teaching metaphor, which may be glossed as "a fraction is a piece of cake". The correct answer is the first one on offer because both diagrams 1 and 2 conventionally represent the fraction $^{3}/_{4}$. However, as I have previously pointed out (Dowling, 1990), this metaphor pedagogically challenges the esoteric domain constitution of a fraction as a number – that is of $^{3}/_{4}$ as a number between 0 and 1. Thus, if we use diagram 1 from Figure 9 to illustrate the sum $^{3}/_{4} + {}^{3}/_{4}$ as in Figure 15, then a perfectly reasonable (though, of course, mathematically incorrect) answer would be $^{6}/_{8}$.[18] The "correct" response to the item in Figure 11 is the second radio button, 14 m. However, this appears to discount the width of the car (and its distance from the building). If the visible side of the car is a little under 2 m from the building, then a viewpoint 7 m away from the car in line with the rear of the car and the lefthand end of the building would make the first option – 18 m – a better answer. The item appears to be testing estimation skills, but the public domain simulation renders it ambiguous.[19] The item in Figure 12 appears to be an esoteric domain text. However, there is a unique answer only if we qualify "relation" with the term "linear". If the nature of the relation is not specified then there is no limitation on what might replace the question mark in the table. We may take the reference to a "missing number" as indicating that the relation is between two

Figure 15. $^{3}/_{4}+{}^{3}/_{4}=?$

[18] This is because the metaphor, "a fraction is a piece of cake", invites the student to take the number of shaded pieces to be the numerator and the total number of pieces to be the denominator. It is also the case that the total amount of shaded cake in Figure 15 is $^{6}/_{8}$ or $^{3}/_{4}$ of the total amount of cake. That we frequently find students making this error does not affirm that they are interpreting the diagrams as I have suggested, but their error is at least consistent with this interpretation.

[19] South Africa – quite easily the lowest scoring country in both mathematics and science – scored 26% answers correct on this item as compared with the 74% international average; It would be interesting to see which responses dominated in South Africa (and, of course, to ask the respondents why).

numerical variables, but, even so, all five offered answers are equally acceptable, mathematically. Here, it is not the construction of a public domain setting that has generated the ambiguity, but a reduction of the complexity of the esoteric domain.[20]

This brief analysis of ten test items[21] suggests that mathematics and science – and the difference between them, here, is not as great as one might suppose – are constructed as laboratorised or, shall we say, laboratorising practices. These laboratorising practices operate on the phenomenal world in much the same way as a hypertext author operates on text, which is to say, by marking that which may legitimately be operationalised; the unmarked, extraneous, subjective regions of the text are methodologically excluded. In both mathematics/science and hyptertext, this marking may often be invisible. In hypertext, however, we are well practiced in scanning the text with the cursor so as to reveal the links; no similar divining rods are to be found in mathematics or science and that is why, of course, my revealing of the ambiguities introduced by the public domain contexts does not challenge the items as suitable for their purpose – I obtained "correct" answers on my first attempt on all of the items, despite my recognition of their "flaws". This is presumably consistent with my standing as a physics graduate and, more to the point, one-time teacher of high school mathematics and science.

So my point is not to criticise the validity or reliability of the test items, but to illustrate the kind of practice that hegemonises the global public educational discourse.[22] To the extent that mathematics and science exhaust this discourse, then we might infer that they define, firstly, the legitimate mode of relationship to the empirical and, secondly, the legitimate form of argumentation. In both cases, legitimacy is established by principles of exclusion that are governed by the esoteric domains of mathematical and scientific practice that exclude, in particular, the subjective and the contingent thus relegating them to the private sphere. As I have suggested above, we may tentatively distinguish between the two esoteric domains by referring to science as a formalising discourse and mathematics as a formalised

[20] A feature that is particularly common in texts directed at lower performing students as is the prevalence of public domain settings (Dowling, 1998).

[21] The site notes that there are about 130 items available, presumably these cover ninth grade civics as well as fourth and eighth grade mathematics and science.

[22] Indeed, critics of multichoice test items tend to limit their criticisms to issues of face and content validity. However, to the extent that the authors of the test have established a strong measure of convergent validity of these items with respect to, shall we say, measures derived from clinical interviews, then there is no reason why they should not be used in large scale surveys, such as TIMSS (see Brown et al. (forthcoming, Dowling & Brown, forthcoming). In their exploration of Piagetian stages, Shayer, Küchemann, & Wylam (1992) precisely did take steps to affirm the convergent validity of their experimental tests in relation to clinical interviews of the type used by Piaget himself. This precaution was ignored by McGarrigle's much cited challenge to Piaget's findings reported in Donaldson (1978). I have not studied the validity tests used by the TIMSS authors, because the point, in this essay, is to examine the workings of this global public discourse and not its convergence with local forms of assessment.

discourse.[23] Given this distinction, we might speculate that science takes the dominant role in respect of the constitution of the first legitimate mode and mathematics in respect of the second. The blurring of the distinction between mathematics and science in their high school forms also blurs this division of discursive labour. In any event, mathematics and science taken together do seem to define the legitimate form of rational action so defining, on a global stage, the bureaucratic public voice,[24] so I'll refer to the public global technology as *mathematicoscience*. Now, clearly, mathematicoscience is not the only public form of discourse. However, apart from the operational matrix[25] of the internet itself, it is arguably the principal form of discourse for which globalised regularity or institutionalisation might be claimed and this is signified by its prominence in the global curricular technology to which I have been referring. Insofar as there is a globally prevalent aspiration for universal schooling and insofar as mathematicoscience, more or less as I have described it here, territorialises the globally public content of schooling, the significance of this discourse should not be understated.

So what are the implications? Well we might begin by considering this essay. I am certainly laying claim to both bureaucratic and traditional authority. My affiliation to the Institute of Education, University of London establishes that I hold an office that authorises me to speak academically about educational matters. This is a very weak claim, however, as the practice of peer review (or clubbing, as I tend to think of it), for example, ensures that the *ex officio* authority of academics is limited, generally to that which they may hold over their students. My recruitment of what I may hope is a familiar academic style and terms also constitutes a bureaucratic action in the way that I (*pace* Max Weber) have defined it: I am, in this sense, allowing (or pretending to allow) the discourse to ventriloquise me. Traditional authority is claimed in terms of my yellowing PhD thesis and also through the community of celebrated academic authors to which I affiliate via my egocentric bibliography

[23] I am reminded here of Foucault's comment on mathematics: "... the only discursive practice to have crossed at one and the same time the thresholds of positivity, epistemologization, scientificity, and formalization. The very possibility of its existence implied that [that] which, in all other sciences, remains dispersed throughout history, should be given at the outset: its original positivity was to constitute an already formalized discursive practice (even if other formalizations were to be used later). Hence the fact that their establishment is both so enigmatic (so little accessible to analysis, so confined within the form of the absolute beginning) and so valid (since it is valid both as an origin and as a foundation); hence the fact that in the first gesture of the first mathematician one saw the constitution of an ideality that has been deployed throughout history, and has been questioned only to be repeated and purified; hence the fact that the beginning of mathematics is questioned not so much as a historical event as for its validity as a principle of history: and hence the fact that, for all the other sciences the description of its historical genesis, its gropings and failures, its late emergence is related to the meta-historical model of a geometry emerging suddenly, once and for all, from the trivial practices of land-measuring" (Foucault, 1972, pp. 188–189).

[24] This seems to be consistent with Max Weber's (1968) remarks on the increasing prevalence of *zweckrationalitat*.

[25] I define "operational matrix" as a technology – a regularity of practice – that incorporates, non-discursively, the principles of its own deployment: a supermarket and the World Wide Web would both be examples.

(clubbing in the imaginary, perhaps). But I am clearly trying to do more than that. Bureaucratic and traditional authority strategies both invoke institutionalised, which is to say, stabilised practices. Such strategies are appropriate in the context of schooling insofar as the authority of the teacher or of the curriculum rests on a training or on a construction that has already been completed. In this respect, at least, schooling is structurally conservative as is illustrated by the recontextualising of set theory, which I mentioned earlier.

The authority of the academic, on the other hand, is established dynamically. The output of research is valued only insofar as it is original (a necessary, but, of course, not sufficient condition for acceptability). Academic discourse, then is structurally dynamic. The academic may rely on traditional authority strategies by, for example, establishing originality only in terms of the empirical setting and not in terms of theoretical framework – replication studies would be of this form. However, work of the highest status must contribute to the development, the construction and/or discovery of the language of the discourse, which is to say, theory.[26] This, of course, entails a destabilising of the institutionalised practice that affirms the two modes of authority action that I have introduced. I need a third mode. This has, fortuitously, also been provided by Max Weber (1964). As with the first two modes, I shall retain his term, but redefine the category: *charismatic* authority is predicated on the closure of the category of author and the opening of the category of practice. In establishing the originality of this essay I am at least in some respects attempting to deploy a charismatic authority action. I am served in this respect by the facility to refer to my own previous publications, establishing myself as an author of already accepted (and so publicly acknowledged as original) practice.

Naturally, there is a general level of resistance in the field to charismatic claims to originality because they must stand in competition with others. My essay, then, must extend, even distort and transform the discourse, but I do not have free license. So how might my essay be challenged? Well, on precisely the principles that are established in the terms of the public global discourse that I am referring to as mathematicoscience – though I have now moved higher up the academic ladder. So: have I deployed appropriate principles of exclusion in my engagement with the empirical and in the construction of my syllogisms; have I deployed an objective methodological apparatus with sufficient rigour to exclude subjective noise or distortion? My critic may point out, for example, that my sampling strategies are inadequate to my grandiose claims and that my analysis and argument are tendentious. Within the context of the public global discourse of mathematicoscience my critic would be entirely justified as I will authoritatively affirm as the co-author of works on research methodology (Brown & Dowling, 1998; Brown, Bryman & Dowling, forthcoming, Dowling & Brown, forthcoming). Insofar as my essay is recognisable in the public sphere, it can be recognised only as heresy.[27]

---

[26] Only theoretical objects may be discovered; an empirical object is merely encountered.

[27] A point illustrated by the Sokal/*Social Text* affair (see http://www.physics.nyu.edu/faculty/sokal/#papers). Sokal complains: "In short, my concern over the spread of subjectivist thinking is both

It is the thrust of my argument, however, that the lance of my quixotic critic cannot penetrate me, precisely because it misses the point, which is as follows. All technologies – including mathematicoscience – are here being regarded as emergent upon the formation of alliances and oppositions in social action; they are the public visibility of these alliances. However we know from our respective experiences that the work that goes into social action is very substantially conducted in private – in the lavatories, not the boardroom. Furthermore, the opening up of private spaces to public scrutiny – ethnography, perhaps, or the ungendered toilets in *Ally McBeal* and the Belgo restaurant in London's West End – will simply resite the private, not eradicate it,[28] just as the zero-tolerance policing paving the way for the gentrification of London's Kings Cross produces assaults on hapless students in Bloomsbury. The private, in other words, is for the most part where, for good or bad, things get done.

Let me complete my schema for authority strategies.[29] I have, in effect, introduced two variables, the category of author and the field of practice and each of these are binary nominal scales, open/closed. The product of these two variables gives rise to the space in Figure 16. It will be apparent that there are now four modes of action, three of which have already been introduced. The fourth mode, which I have termed *liberal*, is essentially a mode of action in which authority is negated. In liberal mode, persons are interchangeable and practice is mutable. Piaget's paradise, perhaps, but a mode of action that does seem to characterise the licence of a private audience: unless you intend or are required to respond to this essay in public, then there are no necessary constraints on the way in which you read and make use

	Field of Practice	
Category of Author	Open	Closed
Closed	*Charismatic*	*Traditional*
Open	*Liberal*	*Bureaucratic*

Figure 16. Modes of Authority Action

intellectual and political. Intellectually, the problem with such doctrines is that they are false (when not simply meaningless). There *is* a real world; its properties are *not* merely social constructions; facts and evidence *do* matter. What sane person would contend otherwise? And yet, much contemporary academic theorising consists precisely of attempts to blur these obvious truths – the utter absurdity of it all being concealed through obscure and pretentious language." (Sokal, 1996a, no page reference in the WWW version). Whilst he may have grounds to complain at the editorial strategies of the journal, *Social Text,* in which he managed to publish his parody of a cultural studies paper (1996b), clearly he just does not understand the positions that he ridicules – this is frequently the case with ridiculers (though I offer no evidence in support of this statement).

[28] *Ally McBeal*, see http://www.imdb.com/title/tt0118254/maindetails. The toilets in the Belgo restaurant actually have gendered sets of cubicles, but in a single space and with communal washbasins.

[29] See Dowling (2004b) for further elaboration of this schema.

of it (or choose not to). The essay stands as a resource or reservoir of resources for recruitment by the audience and, in this aspect, the relationship between author and audience is one of *exchange*. But I will conclude the essay by offering some suggestions.

This essay is written for an international collection, which is managed by an international editorial group. Those of us submitting chapters also had to submit to a peer review process and face the threat of required revision or exclusion. The structure of this practice – also a feature of the most respected academic journals – would appear to militate for some level of adherence to a public discourse which will include, as in this sentence, the genuflections of hedging, because the authority of our utterances must reside, bureaucratically, with the discourse, our mastery of which is yet to be finally affirmed. To read my analysis of the TIMSS test items as literal criticism within the field of the assessment of school science and mathematics would be to sublimate the essay on the level of this public discourse. This would be to render it legitimately open to revision in respect of the necessary exclusion of subjectivity and, incidentally, tricky language which can only be obscuring the clarity (or fallaciousness) of its syllogisms. Interaction in this mode is *equilibration*[30] and, in this mode, an acceptable piece of work must contribute or potentially contribute to the coherence of public rationality to which it stands in synecdochic relation. But if my overall analysis is persuasive (for whatever reason) then, as private intellectuals and teachers, we may be sharpening the sword of our own executioner.

Academic engagement does not always work like this. In the club mode of peer review (including the audiencing of papers at conferences and the recruitment of "the literature" in our own papers) we may also be familiar with the facility to read or listen politely and with at least apparent interest and to withhold equilibrating action on the grounds that contingency insulates us from the other author. I call this mode the *exchange of narratives*. Its inspirational metaphor comes from the telling of stories in a group of holiday friends at a bar in Mombassa (don't ever tell them what they're doing, sociologists are *personae non grata* in bars). Each narrative stands in relation of contiguity – metonymy – to the next. But as an audience this is at best voyeurism (onanism); it passes the time and avoids confrontation.

But the public discourse will not go away. Perhaps the arbitrary nature of public discourses may be made more apparent (or perhaps not) by the introduction of the third set of test items that is made available by clicking the frog on the TIMSS USA website. Perhaps surprisingly, perhaps not, this set of items is from the Civics Education Study (CivEd). The CivEd homepage notes that:

> All societies have a continuing interest in the ways in which their young people are prepared for citizenship and learn to take part in public affairs. At the turn of this new century this has become a matter of increased importance not only

[30] A mechanism that is, interestingly, associated more with the first than the second and third wave of cybernetics. It is the latter two schema that have had greatest influence on the position being developed here giving rise to my preference for autopoiesis and emergence (see Dowling, 2004a; Hayles, 1999).

> in societies striving to establish or reestablish democratic governments, but also in societies with continuous and long established democratic traditions.
> (http://nces.ed.gov/surveys/cived/)

Here is not the place (and I will not be allowed the space) to produce even a brief analysis of the CivEd text items. However, the "International Fast Facts" box in the screen shot of the TIMSS USA home page that I have presented as Figure 1 presents what is presumably a finding from the study:

> In 1999, about 90% of 9th-grade U.S. students reported that it is good for democracy when everyone has the right to express opinions freely.
> Year of the Data: 1999[31]

It would appear that the discourse of liberal democracy is a second key component of the public global technology alongside mathematicoscience. Jean Baudrillard (talking about Saddam Hussain and the first Gulf "War") offers a rather different take on democracy:

> ... as with every true dictator, the ultimate end of politics, carefully masked elsewhere by the effects of democracy, is to maintain control of one's own people by any means, including terror.
> (Baudrillard, 1995, p. 72)

It's not altogether certain that the masking is everywhere very substantial.

Again, here is not the place to engage in an explicit critique – which would, in any event, be quixotic, a quixocritique – of liberal democracy as a universal aspiration and absolute good. All that I should do here is to point to the alignment of discourses associated with the TIMSS site. Alan Sokal (see note 29) would (should he consider an assault on this little piece to be worth the effort) no doubt berate me for making anything at all out of the juxtaposition of the language of democracy with the language of scientific rationality other than that, perhaps, they are *in fact* properly aligned: the one seeking the optimizing of the exigencies of social organization in the context of liberal values; the other seeking the optimizing of our engagement with the empirical world in the face of imperfect knowledge. I am easily defeated in the public discourse that emerges out of social alliances that must overwhelm me. Indeed, even Sokal's far more celebrated public victims must often appear to be skulking back into the privacy of their arcane, alchemic worlds in the face of his dazzling crusade.

The invoking or the awareness of a public/private duality seems to provoke hegemonic or counter-hegemonic, metaphorical action, but to engage in this way is either to play the game of the dominant alliances or to falter. To the extent that the bureaucratized public technology constitutes the language by which expertise

---

[31] It is not helpful to provide a reference as this appeared in a box on the site, the contents of which vary.

is defined, the traditional expert – insofar as their expertise stands in excess of the bureaucratically defined practice – or the charismatic or liberal innovator may participate only as heretics; and heretics always get burned eventually (in this world or a next).

I have introduced three modes of interaction: synecdochic *equilibration*; metonymic *exchange of narratives*; and metaphoric *hegemony*. The first two of these modes presume an alliance of similars – we all speak the same public language. They differ in that equilibration seeks a discursive closure whilst the exchange of narratives deploys contingency to avoid closure. Hegemony contrasts with both in recognition of the public/private partition. Here engagement is between disimilars. But like equilibration, the target is discursive closure. The product of the two variables, alliance (similars/disimilars) and target of discursive action (closure/openness) gives rise to the space shown in Figure 17. As with my analysis of authority action, I am left with a residual category. In this case, the category, *pastiche*, defines an interaction between disimilars – public/private – under conditions of discursive openness. I have offered corresponding tropes for the other modes. The characteristic trope for pastiche is catachresis (see Burbules, n.d.). I want to suggest that it is precisely in this mode that private action in non-bureaucratic mode is most productively elaborated. Here, apostasy in relation to the global public technology of mathematicoscience (and democracy) may be sustained whilst still recruiting from it that which may be of practical value in our local pursuits. We have, in other words, to recognize, that very few of us are going to change the world in any sense at all and that those of us who do may well not welcome the outcome: some people change the world, but not in ways that they themselves choose.

So what does this mean in the context of mathematics and science education? I ought, in righteous exchange mode, to say, "I don't know," but then, I'm a teacher. I suppose that it may well come down to paying close attention to the matter at hand and, in particular, to the nature of the local relations that will tend to dominate any given intervention or interaction. Very little will be served, I think, either by total submission to the hegemony of mathematicoscience or by opposition in quixocritique. The whole point of pastiche interaction is that the integrity of the participating discourses must be maintained – catachresis must not be permitted to degenerate into metaphor or, perhaps worse, the literal discursive identity of equilibration or exchange of narratives. As has been demonstrated by

	Target of Discursive Action	
Alliance	Closure	Openness
Similars	*Equilibration*	*Exchange of Narratives*
Disimilars	*Hegemony*	*Pastiche*

Figure 17. Modes of Interactive Social Action

a wealth of sociological and sociolinguistic work,[32] the predisposition to accept public forms of discourse is itself emergent upon structuration that can be described in socioeconomic terms. As I have demonstrated elsewhere (in relation to school mathematics at least), public forms of discourse necessarily serve to recontextualise and transform and so subordinate private forms where the latter are introduced into the *public domains* of the former (Dowling, 1991b, 1995, 1996, 1998, 2001a). As the bureaucratized spokesperson of mathematicoscience the teacher may draw their students into their own game, but they will not solve any of the problems, address any of the concerns of their students insofar as these problems and concerns are constituted within localized, private discourses—and on suspects that most of them are. Essentially, school is a very bad place to learn anything beyond how to survive as a school student (or teacher).[33] Yet, knowing all of this, my erstwhile[34] mentor, Basil Bernstein had this to say in 1974:

> It is an accepted educational principle that we should work with what the child can offer: why don't we practice it? The introduction of the child to the universalistic meanings of public forms of thought is not compensatory education – it is education.
>
> (Bernstein, 1974, p. 199)

Thirty years and two Gulf "wars" on, you'd think we'd know better. But I fear not; *viva el Don*, it seems.

Acknowledgements

I am grateful to my doctoral students at the Institute of Education for very productive discussion – both individually, and in our fortnightly seminar – on particularly the theoretical issues raised in this chapter.

I am grateful to the International Association for the Evaluation of Educational Achievement (IEA) for permission to reproduce the images in Figures 1–14. These images are screenshots from the Trends in International Mathematics and Science Studies (TIMSS) and National Center for Educational Statistics (NCES) websites taken in September 2004; all of the test items are from TIMSS tests.

References

Baudrillard, J. (1995). *The gulf war did not take place*. Sydney: Power Publications.

Becker, B., & Wehner, J. (2001). Electronic networks and civil society: Reflections on structural changes in the public sphere. In C. Ess (Ed.), *Culture, technology, communication: Towards an intercultural global village* (pp. 65–85). Albany, NY: State University of New York Press.

[32] See, for example: Bernstein (1974, 1999), Bourdieu (1991), Bourdieu & Passeron (1977), Gee et al. (2001), Hasan (1999), Heath (1986), and Moss (2000) – though not all might concur with my formulation of their findings; see also Dowling (2004b).

[33] Cf. Lave & Wenger (1991).

[34] And, despite all, fondly and gratefully remembered – see Dowling (1999, 2001b).

Bernstein, B. B. (1996/2000). *Pedagogy, symbolic control and identity* (1st and Rev. eds). London: Taylor & Francis.

Bernstein, B. B. (1999). Vertical and horizontal discourse: An essay. *British Journal of Sociology of Education, 20*(2), 158–173.

Bernstein, B. B. (1974). *Class, codes and control, Volume I: Theoretical studies towards a sociology of language* (2nd ed.). London: Routledge & Kegan Paul.

Bourdieu, P. (1991). *Language and symbolic power*. Cambridge: Polity Press.

Bourdieu, P., & Passeron, J. -C. (1977). *Reproduction in education, society and culture*. London: Sage.

Brown, A. J., Bryman, A., & Dowling, P. C. (forthcoming). *Educational research methods*. Oxford: Oxford University Press.

Brown, A. J., & Dowling, P. C. (1998). *Doing research/Reading research: A mode of interrogation for education*. London: Falmer Press.

Burbules, N. C. (n.d.). *Web literacy: Theory and practice of reading and writing hypertext*. http://mroy.web.wesleyan.edu/webliteracy/linktropics.htm.

Cooper, B. (1983). On explaining change in school subjects. *British Journal of Sociology of Education, 4*(3), 207–222.

Cooper, B. (1985). *Renegotiating secondary school mathematics*. Lewes: Falmer.

Crotty, M. (1998). *The foundations of social research: Meaning and perspective in the research process*. London: Sage.

Crystal, D. (2003). *English as a global language*. Cambridge: Cambridge University Press.

Deleuze, G., & Guattari, F. (1984). *Anti-Oedipus: Capitalism and schizophrenia*. London: Athlone.

Donaldson, M. (1978). *Children's minds*. Glasgow: Fontana/Collins.

Dowling, P. C. (1990). The shogun's and other curricular voices. In P. C. Dowling & R. Noss (Eds.), *Mathematics versus the national curriculum*. Basingstoke: Falmer, 33–64.

Dowling, P. C. (1991a). A dialectics of determinism: Deconstructing information technology. In H. McKay, M. F. D. Young & J. Beynon (Eds.), *Understanding technology in education*. London: Falmer, 176–192,

Dowling, P. C. (1991b). The contextualising of mathematics: Towards a theoretical map. In M. Harris (Ed.), *Schools, mathematics and work*. London: Falmer.

Dowling, P. C. (1995). Discipline and mathematise: The myth of relevance in education. *Perspectives in Education, 16*(2), 209–226.

Dowling, P. C. (1996). A sociological analysis of school mathematics texts. *Educational Studies in Mathematics*. 31, 389–415.

Dowling, P. C. (1998). *The sociology of mathematics education: Mathematical myths/pedagogic texts*. London: Falmer.

Dowling, P. C. (1999, December). *Basil Bernstein in frame: Oh dear, is this a structuralist analysis*. Presented at the School of Education, Kings College, University of London. http://www.ioe.ac.uk/ccs/dowling/kings1999/index.html.

Dowling, P. C. (2001a). School mathematics in late modernity: Beyond myths and fragmentation. In B. Atweh, H. Forgasz, & B. Nebres, (Eds.), *Socio-cultural research on mathematics education: An international perspective*. Mahwah: Lawrence Erlbaum, 19–36.

Dowling, P. C. (2001b). Basil bernstein: Prophet, teacher, friend. In S. Power et al. (Ed.), *A tribute to basil Bernstein 1924–2000* (pp. 114–116). London: Institute of Education.

Dowling, P. C. (2004a). *Mythologising and organising*. http://homepage.mac.com/paulcdowling/mythologising/index.htm.

Dowling, P. C. (2004b). 'Mustard monuments and media: A pastiche.' Working paper based on 'Who Will Pay the HyperPiper' presentation at the Media Research Centre, Yonsei University, Seoul (October 2003). http://homepage.mac.com/paulcdowling/ioe/publications/mmm/index.htm.

Dowling, P. C. & Brown, A. J (forthcoming). *Doing research/reading research: Re-interrogating education*. London: Routledge.

Dowling, P. C., & Noss, R. (Eds.). (1990). *Mathematics versus the National curriculum*. London: Falmer.

Fish, S. (1995). *Professional correctness: Literary studies and political change*. Cambridge, MA: Harvard University Press.

Flude, M., & Hammer, M. (Eds.). (1990). *The education reform act 1988: Its origins and implications.* Basingstoke: Falmer.

Foucault, M. (1972). *The archaeology of knowledge*. London: Tavistock.

Gee, J. P., et al. (2001). Language, class, and identity: Teenagers fashioning themselves through language. *In Linguistics and Education, 12*(2), 175–194.

Hasan, R. (1999). The disempowerment game: Bourdieu and language in literacy. *Linguistics and Education, 10*(1), 25–87.

Hayles, N. K. (1999). *How we became posthuman: Virtual bodies in cybernetics, Literature and informatics*. Chicago: University of Chicago Press.

Heath, S. B. (1986). Questioning at home and at school: A comparative study. In M. Hammersley (Ed.), *Case studies in classroom research*. Milton Keynes: Open University Press.

Holland, E. W. (1999). *Deleuze and Guattari's Anti-Oedipus: Introduction to schozoanalysis*. London: Routledge.

Joyce, M. (1995). *Of two minds: Hypertext, pedagogy and poetics*. Ann Arbor: University of Michigan Press.

Kogan, M. (1978). *The politics of educational change*. Glasgow: Fontana/Collins.

Lave, J., & Wenger, E. (1991). *Situated learning: Legitimate peripheral participation.* Cambridge: Cambridge University Press.

Moon, B. (1986). *The 'New Maths' controversy: An international story.* Lewes: Falmer.

Moss, G. (2000). Informal literacies and pedagogic discourse. *Linguistics and Education, 11*(1), 47–64.

Shayer, M., Küchemann, D. W., & Wylam, H. (1992). The distribution of Piagetian stages of thinking in British middle and secondary school children. In L. Smith (Ed.), *Jean Piaget: Critical Assessments.* (Vol. 1). London: Routledge.

Sokal, A. (1996a). A physicist experiments with cultural studies. *Lingua Franca*, May/June 1996, pp. 62–64. http://www.physics.nyu.edu/faculty/sokal/lingua_franca_v4/lingua_franca_v4.html.

Sokal, A. (1996b). Transgressing the boundaries: Toward a transformative hermeneutics of quantum gravity. *Social Text, #46/47*, 217–252 (Spring/Summer 1996).

Sosnoski, J. (1999). Hyper-readers and their reading engines. In G. E. Hawisher & C. L. Selfe (Eds.), *Passions, pedagogies and 21st century technologies*. Logan: Utah State University Press.

Symonds, W.C. (2004, March 16). America's failure in science education. *Business week online*. http://www.businessweek.com/technology/content/mar2004/tc20040316_0601_tc166.htm.

Weber, M. (1964). *The theory of social and economic organization*. New York: The Free Press.

Weber, M. (1968). *Economy and society*. New York: Bedminster Press.

Wolf, L. (2002). An environment that encourages change. *IDB América*. http://www.iadb.org/idbamerica/index.cfm?&thisid=353&pagenum=2

11

THE POTENTIALITIES OF (ETHNO) MATHEMATICS EDUCATION: AN INTERVIEW WITH UBIRATAN D'AMBROSIO

Ubiratan D'Ambrosio
Maria do Carmo S. Domite
University of Sao Paulo, Brazil

Abstract: This chapter aims at deepening and founding questions on ethnomathematics in terms of its sociohistorical construction and its influence on the extent of the relation between mathematics education and school knowledge, mathematics education and history of mathematics, education processes and teacher education, pedagogical practice and different sociocultural contexts. The dynamic of the organization of the text, a transforming experience by itself, is a conversation between the authors, beginning with something close to an interview with Professor D'Ambrosio, developing into a teamwork, effectively in the form of a dialogue, even though the authors differ in their history and the knowledge they have amassed in the field of ethnomathematics

Keywords: Ethnomathematics, History of Mathematics Education, Brazil

Domite: I am sure you can imagine how important this conversation is to me, both personally and professionally. I have learned a great deal from you not only about mathematics education, but also about life, so it is with admiration and gratitude that I begin our conversation.
D'Ambrosio: It is a privilege to have this conversation with you. Since the early moments when the ideas behind ethnomathematics were taking shape, the exchanges with you were pleasant and inspiring and I feel now a great identification of viewpoints with you.
Domite: As this interview will be published as a book intended to share mathematics educators' ideas worldwide, I would like to point out your interest in mathematics and mathematics education and ask you to tell us something about your background.

B. Atweh et al. (eds.), Internationalisation and Globalisation in Mathematics and Science Education, 199–208.

D'Ambrosio: Let's talk about my career first and how I got involved with mathematics education. I was born in 1932. In 1949 I was already working as a tutor for people preparing to enter public services in Brazil (mainly teaching Financial Mathematics). I graduated in 1954, with a major in Mathematics (Pure) and taught for some years in high schools. In 1958 I was hired as a full-time instructor and graduate student at the University of São Paulo, USP, Brazil and received my doctorate in 1963, with a dissertation on Calculus of Variations and Measure Theory (very pure!).
Domite: But you also worked in the USA for some years, didn't you?
D'Ambrosio: In 1964 I went to the USA as a research associate at Brown University for one year, but due to the political events in Brazil, I stayed there and became a tenured professor at the State University of New York at Buffalo where I had my first PhD candidate. He wrote his dissertation on Stability of Differential Equations.
During that time, my interest in education was occasional and superficial. In 1972, I returned to Brazil and became the director of the Institute of Mathematics, Statistics and Computer Science of the State University of Campinas (UNICAMP), which became a few years later a major research institution. My first Brazilian doctoral candidate in Campinas wrote a dissertation on Measure Theory and Minimal Surfaces.
Domite: And when did you realize the potentialities of mathematics education, especially its political and social aspect?
D'Ambrosio: From this period onwards I began to realize that mathematics education should be a priority for Brazil. I was motivated by the cultural and social barriers which were responsible for the failing and dropping out of children coming from marginalized groups. They could not compete with children coming from families with better schooling. At the same time, I developed an interest in the history of mathematics and in broader transcultural and transdisciplinarian theories of knowledge. This is my background.
Domite: You are an acknowledged authority in mathematical education and a philosopher of education. You have been active in furthering different movements in mathematics education, predominantly the cultural history of mathematics, ethnomathematics and curriculum policy making. What have been the goals and the focus of your work in recent years? Did something change?
D'Ambrosio: My current concerns about research and practice in math education fit into my broader interest in the human condition as related to the history of natural evolution – from the cosmos to the future of the human species – and to the history of ideas. Particularly, the history of explanations of creation and natural evolution. In the past years – surely much before the last five – my motivation has been the pursuit of peace in all four dimensions: individual, social, environmental and military. I attribute the violations of peace, in all these four dimensions, to the mistaken course of Western civilization. I try to understand the founding myths of Western civilization, and this links to my research on the history of monotheistic religions (Judaism, Christianity, Islam), of techniques, of arts and of how mathematics permeates all this. A great support is gained by looking into non-Western civilizations. I base my research on established forms of knowledge

(communication, languages, religion, arts, techniques, sciences, mathematics) and in a theory of knowledge and behavior which I call the "cycle of knowledge". This theoretical approach recognizes the cultural dynamics of the encounters, based on what I call the "basin metaphor".

Domite: I am delighted to see your vision of mathematics education. It allows us to understand the cultural roots of other social and ethnic groups, as well as a tool that can be used to stimulate or prevent wars. Please, tell us more about those links.

D'Ambrosio: This all links to the historical and epistemological dimensions of the Program Ethnomathematics. It can bring new light into our understanding of how mathematical ideas are generated and how they evolved through the history of mankind.

It is fundamental to acknowledge the contributions of other cultures and the relevance of the dynamics of cultural encounters. Here "culture" is understood in its wider form and includes art, history, languages, literature, medicine, music, philosophy, religion and science. Research in ethnomathematics is necessarily transcultural and transdisciplinary. The encounters are examined in their widest form, to permit exploration of more indirect interactions and influences, and to permit examination of subjects on a comparative basis.

Academic mathematics developed in the Mediterranean basin, expanded to Northern Europe and later to other parts of the World. Nevertheless the codes and techniques to express and communicate the reflections on space, time, classification, comparison – which are proper to the human species – are inherent to the context. Among these codes are measuring, quantifying, inferring and the emergence of abstract thinking.

Domite: I believe we share the same idea that different mathematical relationships and practices can be generated, organized, and transmitted informally to solve immediate needs, as occurs with language. This way mathematics is incorporated into the core of the learning-by-doing processes of the community and thus mathematics is part of what we call culture. From this standpoint, ethnomathematics is concerned not only with the cultural roots of mathematical knowledge, but also quantitative and spatial relationships generated within the cultural community, which often compose mathematics as well.

D'Ambrosio: I agree entirely with you about the way you consider ethnomathematics. Your way of phrasing it summarizes my own trajectory to what I call the Program Ethnomathematics. There has been a controversy in naming it the "Program Ethnomathematics". This name is a way I found to avoid considering ethnomathematics a discipline. The risk of trying to have an ethnomathematics curriculum and to tie ethnomathematics to some rules may distort its transdisciplinary and transcultural character. The wording "Program Ethnomathematics" gives emphasis to this character. It is permanently in process.

Domite: I have noticed you changed your vision on ethnomathematics in favor of a Program Ethnomathematics in recent years. This has had an impact on many substantial ideas in terms of directions to several studies and courses for teachers. Could you talk a little more about this?

D'Ambrosio: At this moment, it is important to clarify that my view of ethnomathematics should not be confused with *ethnic* mathematics, as it is understood by many. This is the reason why I insist on using Program Ethnomathematics. This program tries to explain mathematics, as it tries to explain religion, culinary, dressing, football and several other practical and abstract manifestations of the human species.

Domite: But of course, acknowledging ideas had a role. I mean, initially they inspired the Program Ethnomathematics, didn't they?

D'Ambrosio: Of course, and the ways of doing that reminds us of Western mathematics. What we call mathematics in academia is a Western construct. Although dealing with space, time, classification, comparison, which are proper to the human species, the codes and techniques to express and communicate the reflections on these behaviors are undeniably contextual.

I got an insight into this general approach while visiting other cultural environments. I worked in Africa, in many countries of continental America and the Caribbean, and in some European environments. Later, I tried to understand the situation in Asia and Oceania, although I had no field work there. Thus my approach to cultural anthropology came into being. (Curiously, my first book on ethnomathematics was placed by the publishers in a collection of Anthropology).

To express these ideas, which I call a research program (maybe inspired by Lakatos?), I created a neologism, ethno-mathematics. This gave rise to much criticism, because it does not reflect the etymology of "mathematics". Indeed, the "mathema" root in the word ethnomathematics has little to do with "mathematics" (which is a neologism introduced in the XIV century). The idea of organizing these reflections occurred to me while attending ICM 78 in Helsinki. I played with Finnish dictionaries and was tempted to write *alustapasivistykselitys* for the research program. Bizarre! I believed "ethnomathematics" would be more palatable.

Domite: You have insisted that the Program Ethnomathematics is not ethnic mathematics, as some commentators interpret it. Nevertheless, the mathematics educators involved with ethnomathematics studies work with different cultural environments and work as ethnographers, trying to describe mathematical ideas and practices in other cultures. Is this a style of doing ethnomathematics or is this absolutely necessary?

D'Ambrosio: I would say that it is both a style of doing ethnomathematics and necessary for addressing the questions related to social-cultural studies. But these cultural environments include indigenous populations, labor and artisan groups and marginalized communities in urban environments, farms and professional groups. They develop their own practices, have specific jargons and theorize on their ideas. This is an important element for the development of the Program Ethnomathematics, as important as the cycle of knowledge and the acknowledgement of the cultural encounters.

More recently, I have worked with the preparation of teachers for indigenous communities in the state of São Paulo. Altogether, there are about 100 small tribes (some with different languages), totaling only 8,000 individuals at most.

We are now starting a project on the ethnomathematics of the "quilombolas", which are small communities originated from slaves who fled from the farms in the 17th and 18th centuries and established themselves as small republics in the hinterland of Brazil. The research resulting from all these projects feeds the Program Ethnomathematics.

Domite: In your talk to the state of São Paulo indigenous teachers I could see our mathematics in the context of power. It helps understand what you have always said about the social context discrediting indigenous groups as educated. – This keeps the indigenous groups' crafts from being fully acknowledged as praxis, and limits their empowerment.

D'Ambrosio: And in the face of this, it is difficult for us involved with school education to realize the complexity and richness of the indigenous group's relationship to mathematics. And I go further. It is difficult to understand that a social group or an ethnic group acts with a culture of its own. They have their own jargon, their code of behavior, their values and expectations. Unfortunately, there is not much attention given to this when dealing with school children. And I understand that there are immediate questions facing world societies and education, particularly mathematics education. As a mathematics educator, I address these questions. Hence my links to the study of curriculum, and my proposal for a modern *trivium*: literacy, matheracy and technoracy. This is one of the main lines of work in mathematics education.

Domite: In recent years you have emphasized this modern trivium and the pursuit of peace as urgent needs. I understand that from your point of view both are related to ethical, political and economic issues. What do you say about this?

D'Ambrosio: I agree entirely with you. Both explain my works in Mathematics and Ethics and Mathematics and Citizenship, which have been the themes of several lectures and courses for in-service teachers.

I have published studies on curriculum and oriented several doctoral dissertations with this focus in mind. And these two trends/lines in my current work in mathematics education link naturally to the pedagogical and social dimensions of the Program Ethnomathematics. Of course, they are related and they are long-term concerns. Thus, my links with Future Studies. I have been active – publishing and lecturing – on history and philosophy focusing on the future.

Domite: Besides these important concerns, what are the long term concerns related to them for developing mathematics and science – of course with implications on school mathematics and science?

D'Ambrosio: I see the special role of technology in the human species and its implications in mathematics and science and school mathematics and science. Thus I focus on the history of science (and, of course, of mathematics) trying to understand the role of technology as a consequence of science, but also as an essential element for furthering scientific ideas and theories. Basically, my investigation is geared towards three basic questions: "How are *ad hoc* practices and solution of problems developed into methods?", "How are methods developed into theories?" and "How are theories developed into scientific invention?". Once we've recognized the role of technology in the development of mathematics, reflections about the future of mathematics raise important questions about the role of technology in mathematics education. Hence my line of work in distance education.

Domite: Your vision of the role of technology in modern civilization is related to a variety of emerging fields of knowledge, like Artificial Intelligence and others, especially ones that are different from the current common view. I understand that such personal views are embedded in the broader domain of your type of studies.

D'Ambrosio: Reflections about the presence of technology in modern civilization lead, naturally, to the question about the future of our species. My growing interest in the emerging fields of Primatology and Artificial Intelligence, leads to a reflection about the future of the human species.

Cybernetics and human consciousness lead, naturally, to reflections about cyborgs (a kind of "new" species, *i.e.*, humans with enormous inbuilt technological dependence). Our children will be cyborgs when, around 2025, they become decision-makers and take charge of all societal affairs. Educating these future cyborgs calls, necessarily, for much broader concepts of learning and teaching.

The role of mathematics in the future is undeniable. But what kind of mathematics? Once again, I look for explanations in history. To understand how, historically, societies absorb innovation, is greatly aided by my involvement with world fiction literature, from iconography to written fiction, music and cinema.

I feel it is important to understand the way material and intellectual innovations permeate thinking and myths, and the ways of knowing and doing of non-initiated people. In a sense, how new ideas vulgarize – vulgarize here understood as making abstruse theories and artifacts easier to understand in a popular way. This places me on the side of "post-modernists" in the current science wars.

Domite: Surely this theoretical framework, being different from the various models of thinking, provides a forum for critical academic discussions.

D'Ambrosio: Although this is a difficult position to defend in academic discussions, particularly as a math educator and historian, this summarizes my involvement with current philosophy. The way communities give a character of sacred to space and time has been central to my thinking. One of my students worked for his doctorate on concepts of space and time in a modern popular urbanization program in the Amazon basin. Another is dealing with how the concept of space permeates the medieval emergence of non-Euclidean geometry. And still another student is looking into modern art and mathematics.

I am now working on a paper dealing with the relations of ethnomathematics, ethnomethodology and phenomenology. I look into their evolution and try to identify a common underlying philosophical framework, through a transcultural and transdisciplinary approach.

Domite: Going back to school system subjects: by looking at the development of teacher education and curricula in Brazil, we have observed that even if the foundations for renewal have demonstrated an attempt at relating societal changes – including changes in the school system and general views on teaching and learning – when it comes to the content and its organization within teacher programs, the didactic division between disciplinary and pedagogical knowledge has been a major concern.

We could see in the committee work underlying the making of the Brazilian Curricula Parameters, a regulative framework from the 1990s. Besides the view of teachers' work across school levels and areas expressed, there are attempts at discussing the integration of mathematics to real world situations and other school areas of studies (by looking into Transversal Themes) but the major attempts were just to integrate the traditional four strands of teacher education: subject theory, pedagogy, methods and contents.

My point here is the necessity of creating some substantial links between mathematical ideas and personal/cultural situations – something in the context of ethnomathematics and pedagogy, in the sense of taking into account the students' cultural/social aspects that may positively contribute to their school performance.

D'Ambrosio: In most Latin American countries, curricula are shaped more or less homogeneously all over the country, the major cities serving as models. In these, largely urban populations have expectations with respect to schools either as a step towards social access, hence aiming at college education, or as a need to fulfill basic legal requirements for lower middle class employment in the public services, commerce and banks, among others. For these, it is rarely required to have more than primary education, and post-primary standards are minimal. On the other hand, those aiming at college level degrees look for an education that allows them to pass highly competitive entrance examinations to get into the universities. The program is strongly dictated by what is required in these exams. Secondary education is dominated by these requirements, which are classical and based on training and drilling choice testing.

Domite: But in looking into what is done to change this situation in Brazil, we can find some groups dealing with real life situation and mathematics applications.

D'Ambrosio: Yes, a few examples of attempts to introduce more lively and creative programs can be found. These are isolated cases, and cover interesting applications of the most varied nature. In particular, there are projects that bring the concept of modeling into the very early years of schooling, dealing with real life situations. Problems dealt with in the primary level are, for example, the construction of scale models of houses and cars, among others. The novelty is that "theory" goes together with "doing". Measures are taken, reduction takes place, material is bought and an object is a final product. This goes more in the line of a "project", together with "theoretical" reflections in every step of the process. But these projects are restricted to a few research groups and special training centers and can hardly be seen and known in most of the countries. At the secondary level, the challenge seems to be to make creative and interesting mathematics compatible with what is required on the entrance examination to the universities.

Domite: It seems that the big change lies in teacher education, or in the education of teacher educators, in terms of helping mathematics teachers to be up-to-date with recent developments amid colleges that think the same way.

D'Ambrosio: Yes, I agree. More effective efforts must be concentrated at the graduate level, to prepare educators of teachers. A considerable number of seminars and workshops on applied mathematics should be held for college teachers that are in charge of training prospective primary and secondary school teachers. These can be urged by observing that because of the highly traditional curricula in the teacher training courses, it is unlikely that teachers will be able to stimulate other than routine and trivial applications in their classrooms.
But, as you were looking at, the most effective example of mathematics related to the environment, both socio-cultural and material environment, can be found in ethnomathematics. Although as yet unrecognized, ethnomathematics is being practiced by uneducated people, sometimes even illiterates. Some effort is going on at the research level to identify those ethnomathematics practices and to incorporate them into the curriculum. But there is much ground to cover before the pedagogisation of ethnomathematics takes place – and even before ethnomathematics becomes recognized as valid mathematics.
Domite: To what extent was your work influenced by the current state of mathematics education in your country and/or in the world? Let us talk more specifically with your vision of mathematics education in our country.
D'Ambrosio: The situation of Brazil does not differ much from many other countries that have an intense population dynamics, as for example, the USA. These are countries where much progress is noticed. Never in Brazil have the publishing houses flourished so much, and never was the industrial and agricultural production so intense. This contradicts official assertions, based on the results of tests and national exams, that the school system is a great failure. According to official sources, Brazil is in a terrible shape in reading, writing and mathematics. Any system, with such results, would have been declared irremediably bankrupt.
Domite: This increases our responsibility as educators, parents and administrators, especially as mathematics educators, for mathematics has all to do with the current scenario, does it not?
D'Ambrosio: I am much worried about the cultural dynamics of the encounter of generations (parents, teachers and youth). This encounter is dominated by mistrust and cooptation, reflected in testing and evaluation practices, which dominate our civilization. I use the word cooptation in the sense of attracting to our side, beginning to share our ideas and acting according to our interests. In same way giving up your self! In mathematics education, this is particularly disastrous. Paradoxically, the voices of individual math educators are against this, but the practice of the totality insists on sameness. The result is teaching mathematics in an uninteresting, obsolete and useless way.
Domite: I feel that our role as educators is still very insipient and fragile in order to create more relevant demands for the mathematics curricula proposals. How do you see that?
D'Ambrosio: Yes, our claims of relevance of current math curricula are fragile. Myths surround these claims. From my opinion this fragility is related to evaluation, since resources for testing are the main argument to justify current math curricula.

These have been, since the 60s, the main motivation for my thoughts about math education. In the last years, this has been intensified by my analysis of results of testing in Brazil and elsewhere in the world.

I try to understand the behavior of children and youth and their expectations. History gives us hints on how periods of great changes affect the relationships between generations. Most interesting is the analysis of movements after World War II and the Vietnam war.

Synthesizing, there is more concern with mathematics than with our children. In general, education is dominated by a kind of "corporativist" expectations for the future (this is why education leads to the re-emergence of fundamentalism and fascism) and, regrettably, mathematics and mathematics education have everything to do with this. Tests may be the best instruments to allow this.

Domite: Do you think your work and ideas in terms of culture, society, history, power and peace as related to mathematics education over the last years have impacted on the daily practice and research of mathematics education?

D'Ambrosio: Yes, I believe in all I say and do, and I speak loud about this. This is reflected in the research done by my colleagues and students. They result in a very high level of academic work, leading to different approaches to math education, and realistic enough to get accepted by the system. I learned from a science educator colleague in the early 70sthat changes in education occur by "stealth". I have confirmation of this. When I visit schools, invited by colleagues and former students and I have an opportunity to talk with students, I see how effective the "stealth influence" is.

Although unwritten, unreported, unnoticed by the authorities, the innovation benefits enormously some students. Others do not even notice this, and prepare for testing. What is wrong with testing? Basically, because it penalizes the former, who are creative, and favors the latter, who are co-optable and amenable to sameness. Homogeneous results are unnatural. But this is the supporting argument for standardized testing. (Standardization prevails in evaluation, although it is denied by the theoreticians supporting current testing.)

Domite: Much has been written about globalization and education but only few attempts have been made to conceptualize or develop a framework on globalization in terms of cultural knowledge, to explore the impact on education. For instance, the Latin American countries went through colonization and immigration processes, receiving influences from different cultures, but very little has been discussed about the knowledge produced in its heart, especially among those that constitute the political minorities – Amerindians, blacks and peasants, among other cultural groups. What would you say about this?

D'Ambrosio: The Latin American reality is a source of diversity by itself, with its enormous variety in cultures, linguistics, ethnicity, literature and peoples in addition to different beliefs, ways of life, music, plastic arts, body expression, ideas, dreams and utopias. The reality expresses the cultural plurality of peoples; indigenous and mixed races and immigrants from territories of the whole planet – urban inhabitants as well as peasants – which conform to the multicultural panorama that constitutes our continent. The continent also experiences and resolves in different fashions the tensions and contradictions that are generated in the heart of the ethnic-cultural groups (identity, linguistic varieties, regional and political beliefs), among the groups of a country or between these and the global society.

Domite: The more I read about globalization as a background for discussion on education the more confused I am on the objectives of this discussion. We can say, somehow, that the definition or the implications of globalization for education and the work of educators mean the process of integration of world education systems. There is, maybe, a search for acknowledgement for international curricula for education systems, standard systems for certification as well as professional and academic skills. What are the difficulties faced by ethnomathematics in this context?

D'Ambrosio: In the context of globalization, the diversity is more valued than economy in the different productions of the human being production of knowledge and of its dreams of humanization. The marketplace, the competition, the individualism, the exploration of work, the accumulation, the unilaterality in the relations of international power, the predatory submission of the environment, the strengthening of the controlling systems, the concentration of the economic or political power are components of this hegemonic model. Everything in opposition to our ideal of interculturality as a proposal for human interaction.

Domite: Finally, I would like to thank you and stress that we all know that the task of mathematical education is a constant process. However, so that the changes lead to a more effective practice – in terms of attraction and understanding for children – as well as the responsibility in light of social, political and cultural issues, it is paramount that we have more educators such as you, people who do not just "go with the flow" impregnated by ideological components inherent to education in general, but propose transformative actual models. Thank you so much.

12

ETHNOMATHEMATICS IN THE GLOBAL EPISTEME: QUO VADIS?

Ferdinand Rivera and Joanne Rossi Becker
San José State University, USA

Abstract: This chapter discusses scholarly work in the field of ethnomathematics from three perspectives that seem to encompass much of the current work in the field: challenging Eurocentrism in mathematics; ethnomathematics praxis in the curriculum; and ethnomathematics as a field of research. We identify what we perceive to be strengths and weaknesses of these three perspectives for today's learners who are faced with forces of a global nature. We propose a less traditional view of ethnomathematics that is compatible with postnational, global identities, and exemplify this approach through a professional development program in California. Finally, we raise several issues for future discussions relative to ethnomathematical theory and practice

Keywords: ethnomathematics

According to Habermas (2001), globalisation is still in its emergent state. Currently, we witness various physical and nonmaterial changes in our societies as a consequence of "the increasing scope and intensity of commercial, communicative, and exchange relations beyond national borders" (Habermas, 2001, p. 66). Giddens (1999) also makes sense when he insists that no one group can claim ownership to all the various global forces that are currently influencing the emerging social landscape. As a matter of fact, control takes place at the level of networks that enable globalisation to maintain its multidimensional character. Our intent in this chapter is to confront conceptual and practical difficulties with ethnomathematics and its nuances (herein collectively referred to as "the ethnomathematics program") so that their strengths are articulated and their limitations are surfaced and overcome. Today's learners, irrespective of community and affiliation, are living out the tensions brought about by the reality of globalisation. This social condition implies that various operations, transactions, and interactions that are currently taking place employ disciplinary relations that are not state-specific in the classical sense. They

B. Atweh et al. (eds.), Internationalisation and Globalisation in Mathematics and Science Education, 209–225.

are increasingly performed within a distinctively post-state perspective that has been forged by cosmopolitan solidarity (Habermas, 1998). It is a solidarity that seems to have traversed particular cultures and social filiations or groups and, at the same time, has successfully reconciled the specificity of cultural practices with the generality and universality of lived relations across cultures.

As social theorists of difference, we see some ironies and contradictions that are developing between global and multicultural societies insofar as cultural identities matter. *At the local stage*, immigration has tremendously changed the landscape of nation-states. All prosperous nations that deal with migrants in large numbers experience unanticipated transformations in their societies. Habermas (2001) points out that the "path toward a multicultural society" is a challenge for these nation-states that are confronted with the plurality of lived relationships. A significant issue in education in these multicultural contexts is how to develop good practices of inclusion. Here we note that if by inclusion we mean "a collective political existence [that] keeps itself open for the inclusion of citizens of every background, without enclosing these others into the uniformity of a homogenous community" (ibid, p. 73), what remains unresolved to this day deals with processes and mechanisms that can be effectively institutionalised in schools and in the wider communities so that a more meaningful, harmonious, and productive political integration of different relationships is achieved. According to Habermas,

> (m)ulticultural societies require a "politics of recognition" because the identity of each individual citizen is woven together with collective identities, and must be stabilized in a network of mutual recognition. (Habermas, 2001, p. 74)

Thus, inclusive practices must take into account ways in which different cultural communities with their particular shared traditions and practices can be made to co-exist so that the practices do not produce difficult situations of subcultural formation and marginalisation.

At the global stage, people from around the world develop a shared need or a mass culture for goods, fashion, films, programs, music, books, and other forms of aesthetic expression. The Western influence seems to have produced, Habermas writes,

> [a] "commodified, homogenous culture [that] doesn't just impose itself on distant lands, of course; in the West, too, it levels out even the strongest national differences, and weakens even the strongest local traditions. (Habermas, 2001, p. 75)

Thus, while some critical commentators have pointed out how global forces are driving indigenous cultures to states of moribundity, irrelevance, and homogenisation, they are, as a matter of fact, producing new constellations, new differences, new worldviews, or cosmopolitan identities that celebrate "a new multiplicity of hybridised forms" (ibid, p. 75). In effect, hybridity promotes "new modes of belonging[ness and] new subcultures and lifestyles [that involve] a process [that is] kept in motion through intercultural contacts and multiethnic connections" (p.75). What we now perceive to be the most significant problematic in our schools that are situated in the global cultural economy is not one of inclusion in the worried ethnocentric sense. Rather, it involves finding ways of dealing with new collective experiences, including processes that encourage new individual experiences (ibid, p. 76; Rivera, 2004) and that operate within a sensibility that is compatible with new solidarities or cosmopolitan structures (Habermas, 1998) brought about by emerging global identities.

The chapter is divided into four sections. In section 1, we characterize important aspects of the global episteme that bear on ethnomathematical practices. In section 2, we identify and discuss with some depth three prevailing perspectives (i.e., theory, practice, and research) raised about ethnomathematics. Following Hardt and Negri (2000), since we believe that the construction of a conceptual program is both an epistemological and ontological project – in the sense that the production of knowledge and the construction and deployment of reality are mutually constitutive – we articulate what we perceive to be strengths and weaknesses of the various perspectives that have been proposed and developed about the ethnomathematics program. In section 3, we discuss problems with both the theory and practice of ethnomathematics. Also, we propose a less traditional view of ethnomathematics and propose a hybrid version that is compatible with postnational, global identities. In this section, we draw on insights and tellings from the English Language Development Institute in Algebra, a grant-funded professional development program for in-service mathematics teachers of minority students in California. In Section 4, the conclusion, we raise several issues that are worth considering in future discussions involving ethnomathematical theory and practice.

1. The Global Order Of Things

Hardt and Negri (2000) claim that in the now that is the postmodern, a global concept rules by the name of Empire. Empire deploys a new form of logic that has emerged as a consequence of the globalisation of economic and cultural relations. Further, it produces new modes and conditions of social production. Empire draws its strength from being in control of global capital that is run mainly by networks of transnational corporations and united national and supranational organisms. Networks function around a world market that continues to threaten boundaries and limits imposed by individual nation-states. At the very least, the Empire is "both system and hierarchy, centralized construction of norms and far-reaching production of legitimacy, spread out over world space" (p. 13). Thus, the global market is the site whereby certain binary divisions, generated mostly by nation-states, can no longer be justified since the "new free space" harbours "a myriad of differences" (p. 151) and certain forms of hybridity that enable the market to stay fluid and flexible. Henceforth, individual citizens who live in particular locations witness the decline of the power of their respective countries when confronted with the decentred and deterritorialising rule of the Empire.

At least from the perspective of those of us that are privileged to live in affluent societies and benefit from membership in the top ladder in the global order, today's nation-states have been and, in some cases are being, phased into the postmodern episteme. Hardt and Negri point out that both modernisation and industrialisation represent one and the same economic phenomenon, and that the transition to postmodernisation marks a shift towards an informational economy that "emphasise[s] different kinds of service and different relations between services and manufacturing" (p. 286). Needless to say, such a postmodern condition causes the development of "new mode(s) of becoming human" because advances in "cybernetic intelligence of information and communication technologies" change the manner in which labour is performed in the new global order (p. 289). For instance, individuals are now forced to perform "immaterial labour" by way of manipulating and producing information and knowledge more intensely than ever before. Progress in technological tools has also modified social dynamics as advances in cybernetics have been successful in abstracting important aspects of material, concrete, physical, and bodily labour ("abstract labour") resulting in the further deskilling of work and encouraging "abstract cooperation" in virtual contexts. Consequently, new relations in the division of labour have also taken shape, between creative individuals who are capable of "symbolic-analytic services," that is, "problem-solving, problem-identifying, and strategic brokering activities" (Reich, 1991), and those who can (merely) perform "routine symbolic manipulation" such as data entry and word processing (ibid.).

Analysing the history of the nation-state, Habermas (1998) traces its origin from attempts to organize individuals and communities at a time when the old European feudal order was being phased out, and that nation-states emerged in the period of modernisation and democratisation. For Habermas, traditional notions associated with the nation-state are becoming irrelevant in the global order. Nationality in the usual sense as pertaining to "ethnicity, a common language, or a shared history" (Cronin & De Greiff, 1998, p. xxii) is now being disputed in favour of republicanism that is "founded on the ideals of voluntary association and universal human rights" (ibid.). Further, while loyalties and kinships played an important role in forming national identities in early history, politics and legal institutions also contributed significantly to the constructive process. Thus, a distinction has to be made "between a *civic* and *ethnic* sense of the nation" and "between a *political* and a *majority* culture" (Cronin & De Greiff, 1998, p. xxiii). Cronin and De Greiff capture the differential essence astutely in the following manner:

> Citizens do not have to agree on a mutually acceptable set of cultural practices but must come to a more modest thought still demanding agreement concerning abstract constitutional principles. As with national identity within pluralistic states, Habermas thinks that a supranational identity might evolve around an agreement about political principles and procedures rather than about culture more generally. (Cronin & De Greiff, 1998, p. xxiv)

In summary, the emerging global condition influences the cultural experiences and, consequently, the mathematical education of learners in ways no one can easily predict. As Giddens (1999) has clearly emphasized, we are only at the initial stage of the globalisation process, "at the beginning of a fundamental shake-out of world society, which comes from numerous sources, not from a single source." For some, globalisation is seen in positive terms, while for others, it carries with it some negative elements. A negative instance is worth discussing briefly. While international studies in areas such as school science and mathematics (for example, the Third International Mathematics and Science Study (TIMSS), the International Evaluation of Educational Achievement) drive schools from around the globe to attempt to develop a quality and competitive curriculum, they also create new situations and other externally-induced conditions, such as learned helplessness and relative deprivation, that affect the nature and context of their learners' educational experiences. The TIMSS, including various state-funded examinations that have been informed by results from international assessments, seem to put pressure on schools to develop uniform, standardized, and homogenizing practices without considering their effects on particular cultures. Addressing a positive instance, globalisation has provided the impetus for increased democratisation of life in many countries and, thus, has permitted discussions involving gender, race, and equity, in general. Suffice it to say, globalisation allows individuals to produce new ways of reworking their identities, enabling them to "revolt against traditional forms and styles" and "to create new, more emancipatory ones" (Cvetkovich & Kellner, 1997, p. 10). This observation needs to be articulated considering the fact that many learners from particular cultures show a tendency to value practices other than what their own cultures allow or suggest for them. We tend to view them as comprising the new group of cosmopolites (in Habermas's sense) that value global skills necessary for accomplishing global innovations and activities (Carnoy, 1998).

2. Three Prevailing Perspectives On Ethnomathematics

Ethnomathematics as a field of study has a number of definitions and interpretations. It has evolved significantly from the early, rather narrow definition of Marcia Ascher and Robert Ascher (1997) as "the study of mathematical ideas of non-literate peoples" (p. 26). Powell & Frankenstein (1997) use a broader definition provided by D'Ambrosio, a Brazilian mathematician and mathematics educator whom many consider the intellectual progenitor of the field, that is, ethnomathematics as the mathematics in which all cultural groups engage (D'Ambrosio, 1985). For D'Ambrosio, each group, including "national tribal societies, labour groups, children of a certain age bracket" (pp. 16) has its own mathematics, in contrast to the academic mathematics that is taught in schools. From D'Ambrosio's perspective, ethnomathematics exists at the convergence of the history of mathematics and cultural anthropology.

Eglash (1997) provides a more comprehensive characterization of ethnomathematics. Ethnomathematics is the mathematics of "small-scale or indigenous cultures" (p. 79). It is distinguished from: *non-Western mathematics* (with a focus on contributions from "state empires such as the ancient Chinese, Hindu, and Muslim civilizations" (p. 80) that have developed mathematical methods and theories similar to those of Western mathematics); *mathematical anthropology* (with a focus on "material and cognitive patterns" that are the "structural basis of underlying social forces, or as epiphenomena resulting unintentionally from the nature of the activity itself" (ibid.)); *sociology of mathematics* (with a focus on how mathematics itself is seen as a social construction resulting from the work of professional mathematicians, including the community that validates certain practices), and; *vernacular mathematics* (with a focus on street, situated, folk, informal, and non-standard mathematical practices of individuals that appear not to fall under any of the above categories). An ethnomathematical program strives to see how the mathematical practices and/or social or everyday patterns of minority cultural groups can be shown to be similar or as rigorous and sophisticated as those that have been developed in both Western and non-Western traditions. Further, such practices are not necessarily primitive (i.e., concrete and drawn from nature) and pure (i.e., unsullied by influences from other cultures).

But from its beginnings ethnomathematics has had a decidedly political stance that is not apparent in these definitions. We discuss scholarly work in the field of ethnomathematics from three perspectives that seem to encompass much of the current work in the field: challenging Eurocentrism in mathematics; ethnomathematics praxis in the curriculum; and ethnomathematics as a field of research. By focusing on these conceptions of ethnomathematics, we do not imply discrete categories of work; in fact, various contributions often fit into more than one category. But the categorization does help sort the major points of view represented in the literature.

2.1 Critiques of Eurocentrism

One of the themes of ethnomathematical scholarship is a critique of prevailing views of the history of mathematics as frequently represented as a two-stage development in which the Greeks ($\approx$600 BC to 300 AD) and post-Renaissance Europe and Europeanised countries like the US (16th century to present) were primarily responsible for the development of mathematics. For example, Joseph (1997, 1993) provides an alternative look at the Dark Ages by highlighting the role of Arabs in the history of mathematics, arguing that an Arab renaissance in mathematics between the 8th and 12th centuries provided for a flow of mathematical knowledge *into* western Europe that helped shape the pace of developments for the next five hundred years. Joseph (1997) also stresses that most of the topics taught in school mathematics today are derived from outside Western Europe before the 15th century. So one purpose of this perspective of ethnomathematics is to challenge the Eurocentric foundations of mathematics that ethnomathematics scholars find in many historical treatments of the subject (see, for example, Powell & Frankenstein, 1997).

While colonialism played a critical role in denying the contributions of Arabs and other non-European people of colour to the development of mathematics, the ideology of European superiority arose as an outcome of European political control over vast areas of Africa and Asia. "The contributions of the colonized were ignored or devalued as part of the rationale for subjugation and dominance" (Joseph, 1997, pp. 63). As Walkerdine (1997) points out, the European aristocratic male became the model to which others were compared; all others became inferior. By analysing the mathematics of traditional cultures or others marginalized in mathematics, such as women, scholars have attempted to provide some balance into the historical record (e.g. Gerdes, 1997; Gilmer, 2001 ; Hancock, 2001; Harris, 1997; Zaslavsky, 1973).

2.2 *Ethnomathematics Praxis in the Classroom*

This perspective on ethnomathematics has perhaps engendered the most controversy recently (Adam, Alangui & Barton, 2003; Rowlands & Carson, 2002; Vithal & Skovsmose, 1997). The main goals of proponents of an ethnomathematical approach to curriculum are: to reveal to students the role that mathematics has played throughout human civilization (Gerdes, 1997); to validate students' lived experiences and culture (Zaslavsky, 1997); to capitalize on students' interests and knowledge (Borba, 1997); and to empower students to understand power and oppression more critically (Powell & Frankenstein, 1997). The ultimate aim of an ethnomathematics praxis in the classroom is one of equity. What might such curricular approaches look like?

In their critique of ethnomathematics, Rowlands and Carson (2002) pose four possibilities for an ethnomathematics curriculum and its role relative to formal academic mathematics: replacement for academic mathematics; supplement to academic mathematics; springboard for academic mathematics; or, motivation for academic mathematics. It is clear that supporters of ethnomathematics are promoting much more than cultural adjuncts to lessons: "However, we also stress that we are not advocating the curricular use of other people's ethnomathematical knowledge in a simplistic way, as a kind of 'folkloristic' five-minute introduction to the 'real' mathematics lesson" (Powell & Frankenstein, 1997, p. 254). In their response to Rowlands and Carson, Adam, Alangui, and Barton (2003) propose an "integration of the mathematical concepts and practices originating in the learners' culture with those of conventional academic mathematics" (p. 332). However, their example of perimeter, area and volume within Maldivian culture is so scanty that the reader cannot judge how it answers Rowlands' and Carson's concerns. And despite many fine ethnomathematics articles documenting interesting mathematics arising from real life contexts (for example, Barbie dolls (Kitchen & Lear, 2000); braiding of African American hair (Gilmer, 2001); the mathematics of seamstresses (Hancock, 2001); and the mathematics of carpenters (Millroy, 1992)), we still have few examples of ethnomathematics as educational practice that can serve as stepping stones to formal academic mathematics (Kitchen & Becker, 1998 ; Rowlands and Carson, 2002; Vithal & Skovsmose, 1997).

A further challenge to ethnomathematics and its impact on the school mathematics curriculum is raised by Vithal & Skovsmose (1997) in the South African experience in which ethnomathematics was subverted to provide a justification for apartheid education. Mathematics based on knowledge that students bring from outside school and related to their own situations and culture was used to help justify continued separation of students by racial classification and all the concomitant differences in resources, curricula, and outcomes that would result. So while proponents of ethnomathematics in western countries such as the US consider it as promoting equity (Gilmer, 2001; Secada, 2000), in South Africa during apartheid it helped enable the opposite (Vithal & Skovsmose, 1997).

2.3 Research in Ethnomathematics

Ethnomathematical research seeks to uncover information about various people's mathematical knowledge in both western and non-western contexts, and how that knowledge has been created. This research probes deep epistemological questions, such as what counts as mathematical knowledge? Or, in Eglash's (1997) words: "Once we step outside the acknowledged, professional mathematical community of the west, how will we recognize mathematics when we run into it?" (p. 79). In a western context, Hancock (2001) studied four women seamstresses and the mathematics they used and created while sewing. The four women used mathematics for estimation, problem solving, measurement, spatial visualization, reasoning, geometry, and cost effectiveness. But, according to Hancock (2001), "[b]ecause of their different tools, resources, goals, and thinking, their mathematics rarely resembled school mathematics" (p. 70). The seamstresses not only invented their own language and processes, but created a type of coordinate system on the plane of a fabric that appeared to be different from known, standard systems.

In a non-western context, Knijnik (1997) worked with the Landless People's Movement in Brazil, researching the conceptions, traditions, and mathematical practices of that specific social group and how they codified and interpreted their knowledge in order to solve problems. Gerdes (1995, 1997) has conducted ethnomathematical research in Mozambique starting in the late 1970s, with an aim to ascertain the hidden mathematics of daily life that survived colonization. Gerdes has discovered many examples of use of geometry in daily life in Mozambique, and argues that without colonialism it is possible Mozambicans might have been credited, for example, with the so-called Pythagorean Theorem.

Pinxten (1997) provides an example of how ethnomathematical research might have curricular impact in schools. An anthropologist who studied the Navajo conception of space, Pinxten found that Navajo notions of space are dynamic rather than static, with the emphasis on continuous changes rather than an atomistic structure. This fundamental approach to spatial knowledge creates essential differences in how Navajos approach many concepts, including geometric ones in school. Pinxten proposes an explicit treatment of the Navajo spatial knowledge in geometry courses and in other parts of the curriculum, integrating it with the Western outlook, to improve Navajo children's understandings of spatial concepts.

3. Issues with Theory and Practice, and the English Language Development Institute in Algebra

3.1 Issues with Theory

Researchers who claim Western hegemony in the way mathematics is constructed in our schools today have done quite well in surfacing the contributions of other cultures in the history of mathematics. Equity researchers who advocate widening the space in which to do mathematics by drawing on the cultural practices of learners have as well raised the problematic of contexts in learning mathematics in a more meaningful manner. Various research studies on ethnomathematics (Joseph, 1997; Stapleton, 1996) also show that other early cultures were already familiar with notable mathematical theories such as the Pythagorean Theorem, which only demonstrates the universality of certain mathematical concepts. The question, "What counts as mathematical knowledge?" will remain open and unresolved. Suffice it to say, any response we make to such a foundational question necessitates foregrounding and articulating our favoured paradigms that significantly influence the way we perceive and construct mathematical objects and relationships. Further, what may be nonmathematical to some cultural groups or practitioners may be mathematical to others, and what constitutes the divide between what is and what is not mathematical will remain tied to subscribed epistemes that provide the very "conditions of possibility" (Foucault, 1970).

While we acknowledge the significance of the ethnomathematics program as providing "corrective measures" that may lead to the "redemption of [non-mainstream mathematical] cultures" (D'Ambrosio, 1999, p. 50) we find ourselves echoing Eglash's (1997) predicament: How do we develop alternative ways of thinking about the ethnomathematical practices of small-scale, indigenous groups without imposing the framework of Western mathematics? How might such othered forms of mathematics look if their logic of sense were to remain sophisticated and generally or universally unreasonably effective without being dismissed as primitive? As it were, current conceptualisations of ethnomathematics – as a "history 'from below,' " as the "cultures of the periphery," as "other ways of doing mathematics, proper to different cultures," and as driven by differing cosmovisions that appear opposed to the Western version (D'Ambrosio, 1999) – seem to suggest the view that the mathematical practices of minority groups are culturally-situated and context-dependent.

Barton's (1999) proposal to develop a relativist philosophy further reinforces tensions in ethnomathematical theory. He also suggests renaming ethnomathematics as a QRS system (quantity, relationship, space) to distinguish it from Western mathematics. However, we find that such a philosophy exhibits an epistemo-ontological symptom that Eglash (1997) has described as "western romantic diversions," that is, "illusions of cultural purity and organic innocence [that] are too easily projected onto these traditional cultures" (p. 83). Further, Barton suggests that we view mathematics "as a way of talking" rather than "characterizing mathematical knowledge" (p. 56). Enacting a Wittgenstein move, Barton insists that such talk enables mathematicians to load mathematical objects with real properties. For Barton, however, mathematics "is just a convenient figure of speech – literally" (p. 56). He then articulates that the "real" is at best a human construction that justifies his view that we can set aside judging for correctness (p. 57). While we agree that "talked into existence" is a good thing, however, such an action does not fully take into account how it needs to be evaluated for general and universal effectiveness and usability if at least to assure intergenerational continuity. Also, Barton's QRS system begs the question of a basis for looking at QRS in a way that projects a form or structure that is totally other to Western mathematics. For instance, our current understanding of the weaving patterns of certain indigenous cultures still reflects the use of Western mathematical concepts and processes (e.g., group theory, transformation geometry) in explaining and understanding the patterns. But, how can we begin to understand the patterns in ways that encourage us to look at mathematics differently against/beyond the Western lens?

3.2 Issues with Practice

Despite critiques of assimilation, and anticipating the needs in global times, what is lacking in conversations about ethnomathematics concerns how researchers address the complex issue of ways in which students develop mathematical identities. If certain minority groups in our schools today are known to employ particular ethnomathematical practices, in which case *ethnomathematical practices are viewed as cultural*, should individuals in such groups be bound by those practices? Are those practices too solidified and institutionalised so as not to permit changes that result from developments in their respective societies? Are indigenous mathematical practices not allowed to evolve and expand based on newer forms of social and cultural lives of peoples who engage with others outside their own cultures? From a different lens, if *ethnomathematical practices are seen as socioconstructivist*, should members allow themselves to be continually constructed by those practices that might in effect preclude any consideration of being reconstructed in some other ways? Are members not permitted to improvise based on social, cultural, economic, historical, and material transformations and developments that occur within and outside their societies? Such improvisations are necessary actions especially in situations when traditional practices of the past come into conflict with present needs and circumstances. They are "the openings by which change comes about from generation to generation" (Holland, Lachicotte Jr., Skinner, & Cain, 1998, p. 18).

From our point of view, reconceptualising ethnomathematics involves situating the talk where it is at stake, that is, the formation of students' mathematical identities that go far beyond the confines of traditional conceptualisations (i.e., culturalist, constructivist) oftentimes associated with ethnomathematics. Limiting the scope of the nature of ethnomathematics to those seemingly indigenous practices that define a community tend to essentialise members in ways that effectively close the possibility of multiple and evolving "political" processes relevant to their ways of mathematising. While we acknowledge the benefits that minority groups may acquire from learning more about the mathematical practices of their communities, we also see advantages in broadening their sense of "ethno" to include changes that take place outside their cultures. Moreover, we find it necessary for ethnomathematical researchers who construct what they perceive to be authentic, indigenous mathematical practices of a certain culture to carefully scrutinize the extent to which such practices apply to all individual members that comprise the culture. While a certain cultural community may have developed common practices, it does not simply imply that every member in the group supports the same practices. The formalization of those indigenous practices as an ethnomathematical discourse can in many cases be naively interpreted as applicable to all members despite possible differences in individual, personal, social, and environmental contexts. In other words, we need to be wary of essentialist-driven ethnomathematical programs since there is a possible unintended consequence of categorizing people and their practices in ways that may constrain the manner in which they learn mathematics, and all for the sake of preservation. Similar to Appiah's (1994) cautionary remarks about "tightly scripted identities," it is likely that certain tightly-scripted ethnomathematical practices that have been drawn from a particular culture might curtail individual and personal practices and even prevent members in the same culture from learning a different approach because of the equality assumption that cultural membership also implies shared cultural practices.

3.3 Forging a Hybrid Version of Ethnomathematics

Situating the mathematical education of those minority groups in our classrooms in the positive space of globalisation means providing them with an appropriate mix of past and present mathematical practices that will prepare them to have a better sense of the order in which their immediate and outside worlds are being reorganized in contemporary times. This is "ethno" expanded as a concept that includes all the appropriate "jargons, codes, symbols, myths, and even specific ways of reasoning and inferring" in global times (D'Ambrosio, 1985, p. 45). We emphasise that it does not mean doing away with mathematical practices that learners in particular cultures have come to know by tradition and that have constructed them in some way. However, it does mean reconciling the old with the new and, better still, forging newer practices that enable learners to cope with current modes of living. What we deem to be contemporary ethnomathematical practices involve the development of a hybrid set of altered practices and an assemblage of new collective mathematical registers that enable minority learners to cope with the global imaginary. Such practices and meaning systems should bridge the divide between the abstract, universal, and decontextualised nature of Western mathematics and the situated, local, and contextualised nature of ethnomathematics.

In 2001–2002, data from the U.S. Department of Education shows that close to 4 million students in public schools throughout the country obtained some level of assistance to learn English, with about three quarters of the students speaking Spanish as their first language. In the state of California, there has been a steady growth of English learners from 1995 to 2003. California has a "higher concentration of English learners than anywhere else in the US" (Gándara, Maxwell-Jolly, & Driscoll, 2005). In the 2003–2004 Language Census, data from the California Department of Education reveals that 85% of English learners spoke Spanish, while the remaining ones spoke any one of fifty-five different languages. Efforts have been established to assist these students to acquire proficiency in the official language (i.e., English) at both conversational and academic levels. The English Language Development Institute in Algebra (ELDI-A) was one of several efforts. It has both ethno- and Western-mathematical components integrated in its program for in-service and certified middle school and high school teachers. ELDI-A works within a premise that English learners' mathematical identities are never pregiven to them. That is, while it is true that they come to American classrooms after having been already exposed to levels of ethnomathematical practices in their respective home countries, they are still capable of acquiring knowledge about (Western) mathematics. What the ELDI-A seeks to accomplish is for teachers to provide a hybrid space in which English learners acquire Western mathematics by grounding their knowledge on what they know about their ethnomathematical practices. This perspective shadows Cummins's (1994) common underlying proficiency thesis whereby linguistic elements in a student's native academic language share syntactical, semantical, and structural commonalities with the elements in the new academic language. In the case of school mathematics, using the mathematical knowledge that students bring with them and then connecting that knowledge with the appropriate mathematical knowledge in English will enable learners to achieve some level of success in learning academic, formal mathematics.

Because the ELDI-A focuses on implementing a mathematical discourse that is drawn from activities from various traditions, what is constructed for learners is a discourse in which various frames of reference for meaning have not been drawn from a single source (i.e., Western). Further, the pedagogical strategies appropriate for English learners, called Specially Designed Academic Instruction in English (SDAIE), are sensitive to similarities and differences in cultural practices. Thus, English learners' knowledge of concepts, skills, and processes has been generated from a diverse set of mathematical practices. Because the ELDI-A Program represents a collective discourse from several cultures, students' mathematical practices evolve out of such a hybrid condition.

Barton (1999) provides some evidence about a possible relationship between the manner in which cultural groups use and practice mathematical language and their conceptions of quantity. For instance, the traditional Maoris in New Zealand and some American Indian groups consider "number words [as] action words, they act like verbs" (p. 57; see, also, Denny (1986)). Barton laments that such linguistic practices have "been talked out of existence, or, at the least, [they have] been talked out of existence as mathematics" (ibid.). In the ELDI-A, every effort is made to bridge such differences in mathematical practice. What mathematics teachers acquire is that an understanding of English and relevant discourse and linguistic patterns reflect cultural traditions and practices. Further, it is generally acknowledged that the English language relevant to mathematics has a structure that is not shared by other cultures. In other terms, there are variations in the manner in which language is used and practiced by, say, American Indians, Native Hawaiians, Puerto Ricans, and African Americans, which tend to significantly influence the way mathematics is learned. Also, Fillmore & Snow (2000) point out that if teachers are aware of the grammatical and extra-linguistic (cultural) structures that different minority groups employ to convey their thoughts and processes, then they can at least "see the logic behind [their students'] errors" (p. 15). Thus, in ELDI-A, it is not the case that certain ethnomathematical practices are effaced or talked out of existence. In fact, they serve as the basis for assisting students to acquire competence in the academic, formal language in which mathematics is represented (which happens to be English in the case of the U.S.). Various SDAIE strategies attempt to integrate ethnomathematical practices with those used in the mainstream.

4. Provisional Closure

D'Ambrosio (1985) claims that the field of ethnomathematics is about acknowledging how "different modes of thought may lead to different forms of mathematics" (p. 44). We are fortunate that there is now a strong research base that shows the mathematical capabilities of quite a number of cultural groups that have developed particular "quantitative and qualitative practices, such as counting, weighing and measuring, comparing, sorting, and classifying" (D'Ambrosio, 1999, p. 51). D'Ambrosio (1999) points out as well how tellings in cognitive theories suggest a strong connection between culture and cognition. While his early views are worth considering in our efforts to theorize mathematical practice based on cultural specificities and necessities, there is also a need to consider how promoting such differences in thought and context will benefit minority learners in the long haul. While we possess a wealth of information about the mathematical systems and discursive and symbolic representations of different cultural groups, the most significant question for ethnomathematical theory and practice is: What now? Restivo (1983) has astutely articulated how transformations "in the social, economic, and political conditions of [and relationships in] our lives" would inevitably necessitate transformations in "the material bases and social structure of mathematics" (p. 178). Considering the global episteme, how can teachers use ethnomathematical knowledge that will enable their students, especially those individuals that come from "cultures of the periphery" (D'Ambrosio, 1999, p. 51), to meet the demands of a changing global society?

Bracketing unresolved conceptual issues with the ethnomathematics program, we believe that all learners' mathematical experiences will be enriched if every effort is made to reconcile the traditions of both Western mathematics and ethnomathematics, including other types of mathematical systems such as non-Western and vernacular mathematics (see Eglash's (1997)). Drawing on Habermas (2001, 1998), this reconciliatory view stems from our belief that it is possible to have shared mathematical practices in spite of cultural differences. Western mathematics for us represents those corpora of disembodied, universal, and institutionalised mathematical knowledge and practices that continue to impose its hegemony as a result of centuries of shared thinking across cultures. For instance, contemporary school algebra reflects an interesting history of shared knowledge as a result of early mathematicians who have engaged in trade and commerce, and at the same time, have acquired knowledge of mathematical systems in other cultures. In Section 3, we briefly discussed how the ELDI-A program that we offer our in-service mathematics teachers in California was an attempt to resolve certain linguistic and extra-linguistic (cultural, social) differences and difficulties. Thus, we see a complementary relationship between Western mathematics, the mainstream discourse that is implemented in almost all schools around the globe, and the contextual nature of ethnomathematics.

Ethnomathematics researchers are also not exempt from criticisms that in effect claim they are imposing ethnomathematical traditions onto learners who may favour or benefit from other ways of learning mathematics. We believe that a more powerful ethnomathematics program in contemporary times involves understanding the structure of complexity of cultures in ways that explain how members in such cultures are able to preserve valuable mathematical practices and might overcome those that constrain them from fully participating globally. Holland, Lachicotte Jr., Skinner, and Cain cogently capture what we envision to be the next phase in the ethnomathematics agenda in the sentences below.

> The very conceptions of culture have changed drastically. Anthropology no longer endeavours to describe cultures as though they were coherent, integrated, timeless wholes. ... Anthropology is much less willing to treat the cultural discourses and practices of a group of people as indicative of one underlying cultural logic or essence equally compelling to all members of the group. Instead, contest, struggle, and power have been brought to the foreground. *The objects of cultural study are now particular, circumscribed, historically and socially situated "texts" or "forms" and the processes through which they are negotiated, resisted, institutionalised, and internalised.*
> (Holland, Lachicotte Jr., Skinner, & Cain, 1998, pp. 25–26; emphasis added).

Below we raise four issues that need to be addressed in future discussions on ethnomathematics.

(1) In constructing knowledge about the ethnomathematical practices of indigenous groups, how were those practices institutionalised? What were the social, economic, and political conditions that have allowed those practices to be taken as shared? Are those conditions still evident in their societies?
(2) To what extent do individual members within indigenous groups subscribe to the same ethnomathematical practices? How do they negotiate and internalise such practices?

(3) Are there members within indigenous groups who do not subscribe to the same ethnomathematical practices? Why do they resist the practices?
(4) Considering the fact that the ethnomathematical practices of minority groups have been developed and influenced by specific cosmovisions, epistemologies, and ontologies, how can teachers and learners be assisted in reconciling possible conceptual and praxiological differences between mainstream and minoritarian views and practices?

References

Adam, S., Alangui, W., & Barton, B. (2003). A comment on Rowlands' and Carson's "Where would formal, academic mathematics stand in a curriculum informed by ethnomathematics? A critical review." *Educational Studies in Mathematics, 56*(3), 327–335.

Appiah, A. (1994). Identity, authenticity, survival: Multicultural societies and social reproduction. In A. Gutman (Ed.), *Multiculturalism: Examining the politics of recognition* (pp. 149–163). Princeton, NJ: Princeton University Press.

Ascher, M., & Ascher, R. (1997). Ethnomathematics. In A. Powell & M. Frankenstein, (Eds.), *Ethnomathematics: Challenging Eurocentrism in mathematics education* (pp. 25–50). New York: SUNY Press.

Barton, B. (1999). Ethnomathematics and philosophy. *Zentralblatt für Didaktik der Mathematik, 31*(2), 54–58.

Borba, M. (1997). Ethnomathematics and education. In A. Powell & M. Frankenstein, (Eds.), *Ethnomathematics: Challenging eurocentrism in mathematics education* (pp. 261–272). New York: SUNY Press.

Carnoy, M. (1998). The changing world of work in the information age. *New Political Economy, 3*(1), 123–128.

Cronin, C., & De Greiff, P. D. (1998). Introduction. In J. Habermas, (Ed.), *The inclusion of the other: Studies in political theory* (pp. vii–xxxiv). Cambridge, MA: MIT Press.

Cummins, J. (1994). Primary language instruction and the education of language minority students. *Schooling and language minority students: A theoretical framework* (pp. 3–46). Los Angeles, CA: Evaluation, Dissemination, and Assessment Centre of the California State Department of Education.

Cvetkovich, A., & Kellner, D. (Eds.). (1997). *Articulating the global and the local: Globalisation and Cultural Studies*. Boulder, CO: Westview Press.

D'Ambrosio, U. (1985). Ethnomathematics and its place in the history and pedagogy of mathematics. *For the Learning of Mathematics, 5*(1), 44–50.

D'Ambrosio, U. (1999). Ethnomathematics and its first international congress. *Zentralblatt für Didaktik der Mathematik, 31*(2), 50–53.

Denny, P. (1986). Cultural ecology of mathematics: Ojibway and Inuit hunters. In M. P. Closs (Ed.), *Native American Mathematics* (pp. 129–180). Austin, TX: University of Texas Press.

Eglash, R. (1997). When math worlds collide: Intention and invention in ethnomathematics. *Science, Technology, and Human values, 22*(1), 79–97.

Fillmore, L.W., & Snow, C. (2000). What teachers need to know about language. *ERIC Clearinghouse on Language and Linguistics: Special Report.* http://faculty.tamu-commerce.edu/jthompson/Resources/FillmoreSnow2000.pdf

Foucault, M. (1970). *The order of things: An archaeology of the human sciences.* New York: Random House-Vintage.

Gandara, P., Maxwell-Jolly, J., & Driscoll, A. (2005). *Listening to teachers of English language learners.* Santa Cruz, CA: Centre for the Future of Teaching and Learning.

Gerdes, P. (1995). *Ethnomathematics and education in Africa.* Stockholm: University of Stockholm Institute of International Education.

Gerdes, P. (1997). Survey of current work in ethnomathematics. In A. Powell & M. Frankenstein, (Eds.), *Ethnomathematics: Challenging eurocentrism in mathematics education* (pp. 331–372). New York: SUNY Press.

Giddens, A. (1999). Excerpts from a keynote address at the UNRISD conference on globalisation and citizenship. *UNRISD News* #15.

Gilmer, G.F. (2001). Ethnomathematics: A promising approach for developing mathematical knowledge among African American women. In J. E. Jacobs, J. R. Becker, & G. F. Gilmer (Eds.), *Changing the faces of mathematics: Perspectives on gender* (pp. 79–88). Reston, VA: National Council of Teachers of Mathematics.

Habermas, J. (1998). *The inclusion of the other: Studies in political theory* (C. Cronin & P. D. Greiff, Eds.). Cambridge, MA: MIT Press.

Habermas, J. (2001). *The postnational constellation: Political essays* (M. Pensky, trans. & ed.). Cambridge, MA: MIT Press.

Hancock, S. J. C. (2001). The mathematics and mathematical thinking of four women seamstresses. In J. E. Jacobs, J. R. Becker, & G. F. Gilmer (Eds.), *Changing the faces of mathematics: Perspectives on gender* (pp. 67–78). Reston, VA: National Council of Teachers of Mathematics.

Hardt, M., & Negri, A. (2000). *Empire*. Cambridge, MA: Harvard University Press.

Harris, M. (1997). An example of traditional women's work as a mathematics resource. In A. Powell & M. Frankenstein, (Eds.), *Ethnomathematics: Challenging eurocentrism in mathematics education* (pp. 2215–222). New York: SUNY Press.

Holland, D., Lachicotte Jr. W., Skinner, D., & Cain, C. (1998). *Identity and agency in cultural worlds*. Cambridge, MA: Harvard University Press.

Joseph, G. G. (1993). A rationale for a multicultural approach to mathematics. In D. Nelson, G. G. Joseph, & J. Williams (Eds.), *Multicultural mathematics: Teaching mathematics from a global perspective* (pp. 1–24). Oxford: Oxford University Press.

Joseph, G. G. (1997). Foundations of eurocentrism in mathematics. In A. Powell & M. Frankenstein, *Ethnomathematics: Challenging eurocentrism in mathematics education* (pp. 61–82). New York: SUNY Press.

Kitchen, R. S., & Becker, J. R. (1998). Mathematics, culture, and power–A review of "Ethnomathematics: Challenging eurocentrism in mathematics education." *Journal for Research in Mathematics Education, 29*(3): 357–363.

Kitchen, R. S., & Lear, J. M. (2000). Mathematising Barbie: Using measurement as a means for girls to analyse their sense of body image. In W. G. Secada (Ed.), *Changing the faces of mathematics: Perspectives on multiculturalism and gender equity* (pp. 67–74). Reston, VA: National Council of Teachers of Mathematics.

Knijnik, G. (1997). An ethnomathematical approach in mathematical education: A matter of political power. In A. Powell & M. Frankenstein, (Eds.), *Ethnomathematics: Challenging eurocentrism in mathematics education* (pp. 403–410). New York: SUNY Press.

Millroy, W. L. (1992). *An ethnographic study of the mathematical ideas of a group of carpenters*. Reston, VA: National Council of Teachers of Mathematics.

Pinxten, R. (1997). Applications in the teaching of mathematics and the sciences. In A. Powell & M. Frankenstein, (Eds.), *Ethnomathematics: Challenging eurocentrism in mathematics education* (pp. 373–402). New York: SUNY Press.

Powell, A., & Frankenstein, M. (1997). (Eds.). *Ethnomathematics: Challenging eurocentrism in mathematics education*. New York: SUNY Press.

Reich, R. (1991). *The work of nations: Preparing ourselves for 21st century capitalism*. New York: Knopf.

Restivo, S. (1983). *The social relations of physics, mysticism, and mathematics*. Boston, MA: D. Reidel Publishing Company.

Rivera, F. (2004). In Southeast Asia (Philippines, Malaysia, and Thailand): Conjunctions and collisions in the global cultural economy. In W. Pinar (Ed.), *International handbook of curriculum research* (pp. 553–574). Mahwah, NJ: Erlbaum.

Rowlands, S., & Carson, R. (2002). Where would formal, academic mathematics stand in a curriculum informed by ethnomathematics? A critical review of ethnomathematics. *Educational Studies in Mathematics, 50*(1), 79–102.

Secada. W. G. (Ed.). (2000). *Changing the faces of mathematics: Perspectives on multiculturalism and gender equity*. Reston, VA: National Council of Teachers of Mathematics.

Stapleton, R. (1996). *Ancient Chinese Mathematics*. Ethnomathematics Digital Library. Available online: http://www.unisanet.unisa.edu.au/07305/chinese.htm.

Vithal, R., & Skovsmose, O. (1997). The end of innocence: A critique of ethnomathematics. *Educational Studies in Mathematics, 34*(2), 131–157.

Walkerdine, V. (1997). Difference, cognition, and mathematics education. In A. Powell & M. Frankenstein, (Eds.), *Ethnomathematics: Challenging eurocentrism in mathematics education* (pp. 201–214). New York: SUNY Press.

Zaslavsky, C. (1973). *Africa Counts: Number and pattern in African culture*. Boston: Prindle, Weber & Schmidt.

Zaslavsky, C. (1997). World cultures in the mathematics class. In A. Powell & M. Frankenstein, (Eds.), *Ethnomathematics: Challenging eurocentrism in mathematics education* (pp. 207–320). New York: SUNY Press.

13

POP: A STUDY OF THE ETHNOMATHEMATICS OF GLOBALIZATION USING THE SACRED MAYAN MAT PATTERN

Milton Rosa[1] **and Daniel Clark Orey**[2]
[1]*Mathematics Department, Encina High School, SJUSD, 1400 Bell Street, Sacramento, California, USA, milrosa@hotmail.com*
[2]*California State University, Sacramento orey@csus.edu*

Abstract: There exists a belief that mathematics produced by non-Western cultures is irrelevant for both the economic and technological development of our modern globalised world. From a global perspective, ethnomathematics can be considered an academic counterpoint to globalization, and offers a critical perspective of the internationalism of mathematical knowledge through attempts to connect mathematics and social justice. It is also possible to perceive ethnomathematics as a form of academic articulation between cultural globalization and the mathematical knowledge of diverse non-Western cultural groups. Through a study of the mathematical practices found in the sacred mat and geometric diamond patterns of the Maya, it is possible to use an ethnomathematical, anthropological, and global perspective, to demonstrate one way in which we might preserve a portion of the wisdom and knowledge of these unique and resilient peoples

Keywords: globalization; ethnomathematics; Mayan civilization; anthropology; mathematical knowledge; mat patterns; cultural groups; non-Western cultures, mathematical modelling

1. Introduction

Before the present era of globalization, the world's continents were separated by vast expanses of ocean and sea. Ancient peoples knew of the existence of others only through myth, legend, and the stories of conquerors or travellers. Most of humanity lived in isolated and self-sufficient cultural groups and lived and died in the same place (Toffler, 1980). Recently, the world's peoples have been linked together through extensive systems of communication, migration, trade, and production.

B. Atweh et al. (eds.), Internationalisation and Globalisation in Mathematics and Science Education, 227–246.

2. Globalization

Globalization is an ongoing historical process that has, at its roots, the very first movement of peoples from their original homelands. Explorers, conquerors, migrants, adventurers, and merchants have always taken their own ideas, products, customs, and mathematical practices with them in their travels. The analysis of the great events of human history such as the conquests by Caesar, Alexander, Cortez; the adventures of Marco Polo, the Portuguese Naval School of Dom Enrique, and the navigation of Columbus, all occurred primarily for economic reasons. Imperialistic adventures determined the colonial social-cultural characteristics through the imposition of non-native customs on local and diverse indigenous peoples. This form of colonialism was practiced primarily by European nations and is often referred to as the *Europeanization* (Featherstone & Radaelli, 2003) of the world.

In order to maintain and govern their possessions, European colonizers required enormous amounts of capital and power, and settled most questions of cultural difference by force. This increased a certain amount of awareness of non-Western cultures by the colonizers, and has raised new questions for scholars about the nature of society, culture, language, and knowledge. For example, Recinos (1978) stated that "from the first years of the colonization, the Spanish missionaries were aware of the need to learn the languages of the Indians in order to communicate with them directly and to instruct them in the Christian doctrine" (p.30).

Emerging theories of social evolution allowed Europeans to organize this new knowledge in a way that justified the political and economic domination of others. Colonized people were considered less-evolved, thus giving the powerful sense of justification to the colonizers as they came to believe themselves more evolved. Nevertheless, an effective administration required some degree of understanding of other cultures.

Colonial powers built educational institutions based on their own educational paradigms and systems. Because of this, it is possible to identify the early processes of globalization and internationalization of scientific-mathematical knowledge through the very establishment of the school systems that were built and adapted in colonies on Asia, the Americas, and Africa. Worldwide, the concept of higher education generally takes on the system, titles, and structures of a medieval European design that was passed around the world through colonial expansion.

With Guttenberg's invention of the printing press in 1455, some European cultural groups were quickly empowered over others and began to expand their culture, values, thoughts and civilization. Initially, this process of globalization developed relatively slowly, however, with the onset of Industrial Revolution, and the subsequent rise of materialism and capitalism, globalization has rapidly expanded.

3. The Globalization of Mathematical Knowledge

We do not really know when an interest in the mathematical practices of other cultures was first expressed. The earliest observations of distinct mathematical practices probably occurred in tandem with the first travels to different regions of the world, made by those who came in contact with local cultures. They observed different customs that no doubt included different mathematically-related practices such as counting and measuring. Even though an absence of early records has hindered true understanding, observations of those practices allowed early scientists, philosophers, and mathematicians to apply many mathematical concepts and ideas that were brought back from their travels. The development of writing allowed historians of mathematics to piece together knowledge accumulated by early civilizations.

In the light of these facts, the globalization of mathematical, scientific, and technological knowledge brought accelerated technological *progress* to various parts of the world. For example, when in the 7th century the Arabs invaded Europe, they brought with them the mathematical knowledge that they acquired from India (thus the term Hindu-Arabic numeration system). They also influenced Medieval Europe by exchanging food, customs, culture, science and technology. In turn, when they conquered and colonized the peoples who lived there, Europeans introduced this system into the New World.

The mathematical discoveries made by the Hindus around the 9th century were transmitted to the Arab peoples through religious expansion and commercial activities, war, and conquest. At this time, the number system used by the Greeks and Romans was cumbersome and impractical for many uses and the adoption of the decimal number system used by the Hindus and brought to Europe by the Arabs made perfect sense. This improved ability to calculate allowed for growth in the western sciences.

The Hindus also took advantage of this same cultural interchange by learning important concepts of Greek mathematics by way of the Arabs. Despite this "Eastern" globalization, the earliest systematic use of a symbol for zero in a place value system was used by the Mayans centuries before the Hindus began to use a symbol for zero (Cajori, 1993; Diaz, 1995; Jr. Merick, 1969). It is very important to note that the Mayan number system was in use in Mesoamerica while the Europeans were still struggling with the Roman numeral system which suffered from serious defects because there was no zero and the numbers were entirely symbolic with no direct connection to the number of items represented.

In the Mayan number system the symbol for zero was used to indicate the absence of any units of the various orders of the modified base-twenty system. In this context, Ifrah (1998) stated "What is quite remarkable is that Mayan priests and astronomers used a numeral system with base 20 which possessed a true zero and gave a specific value to numerical signs according to their position in the written expression" (p. 308). Further evidence of this phenomenon resulted from Ifrah's study of Mayan achievement in mathematics:

> So we must pay homage to the generations of brilliant Mayan astronomer-priests who, without any Western influence at all, developed concepts as sophisticated as zero and positionality, and despite having only the most rudimentary equipment, made astronomical calculations of quite astounding precision
>
> (p. 322).

In the 11th century, the internationalization of scientific, technological, and mathematical knowledge was not only influenced by Western cultures, because the agents of globalization were located in other regions of both the known and unknown world (Sen, 2002). Powerful technological items such as paper, gunpowder, the magnetic compass, and the iron–chain suspension bridge were used in China, but other cultures around the world had little if any knowledge of these technologies (Sen, 2002).

In the 14th century, the Arab historian and philosopher Ibn Khaldun (1332–1406) examined social, psychological, economic, and environmental factors that affected the development, ascension and fall of different civilizations. In his study, Khaldun analyzed several economic policies and demonstrated the consequences for both local and distant communities (Oweiss, 1988). These facts accompanied a mathematical knowledge that strongly contributed to the defence of communities against the injustice and oppression of the ruling class.

At the end of 15th and the beginning of the 16th centuries, explorers provided descriptions of different aspects of the "exotic" cultures they encountered in Asia, Africa, and the Americas. Early chroniclers of the Americas reported observations and registered data collected in relation to the cultures they encountered in their explorations. Using a process that can be considered ethnomathematical in nature, Juan Diaz Freyle published, in 1556, the first book of arithmetic of the new world entitled *Sumario compendioso de las quentas de plata y oro que en los reinos del Pirú son necessarias a los mercaderes y todo genero de tratantes: Con algunas reglas tocantes al arithmética*[1]. In this book, Freyle described the arithmetic practiced by the indigenous people. It is important to observe that this book described the process of the indigenous people's assimilation of the conquering people's mathematical knowledge. This can be perceived as a transformation of the native mathematical system through a global and cultural dynamic perspective.

According to Grattan-Guinness (1997), when Europeans invaded and conquered the northern part of the Americas during the early 16th century, they "began to apply commercial arithmetic to the purchase of citizens in North America from local chiefs and kings, and the later sale of those still alive, to entrepreneurs and landowners across to the Americas" (p. 112). He also affirmed:

> They too made little effort to conserve the culture of either slaves or of the indigenous tribes. Nevertheless, the latter have managed to maintain a repertoire of mathematical theories, not only in arithmetic, geometry and astronomy but especially in connection with skills such as archery and in games of chance involving the throwing down of rods and sticks decorated in various ways
>
> (p. 113).

---

[1] Translation: *A Compendium Summary of the Accounts of Silver and Gold that in the Kingdoms of Peru are Necessary to Merchants and All Kinds of Dealers: With Some Rules Concerning Arithmetic.*

The ascension of the Portuguese, Spanish, French, Dutch, English, and Belgian Empires in 18th and 19th centuries contributed to increasing contact with the cultures they colonized. This context allowed for an increased development of global commerce, a greater spread of the growing capitalist economy, and the industrialization of Europe.

The newly industrialized countries continued their search for new lands as sources of supply, cheap manpower, and the raw materials to be manufactured at low costs. At the same time, millions of Europeans from the lower classes were encouraged to immigrate to the newly established colonies in promise of better lives. These cultural exchanges allowed for a continued accumulation of data and information of distinct cultural groups that were "found" and subjugated in the colonies.

In the 19th century, the first forms of what would become modern anthropology began to be systematized. According to some experts, as different cultures were studied during the ongoing processes of assimilation and colonization, the customs and mathematical practices of diverse cultural groups also became objects of study by many early European anthropological societies. In the 20th century, a growing and increasingly sensitive understanding of mathematical practices and ideas from diverse cultural groups became increasingly available through the growth of the fields of ethnology, culture, history, anthropology, linguistics, and the development of ethnomathematics. Insights from many theoretical studies signal the possibility of the sensitive internationalization of mathematical practices and ideas expressed in different cultural contexts.

4. The Perspective Offered by Ethnomathematics

Ethnomathematics recognizes that all cultures and all people develop unique methods and sophisticated explications to understand and to transform their own reality. It also recognizes that the accumulated methods of these cultures are engaged in a constant, dynamic, and natural process of evolution and growth in every society. In this context, culture is a complex whole that includes knowledge, beliefs, art, laws, morals, customs, and any other practices and habits assured by a member of a society. Lindsey, Robins, & Terrel (2003) define culture as "a group of people identified by their shared history, values, and patterns of behaviour" (p. 41). Lindsey et al (2003) also believe that "culture is a problem-solving resource we need to draw-on, not a problem to be solved" (110). Ethnomathematics looks at the mathematics of this problem-solving resource.

Another presupposition of ethnomathematics is that it validates all forms of mathematical explaining and understanding formulated and accumulated by different cultural groups (Rosa, 2000). This knowledge is regarded as part of an evolutionary process of change that is part of the same cultural dynamism present as each group comes into contact with each other in this new global reality. In this perspective, all cultures have, by necessity, evolved unique ways to *quantify, count, classify, measure, explain and model* the phenomena of their own daily occurrences (Borba, 1997). A study of the different ways in which people resolve problems and the practical algorithms on which they base these mathematical perspectives becomes relevant for any real comprehension of the concepts and the practices in the mathematics that they have developed over time. For example, when we speak of patterns and sequences, we know that humanity utilized different numeric and geometric patterns to make music, dance, or create basketry, ceramics, rugs, and fabric. Many times, these patterns possessed religious and spiritual aspects that sought to connect their own human perspective with the "divine" around them.

5. Ethnomathematics and Anthropology

One of the most important concepts of ethnomathematics is the association of the mathematics found in distinct cultural forms. Ethnomathematics as a program is much wider than traditional concepts of mathematics and ethnicity. In this case, D'Ambrosio (1990) refers to "ethno" as that related to distinct cultural groups identified by cultural traditions, codes, symbols, myths, and specific ways of reasoning and inferring.

The focus of ethnomathematics consists essentially of a serious and critical analysis of the generation and production of knowledge (creativity), intellectual processes in the production of this knowledge, the social mechanisms in the institutionalization of knowledge (academic ways), and the diffusion of knowledge (educational ways). In this holistic context, the study of the systems that form reality and look to reflect, understand, and comprehend extant relations among all of the components of the system require constant analysis of their reality. Rosa (2000) has defined ethnomathematics as the intersection of cultural anthropology, mathematics, and mathematical modelling which is used to translate diverse mathematical practices.

All as shown in Figure 1 individuals possess both anthropological and mathematical concepts; these concepts are rooted in the universal human endowments of curiosity, ability, transcendence, life, and death. They characterize our very humanness. Awareness and appreciation of cultural diversity that can be seen in our clothing, methods of discourse, our religious views, our morals, and our own unique world view allow us to understand each aspect of the daily life of humans.

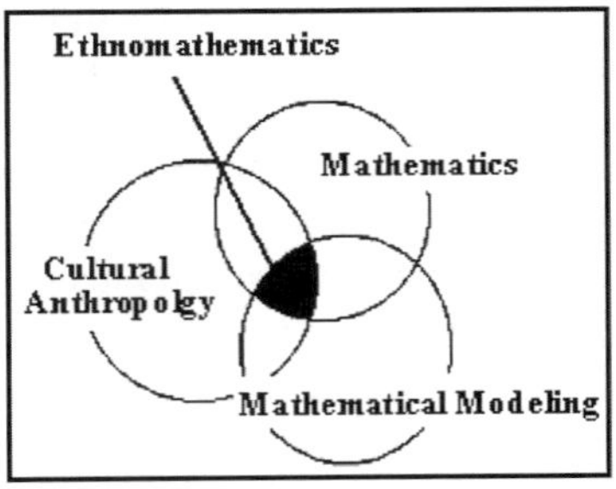

Figure 1. Ethnomathematics as an Intersection of Three Disciplines

The culture of each group represents a set of values and the unique way of seeing the world as it is transmitted from one generation to another. The principal focus of anthropology that is relevant to our work in this chapter includes such aspects of culture as language, economy, politics, religion, art, and our daily mathematical practices. Since, cultural anthropology gives us the tools to increase our understanding of the internal logic of a given society; an anthropological study of distinct cultural groups allows us to further our understanding of the internal logic and beliefs of different peoples.

6. Ethnomathematics in the Process of Globalization

Knowledge is generated and intellectually organized by individuals in response to their own social, cultural, and natural environment. This knowledge is socially organized and used to recognize and explain activity in the daily lives of people. According to D'Ambrosio (2002), observers, chroniclers, theoreticians, sages, and professionals expropriated this knowledge, and then classified, labelled, diffused, and transmitted it across generations.

There are structured forms of knowledge such as language, religion, the culinary arts, medicine, dress, values, sciences, and forms of mathematical thinking that are interrelated and respond to the way reality is perceived through the unique social, cultural, and local environment of an individual (D'Ambrosio, 2002). These forms of knowledge are structured differently because cultural dynamics increasingly plays a role in the broadening perception of reality which, as a consequence, modifies responses to these cultural structures that are in a dynamic state of change as shown in Table 1. Some individuals, groups, societies, or nations freeze these forms of knowledge (Gerdes, 1985). The frozen knowledge becomes accepted and energy is directed towards keeping these forms of knowledge static.

Table 1. Mathematical Practices as Diverse Cultural Forms of Knowledge

Mathematical Practices	are	Diverse Cultural Forms of Knowledge
• measurement • comparison • classification • quantification • ordering • selecting • cipering • memorization of routines • counting • inference • modelling	• general • product • organized • diffused • transmitted formally and informally • symbols • values • beliefs	• languages • communication • jargon • mathematical ideas • codes of behaviours • myth

D'Ambrosio (1985) has stated that there are many different kinds of ethnomathematics. Each one of them responds to different cultural, social, and natural environments. One of these environments originated in the Mediterranean basin and gave origin to a form of ethnomathematics called "mathematics." Through the subsequent processes of conquest and colonization, and now a corporately forced globalization, this "mathematics" has been imposed across the world at large. It has been accepted because of its tremendous scientific success and its ability in dealing with space and time which accompanied the colonial world view of property ownership, production, labour, consumption, and subsequent capitalistic values. Mediterranean-based mathematics has come to be known as "Western" mathematics, is often referred to as "universal mathematics", but in reality it can be seen as a subset of our overall basic human endowment.

Ethnomathematics is the mathematics practiced by identifiable cultural groups (D'Ambrosio, 1990) such as national and tribal societies, labour groups, children of a certain age, and professionals. Borba (1997) agrees with this point of view and stated that "even the mathematics produced by professional mathematicians can be seen as a form of ethnomathematics because it was produced by an identifiable cultural group" (p. 40). Mtetwa (1992) found that "some people have misunderstood the term, using it exclusively to refer to mathematical forms created and practiced by and for a specific ethnic group" (p. 1). This interpretation of ethnomathematics does not represent the broader definition given by D'Ambrosio (1993) to this program.

Ethnomathematics includes, but is certainly not limited to academic or school mathematics, and the kinds of mathematics conceived and practiced by the professional scientific community (Orey & Rosa, 2003). Powell and Frankenstein, (1997) stated that it "is the informal and ad hoc aspects of ethnomathematics that broaden it to include more than academic mathematics" (p. 7). Western mathematics can and should be considered as a subgroup of ethnomathematics because there is a relationship between ethnomathematics and academic mathematics through mathematical modelling (Borba, 1997; D'Ambrosio, 1993; Orey & Rosa, 2003). In this perspective "school mathematics is an outgrowth and subset of ethnomathematics" (Mtetwa 1992, p. 3).

However, the distinction between western and non-western mathematics is weakened by a theory of knowledge that is supported by cultural dynamics which occurs through encounters among different cultural groups, and produces cycles of generation, organization, transmission, and the diffusion of knowledge. Traditional or academic mathematical practices are a form of ethnomathematics defined by the cultural background and patterns of individuals that practice them. They translate this knowledge in a form of academic language and incorporate it as mathematical practice in their daily lives. This cultural dynamics is defended and described by D'Ambrosio (2000).

In an increasingly globalised world, the subtlest weapon of the colonizer[2] has been the institution of "formal" education. By instituting the use of formal education, the colonizer removes the roots of the colonized and facilitates the cultural process of submission. In this case, one of the main functions of ethnomathematics is the *decolonization* of individuals and communities (Orey & Rosa, 2003). Because of this, ethnomathematics is frequently criticized as being political, since it seeks the liberation of the oppressed in the ongoing process of colonization and globalization. By doing this, ethnomathematics seeks to raise the self-confidence, to enhance creativity, and to promote cultural dignity of diverse cultural groups. These factors are essential to value an individual's cultural background.

D'Ambrosio (1999) gives a global dimension to ethnomathematics without being colonialist or imperialistic. According to his theory, ethnomathematics is international in its own rite. In his perspective, it is possible to internationalize and value different mathematical practices through mathematical modelling[3]. Modelling acts as a bridge between mathematical ideas and academic mathematics. This environment allows us to internationalize diverse mathematical practices by including mathematical modelling in the mathematics curriculum which will enable students to understand and function in a globalised world (Atweh & Clarkson, 2002). This process shows that mathematics is a cultural endeavour, and is rooted in tradition, and which considers all systems of mathematical ideas developed by every civilization as valid (Rosa, 2000).

Most notably, unlike much of modern sciences and mathematics, the program of ethnomathematics as defined by D'Ambrosio, emerged from the unique conditions of the Brazilian socio-political-economic reform movement in the late 20th century. For all these reasons, we believe that an ethnomathematics program cannot be viewed as a neo-colonial approach in mathematics education.

Both ethnomathematics and a new globalised mathematics must take care not to trivialize other cultures based on the misrepresentations of their scientific and mathematical ideas or structures. It is also important to uphold a balanced analysis that maintains a group's cultural integrity while accurately portraying its scientific, mathematical, and technological contributions. We have outlined here, one example of how this future scholarship might proceed.

7. The Great Architect

Many cultures share the belief that a "Great Architect" of the universe possessed certain mathematical characteristics. This "Great Architect" is, according to many Mediterranean traditions, God, Yahweh, Allah, and, according to the Mayan tradition, named *Tzakol* (Recinos, 1978). The knowledge of this "Great Architect"

[2] The colonizer can be any government, religion, individual, or corporation.

[3] Mathematical modelling is a tool that provides a translation of different mathematical practices into academic mathematics.

was learned and captured by many Mediterranean and ancient non-Western civilizations (D'Ambrosio, 2000). Since there is more than one religious practice, more than one system of values, more than one name for the "Great Architect", there is, perhaps more than one way of *explaining, knowing,* and *understanding* these diverse realities (D'Ambrosio, 2001).

A study of the mathematics of indigenous peoples who were "discovered" and colonized by Europeans allows us to introduce mathematical ideas of cultural groups who have been excluded from traditional mathematical discourse. It is in this context that an ethnomathematical perspective can be used to challenge what is often known as an ethnocentric view of diverse cultural systems. Complex social organizations are typically thought of as having advanced technology and thus, a more "complicated" mathematical system; yet, indigenous cultures such as the Mayans, developed equally complicated mathematics which had an equally conscious effect on the world around them.

8. The Mayan Civilization

Mayan civilization has survived for more than 3000 years in the region now called Central America. The Mayan people are best-known by their distinct architecture, the patterns they found in their observations about the universe, the development of mathematical relationships, and a symbolic and sacred system that they developed to represent these patterns. About 7 million Mayan people are dispersed in urban and rural communities in Southern México, Belize, Guatemala, Honduras and El Salvador. With centuries of persecution, cultural insulation, and disrespect of Mayan traditions, beliefs and religion, most Mayan people now live in crushing poverty.

9. The Mayan Process of Globalization

For indigenous Mayan people, the violent encounter with globalization began in 1524 with the arrival of the Spanish conqueror Pedro de Alvarado. With the invasion of the Americas by Europeans, the world of the Mayans, Incas, and Aztecs, like all the other Indigenous societies in the New World, came to an abrupt and extremely brutal end. On other hand, according to Ascher & Ascher (1981), the Incas did not destroy and replace the cultures they conquered, "An Inca deity was added to, not substituted for, the local gods... locally important people continued to be important..." (p. 5–6).

Although medieval Europe was in many ways less developed than the Mayans, the conquerors arrived with an enormous military advantage such as gunpowder, steel swords, and horses. At the same time, indigenous societies were weakened by diseases against which they had no immunity. It was the superior European technology and firearms that proved a vital factor to the success of the conquest of the Americas. They justified their "destructive acts on the basis of cultural superiority" (Ascher, 1991, p. 17). In a quest for riches, the European invaders defeated the Mayans. In so doing, they destroyed libraries that were possibly the greatest repositories of indigenous science in the Western Hemisphere. Some surviving texts were carried to safety by Mayan priests. Among them was the hieroglyphic source for the *Popol Vuh*, which is considered by some to be the "Mayan Bible", and the Dresden Codex, which reveals the sophistication of Mayan knowledge of astronomy and mathematics. Knowledge about the Mayan world in these texts is just a small fraction of knowledge that they accumulated during thousands of years (Coe, 1992).

When Mayan cities were decimated by disease, burned and sacked, their religion and culture were banned and forced underground. Within a short time; the Mayans had become slaves in their own homeland and were deprived of their land, their rights, and any kind of political or social representation. The once proud Mayan kingdoms were subjugated and colonized. Yet, despite this, they have continued to maintain much of their heritage, religion, mathematical knowledge, and languages.

However, the Mayans did not accept this fate lightly; a study of Mayan history shows that in every generation since the initial invasion by Spain, the Mayans have risen-up in rebellion (Wilkinson, 2002). The Mayan peoples have never forgotten their cultural identity. Despite centuries of oppression and prejudice, they continue to celebrate their own cultural and religious ceremonies, and maintain and speak their own languages. There is no doubt that Mayan culture has been weakened due to the processes of disease, slavery, colonization, conquest, and globalization. The early greed and ambition of colonizers has recently been replaced by the phenomenon of globalization, resulting in social deprivation and degradation for Mayan peoples. Yet, Mayan culture survives despite a brutal history of religious repression, racism, inequality, and exclusion (Wilkinson, 2002).

10. The Geometric Pattern of the Mayan Diamond

The Mayans made use of a series of sacred geometric-numeric patterns that they transmitted from generation to generation. The utilization of these patterns probably originated with a species of rattlesnake *Crótalus durissis* (Figure 2), found in the region (Nichols, 1975; Diaz, 1995; & Grattan-Guinness, 1997). Rattlesnake skins possess a unique diamond pattern (Figure 3); this particular species is called the "diamond backed rattle snake" in English. The contemplation of this form and geometric pattern inspired Mayan art, geometry, and architecture (Diaz, 1995, & Grattan-Guinness, 1997). The images of rattlesnakes are found in many aspects of Mayan culture. They symbolize the birth and life changes of the ancient Mayans because *Crótalus durissis* enlivens and crawls its way across time.

The significant and purely abstract, patterns found in geometric rattlesnake forms are found in fabrics and in façades of numerous ancient buildings, monuments and architectural structures though out the ancient Maya territories.

In Figure 4, it is possible to observe that the degrees of slope of Mayan pyramids are extremely steep and are difficult to climb comfortably. The easiest and most comfortable way to climb Mayan pyramid stairs is to climb the steps in a zigzag.

Figure 2. Crotálus durissus

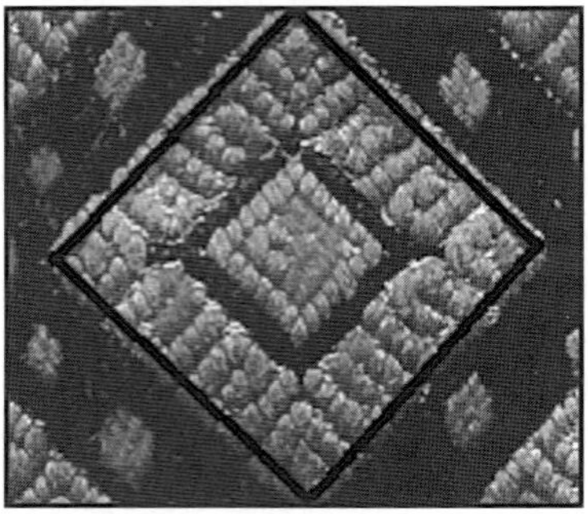

Figure 3. Rhombus representing the geometric from of the skin of the rattlesnake

Figure 4. EI Castillo in Chichen Itza

The trajectories formed by the movement of the priests ascending and descending of the pyramids have the same form and geometric patterns found in the rattlesnake skin (Diaz, 1995; Grattan-Guinness, 1997). In this case, Mayan priests ascended and descended pyramids in a criss-cross ritual that reproduced the diamond pattern of the rattlesnake.

11. The Sacred Mayan Mats

The word *Popul* present in the title in of the sacred book *Popul Vuh* contains the prefix *Pop (Ahpop)*, that is, the Maya Quiché word for *mat* (Recinos, 1978). The Gods that were represented in the monuments of numerous Mayan pyramids sat on top of *Pop* patterns built over sacred mat patterns. The monuments themselves were constructed over *mats* that had magic or mystical power and used number values to provide a spiritual foundation to accompany the physical buildings.

Diaz de Castillo (1983) affirmed that the priests and the Mayan nobility also sat on top of sacred *mats* for ceremonies and festivities. He also described that in the time of the conquest of the Mayans by the Spanish, important meetings were made between Spanish leaders and the Mayan nobility and priests. In these meetings, the Spanish leaders sat on sacred mats that were offered by the Mayan nobility. However, they covered the mats with cloth that contained values that neutralized any mystical power and blessing that emanated from the numbers presented in the geometric patterns in the mats (Figure 5).

These patterns were sculpted in stones and used in jewellery and cloth. They are still used in the clothing of 21st century Maya descendents (Figure 7).

Through much of their weaving, the present *magic* of the designs in the vestments are connected with ceremonies that were promoted by their ancestors.

In the universal diamond (Figure 6), the four fields represent the frontiers between space and time in the Mayan universe. The small diamonds that are in each field represent the cardinal points of this universe; the east is placed where the sun rises, the west is placed below and represents the end of the day, the north is placed on the left and the south on the right.

The Mayan spatial orientation of the four corners of their universe is not based on the cardinal points of the western compass (Morales, 1993). Frequently, the diamonds are placed so eastern and western fields are coloured blue to represent the Caribbean on the east and Pacific Ocean on the west. The centre of each large diamond is placed so that a small diamond represents the sun. Sometimes, a fine line is placed on the design that connects the east and west and represents the trajectory of the sun across the sky.

Figure 5. Different Geometric Patterns of the Mayan Sacred

Figure 6. The Universal Diamond

Figure 7. Huiple-Traditional Maya Dress

Figure 8. Wall of a Mayan Temple in Yucatan, Mexico

Many present-day Mayans weave and sew many of the same designs and motifs that have been popular since the classic period of Mayan culture between 3rd and 10th centuries (Deuss, 1981; Rowe, 1981). Many of the pictures found on ceramics, lintels, stela and murals also contain the same patterns and geometric forms that are utilized in the Mayan weavings (Figure 8).

The diamond shape was considered extremely important, indeed sacred because it represented the light reflected with brilliance in a polished diamond. This diamond shape brought a sense of order and light, and reminded them that all need to live in harmony.

The attraction of the diamond form was in concord with the sacred numbers of the Gods; it was divine power that implied the numbers of 1 to 9 (Nichols, 1975, Orey, 1982). This context allowed the Mayans to use these numbers which were based on the snakeskin and diamond patterns for a type of numerology because they could have had a sacred value and a specific significance (Coe, 1966, Coe & Kerr, 1988; Nichols, 1975, Orey, 1982).

12. Decoding Mayan Messages

According to Nichols (1975), the patterns X's or XX's[4] found on many Mayan mats (Pop) contained information. The numbers placed on these mats progressed sequentially and zigzagged diagonally as shown in Figure 9. The first number is positioned on the right vertice of the first square that composed the mat. For example, on a mat of 3 lines by 2 columns, the numbers are placed as in the diagram below:

The final numerical number of this matrix might be calculated in the following manner:

1. We add the corresponding numbers of each line of the matrix.

$$1+6=7$$

$$5+2=7$$

$$3+4=7$$

Consulting the table 2, the result 7 has the value: God in Divine Power.

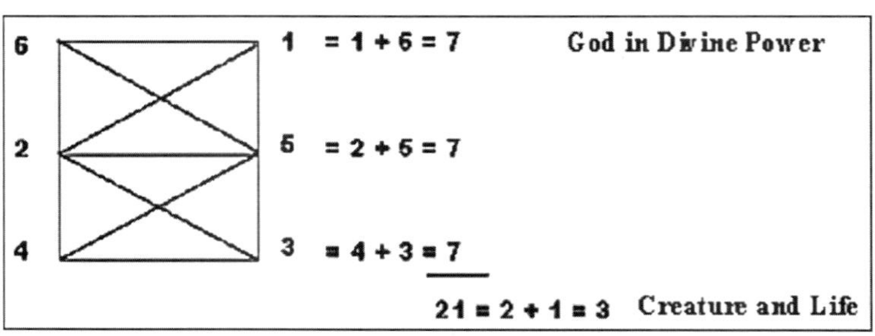

Figure 9. Decoding Mayan Messages

[4] According to Girard (1979), "when the King spreads his legs and lifts his arms over his head, he assumes a posture that can be called a cross and which is nothing more nor less than the representation" (p. 293) of the glyph of kin or glyph of the sun.

Table 2. The Sacred Significance of the Numbers

1. God or Goddess
2. The Creator: Parents
3. Creature and Life
4. Venus, called Kulkulkan
5. Priest : The Hand of God
6. Life and Death
7. God in Divine Power
8. Body and Soul
9. The Nine Drinks

2. Adding all the results we get:

$$7+7+7=21$$

3. We then add the digits resulting in the ultimate value of: 2 + 1 = 3
4. According to the table 2, the number 3 corresponds to Creature and Life.

A possible interpretation of the message of this result can then be: *God utilizes His Divine Power to give life to all creatures in the world.*

Objects found in some of the most important archaeological sites of Guatemala such as Tikal and Quirigua reveal that Mayan priests made certain decisions based on sacred mats because they contained significant sacred numbers that were based on *ultimate values* for each pattern. For example, to find a solution for a given situation, a priest needed to make a decision towards codifying a mat that contained the ultimate value 6 which signifies "Life and Death." In this perspective, the Mayan priests were charged with maintaining the spiritual, religious, scientific, and mathematical knowledge of Mayan civilization.

13. The Mayan Number System of the Divine Creation

According to Mayan theosophy, the creation of the world was closely associated with mathematical concepts. In accordance with this perspective, Girard (1979), states:

> The Quiché codex begins by referring to the creation of the universe. Divinity – pre-existent to its works – creates the cosmos, which extends through two superimposed, quadrangular planes – heaven and earth – their angles delimited and their dimensions established. Thereby is established the geometric pattern from which will derive the rules for cosmology, astronomy, the sequential order in which events occur, and the marking out and use the land, which for the Maya are all reckoned from that space-time scheme.
>
> (p. 28).

Diaz (1995) stated that the creation of the four corners of the Mayan universe was governed by the geometric pattern of the rhombus which represents the geometric pattern on the skin of the rattlesnake *Crótalus durissus*. In the creation of the Mayan universe, the god Tzakol's[5] used his supernatural intervention in the creation process by applying the sacred-symbolic power of the numbers as described in the book *Popol Vuh* (Recinos, 1978). This can be interpreted by the following mathematical pattern:

Number 0: "This is the first account, the first narrative. There was neither man, nor animal, birds, nor forests; there was only the sky. ... Nothing existed." (Recinos, 1978, p. 81). It was like a seed phase because all was in suspense, all calm, in silence, all motionless, and the expanse of the sky was empty. Thus, the Mayans used a seed symbol  for zero.

Number 1: *Tzakol*, known as Huracán, is the first hypostasis of God. He planned the creation of the universe, the birth of life, and the creation of man (Recinos, 1978).

Number 2: The Creator brought the Great Mother (*Alom*) and the Great Father (*Qahalom*). *Alom* is the Great Mother and represents the essence of everything that is conceived. *Qahalom* is the Great Father who gives breath and life.

Number 3: Then came the three: *Caculhá Huracán* (the lightning), *Chipi-Caculhá* (the small flash) and *Raxa-Caculhá* (the green flash) that represent life and all creatures.

Number 4: Diaz (1995) states that the Venus Goddess, called *Kukulkan* is represented by number 4 because it corresponds to the four sides of the rhombus. His view is that the number 4 is "in the design on the skin of the *Crótalus*" (p. 8).

Number 5: The gods delegated their power to the priests. The priests were considered as the hands of the god because they gave to the Mayan people the gods' answers to their prayers. In Mayan ceremonies, the priests held ceremonial rods decorated with rhombuses in the centre and a snake head on top and they were "the mathematical insignias of the wise priests that ordered the construction of the Mayan temples" (Diaz, 1995, p. 8).

Number 6: In Mayan cosmology, bones are like seeds because everything that dies goes in the Earth and then new life emerges from the Earth in a sacred cycle of existence.

Number 7: The Mayans believed that the divine power of the gods reorganizes the order of the cosmos and reunites the human world with the supernatural and mystical worlds.

Number 8: Everything on and of the Earth relates to material reality (the body) and spiritual reality (the soul).

Number 9: *Alom* made nine drinks with the milling of yellow and white corn. With these drinks she created the muscular body and the robustness of men.

---

[5] According to Diaz (1995), "the root of Tz'akol is Tsa or Tza, that is Tzamná or Itzamná, which comes from Tzab, rattlesnake, which is onomatopoeic with the sound of the rattle" (p. 8).

14. The Symbolism of Maya Numerology

Mayans perceived that natural events occurred in accordance with numerical patterns, as in the annual sequence of the lunar cycles. Numbers were related to the manifestations of nature and for this reason it was possible to determine that the universe obeys laws that allowed them to measure and anticipate certain forms of natural events. Because of these observations, "the Maya are said to have "mathematized" time, and, through it, their religion and cosmology" (Ascher 2002, p. 63).

Despite advanced mathematical knowledge of the Mayan people, they incorporated concepts of theogony[6] with concepts of numbers by utilizing symbolic elements to express their ideas about the creation of the universe. See in this context, the Mayan theology posits nine cosmic manifestations that are perceived in nature and through which the Mayan people infer the abstract manifestations of God.

The theogonic philosophy of the Mayans exceeds the limits of mathematical knowledge because it relates to the numbers of the abstract manifestations of The Great Architect, with the objective of explaining, understanding, and comprehending the organizational principles of the creation of the universe.

15. Final Considerations

This study focuses here on the ethnomathematics of the Mayan, aiming to understand how they knew, understood, and organized part of their mathematical knowledge to comprehend and explain the creation of the universe according to their believes. In this perspective, the Mayans developed a sacred and magical numbers system through the construction of mats that were elaborated in divine patterns.

Mayan people possessed a sophisticated geometric and numerical creation story of their universe, whose first record is related to sacred numerical values. From what we understand of the Mayan cultural perspective, numbers, symbols, and words could direct the priests to deities of corresponding numerical values. The Mayans were not the only Americans to use this perspective, it is important to highlight a study of the Inca quipu by Ascher & Ascher (1981) who found "... the quipu could be a demonstration of interest in the number itself; or the number could have significance because it has been invested with some meaning beyond its numerical value" (p. 140).

The study of mathematical practices as found in the Mayan sacred mat and geometric diamond patterns serve as a tool to understand and analyze the sacred power of numbers, from 1 to 9, which can be considered as a useful numerological system used by the priests to codify and interpret messages. This aspect of the Mayan culture helps us to demonstrate one use of an ethnomathematical, anthropological, and global perspective in which we might recreate, internationalize, study, and preserve a portion of the wisdom and knowledge of these unique and resilient peoples.

---

[6] The genealogical account of the origin of the gods.

In this context, from a global perspective, ethnomathematics can be considered an academic counterpoint to globalization, and offers a critical perspective of the internationalization of mathematical knowledge through attempts to connect mathematics and social justice. It is also possible to perceive ethnomathematics as the academic articulation between cultural globalization of mathematical knowledge and diverse non-Western cultural groups. In this ethnomathematical perspective, it is important that individuals in different cultural groups understand the overall importance of their own mathematical knowledge. They may also need to extend the scope of this knowledge through collaboration with diverse cultural groups (other than their own) by sharing different mathematical practices that are part of a developing new context of globalization.

See in the above context, when discussing, sharing, and internationalizing mathematical practices and the ideas used by other cultures, it is necessary to recast them into an individual's Western mode, modelling allows us to translate these practices into western mathematics. In this cultural dynamism it is possible to distinguish between the mathematical practices and ideas which are implicit and those which are explicit, between western mathematical concepts and non-western mathematical concepts which are used to describe, explain, understand, and comprehend the knowledge generated, accumulated, transmitted, diffused, internationalized, and globalised by people in other cultures.

References

Ascher, M. (1991). *Ethnomathematics: A multicultural view of mathematics ideas.* Belmont, CA: Wadsworth, Inc.

Ascher, M. (2002). *Mathematics elsewhere: An exploration of ideas across cultures.* Princeton, NJ: Princeton, University Press.

Ascher, M., & Ascher, R. (1981). *Mathematics of the incas: Code of the quipu.* Mineola, NY: Dover Publications, Inc.

Atweh, B., & Clarkson, P. (2002). Mathematics educator's views about globalization and internationalization of their discipline: Preliminary findings. In P. Valero & O. Skovsmose, (Eds.), *Proceedings of the 3rd International MES Conference,* (pp. 1–10). Copenhagen: Centre for Research in Learning Mathematics,

Borba, M. (1997). Ethnomathematics and education. In A.B. Powell & M. Frankenstein, (Eds.), *Ethnomathematics: Challenging eurocentrism in mathematics education* (pp. 261–272). New York: State University of New York.

Cajori, F. (1993). *A history of mathematical notations: Two volumes bound as one.* New York: Dover Publications, Inc.

Coe, M. D. (1966). *The maya.* New York: Praeger Publishers.

Coe, M. D. (1992). *Breaking the maya code.* New York: Thames and Hudson.

Coe, M. D., & Kerr, J. (1988). *The art of the maya scribe.* New York: Harru N. Abrams.

D'Ambrosio, U. (1985). Ethnomathematics and its place in the history and pedagogy of mathematics. *For the learning of mathematics, 5*(1), 44–48.

D'Ambrosio, U. (1990). *Etnomatemática* [Ethnomathematics], Editora Ática. Brazil: São Paulo, SP.

D'Ambrosio, U. (1993). Etnomatemática: um programa [Ethnomathematics: A program], *A Educação Matemática em Revista, 1*(1), 5–11.

D'Ambrosio, U. (1999). *Educação para uma Sociedade em Transição* [Education for a society in transition], Papirus Editora. Brazil: Campinas, SP.

D'Ambrosio, U. (2000). Ethnomathematics: A step toward peace, *Chronicle of Higher Education, 12*(2), 16–18.
D'Ambrosio, U. (2001). *Etnomatemática: Elo entre as tradições e a Modernidade* [Ethnomathematics: A link between traditions and modernity], Editora Autêntica. Brazil: Belo Horizonte, MG.
D'Ambrosio, U. (August, 2002), *Etnomatemática* [Ethnomathematics], Closing lecture delivered to the II Congress on Ethnomathematics, UFOP, Ouro Preto. Brazil: Minas Gerais.
Deuss, K. (1981). *Indian costumes from guatemala.* Great Britain: CTD Printers Ltd..
Diaz de Castillo, B. (1983). *Historia Verdadera de la Conquista de La Nueva España* [True History of the Conquest of New Spain], Porrúa. México: Ciudad de México,
Diaz, R. P. (1995). The mathematics of nature: The canamayté quadrivertex, *ISGEm Newsletter, 11*(1), 5–12.
Featherstone, K. & Radaelli, C. (Eds.). (2003). The politics of Europeanization, http://www.oxfordscholarship.com/oso/public/content/politicalscience/0199252092/toc.html,' ISBN-10: 0-19-925209-2. Oxford, UK: Oxford Scholorship.
Gerdes, P. (1985). How to recognize hidden geometrical thinking? A contribution to the development of anthropological mathematics. *For the learning of mathematics, 6*(2), 10–12, 17.
Girard, R. (1979). *Esotericism of the popol vuh: The sacred history of the Quiché-Maya.* Pasadena, CA: Theosophical University Press.
Grattan-Guinness, I. (1997). *The rainbow of mathematics: A history of the mathematical sciences.* London, Great Britain: W. W. Norton & Co.
Merick Jr., L. C. (1969). Origin of zero, in National Council of Teacher of Mathematics. In J. K. Baumgart, D. E. Deal, B. R. Vogelt & A. E. Hallerberg (Eds.), *Historical topics for the mathematics classroom.* Washington, DC: NTCM.
Ifrah, G. (1998). The universal history of numbers: From prehistory to the invention of the computer. New York: John Wiley & Sons, Inc.
Lindsey, R. B., Robins, K. N., & Terrel, R. D. (2003). *Cultural proficiency: A manual for schools leaders.* Thousand Oaks, CA: Corwein Press, Inc.
Morales L. (1993). Mayan geometry, *ISGEm Newsletter, 9*(1), 1–4.
Mtetwa, D. K. (1992). Mathematics & ethnomathematics: Zimbabwean student's view. *ISGEm Newsletter*, 7(1), 1–4.
Nichols, D. (1975). *The lords of the mat of tikal.* Antigua, Guatemala: Mazda Press.
Orey, D., (February, 1982), Mayan math, *The Oregon Mathematics Teacher, 1*(1), 6–9.
Oweiss, I. M. (1988). *Arab civilization.* New York: State University of New York Press.
Powell, A. B. & Frankenstein, M., (1997), Ethnomathematics praxis in the curriculum. In A. B. Powell & M. Frankenstein, (Eds.), *Challenging eurocentrism in mathematics education,* (pp. 249–259). New York: SUNY.
Recinos, A. (1978). In D. Goetz & S. G. Morley, (Trans.), *Popul Vuh: The sacred book of the Ancient Quiché Maya.* Oklahoma: Norman University of Okalahoma Press, (Original work published 1960).
Rosa, M. (2000). *From reality to mathematical modelling: A proposal for using ethnomathematical knowledge*, unpublished master's thesis. Sacramento: California State University.
Orey D. C., & Rosa, M. (2003). Vinho e queijo: Etnomatemática e Modelagem! [Wine and cheese: Ethnomathematics and modelling!]. *BOLEMA, 16*(20), 1–16.
Rowe, A. P. (1981). *A century of change in guatemalan textiles.* New York: The Centre for Inter-American Relations.
Sen, A. (2002), *How to judge globalism,* [33 paragraphs], The American prospect [On-line serial], 13(1); http://www.propsect.org/print/V13/1/sen-a.html.
Toffler, A. (1980). *The third wave.* New York: William Morrow and Company, Inc.,
Wilkinson, D. (2002). Silence on the mountain: Stories of terror, betrayal, and forgetting in Guatemala. New York: Houghton Mifflin Company.

14

INTERNATIONALISATION AS AN ORIENTATION FOR LEARNING AND TEACHING IN MATHEMATICS

[1]Anna Reid and [2]Peter Petocz
[1]Centre for Professional Development and [2]Department of Statistics, Macquarie University, North Ryde, NSW 2109, Australia

Abstract: In this chapter, we put forward the claim that any specific view of internationalisation corresponds to a particular orientation for learning and teaching in mathematics. We use a critical discourse perspective to explore variation in the intentions and outcomes of an 'internationalised curriculum' and apply the results to the discipline of mathematics. We support the discussion with reference to several components of our research: in particular, a study on students' conceptions of mathematics and learning in mathematics, and another study reporting on lecturers' understanding of the intersections between teaching and sustainability, an important correlate of internationalisation. Our aim is to consider the way in which internationalisation contributes as a 'value' orientation for our students' approaches to their study and indeed to their whole lives. We then apply our model to a practical discussion of the construction of learning environments that support a focus on students' professional formation and the development of their global perspectives

Keywords: Internationalised curriculum, conceptions, values, sustainability

1. Introduction

Mathematicians have often considered internationalisation to be a core feature of their subject, acknowledging its rich multicultural heritage and the global endeavour of mathematical research. In many discussions in the general tertiary context, however, internationalisation is seen as an umbrella term for the enticement of students from around the globe to study in another country. Between these views lies a range of different ways of thinking about internationalisation, which seems to be becoming a necessary component of any learning environment. We claim that

B. Atweh et al. (eds.), Internationalisation and Globalisation in Mathematics and Science Education, 247–267.

any specific view of internationalisation corresponds to a particular orientation for learning and teaching in mathematics.

In this chapter, we use a critical discourse perspective to explore variation in the intentions and outcomes of an internationalised curriculum in any discipline, and use this exploration to suggest appropriate reconceptualisations of teaching, learning and researching in mathematics. We support the discussion with reference to several components of our research: in particular, a recent large cross-national research study on students' conceptions of mathematics and learning in mathematics, and an empirical study reporting on lecturers' understanding of the intersections between teaching and sustainability, an important correlate of internationalisation. Our aim is to consider the manner in which internationalisation contributes as a 'value' orientation for our students' approaches to their study and indeed to their whole lives.

The chapter concludes with a practical discussion of the construction of learning environments that favour a questioning and interactive approach, curriculum that focuses on professional formation and provides support for the development of students' global perspectives, and ways in which we as lecturers can become aware of our students' expectations for their learning. As the nature of knowledge and the opportunities for professional work change, traditional disciplinary approaches to teaching can be modified. The focus of the learning environment should be on the needs of all our students: indeed, we should consider that all students are 'international' in the 21st Century.

2. Mathematics – an International Subject

Mathematicians have often considered internationalisation to be a core feature of their subject, acknowledging its rich multicultural heritage and the global endeavour of mathematical research. Firstly, of course, mathematics is an international language – maybe the only truly international one! A result written in mathematical notation can be read by mathematicians in any country in the world. Indeed, it may be more than international: in Carl Sagan's novel Contact (1985), the initial communication received from extra-terrestrial sources is a mathematical one – pulses representing the sequence of primes. Various projects have been undertaken to construct and send mathematically-based messages in an attempt to establish communication with other intelligences (SETI, 2004; Vakoch, 2002).

Then, the development of mathematics has always been an international and multicultural affair. Classical histories of the subject (such as Boyer & Merzbach , 1991) highlight the contribution of the Egyptians, the Babylonians, the Greeks, the Chinese, Indians and Arabs, and the European Renaissance. Other books focus specifically on the contribution of non-European cultures (Joseph, 1991; Ifrah, 1998). Writers in the area of ethnomathematics (D'Ambrosio, 2001; Frankenstein & Powell, 1997) investigate the mathematical practices of various (often indigenous) groups of people and apply the same methods to explicating the mathematical practices of specific groups of people in specialised contexts (D'Ambrosio, 2001, gives several examples). Current developments and applications of mathematics are made by an international group of research mathematicians with contributions coming from around the globe. International conferences, journals and professional exchanges, and global communication via e-mail and the internet, allow this group of research mathematicians to work together. While it is true that 'western' mathematics plays a dominant role in today's world, alternative traditions and approaches are also represented.

In the area of mathematics education, there is a similar international dimension. Conferences such as the International Congress on Mathematics Education (ICME) bring together educators from around the globe, journals such as the International Journal of Mathematical Education in Science and Technology (iJMEST) and the Statistics Education Research Journal (SERJ) provide forums for exchange of ideas, and publications such as the Second International Handbook of Mathematics Education (Bishop et al., 2003) summarise debate on current problems world wide. International comparative studies such as the Trends in International Mathematics and Science Study (TIMSS, US Department of Education , 2003) give countries an opportunity to benchmark their standards and their curricula against other countries in the World. Large numbers of students, particularly at the postgraduate level, study mathematics in countries other than their country of origin, and gain experience in mathematics and pedagogical methods that they can take back to broaden the approaches in their own countries. This movement of students between different countries in the world makes mathematics a truly internationalised subject.

But this is not to say that there are no problems! Atweh et al. (2003) discuss some of these problems: issues about the role of international organisations such as the International Commission on Mathematical Instruction (ICMI) and Psychology of Mathematics Education (PME) and the difficulties faced by participants from under-developed countries; the negative political and social effects of international comparative studies (such as TIMSS) that are dominated by countries such as the USA; the marginalisation of ethnomathematics and concerns over power and voice; the notion of global curriculum that is really a western curriculum, propped up by textbooks written for US markets and focusing on US concerns. Other problems that are apparent include the domination of English-language journals from the US and the UK, even over people from other countries writing in English; international students in mathematics who are seen primarily as a source of income for the host country; and the one-way nature of international student exchanges. Asking mathematicians and mathematics educators about internationalisation results most commonly in the response that 'mathematics is already internationalised' (Atweh & Clarkson, 2002, is an exception). However, the reality does not bear this out.

3. Internationalisation as a Value For Learning

Mathematics and mathematics education are situated in the wider discourse of internationalisation. Various notions of internationalisation impact on the professional preparation of our mathematics students and the range and variety of these components influence the manner in which internationalisation may contribute as a 'value' orientation for our students' approaches to study and indeed to their whole lives. In this section, we investigate current thinking surrounding the notion of internationalisation using a critical discourse analysis of recent writings on the subject, not limited to the area of mathematics. This will enable us to appreciate the broader context when we return our focus to mathematics education in the following sections.

The notion that internationalisation can be seen as a value is at odds with much of the current discourse surrounding internationalisation. The exigencies of student and faculty mobility in the 21st Century have demanded that attention be focused on the management, and quality issues associated with that management, of the recruitment of students from one nation to another, and the delivery of courses in countries other than the one in which they were developed (Fallshaw, 2003). This demand, which will continue to increase, subtly draws our attention towards these managerial concerns at the expense of what is truly the core of internationalisation – appreciation of diversity and the personal ability to focus on fostering inclusive attitudes and practice within our professional lives. However, it is not possible to consider internationalisation as a value without any consideration of its managerial aspects. Students who cross national borders for their education have experienced a range of conflicting situations as they negotiate the location of their course of study, the entry requirements of any particular university, the accommodation and student support opportunities in their host country, and the host country's visa requirements. All these components, and their previous educational and cultural experiences, become a part of the students' perceptions of their learning environment.

Of course, this is a picture of only one sort of student mobility. Others experience learning in courses that are taught by academics from institutions in another country using face-to-face, block or e-learning approaches. Still others are part of situations where courses have been adapted from those delivered elsewhere to incorporate a regional flavour. The commonality between these different learning situations is that all of the students have an experience of the globalisation of learning which not only intrudes upon their own personal intentions for learning, but also enables them to benefit from learning situations which are different from their previous experiences. Most students learning in any country are in a situation where their cohort consists of students from many countries. There is rarely a notion of a completely homogenous group where cultural and pedagogical values are shared in common. Instead, there is an expectation that all students will need to adjust to different learning cultures (Scheyvens et al., 2003) and negotiate language in diverse social and academic contexts (Volet & Ang, 1998; Montgomery & McDowell, 2004). We suggest that these conditions have set up an opportunity for the students of the 21st Century to have learning experiences that prepare them to take part in a world where professional knowledge and inter-cultural relations are changing at a rapid pace.

In addition to the issues of mobility (and the inclusion of international students in a 'domestic' environment), many students will have had previous educational experiences where knowledge is packaged for them as a series of irrefutable facts. This epistemology suggests that knowledge is a bounded concept in that it is absolute and can be learned through acquisition and repetition. An alternate experience that some students may have had is that knowledge is a construction of a particular social situation and is thus subject to critique from differing positions. Student learning in this situation involves active deconstruction and reconstruction of the knowledge encountered. Rizvi (2000) suggests that internationalisation must look at the 'global-local relationship' (that is, the situatedness of knowledge): the 'changing nature of the knowledge economy has identified the important issues of tension between domestic and global education and market agendas, and the acknowledgement that education has become a marketable commodity'. Indeed, the marketing of knowledge is an obvious and relatively easy objective. The very idea that knowledge can be marketed also emphasises the notion that knowledge is finite and in some way unchanging. This epistemology is sustained by educational practices that encourage memorisation and recall of facts. Educational practices that focus rather on students' application, integration and creation of knowledge are the antithesis of this, and are related to higher quality learning outcomes (Biggs, 1999). Both situations have been described in the context of mathematics and science education. The large literature that describes ethnomathematics and ethnoscience attempts to raise awareness about the value of cultural knowledge that approaches various concepts from perspectives other than the western, empiricist standpoint.

The dominant 'marketing and quality assurance' paradigm encourages a neglect of the central focus of universities – that is, knowledge, its discovery through research, and its dissemination through teaching and learning. Australian universities have been directing their attention towards different aspects of internationalisation for some time. Adams & Walters (2001) suggest a particular orientation to internationalisation, writing that 'International education in Australian universities has been dominated by a single paradigm. This paradigm can be described as the recruitment to home campuses of international students via differentiated regional and country strategies, conventional marketing techniques, and commission agents' (p. 269). This idea is extended by Ball (1998) through the identification of a tension between national educational and economic development with international focuses. He indicates that the inevitable results of this 'policy dualism' are 'changes in the way education is organised and delivered, also … changes [in] the meaning of education and what it means to be educated and what it means to learn' (p. 128). There is an obvious tension between these two focuses, as economic development though an educational market may (and often does!) result in the recruitment of students who may not yet be ready to take advantage of the opportunities presented by international study. Receiving institutions respond to this challenge through the provision of social and academic support systems, but the students are still often unable to receive the maximum benefit (Leask, 2003; Curro & McTaggart, 2003).

Discussion amongst faculty members suggests that internationalisation has a significant role to play in the relationships between learning and teaching, universities and professional bodies, traditional disciplinary approaches and the integration of new research-based knowledge, and the development of new ways of thinking about teaching and learning support roles (Reid & Loxton, 2004). For students, internationalisation can mean an experience of visiting a different country, learning about contrasting ways of living and thinking, and perhaps integrating these experiences into their own value systems. Beyond this, their chosen areas of study will help them develop and refine their understanding of the world. Domestic students also have the opportunity to value the variation found in their classes which brings to their learning multiple perspectives on living, learning and attitudes. What students encounter and grapple with while they are at university plays an important role in their orientation to life and work at the conclusion of their formal study (Haigh, 2002; Ryan & Hellmundt, 2003).

The diverse nature of the student population brings with it a series of implications for the development of supportive learning environments. Diversity can be considered on a number of levels: it can be seen as the range of intellectual and physical abilities within the student group, the range of cultural backgrounds and experiences, language expertise, work or professional experience, or simply as the range of ways that students learn. As student mobility increases so too does the level of engagement with international issues. Buenfil-Burgos (2000) alerts us to the misuse of diversity in the classroom when internationalisation is seen as 'universalisation, homogenisation, integration and centralisation' (p. 4). From this perspective, the implication of the internationalisation of education is the loss of cultural diversity if we as academics attempt to run courses that emphasise predominantly western views. Buenfil-Burgos says 'from a cross-cultural perspective, globalisation has been construed as a cultural catastrophe that is harassing cultural minorities, and therefore a cultural and educational strategy becomes crucial to defend minority cultures' (p. 4).

From an Australian context, this 'loss of cultural diversity' can be experienced when one philosophy of education is privileged above another. This is the case when a largely western education system favours and fosters the ideas of 'individualism' at the expense of 'collectivism' which is favoured by students with a Confucian education experience (Chalmers & Volet, 1997; Triandis, 2002). However, acknowledging that these differences exist within a classroom and are also characteristics (amongst others) of professional work can enable a powerful integration of ideas where each orientation is valued (Ryan & Hellmundt, 2003). Haigh (2002) suggests that this approach can lend itself towards the development of an inclusive curriculum, and Ryan & Hellmundt (2003) expand on this notion with concrete suggestions for the enactment of such a curriculum: 'Many lecturers recognised that common unit objectives (attitudes, knowledge and skills) assume homogeneity of learners and the desirability of homogeneous learning outcomes. They instead offered flexible and negotiable learning objectives and outcomes and learning contracts for individual needs and interests. Some lecturers recognised that students may have no previous experience of some types of assessment and may need training and 'scaffolding' until they are familiar with these approaches. Some were also aware that their own perceptions of ability were influenced by cultural assumptions, and that students were often being assessed for their mastery of academic discourse rather than for critical or original thinking' (p. 5).

The need for educational strategies that assists students to develop their understanding of different cultures and to empathise with others with different experiences and perspectives is apparent. Marton and Trigwell (2000) suggest that it is awareness of variation that enables students to learn tolerance: 'When it comes to preparing students for an unknown future, the nature of variation is of decisive importance… If you want to contribute to enabling students to participate in the yet unknown learning communities of the future, you have to let them participate in the learning communities of today, which keep changing and which differ quite significantly from each other' (p. 394). It is possible to use the breadth of experience found within a class to challenge assumptions about the nature of learning, the subject and the world, thereby benefiting from the variation found within a group and enhancing the quality of learning. Yet, it appears that it is much more difficult to develop and act upon pedagogical strategies to promote internationalisation as a concept for learning than it is to develop and promote strategies that deal with the quality provision of recruitment and academic management.

4. Intersections Between Research and Internationalisation

Returning to the particular discipline of mathematics, we focus on the various experiences of learning and teaching that can be experienced by students and lecturers. Looking at the nature of mathematics learning at tertiary level from the viewpoints of the participants allows us to explore the connections between views of mathematics and internationalisation. Mathematics lecturers' views of 'diversity' as shown by their writings in recent forums, including papers presented at a conference whose title included The Challenge of Diversity, demonstrate the ways in which they are grappling with ideas of internationalisation. Empirical research on students' conceptions of mathematics, statistics and learning, derived from interviews that we have carried out over the previous four years, is also summarised in the following section. We follow this with a discussion of academics' conceptions of sustainability, an important correlate of internationalisation, again obtained from an interview-based study. Using these research studies and our previous investigations of the intersection between mathematics and sustainability, we suggest that a similar framework can be used to explore the relationship between views of mathematics and internationalisation.

In 1999, the annual Delta Symposium on Undergraduate Mathematics was specifically focused on The Challenge of Diversity (Spunde et al.,1999). An analysis of the content of the papers presented shows the ways in which the idea of diversity was understood by the participants at the conference. The papers describe diversity in students' mathematical preparedness for tertiary study, diversity in teaching methods, mathematics as a component of diversity for other areas of study such as engineering, and diversity in the range of assessment methods. Only three papers at that conference suggested that diversity was related to students' experiences and expectations. McIntyre and Pfannkuch (1999) indicated that the needs of students with different life and pedagogical experiences are actively considered as part of curriculum change activities, and these student characteristics are possibly the reason why some students think that mathematics 'lack(s) relevance to the real world' (Oldknow, 1999). Motivated by a notion of justice (which appears again in our investigations of sustainability), Snyders (1999) proposed that social and cultural equity could be implemented through the systematic development of different entry points to mathematical study that reflect the different levels of student preparedness which result from their previous experiences.

More recently, the tenth International Congress on Mathematical Education (held in 2004 in Copenhagen, Denmark) attracted a large number of participants from around the world. One of the discussion groups (DG5) focused on 'International cooperation in mathematics education', and a discussion paper on the website formulated a set of relevant questions which illustrates the range of the discussions: What are the goals for international collaborations? Should cooperation be regional or global? What are the barriers to genuine and equitable international cooperation? What forms could such cooperation take, and how could it be organised and implemented? How can a cooperative preparation of researchers in mathematics education contribute to the development of a genuine and equitable cooperation? Is there a danger that international cooperation may lead to excessive homogenisation of mathematics education? Aside from this, only a small proportion of the conference papers focused on social, cultural or professional issues relevant to internationalisation, and most of those were oriented towards ethnomathematical approaches (e.g. Matang & Owens, 2004; Mosimege & Ismael, 2004).

More generally, there is a large body of research that investigates students' ideas of learning in a variety of subject areas (see Marton & Booth, 1997, for a summary). Other studies have investigated academics' notions of teaching (Kember, 1997, summarises these), as well as students' and academics' views of their subjects. These research studies help us to reorient our thinking about internationalisation as we examine the complexity of teachers' and students' experience of learning in specific situations. In some of our own research, we have investigated students' conceptions of mathematics (Reid et al., 2003, 2005) and statistics (Petocz & Reid, 2001, 2003a; Reid & Petocz, 2002a, 2002b). These studies were carried out using several series of in-depth interviews with students who were studying the mathematical sciences and planning to become professionals in some area of mathematics, statistics, mathematical finance or operations research. The interviews were designed to encourage our participants to think about the discipline and their learning in the mathematical sciences, and to describe the ways that they constituted meaning from their experiences. We asked students questions such as: What is statistics?, What do you understand mathematics to be about?, What do you aim to achieve when you are learning in mathematics?, How do you know when you have learned something in statistics? and What do you think it will be like to work as a qualified mathematician? These questions were followed by further probing questions which responded to their answers: for example, general questions such as: Can you give me an example of that? and What do you mean by 'understand'? or specific questions such as: So how does maths help you find things out?

The interviews were analysed using a phenomenographic approach (Marton & Booth, 1997). The results from these studies are summarised in the 'outcome spaces' shown in Table 1 showing the essential aspects of qualitative differences between them (and supporting quotes from the interviews are given in the papers referenced). Some students held conceptions of mathematics or statistics as being concerned with techniques and components, and learning as being focused on doing required activities in order to acquire these techniques. For other students, mathematics was concerned with models (including models derived from data in the statistical context) and learning was focused on constructing and applying such models in order to understand the discipline. Some students viewed mathematics or statistics as an approach to life and a way of investigating problems, and learning as a process of developing mathematical or statistical ways of thinking and changing their view of the world. In common with other phenomenographic outcome spaces, these conceptions were hierarchical and inclusive. Those students who talked about mathematics in terms of techniques did not include or appreciate elements of any of the other views. However, those students who talked about mathematics as an approach to life could also discuss mathematical models and the techniques of mathematics: for this reason, we refer to such conceptions as 'broad' or 'holistic', as opposed to the 'narrow' or 'limited' conceptions described earlier.

In the second phase of the mathematics study (described in Petocz et al., 2007), we used an open-ended questionnaire to verify and amplify the conceptions found previously, investigate their distribution, and add an international dimension by including students from four other countries, South Africa, Brunei, Canada and Northern Ireland (as well as Australia). The open-ended questions were based on those used in the interviews: What is mathematics? What part do you think mathematics will play in your future studies? What part do you think mathematics will play in your future career? Of course, the answers were written, were relatively brief, and did not afford the opportunity of further probing or follow-up: nevertheless, they established the existence of the three levels of conceptions described earlier. The questionnaire also included a number of closed questions asking about academic, linguistic and demographic background: these were used to investigate the distribution of the conceptions and their variation across countries, years of study and language groups (see Wood et al., 2006 for a summary).

Considering the divergent views of internationalisation described by our colleagues in a range of disciplines (including mathematics, Atweh & Clarkson, 2002), it is reasonable to expect that no one way of tackling internationalisation as a core value is appropriate. Maybe we can find more agreement on various components of internationalisation, and one important correlate is the idea of sustainability. Both ideas can be described as values or dispositions, and they each contain elements that are important to the other. A look at the aims of the recent Johannesburg Earth Summit (United Nations, 2002) shows that sustainability shares a concern about economic and cultural diversity with those who support internationalisation. As

Table 1. Conceptions of mathematics and statistics and learning in these areas

Mathematics and statistics		Learning in mathematics and statistics	
(1) Techniques and Components	Mathematics and statistics consist of individual techniques and components, and students focus their attention on disparate mathematical and statistical activities, including the notion of calculation (in the widest sense).	(A) Focus on Techniques	Learning in mathematics and statistics is doing required activities, collecting methods and information, in order to pass or do well in assessments, examinations or future jobs.
(2) Models and Data	Mathematics and statistics consist of models of some aspect of reality, and students focus their attention on setting up models of a specific situation (a production line, a financial process) or a universal principle (the law of gravity).	(B) Focus on Subject	Learning in mathematics and statistics is about applying mathematical and statistical methods, linking theory and practice, in order to understand mathematics and statistics and areas where they are applied.
(3) Meaning and Life	Mathematics and statistics are an approach to life, a way of thinking, an inclusive tool to make sense of the world, and students make a strong personal connection between the subject and their own lives.	(C) Focus on Student/Life	Learning in mathematics and statistics is about acquiring mathematical and statistical ways of thinking, using this to change ones view of the world and satisfying ones intellectual curiosity.

Wals and Jickling (2002, p. 227) put it, 'teaching about sustainability includes deep debate about normative, ethical and spiritual convictions'.

In another research project (Reid & Petocz, 2006), we undertook an empirical study reporting on lecturers' understanding of the intersections between teaching and sustainability. Again, the research was carried out using in-depth interviews, in this case with lecturers at Macquarie University who were involved with postgraduate teaching. Participants were asked about their views of sustainability, teaching and the relations between them using the key questions: What do you understand sustainability to be about? and How do you include the ideas of sustainability in your teaching? These were followed with further questions that explored the responses in depth, for example, When you talk about resources what are you referring to? and You mentioned two terms, ethical investment and triple bottom line, can you explain to me what those two terms mean? An important feature was that participants were allowed to come up with their own definitions of sustainability and the interviews explored those definitions: this is in contrast to an approach where researchers supply their own definition or one from the literature. The results are summarised in the phenomenographic outcome space (Table 2) which shows how lecturers adopt different levels of responsibility for the enactment and integration of specific philosophies into their teaching practice (supporting quotes from the interviews are given in the paper).

Teaching (in the context of sustainability) represents a 'structural dimension', since it describes aspects that the academics have control over, i.e. themselves. Here, there are three hierarchical conceptions: disparate, overlapping and integrated. Sustainability (in the context of teaching) gives the 'referential dimension', since it focuses on ideas or thinking – what sustainability means, rather than the actions that comprise it. Here again there are three conceptions: distance, resources and justice. Again, the conceptions are hierarchical and inclusive: a person who holds the 'justice' view of sustainability is also aware of and able to use the 'resources' view and is able to give the sorts of definitions that might be used by people with the 'distance' view. However, this does not happen in the other direction: so a person who holds a 'disparate' view of teaching sustainability will not easily understand or have sympathy with the 'overlapping' view and may have no idea at all about the 'integrated' conception. The implications of this range of ways of seeing teaching and sustainability will be important to our discussion regarding internationalisation.

We have also investigated the problem of integrating issues of sustainability into mainstream curriculum in mathematics (Petocz & Reid, 2003b). Our approach was to compare and combine (students') conceptions of mathematics as a discipline with (lecturers') conceptions of sustainability, looking at the possibilities afforded by the various levels of conceptions of mathematics and sustainability. We conjectured that the narrowest views of mathematics (as techniques or components) are likely to coexist with a 'distance' approach to sustainability carried out using a 'disparate' teaching approach. On the other hand, the broadest views of mathematics (as an approach to life and a way of thinking) give scope for views of sustainability that include the idea of 'justice' and can be implemented with an 'integrated' teaching

Table 2. Conceptions of sustainability and teaching

Sustainability (in the context of teaching)		Teaching (in the context of sustainability)	
(A) Distance	Sustainability is approached via a definition (maybe a dictionary definition of 'keeping something going') but essentially to keep the concept at a distance and avoid engagement with it.	(1) Disparate	Teaching and sustainability are seen as unrelated ideas. Teaching focuses on the course content and 'covering' a syllabus, sustainability is seen as keeping something going, the 'green' approach.
(B) Resources	Sustainability is approached by focusing on various resources, either material (minerals, water, soil), or biological (fish, crops), or human (minority languages, populations, economies).	(2) Overlapping	The notion of sustainability overlaps with the activity of teaching. Teaching is seen as ensuring that students understand the course content. Ideas of sustainability can be incorporated (as examples, etc) to the extent that the situation allows.
(C) Justice	Sustainability is approached by focusing on the notion of 'fairness' from one generation to the following one, or even within one generation. The idea is that sustainability can essentially only happen under these conditions.	(3) Integrated	Sustainability is an essential component of teaching. Teaching is seen as encouraging students to make a personal commitment to the area represented by course content, including sustainability as part of that.

approach. The narrowest conceptions seem to limit the opportunities to incorporate sustainability into mathematics classes, while the broadest conceptions allow scope for this to happen. Although these conclusions are speculative, they are supported by evidence in the form of a small number of interviews on sustainability with lecturers in mathematics, and also by a large amount of experience with our own and our colleagues' teaching.

Here we would like to extend our speculation by replacing the notion of sustainability with that of internationalisation, and once again comparing and combining (students') conceptions of mathematics with (lecturers') ideas about internationalisation. In carrying out this process, we are relying on the connections between sustainability and internationalisation that we discussed earlier, and we are replacing our interview-based evidence on conceptions of sustainability with analysis of the discourse on internationalisation that is found in the published literature (and summarised earlier, in section 2). Most of this literature is not specifically concerned with mathematics, although there are a few articles that discuss internationalisation and mathematics (see for example, Atweh & Clarkson, 2001, 2002; Atweh et al., 2003; Atweh, 2004). The success of our speculation should be judged by the degree to which the ideas that we combine are useful in elucidating the nature of internationalisation as an orientation for learning mathematics.

Exploring the possibility of a correspondence between the conceptions of sustainability that we described previously and views of internationalisation as a value that appear in the published literature (see Table 3) enables us to expand the way in which research can illuminate the pedagogical issues. First, the 'distance' conception of sustainability is paralleled by a 'distance' view of internationalisation (Adams & Walters, 2001; Reid & Loxton, 2004): this is evident in the many statements that 'maths is already international' and also by the view that internationalisation is all about marketing. Next, the 'resources' conception of sustainability seems to correspond to a 'curriculum' view of internationalisation that focuses on the content (examples, issues, subject matter), the methods (pedagogy, epistemology) and the student body (experience, mobility, heterogeneity) (Ryan & Hellmundt, 2003; Haigh, 2002): in each case, the idea is applied to the practical manifestations of the theoretical value. Finally, the 'justice' conception of sustainability has its parallel in the 'justice' view of internationalisation that is implicit in some writings (Jackson, 2003) and made explicit in Atweh (2004). The conceptions of teaching in the context of sustainability are equally applicable to views of teaching in the context of internationalisation, and the translation is straightforward.

A recent report by Wihlborg (2004) gives some interesting information about Swedish student nurses' conceptions of internationalisation, and is one of the few studies in this area. Using a phenomenographic approach, Wihlborg identified three levels of conceptions. The first was referred to as 'competition, formal validity': nursing gives an internationally recognised qualification that can be used to get jobs in other countries. The second was a 'Swedish perspective on the nurse education program': inserting international content into the Swedish nursing curriculum so that nurses can better deal with people from other national backgrounds. The third

Table 3. Views of internationalisation and teaching

Internationalisation (in the context of teaching)		Teaching (in the context of internationalisation)	
(a) Distance	The discipline (whatever it is) is already international. Focus on marketing aspects (e.g. international qualification) ensures that internationalisation is of only peripheral concern.	(i) Disparate	Teaching and internationalisation are seen as unrelated ideas. Teaching focuses on the course content and 'covering' a syllabus, internationalisation is the job of marketers and administrators.
(b) Curriculum	Internationalisation can be approached via content (examples, issues, subject matter), methods (pedagogy, epistemology) and the characteristics of the student body (experience, mobility, heterogeneity).	(ii) Overlapping	The notion of internationalisation overlaps with the activity of teaching. Teaching is seen as ensuring that students understand course content. Ideas of internationalisation can be incorporated (as examples, etc) to the extent that the situation allows.
(c) Justice	Internationalisation is approached by focusing on the notion of 'fairness' of contacts between educators and students in different countries, and can essentially only occur under such conditions.	(iii) Integrated	Internationalisation is an essential component of teaching. Teaching is seen as encouraging students to make a personal commitment to the area represented by course content, including internationalisation as part of that.

was called 'socio-cultural knowledge': knowledge of the similarities and differences between cultures, which went far beyond the current syllabus in nursing. This outline seems to be broadly parallel to our 'views of internationalisation', although taken from the viewpoint of students rather than lecturers.

Now we have set up this correspondence and replaced conceptions of sustainability with views of internationalisation, we can take the final step in our speculation. We can postulate, as before, that the narrowest views of mathematics (as techniques or components) are likely to coexist with a 'distance' approach to internationalisation carried out using a 'disparate' teaching approach. Broader views of mathematics (as models and data) present opportunities for a 'curriculum' view of internationalisation realised through an 'overlapping' teaching approach. Finally, the broadest views of mathematics (as an approach to life and a way of thinking) give scope for views of internationalisation that include the idea of 'justice' and can be implemented with an 'integrated' teaching approach. As with sustainability, the narrowest conceptions seem to limit the opportunities to incorporate internationalisation into mathematics classes, while the broadest conceptions allow plenty of scope to do this.

Do we have any evidence to support this last step? The only relevant paper we can find is Atweh & Clarkson (2002) (and the study that it comes from) where focus groups of mathematics educators with substantive international contacts in Australia/New Zealand, Mexico and Colombia (also Brazil, not then analysed) discussed terms and issues presented to them by the researchers. Much of the discussion revolved around the definitions of the terms 'internationalisation' and 'globalisation' presented to participants by the researchers, and alternative conceptualisations put forward by the participants themselves, involving notions such as colonialism, assimilation and integration, universalism and even 'Americanisation' (referring to the United States of America). However, in the report of the focus groups, we do find discussion of economic and marketing considerations, reasons for the internationalisation of mathematics curricula, and humane, ethical and equity reasons for becoming involved in international projects of development and research.

5. Internationalisation as a Value for Teaching and Learning Mathematics

In this last section, we look at the notion of internationalisation as a value orientation for mathematics education, focusing on the broadest level of the hierarchical structure that we have postulated. We investigate how the theoretical approach can be translated into the design of appropriate learning environments that encourage lecturers and students towards the broadest and most holistic approaches.

As an initial step, our model can be used to interrogate the content and pedagogy of any university mathematics curriculum to indicate the extent to which it may be sympathetic to the idea of internationalisation as a value for learning and teaching.

A curriculum and corresponding pedagogy strongly focused on specific mathematical content – techniques and components – is unlikely to allow much room for including notions of internationalisation. A broader curriculum that includes a modelling approach is more likely to incorporate internationalisation in examples and approaches. At the broadest level, a curriculum that includes a focus on mathematics as a way of thinking and an approach to problems, and is supported by appropriate pedagogy, is most likely to include discussion of values in general and aspects of internationalisation in particular. Of course, it is possible for any teacher to 'broaden' the written curriculum from which they are teaching, although this is less likely to occur with a curriculum that is loaded up with specific mathematical content that needs to be 'covered'.

Many studies, including the ones we have discussed in earlier sections, demonstrate that there is a wide range of variation in students' and teachers' ideas about their subject and their learning. This finding, while not startling in itself, provokes us to explore the implications of this wide variety for any pedagogic situation. An awareness of the range of ways people experience and understand a situation enables us to develop learning environments where implicit understandings are made explicit and thus subject to critique and possible change. In the specific context of mathematics education, we suggest that internationalisation can be considered as a value orientation: any specific view of internationalisation corresponds to a particular conception of mathematics and learning and teaching in mathematics (and vice versa). Since the conceptions are hierarchical, we aim to encourage lecturers and students towards the broadest ones, aware that aspects of the narrower views will also be included.

Our students will have a wide range of pedagogical experiences before they encounter a tertiary environment, they will come from a variety of cultural, religious and linguistic environments, and they will have different expectations for their futures. In the course of their studies, tertiary mathematics students will encounter a range of different focuses for their thinking – mathematics itself, ethics, philosophy, communication, for instance – which all contribute towards preparing them for their professional and personal lives in the 21st century. We as teachers have also had a range of different experiences and hold different expectations for our mathematics lectures, tutorials and laboratory classes. We are aware that our discipline-specific, nationally-specific courses do not meet the needs of many of our students, recognising that the learning environment has shifted dramatically, that our students' expectations for their learning outcomes have changed, that our assumptions about teaching result in flawed pedagogical practice (and even that the jokes we habitually use are falling flat). For lecturers, this suggests an important challenge: dealing with the diversity of students' (and colleagues') experiences within a mathematical and scientific context.

Our theoretical investigation of conceptions of mathematics and views of internationalisation has several immediate pedagogical implications. First, it seems that students are generally unaware of the range of variation in their fellow students' conceptions, and making them aware of this range gives them the opportunity to broaden their views (see, for example, Ho Watkins & Kelly, 2001, where the students were lecturers). A discussion in class early in the course, or a description of different approaches in a course hand-out can provide students with an opportunity for exploration and change. It will also set up an expectation that the mathematics course will have a broader scope than the immediate techniques and methods that form the 'subject matter'.

Reid (2001) has shown that there are strong relations between teachers' conceptions of their discipline and the way that they go about teaching. Moreover, the learning environment that they set up in their classes can encourage those students who identify with the narrower levels to engage with their learning at a broader level. However, this can also work the other way: if a teacher sets students tasks that are best carried out using the narrower conceptions of learning, evidence in our interview transcripts shows that students who are aware of the more inclusive levels consider learning approaches that relate to the narrower levels (Reid & Petocz, 2002a). Thus, the way that we set up the learning environment in our mathematics classes can influence the conceptions of mathematics that our students have and use.

Learning materials that are set in real contexts, and that encourage students to investigate the mathematical approach and the role of mathematics in their professional and personal lives will lead them towards the broadest conceptions of mathematics. Two examples explicitly incorporating the results of our research are Advanced Mathematical Discourse (Wood & Perrett, 1997), a textbook for a first-year course in Mathematical Practice – thinking, communicating and working mathematically, and Reading Statistics (Wood & Petocz, 2003), a book which asks students to 'read' and engage with research articles in a variety of areas of application, and to communicate the statistical meaning in a range of professional situations. Assessment that is based on learning and reproducing mathematical theory will tempt even those students with the broadest conceptions of mathematics towards the narrowest views. In summary, curriculum needs to accommodate variation in students' conceptions of mathematics, both because this variation exists and in order to help students broaden their views of their subject. It can also be important to explore the nature of work as a professional mathematical scientist and to demonstrate the applicability of students' studies to their future professional roles.

We have argued that the broader and more holistic conceptions of mathematics allow room for integration with the broader conceptions of internationalisation, while the narrower and more limiting conceptions seem to preclude any serious engagement with the ideas. If students hold the narrowest views of mathematics in terms of techniques and components, they will focus on acquiring and perfecting these techniques, and ideas of internationalisation will be seen as irrelevant. If lecturers view mathematics in this way, they will focus their teaching on practice of techniques, covering a syllabus and making sure that students can reproduce the material in an examination. The broadest views of mathematics are as an approach to life and a way of thinking, a creative, human endeavour that includes intuitive and aesthetic components, and makes important connections with areas of social concern such as ethics, peace and democracy: in such a context, notions of internationalisation can flourish. There is evidence that many academic research mathematicians do view their subject in this way, although sadly this approach does not seem to be carried through into their teaching (Burton, 2004).

Internationalisation as an orientation for learning implies a value for mathematics and suggests that we, as a community of educators, may imagine a complex and different future. Considering the 'justice' view as a central component, learning organisations need to refocus their (our) attention from the dominant marketing recruitment paradigm towards an attention that acknowledges the diversity found amongst faculty and students. Learning organisations for this future may focus on equitable learning situations where students' mathematical experiences are seen as part of a whole leading to their future professional, social and cultural roles. Curriculum that encourages a range of different learning outcomes as a valid practice may be one way of doing this. For internationalisation to be recognised as a value for learning it must be seen as a means of developing relationships amongst people and, inter alia, their (mathematical) ideas.

6. References

Adams, T., & Walters, D. (2001). Global reach through a strategic operations approach: An Australian case study. *Journal of Studies in International Education*, *5*(4), 269–290.

Atweh, B. (2004). Towards a model of social justice in mathematics education and its application to critique of international collaborations. In I. Putt, R. Faragher & M. McLean, (Eds.), *Mathematics education for the third millennium: Towards 2010*, (pp. 47–54). Proceedings of the Annual Conference of the Mathematics Education Research Group of Australasia, Townsville, Australia: MERGA, James Cook University.

Atweh, B., & Clarkson, P. (2001). Issues in globalisation and internationalisation of mathematics education. In B. Atweh, H. Forgasz & B. Nebres, (Eds.), *Sociocultural research on mathematics education: An international perspective*, (pp. 77–94). Mahwah NJ: Lawrence Erlbaum.

Atweh, B., & Clarkson, P. (2002). Mathematics educators' views about globalization and internationalization of their discipline: Preliminary findings. In P. Valero & O. Skovsmose, (Eds.), *Proceedings of the 3rd International MES Conference*, (pp. 1–10). Copenhagen: Centre for Research in Learning Mathematics.

Atweh, B., Clarkson, P., & Nebres, B. (2003). Mathematics education in international and global contexts. In A. J. Bishop, M. A. Clements, C. Keitel, J. Kilpatrick & F. K. S. Leung, (Eds.), *Second international handbook in mathematics education,* (pp. 185–229). Kluwer, Dordrecht: The Netherlands.

Ball, S. (1998). Big policies/small world: An introduction to international perspectives in educational policy. *Comparative Education*, *34*(2), 119–130.

Biggs, J. (1999). *Teaching for quality learning at university*, Society for research in higher education, Great Britain: Open University Press.

Bishop, A. J., Clements, M. A., Keitel, C., Kilpatrick, J., & Leung, F. K. S. (Eds.). (2003). *Second international handbook of mathematics education*. Kluwer, Dordrecht: The Netherlands.

Boyer, C. B., & Merzbach, U. C. (1991). *A history of mathematics*, (2nd ed.), New York: Wiley.

Buenfil-Burgos, R. (2000). Globalisation, education and discourse political analysis: Ambiguity and accountability in research. *Qualitative Studies in Education*, *13*(1), 1–24.

Burton, L. (2004). *Mathematicians as enquirers – Learning about learning mathematics*. Kluwer, Dordrecht: The Netherlands.

Chalmers, D., & Volet, S. (1997). Common misconceptions about students from South-East Asia studying in Australia. *Higher Education Research and Development*, *16*(1), 87–97.

Curro, G. & McTaggart, R. (2003). Supporting the pedagogy of internationalization, presented at 17th IDP Australian International Education Conference, Melbourne, Australia, October (June 6, 2006); http://www.jcu.edu.au/office/tld/teachingsupport/ documents/Supporting_pedagogy_internat.pdf.

D'Ambrosio, U. (2001). General remarks on ethnomathematics, *Zentralblatt fur Didaktik der Mathematik, 33*(3), 67–69.

Fallshaw, E. (2003). Overseas partnerships – a case study in quality, presented at Higher Education Research and Development Society of Australasia Annual Conference, Canterbury, New Zealand, July (June 6, 2006); http://surveys.canterbury.ac.nz/herdsa03/pdfsref/Y1052.pdf.

Frankenstein, M. & Powell, A. B. (Eds.) (1997). *Ethnomathematics: Challenging eurocentrism in mathematics education.* Albany, New York: State University of New York Press.

Haigh, M. J. (2002). Internationalisation of the curriculum: Designing inclusive education for a small world. *Journal of Geography in Higher Education, 26*(1), 49–66.

Ho A., Watkins D., & Kelly M. (2001). The conceptual change approach to improving teaching and learning: An evaluation of a Hong Kong staff development programme. *Higher Education, 42*, 143–169.

Ifrah, G. (1998). *The universal history of numbers.* London: The Harvill Press.

Jackson, M. (2003). Internationalising the university curriculum. *Journal of Geography in Higher Education, 27*(3), 325–340.

Joseph, G. G. (1991). *The crest of the peacock.* Princeton NJ: Princeton University Press.

Kember, D. (1997). A reconceptualisation of the research into university academics' conceptions of teaching. *Learning and Instruction, 7*, 255–275.

Leask, B. (2003). Beyond the numbers – levels and layers of internationalisation to utilise and support growth and diversity, presented at 17th IDP Australian International Education Conference, Melbourne, Australia, October (June 6, 2006); http://www.idp.com/17aiec/ selectedpapers/Leask%20-%20Preparing%20students%20for%20life%2022-10-03.pdf.

Marton, F., & Booth, S. (1997). *Learning and awareness.* New Jersey: Lawrence Erlbaum.

Marton, F., & Trigwell, K. (2000). Variatio est mater studiorum, *Higher Education Research and Development, 19*(3), 381–395.

Matang, R., & Owens, K. (2004). Rich transitions from indigenous counting systems to English arithmetic strategies: implications for mathematics education in Papua New Guinea, in: *Proceedings of International Congress on Mathematics Education ICME10*, Copenhagen, July (June 6, 2006); http://www.icme-organisers.dk/dg15/ DG15_RMandKO_final_ed.pdf.

McIntyre, D., & Pfannkuch, M. (1999). Case study: evaluation of a university mathematics program. In W. Spunde, P. Cretchley & R. Hubbard, (Eds.), *The challenge of diversity: Proceedings of the delta '99 symposium on undergraduate mathematics*, (pp. 126–130). Toowoomba: University of Southern Queensland Press, (June 6, 2006); http://www.sci.usq.edu.au/staff/spunde/delta99/deltaon2.htm.

Montgomery, C., & McDowell, L. (2004). Social networks and learning: A study of the socio-cultural context of the international student. In C. Rust, (Ed.), *Improving Student Learning 11: Theory, Research and Scholarship*, (pp. 66–79.) UK: OCSLD.

Mosimege, M., & Ismael, A. (2004). Ethnomathematical studies on indigenous games: examples from Southern Africa. In *Proceedings of international congress on mathematics education ICME10*, Copenhagen, July (June 6, 2006); http://www.icme-organisers.dk/dg15/DG15_MMandAI_final_ed.pdf.

Oldknow, A. (1999). Diverse students – diverse solutions? In W. Spunde, P. Cretchley & R. Hubbard, (Eds.), *The Challenge of Diversity: Proceedings of the Delta '99 Symposium on Undergraduate Mathematics*, (pp. 11–16).Toowoomba: University of Southern Queensland Press, (June 6, 2006); http://www.sci.usq.edu.au/staff/spunde/delta99/deltaon2.htm.

Petocz, P., & Reid, A. (2001). Students' experience of learning in statistics, *Quaestiones Mathematicae, Supplement 1*, 37–45.

Petocz, P., & Reid, A. (2003a). Relationships between students' experience of learning statistics and teaching statistics. *Statistics Education Research Journal*, 2(1), 39–53.

Petocz, P., & Reid, A. (2003b), What on earth is sustainability in mathematics?, *New Zealand Journal of Mathematics, 32* Supplementary Issue: 135–144.

Petocz, P., Reid, A., Wood, L. N., Smith, G. H., Mather, G., Harding, A., Engelbrecht, J., Houston, K., Hillel, J., and Perrett, G. (2007). Undergraduate students' conceptions of mathematics: an international study. International Journal of Science and Mathematics Education, *5*, 439–459.

Reid, A. (2001). Variation in the ways that instrumental and vocal students experience learning in music. *Music Education Research*, *3*(1), 25–40.

Reid, A., & Loxton, J. (2004). Internationalisation as a way of thinking about curriculum development and quality, presented at the Australian Universities Quality Forum, Adelaide, July (June 6, 2006); http://www.auqa.edu.au/auqf/2004/program/papers/ Reid.pdf.

Reid, A., & Petocz, P. (2002a). Students' conceptions of statistics: a phenomenographic study. *Journal of Statistics Education*, 10(2), (June 6, 2006); http://www.amstat.org/publications/jse/v10n2/reid.html.

Reid, A., & Petocz, P. (2002b). Learning about statistics and statistics learning, in: *Australian Association for Research in Education 2002 Conference Papers*, P. Jeffrey, comp., AARE, Melbourne (June 6, 2006); http://www.aare.edu.au/02pap/rei02228.htm.

Reid, A., & Petocz, P. (2006). University lecturers' understanding of sustainability. *Higher Education*, *51*(1), 105–123.

Reid, A., Petocz, P., Smith, G. H., Wood, L. N., & Dortins, E. (2003). Maths students' conceptions of mathematics. *New Zealand Journal of Mathematics*, *32* Supplementary Issue, 163–172.

Reid, A., Smith, G. H., Wood, L. N., & Petocz, P. (2005). Intention, approach and outcome: university mathematics students' conceptions of learning mathematics. *International Journal of Science and Mathematics Education, 3*(4), 567–586.

Rizvi, F. (2000). Internationalisation of curriculum. In RMIT teaching and learning strategy, RMIT website (June 6, 2006); http://www.pvci.rmit.edu.au/ioc/.

Ryan, J., & Hellmundt, S. (2003). Excellence through diversity: Internationalisation of curriculum and pedagogy, presented at 17th IDP Australian International Education Conference, Melbourne, Australia, October (June 6, 2006); http://www.idp.com/17aiec/selectedpapers/Ryan%20-%20Excellence%20through%20diversity%2024-10-03.pdf.

Sagan, C. (1985). *Contact*, New York: Simon and Schuster.

Scheyvens, R., Wild, K., & Overton, J. (2003). International students pursuing postgraduate study in geography: Impediments to their learning experiences. *Journal of Geography in Higher Education*, 27(3), 309–323.

SETI Institute (2004). The Arecibo broadcast (June 6, 2006); http://www.seti.org/site/pp.asp?c=ktJ2J9MMIsEandb=179070.

Snyders, A. (1999). Foundation mathematics for diversity: Whose responsibility and what context? In W. Spunde, P. Cretchley & R. Hubbard, (Eds.), The challenge of diversity: *Proceedings of the delta '99 Symposium on undergraduate mathematics*, (pp. 200–205). Toowoomba: University of Southern Queensland Press, (June 6, 2006); http://www.sci.usq.edu.au/staff/spunde/delta99/deltaon2.htm.

Spunde, W., Cretchley, P., & Hubbard, R., (Eds.). (1999). *The challenge of diversity: Proceedings of the delta '99 symposium on undergraduate mathematics.* Toowoomba: University of Southern Queensland Press, (June 6, 2006); http://www.sci.usq.edu.au/staff/spunde/delta99/deltaon2.htm.

Triandis, H. C. (2002). Generic individualism and collectivism. In M. J. Gannon & K. L. Newman, (Eds.), *The blackwell handbook of cross-cultural management*, (pp. 43–44). Oxford: Blackwell Publishers.

United Nations. (2002). *Report of the world summit on sustainable development* (June 6, 2006) http://www.johannesburgsummit.org/html/documents/documents.html.

US Department of Education. (2003). *Teaching mathematics in seven countries: Results from the TIMSS 1999 video study*, National Center for Education Statistics, NCES 2003–2013.

Vakoch, D. A. (2002). The art and science of interstellar message composition (June 6, 2006); http://www.seti.org/atf/cf/%7BB0D4BC0E-D59B-4CD0-9E79-113953A58644%7D/art_science_special_intro_2002.pdf.

Volet, S. E., & Ang, G. (1998). Culturally mixed groups on international campuses: an opportunity for inter-cultural learning. *Higher Education Research and Development*, *17*(1), 5–23.

Wals, A., & Jickling, B. (2002). Sustainability in higher education: From doublethink and newspeak to critical thinking and meaningful learning. *International Journal of Sustainability in Higher Education, 3*(3), 221–232.

Wihlborg, M. (2004). Student nurses' conceptions of internationalization. *Higher Education Research and Development, 23*(4), 433–454.

Wood, L. N., & Perrett, G. (1997). *Advanced mathematical discourse*, Sydney: University of Technology.

Wood, L. N., & Petocz, P. (2003). *Reading statistics*, Sydney: University of Technology.

Wood, L. N., Smith, G. H., Mather, G., Harding, A., Engelbrecht, J., Houston, K., Perrett, G., Hillel, J., Petocz, P. & Reid, A. (2006). Student voices: Implications for teaching mathematics. *Proceedings of the 3rd International Conference on the Teaching of Mathematics at the undergraduate level*, Istanbul, Turkey. John Wiley and Sons [CD], (June 16, 2007); http://www.tmd.org.tr/sites/ICTM3/uploads/documents/paper-256.pdf.

15

CONTRIBUTIONS FROM CROSS-NATIONAL COMPARATIVE STUDIES TO THE INTERNATIONALIZATION OF MATHEMATICS EDUCATION: STUDIES OF CHINESE AND U.S. CLASSROOMS

Jinfa Cai[1] **and Frank Lester**[2]

[1]*Department of Mathematical Sciences, University of Delaware;*
[2]*School of Education, Indiana University*

Abstract: Cross-national studies offer a unique contribution to the internationalization of mathematics education. In particular, they provide mathematics educators with opportunities to situate the teaching and learning mathematics in a wider cultural context and to reflect on generalization of theories and practices of teaching and learning mathematics that have been developed in particular countries. In this chapter, we discuss a series of cross-national studies involving Chinese and U.S. students that illustrate to how cultural differences in Chinese and U.S. teachers' teaching practices and beliefs affect the nature of their students' mathematical performance. We do this by showing that the types of mathematical representations teachers present to students strongly influence the choice of representations students use to solve problems. Specifically, the Chinese teachers overwhelmingly used symbolic representations of instructional tasks, whereas the U.S. teachers relied almost exclusively on verbal explanations and pictorial representations, illustrating that mathematics teaching is local practice which takes place in settings that are both socially and culturally constrained. These results demonstrate the social and cultural nature of teachers' pedagogical practice

Keywords: cross-national studies; internationalization; Chinese and U.S. Classrooms, mathematical problem solving; solution representations; pedagogical representations

1. Introduction

The growing number of cross-national studies in mathematics during the past two decades, particularly those related to TIMSS, offers a unique contribution

B. Atweh et al. (eds.), Internationalisation and Globalisation in Mathematics and Science Education, 269–283.

to the internationalization of mathematics education. The special contribution of cross-national studies to the development of international perspectives can be seen in various ways. First, cross-national studies are generally conducted collaboratively and involve researchers from several nations. Second, cross-national studies introduce cultural and social dimensions of mathematics education that can be acquired in no other way. In particular, cross-national studies provide opportunities to: (1) improve the ability to measure students' educational achievement, (2) enhance the possibility of generalisability of studies that explain the factors important in educational achievement, and (3) increase the probability of the dissemination of new ideas to improve the quality of classrooms and schools. Third, cross-national studies in mathematics provide a valuable context within which to understand major aspects of the educational systems in different cultures. In strongly culture-related content areas, such as history, not only does the context of learning vary across cultures, but also the content of what is being learned varies. However, in mathematics, the content remains similar even though the cultures vary.

A fourth and perhaps most important contribution of cross-national studies is related to the universally perceived importance and usefulness of mathematics. Mathematics is viewed no longer simply as a prerequisite subject but rather as a fundamental aspect of literacy for a citizen in contemporary society (Mathematics Sciences Education Board, 1998; National Council of Teachers of Mathematics, 2000; Organization for Economic Cooperation and Development, 2004). In view of the importance of mathematics for society and for individual students, the efficacy of mathematics teaching and learning in schools deserves sustained scrutiny. Cross-national studies provide mathematics educators with opportunities to identify effective ways of teaching and learning mathematics in a wider cultural context. Examination of what is happening in the learning of mathematics in other societies helps researchers and educators understand how mathematics is taught by teachers and is learned and performed by students in different cultures. It also helps them to reflect on theories and practices of teaching and learning mathematics in their own cultures.

The discussion of the cross-national performance differences in mathematics is currently a hot topic in education. However, some of the recent discussion has shifted from focusing on the international ranking of mathematical performance to what we can learn from cross-national studies to improve students' learning (e.g., Cai, 2001; Stigler & Hiebert, 1999). Cross-national studies not only provide information on students' mathematical performance examined in the context of the world's varied educational systems, but also help to identify the factors that do and do not promote mathematics learning. In particular, researchers and educators have begun to explore the nature of differences in students' mathematical performance, but also cultural and educational differences in an attempt to understand why cross-national differences in mathematics exist.

The purpose of this chapter is to promote ways we, as researchers, can use international comparisons of mathematics learning to go beyond simply ranking countries. Indeed, we take the position that an overemphasis on the rankings of countries by achievement may sidetrack the search for what causes cross-national differences in mathematics. As an alternative we will show how such comparisons can enhance our understanding of cultural differences and lead to improved student learning.

2. Chinese and U.S. Students' Solution Representations

Representations are both an inherent part of mathematics and an instructional aid for making sense of mathematics (Ball, 1993; Dufour-Janvier, Bednarz, & Belanger, 1987; Goldin, 2002, 2003; Leinhardt, 1993; NCTM, 1991, 2000; Pimm, 1995). In mathematics, some form of representation must necessarily be used to express any mathematical object, statement, concept, or theorem (Dreyfus & Eisenberg, 1996). Representations are tools for solving mathematical problems. In solving a problem, a solver needs to establish representations of the problem not only to help her or him organize and make sense of the problem, but also to communicate her or his thinking to others. Solution representations are the visible records generated by a solver to communicate her or his thinking about how the problem was solved.

2.1 Differences Between Chinese and U.S. Students' Representations

Previous cross-national studies have used "open-ended tasks" to examine thinking and reasoning involved in U.S. and Asian students' mathematical problem solving in addition to examining the correctness of answers (e.g., Becker, Sawada, & Shimizu, 1999; Cai, 1995, 2000a, 2000b; Silver, Leung, & Cai, 1995). To solve an open-ended problem, students are expected to produce both a correct answer and a written record of the thinking and reasoning involved to obtain the correct answer. These studies have revealed a striking difference between U.S. and Asian students' solution representations. Asian students have tended to use symbolic representations (e.g., arithmetic or algebraic symbols); while U.S. have students tended to use visual representations (e.g., pictures).

In a recent study, Cai and Lester (2005) examined students' solution representations on 12 open-ended assessment tasks and how the use of representations was related to their overall performance. The representations were classified into three categories: (1) verbal (i.e., using written words), (2) visual (i.e., using drawing or pictures), and (3) symbolic. To examine the representational differences across the 12 tasks, each student was assigned a representational score. Table 1 shows mean scores for both Chinese and U.S. students for each representation. Overall, there was a significant difference between Chinese and U.S. students' representational mean scores ($\underline{F}(2, 542) = 318.71$, $\underline{p} < .001$). A post-hoc analysis using t-tests showed that U.S. students had significantly higher representational mean scores than Chinese students in using written words ($\underline{t} = 6.88$, $\underline{p} \leq .01$) and visual drawings ($\underline{t} = 16.14$, $\underline{p} \leq .001$). However, Chinese students had a significantly higher representational mean score than U.S. students in using mathematical symbols ($\underline{t} = 15.85$, $\underline{p} < .001$).

Table 1. Mean score in each of the Chinese and U.S. students' dominant representations

	Verbal	Symbol	Pictorial	No explanations
CHINA (n=310)	.23 (.11)[a]	.70 (.21)	.01 (.02)	.06 (.04)
U.S. (n=232)	.43 (.20)	.37 (.26)	.17 (.10)	.03 (.03)

[a] Numbers in parentheses are standard deviations.

The overall representational differences were consistent with those found for each individual task: U.S. students are much more likely to use pictorial representations and Chinese students are much more likely to use symbolic representations (Cai, 2000a; Cai & Lester (2005)). As an example, Table 2 shows the percentages of Chinese and U.S. students' solution representations for a particular problem, the *Block Pattern Problem*. Chi-square analyses showed that the frequencies of Chinese and U.S. students who used each of the representations differed significantly on both the 5-step question of the *Block Pattern Problem* [χ^2 (2, $\underline{N}$ = 542) = 220.43, $\underline{p} < .001$], and the 20-step question of the *Block Pattern Problem* [χ^2 (2, $\underline{N}$ = 509) = 165.55, $\underline{p} < .001$]. In particular, in answering the 20-step question, about 38% of the U.S. students used pictorial representation, but only about 9% of the Chinese students used pictorial representation. In contrast, about 67% of the Chinese students used symbolic representations, but only 11% of the U.S. students used symbolic representations.

2.1.1 Relatedness Between Representation and Performance

How would students' use of representations be related to their performance? A few studies have examined the relatedness between representation and performance (Cai, 2000b; Cai & Hwang, 2002; Cai & Lester, 2005). Cai (2000b) showed that if the analysis is limited to U.S. students using symbolic representations, there is no performance difference between U.S. and Chinese students. In addition, if the comparative analysis is limited only to those Chinese and U.S. students who used pictorial representations or concrete strategies, the two groups' performance is quite similar (Cai & Hwang, 2002). These findings not only support the argument that the representations students use can serve as an index of how well they might solve problems (Dreyfus & Eisenberg, 1996; Janvier, 1987; Larkin, 1983), they also suggest that Chinese students' superior performance on some problems may be due, in part, to their use of more sophisticated representations.

To further examine the relation between representation and performance, Cai & Lester (2005) calculated the correlation coefficients between Chinese and U.S. students' representational scores and their performance on the three types of tasks. Table 3 shows these correlation coefficients. For both samples, students' scores on symbolic representations are highly correlated with their performance on each type of the assessment problems ($\underline{p} < .01$). This means that those students who

Table 2. Percentage of Chinese and U.S. students' solution representations on the block pattern problem

Block Pattern Problem
Look at the figures below.

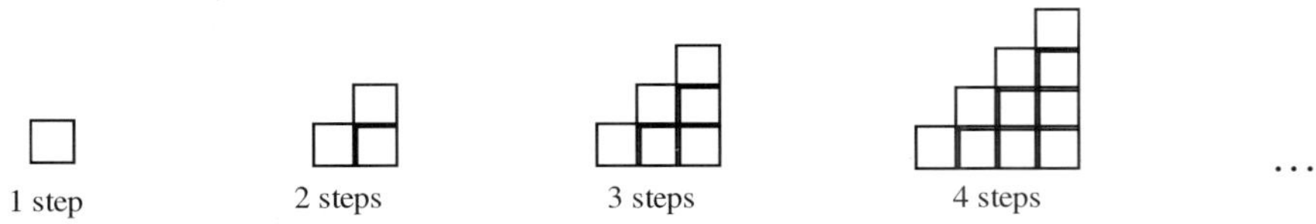

A. How many blocks are needed to build a staircase of 5 steps?
Explain how you found your answer.
B. How many blocks are needed to build a staircase of 20 steps?
Explain how you found your answer.

	5-steps		20-steps	
	US (n=232)	CHINA (n=310)	US (n=218)	CHINA (n=291)
Symbolic	18	82	11	67
Pictorial	59	15	38	9
Verbal	23	3	51	24

Table 3. Correlation coefficients between the representational scores and task scores

	Computation		Simple Problem Solving		Non-routine Problem Solving	
	US	CHINA	US	CHINA	US	CHINA
Verbal	.067	-.010	.013	-.04	.041	-.004
Symbolic	.352**	.194**	.269**	.162**	.574**	.382**
Pictorial	.217**	.041	.098	.059	.278**	.061

** p<.01

use symbolic representations generally had better scores on the tasks. This finding supports the hypothesis that students who used symbolic representations have higher mean scores than those who used other representations, such as visual representations. For the U.S. sample, students' scores on pictorial representations are highly correlated with their performance on both computation tasks ($p < .01$) and non-routine problem solving ($p < .01$), but this was not the case for the Chinese sample. For both samples, students' scores on verbal representations were not correlated with their performance on each type of problems.

Why do Chinese and U.S. students solve problems so differently? In this section, we have examined Chinese and U.S. students' solution representations. In the following section, we look at Chinese and U.S. teachers' beliefs about teaching and representations.

3. Chinese and U.S. Teachers' Beliefs About Teaching and Representations

In Cai and Lester, 2005 study, although Chinese students were more likely than the U.S. sixth grade students to construct symbolic expressions for their solutions, a considerable number of the Chinese students were not able to construct mathematical expressions or algebraic equations for their solutions. This raises the question: Why did these students not use concrete or pictorial approaches to solve the problems as U.S. students did? Concrete and pictorial approaches may provide an entry-level facility to solve the problems. In fact, for those Chinese students who did not use mathematical expressions, a drawing strategy might have enabled them to solve more of the problems. In a recent study, Cai (2004) examined Chinese and U.S. teachers' views of solution representations and strategies by analyzing teachers' scoring of 28 student responses. The results showed that overall U.S. teachers are much more accepting than Chinese teachers of a variety of representations and strategies. However, the U.S. teachers' acceptance was not reflected in their evaluation of students' responses involving conventional approaches, such as using algebraic equations and other mathematical expressions. Table 4 shows the mean rating of Chinese and U.S. teachers to three solution strategies for the following problem: *Angela is selling hats for her Mathematics Club. She sold 9 hats in the first week, 3 hats in the second, and 6 hats in the third. How many hats does she have to sell in the fourth week so that the average number of hats sold in four weeks is 7?*

Both U.S. and Chinese teachers gave responses with algebraic and arithmetic approaches similar scores (Cai, 2004). In particular, almost all U.S. and Chinese teachers valued the responses with algebraic approaches the highest when compared to other responses in the same problem. Almost all the U.S. and Chinese teachers liked the responses with the algebraic approach the best when compared to other responses for the problem. However, the U.S. teachers seem to have different expectations than the Chinese teachers. All the U.S. teachers except one believed that sixth grade students in the United States in general are not expected to solve problems using algebraic approaches. For example, one of the U.S. teachers interviewed explicitly said, "I wish my sixth graders could do this. But in our school, only seventh grade students are taught algebraic concepts. Sixth-graders are only learning pre-algebra and are not expected to solve them using this kind approach involving x's." On the other hand, all Chinese teachers interviewed expected their sixth graders to solve problems using algebraic approaches.

The U.S. teachers' acceptance of a wider variety of strategies was especially evident in their rating of responses involving visual strategies. If a response involved a visual or concrete strategy, Chinese teachers usually gave a relatively lower score even though the strategy is appropriate for a correct answer. By contrast, while U.S.

Table 4. Mean rating of Chinese and U.S. teachers on three responses

Descriptions of Responses	Chinese Teachers (n=59)	US Teachers (n=52)
Algebraic Approach: Let x be the number of hats sold in Week 4. $9+3+6+x=4\times7$. $x=10$ by solving the equation for x. So Angela should sell 10 hats in Week 4.	3.37	3.77
Drawing Approach: The average (7) was viewed as a leveling basis to "line up" the numbers of hats sold in the week 1, 2, and 3. Since 9 hats were sold in week 1, it has two extra hats, which were moved to week 2. Since 3 hats were sold in week 2, 4 additional hats are needed in order to line up the average. Week 2 got 2 hats from Week 1 so only 2 more were needed. Since 6 hats were sold in week 3, it needs 1 additional hat to line up the average. In order to line up the average number of hats sold over four weeks, 10 hats should be sold in Week 4.	3.17	3.85
Arithmetic Approach: $9 + 3 + 6 = 18$. $4 \times 7 = 28$. $28–18 = 10$. Therefore, Angela should sell 10 hats in Week 4 so that the average number of hats sold in four weeks is 7.	3.73	3.63

Note . The maximum possible score was 4.

teachers realized that the drawing strategy is not a sophisticated strategy and can be very time consuming, they recognized that drawing is a viable approach that can produce correct answers. Therefore, in their view, a response containing a visual or concrete strategy should not be penalized. Chinese teachers seemed to have a clear goal that students should learn more generalized strategies. However, there is no evidence from the interview that U.S. teachers have such a clear goal. Instead, the U.S. teachers' goal is to have students solve a problem correctly no matter what approaches they use.

Chinese teachers also gave lower scores for responses involving estimates than did U.S. teachers. For example, in evaluating an estimation response to the following problem, U.S. teachers awarded significantly higher scores than Chinese teachers (Mean for U.S. Teachers = 2.43 and Mean for Chinese teachers = 1.92).

Problem: The average of Ed's ten test scores is 87. The teacher throws out the top and bottom scores, which are 55 and 95. What is the average of the remaining set of scores? Show how you found your answer.

Response: I think that the average for the remaining set of scores is between 55 and 95. But 87 is closer to 95 than 55. So the average for the remaining must be about 90.

For the Chinese teachers, if a problem included all of the information to provide an accurate answer, they did not consider it desirable to give an estimate for the answer. Some U.S. teachers seemed to hold a similar view about responses involving estimates of answers. However, other U.S. teachers focused more on the processes of solving a problem. For them, if the process of solving a problem is sound and the process shows an understanding of the concept involved, the student response should receive high scores even though only estimates of answers were provided.

In addition, Chinese teachers seemed to be much more concerned about details of the writing format and inclusion of units for answers than were U.S. teachers. Chinese teachers believed that the use of an appropriate writing format and units in problem solving can help students develop their abilities to think logically. Such a requirement is also demanded in examinations. The Chinese teachers' concerns about the details of the written format and the inclusion of units in answers are also related to their beliefs about understanding mathematics. Chinese teachers seemed to believe that the use of an appropriate written format and the inclusion of units in problem solving can help students develop their abilities to think logically.

In contrast, U.S. teachers were not as concerned with details about written formats. For example, in a response, the following expressions were included to find the number of blocks needed to build a 20-step staircase (refer to the *Block Pattern Problem* in Table 2): $1 + 2 = 3 + 3 = 6 + 4 = 10 + 5 = 15 + 6 = 21 + 7 = 28 + 8 = 36 + 9 = 45 + 10 = 55 + 11 = 66 + 12 = 78 + 13 = 91 + 14 = 105 + 15 = 120 + 16 = 136 + 17 = 153 + 18 = 171 + 19 = 190 + 20 = 210$. The mean rating for U.S. teachers was 3.62, but the mean rating for Chinese teachers was 2.39. Such an imprecise, in fact incorrect, expression did not seem to bother U.S. teachers at all. As a U.S. teacher interviewed commented, "No, it does not bother me. The little fellow just put down what he thinks in his head. Isn't it the way we think?" For a response like this, Chinese teachers were less willing to give higher score. The explanation provided by a Chinese teachers interviewed was typical: "The result is correct, but there are some mistakes in the writing. Two sides of an equal sign should be equal."

In summary, the findings of this study clearly showed that the differences between Chinese and U.S. students' thinking may well be due to differential beliefs of teachers in two nations. Chinese and U.S. teachers not only have different learning goals, but also place different emphasis on their teaching of problem solving. In particular, U.S. teachers hold a much higher value for responses involving concrete strategies and visual representations than do Chinese teachers (Cai, 2004).

4. Pedagogical Representations in Chinese and U.S. Classrooms

Cultural beliefs do not dictate what teachers do, but teachers do draw upon their cultural beliefs as a normative framework of values and goals to guide their teaching (Bruner, 1996). It is quite possible that Chinese and U.S. students' uses of different solution representations may be related to how Chinese and U.S. teachers view and construct various representations to help students learn mathematics and solve problems in classroom instruction. To investigate this possibility, Cai and Lester (2005) analyzed the videotaped lessons of four U.S. and five Chinese teachers.

4.1 General Descriptions of Instruction in Chinese and U.S. Classrooms

Lesson structure and organization across the five Chinese teachers were quite similar. They usually started with a brief review of what had been taught in the previous lesson, then they clearly stated the topic for the lesson and wrote down the topic on the blackboard. Each of the Chinese lessons consisted of a combination of whole-class discussion and individual student seatwork. In Chinese classrooms, individual students were frequently asked to write down their solutions to problems on the blackboard and explain them to the rest of the students.

During the instruction, Chinese students were also frequently asked to read important points together. Chinese teachers seemed to believe that the more students read aloud, the more they would be able to remember and understand the important points. At the end of each lesson, Chinese teachers always tried to summarize the main or important points related to the lesson. Some teachers had already written these important points on a small blackboard in advance. They just hung up the small blackboard and pointed out the important points to students. Other teachers referred to what had been written on the large blackboard while they were pointing out these important points. We should note that Chinese teachers rarely erased writings on the blackboard during their instruction because the design of the blackboard presentation is considered part of the lesson plan. No group work was involved in any of the Chinese lessons.

The four U.S. teachers demonstrated two kinds of teaching styles. Two of the teachers used a combination of whole-class discussion and group work. The other two teachers mainly used whole class discussion. Every U.S. teacher mainly used an overhead projector in her or his instruction. There were two lessons from two different teachers in which U.S. students were asked to display their solutions on the overhead, but they were not asked to explain their solutions to the rest of the students in the room. Like Chinese teachers, U.S. teachers tended to begin their lessons by reviewing what had been taught previously. Unlike Chinese teachers, U.S. teachers rarely summarized main or important points at the end of each lesson. Our general impression is that the Chinese lessons were well organized; a claim we cannot make about the U.S. lessons.

4.2 Pedagogical Representations of Instructional Tasks

Pedagogical representations are the representations teachers and students use in their classroom as carriers of knowledge and thinking tools to explain a concept, a relationship, a connection, or a problem-solving process. The topics covered in U.S. lessons were very different from those covered in Chinese lessons. We assumed that different pedagogical representations would be used in teaching different topics, and also that the use of pedagogical representations is related to the kinds of instructional tasks. Therefore, we classified the instructional tasks into eight categories:(1) judgment tasks, (2) tasks requiring changing how numbers are represented, (3) construction tasks, (4) recall tasks, (5) Tasks requiring show and observe examples, (6), Tasks requiring comparisons of two numbers, (7) decontextualised computation tasks, and (8) contextualized problem-solving tasks. We made this classification for the convenient purpose of comparisons of pedagogical representations in Chinese and U.S. classrooms. Table 5 shows examples of each type of instructional tasks and percentages of each type of from both Chinese and U.S. classrooms.

The percentage distributions of instructional tasks are quite different between China and U.S. classrooms. In U.S. classrooms, the most frequently used instructional tasks were tasks requiring comparisons of two numbers (about 29%), whereas only 1% of the tasks from Chinese classrooms involved comparing two numbers. For Chinese teachers, the most frequently used tasks were the contextualized problem-solving tasks, which also occurred about 29% of the time. The contextualized problem-soling tasks from Chinese classrooms were pretty evenly distributed among the five Chinese teachers. About a quarter of the tasks from U.S. classrooms involved contextualized problem solving. Chinese teachers quite often used judgment tasks, whereas none of the US teachers did so. Through using judgment tasks, Chinese teachers intended to clarify common misconceptions and to help students better understand related concepts or properties. For example, in order to help students understand the concept of percent as a ratio, Chinese teachers used a number of judgmental tasks, such as, "Is it correct to say that a building uses 50% of a ton of coal for heating?" None of the U.S. teachers used judgment tasks in their instruction.

About a quarter of the instructional tasks in Chinese classrooms required changes in representations for numbers, but only 11% of the instructional tasks in U.S. classroom were this sort of task. For both groups of teachers, a considerable number of tasks involve decontextualised computation and only a few tasks required students to recall some facts or formulas. There were a few instructional tasks used in Chinese classroom that require students to show and observe examples, but none of the U.S. teachers used such tasks in their lessons.

We compared the teachers' pedagogical representations to the following five types of instructional tasks: tasks requiring changes in representations of numbers, construction tasks, recall tasks, decontextualised computation, and contextualized problem-solving tasks. We found that across these different types of tasks, U.S.

Table 5. Sample instructional tasks and percent of each type in Chinese and U.S. classrooms

Types of Tasks	% of Tasks	
	US (n=96)	CHINA (n=88)
Judgment tasks (a) In each of the following figures, which one shows a diameter? 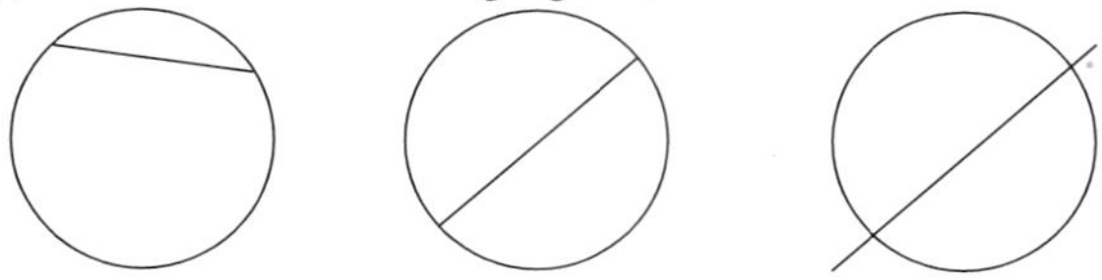 (1) (2) (3) (b) Is it correct to say that a building uses 50% of a ton of coal for heating.	0	14
Tasks requiring change representations of numbers (a) Change 2.5% into a fraction. (b) Using fraction pieces to demonstrate the meaning of 4/3	11	24
Construction tasks (a) Make a new figure that has a perimeter of 18 by adding on tiles along the edges. What are the area and perimeter for the figure? (b) fold a paper circle so that two semicircles are completely overlap from different directions and make observations.	11	8
Recall tasks (a) How can you find the perimeter of a rectangle and a square? (b) What shapes have we learned so far? How many segments are needed to form these figures?	2	3
Tasks requiring show and observe examples (a) Find three symmetric figures around you. (b) Show several examples of circles.	0	5
Tasks requiring comparisons of two numbers (a) Which of the following two fractions has the larger value? 72/108 vs. 84/126 (b) Which of the following fractions is the nearest 2? 1(15/16) vs. 63/32 vs. 17/8	29	1
Decontextualised computation tasks (a) In a circle, if d = 3.5, what is r ? (b) 5/6 + 3/8 = ?	23	16
Contextualized problem-solving tasks (a) Create a story problem for which requires you to compare 9/15 and 20/24. (b) In 100 fifth grade students, there are 20 honor students. In 200 fourth grade students, there are 30 honor students. Which one has a higher rate of honor students?	24	29

teachers were much more likely to use visual representations or physical manipulatives in their instruction, but Chinese teachers were much more likely to use symbolic representations along with verbal explanations in instruction. For example, in the vast majority of the instructional tasks requiring changing representations of numbers, visual representations or physical manipulatives were used as part of the pedagogical representations in solving these tasks in the U.S. classrooms. In contrast, Chinese teachers almost exclusively used symbolic representations combined with verbal explanations for the instructional tasks requiring changing representations of numbers. In each of the construction tasks from the U.S. classrooms, manipulatives were used in the process of solving all these tasks. In each of the construction tasks used in the Chinese classrooms, visual drawings were used in the process of solving all these tasks. However, no physical manipulatives were used to solve these tasks.

5. Lessons for Improving Student's Learning

A main theme of this volume is to consider international comparisons of mathematics and science performance through cultural lenses, looking for similarities and differences (Atweh & Clarkson, 2001). The results from the Third International Study in Mathematics and Science Study (TIMSS) and subsequent studies have had both positive and negative consequences. Two positive consequences are readily apparent. First, researchers from all corners of the world have come together not only to discuss what school mathematics is worth learning, but also to learn from each other about curricular, pedagogical, policy, and research issues. Second, through the international studies, we have come to realize in a very direct way that the idea that mathematics learning is "culture free" is clearly illusory. Indeed, the results of the studies discussed here provide strong evidence that the culture of the classroom established by the teacher has a strong influence both on what mathematics is learned and how it is learned. Furthermore, we suspect that the classroom culture is influenced by the larger societal culture outside the classroom. Our suspicion is supported by the fact that Chinese and U.S. teachers have such different beliefs about what constitutes an appropriate solution and use such different sorts of tasks and representations in their instruction. Sekiguchi (1998) points out that 'it does not make much sense to talk about "domestic" or "regional" mathematics. But in mathematics education, each country has its domestic educational problems and interests, and they shape its research practice significantly' (p. 391). To Sekiguchi's observation, we would add that the same is true of the practice of teaching mathematics; like research, the very nature of teaching is shaped by domestic educational problems, interests, and values.

There are, however, serious barriers to be overcome before we can truly claim that mathematics education is an international field of study. The wide variety of cultural differences among countries make the problem of understanding why one country performs differently from another extremely difficult, if not impossible. So, what can we learn from studies such as the ones we have discussed in this paper?

In the remainder of the chapter, we discuss the lessons we can learn to improve students' learning.

In this chapter, we synthesized findings about Chinese and U.S. students' representational differences in problem solving, as well as their teachers' beliefs about representations and the types they use during their instruction. The results from previous studies clearly show that Chinese and U.S. students prefer to use different types of solution representations. In particular, the mean score for using symbolic representations for Chinese students was significantly higher than that of the U.S. students; while the mean score of using pictorial representations for U.S. students was significantly higher than that of the Chinese students. The results from the correlation analyses showed that for both samples, students' use of symbolic representations had a strong association with their performance on computation, simple problem solving, and non-routine problem solving. These findings suggest the value of using symbolic representations to improve students' learning of mathematics. If a central aim of classroom instruction is to foster the transfer of acquired knowledge beyond the initial learning context to new circumstances, classroom instruction should support students' use of symbolic representations.

The analyses of the video-taped Chinese and U.S. lessons as well as their evaluation of students' responses showed a clear connection between students' use of solution representations and teachers' conceptions and use of pedagogical representations. Teachers in U.S. classrooms were much more likely to use concrete and visual representations than teachers in Chinese classrooms. This finding suggests that U.S. teachers less frequently encourage students at this grade level to move to more abstract representations and strategies in their classroom instruction. One of the common misconceptions held by many U.S. teachers is that concrete representations or manipulatives are the basis for all learning (Burrill, 1997). However, research has shown that manipulatives or concrete representations do not guarantee students' conceptual understanding (e.g., Baroody, 1990). Some researchers (e.g., Dreyfus & Eisenberg, 1996; Smith, 2003) have pointed out that concrete representations and strategies have limitations since they are context- or task-specific strategies in problem solving. Concrete representations may limit students' thinking and further learning unless they can be helped shift to more generalized approaches. The purpose of using concrete and visual representations is to mediate students' conceptual understanding of the abstract nature of mathematics, but concrete experiences do not automatically lead to generalization and conceptual understanding. Unfortunately, the U.S. teachers made little attempt to help students make the transition from concrete and visual representations to symbolic ones.

On the other hand, findings from these studies also suggest that Chinese teachers may overemphasize symbolism and abstract thinking in the classroom. In fact, many Chinese teachers only view symbolic and numerical solutions as "mathematical solutions." Most of the Chinese teachers do not regard pictorial solutions of a problem as "mathematical" (Cai, 2004). Although we should expect students to have an understanding that goes beyond "concreteness" since concrete visual representations may impose limitations on students' development of mathematical reasoning abilities, students should also be given the opportunity to construct their own representations of mathematical concepts, rules, and relationships. Overemphasizing symbolism may promote rote memorization of procedures.

One of the practical implications of the findings from these studies is that students should be given the opportunity to construct their own representations of mathematical concepts, rules, and relationships. However, we should expect them to have an understanding that goes beyond concrete cases. For example, teachers may start with concrete representations or physical manipulatives to encourage students to use their own strategies for solving problems and making sense of mathematics. But, the students' further conceptual development requires that teachers help students develop more generalized solution representations and strategies.

It is clear that mathematics teaching is local practice that for the most part takes place in settings that are both socially and culturally constrained. From the findings we reviewed, the representational differences between Chinese and U.S. students helped interpret cross-national performance differences found in large-scale international studies, such as TIMSS and PISA (Organization for Economic Cooperation and Development, 2004). The examination of Chinese and U.S. teachers' conceptions and constructions of representation through analyzing video-taped lessons and their scoring of students' responses helped us understand why there are representational differences in Chinese and U.S. students' problem solving. The fact that the Chinese teachers overwhelmingly used symbolic representations of instructional tasks, whereas the U.S. teachers relied almost exclusively on verbal explanations and pictorial representations must be taken as evidence of the social and cultural nature of the pedagogical practice.

6. References

Atweh, B., & Clarkson, P. (2001). Internationalization and globalization of mathematics education: Toward an agenda for research/action. In B. Atweh, H. Forgasz, & B. Nebres, (Eds.), *Sociocultural research on mathematics education: An international perspective*, (pp. 77–94). Mahwah, NJ: Erlbaum.

Ball, D. L. (1993). Halves, pieces, and troths: Constructing and using representational contexts in teaching fractions. In T. P. Carpenter, E. Fennema, & T. A. Romberg, (Eds.), *Rational numbers: An integration of research,* (pp. 328–375). Hillsdale, NJ: Erlbaum.

Baroody, A. J. (1990) How and when should place-value concepts and skills be taught? *Journal for Research in Mathematics Education, 21*, 281–286.

Becker, J. P., Sawada, T., & Shimizu, Y. (1999). Some findings of the U.S.-Japan cross-cultural research on students' problem-solving behaviors. In G. Kaiser, E. Luna, & I. Huntley, (Eds.), *International comparisons in mathematics education,* (pp. 121–139). London: Falmer Press.

Bruner, J. (1996). *The culture of education.* Cambridge, MA: Harvard University Press.

Burrill, G. (1997). The NCTM standards: Eight years later. *School Science and Mathematics, 97*(6), 335–339.

Cai, J. (1995). *A cognitive analysis of U.S. and Chinese students' mathematical performance on tasks involving computation, simple problem solving, and complex problem solving*, (*Journal for Research in Mathematics Education* monograph series 7), National Council of Teachers of Mathematics, Reston, VA.

Cai, J. (2000a). Mathematical thinking involved in U.S. and Chinese students' solving process-constrained and process-open problems. *Mathematical Thinking and Learning, 2*, 309–340.

Cai, J. (2000b). Understanding and representing the arithmetic averaging algorithm: An analysis and comparison of U.S. and Chinese students' responses. *International Journal of Mathematical Education in Science and Technology, 31*, 839–855.

Cai, J. (2001). Improving mathematics learning: Lessons from cross-national studies of U.S. and Chinese students. *Phi Delta Kappan, 82*(5), 400–405.

Cai, J. (2004). Why do U.S. and Chinese students think differently in mathematical problem solving? Exploring the impact of early algebra learning and teachers' beliefs. *Journal of Mathematical Behavior, 23*, 135–167.

Cai, J., & Hwang, S. (2002). U.S. and Chinese students' generalized and generative thinking in mathematical problem solving and problem posing. *Journal of Mathematical Behavior, 21*(4), 401–421.

Cai, J., & Lester, F. K. (2005). Solution and pedagogical representations in Chinese and U.S. classrooms. *Journal of Mathematical Behavior, 24*(3–4), 221–237.

Dreyfus, T., & Eisenberg, T. (1996). On different facets of mathematical thinking. In R. J. Sternberg & T. Ben-Zeev, (Eds.), *The nature of mathematical thinking,* (pp. 253–284). Hillsdale, NJ: Erlbaum.

Dufour-Janvier, B., Bednarz, N., & Belanger, M. (1987). Pedagogical considerations concerning the problem of representation. In C. Janvier, (Ed.), *Problems of representation in the teaching and learning mathematical problem solving* (pp. 109–122). Hillsdale, NJ: Erlbaum.

Goldin, G. A. (2002). Representation in mathematical learning and problem solving, In L. D. English (Ed.), *Handbook of international research in mathematics education* (pp. 197–218). Mahwah, NJ: Erlbaum.

Goldin, G. A. (2003). Representation in school mathematics: A unifying research perspective. In J. Kilpatrick, W. G. Martin, & D. Schifter, (Eds.), *A research companion to principles and standards for school mathematics* (pp. 275–285). Reston, VA: National Council of Teachers of Mathematics.

Janvier, C. (1987). *Problems of representation in the teaching and learning of mathematical problem solving.* Hillsdale, NJ: Erlbaum.

Larkin, J. H. (1983). The role of problem representation in physics. In D. Genter & A. L. Stevens (Ed.), *Mental models* (pp. 75–98). Hillsdale, NJ: Erlbaum.

Leinhardt, G. (1993). On teaching. In R. Glaser (Ed.), *Advances in instructional psychology* (Vol. 4, pp. 1–54)., Hillsdale, NJ: Erlbaum.

Mathematical Science Education Board. (1998). *The nature and role of algebra in the K-14 curriculum: Proceedings of a national symposium.* Washington DC : National Research Council.

National Council of Teachers of Mathematics. (1991). *Professional standards for school mathematics.* Reston, VA: The Author..

National Council of Teachers of Mathematics. (2000). *Principles and standards for school mathematics.* Reston, VA: The Author.

Organization for Economic Cooperation and Development. (2004). *Learning for tomorrow's world: First results from PISA 2003.* Paris: Author.

Pimm, D. (1995). *Symbols and meanings in school mathematics.* London: Routledge.

Sekiguchi, Y. (1998). Mathematics education research as socially and culturally situated. In A. Serpinska & J. Kilpatrick (Eds.), *Mathematics education as a research domain: A search for identity* (pp. 391–396). Dordrecht, The Netherlands: Kluwer Academic Publishers.

Silver, E. A., Leung, S. S., & Cai, J. (1995). Generating multiple solutions for a problem: A comparison of the responses of U.S. and Japanese students. *Educational Studies in Mathematics, 28*(1), 35–54.

Smith, S. P. (2003). Representation in school mathematics: Children's representations of problems. In J. Kilpatrick, W. G. Martin, & D. Schifter (Ed.), *A research companion to principles and standards for school mathematics* (pp. 263–274). Reston, VA: National Council of Teachers of Mathematics.

Stigler, J. W., & Hiebert, J. (1999). *The teaching gap: Best ideas from the world's teachers for improving education in the classroom.* New York: The Free Press.

16

INTERNATIONAL PROFESSIONAL DEVELOPMENT AS A FORM OF GLOBALISATION

Hedy Moscovici[1] **and Gary F. Varrella**[2]

[1] *California State University – Dominguez Hills, College of Education, 1000 East Victoria St., Carson, California 90747-0005, USA*

[2] *Washington State University, Spokane County Extension, 222 N. Havana, Spokane, Washington, 99202, USA.*

Abstract: This chapter examines models of professional development (PD) for science teachers in four different countries (the United States, Romania, Israel and the Republic of Armenia), and using this comparison as a launch point, considers the notion of professional development as a form of globalization. Comparing and contrasting familiar PD models for established democracies (the US and Israel) with less familiar PD activities (from countries exiting socialism/communism such as Romania and the Republic of Armenia) help us extend our understanding of the roots, strengths and flaws of the familiar PD and readdress the role of cross-cultural PD providers who think and act globally

Keywords: professional development (PD) models; international PD; globalisation of PD; international educational reform; educational reform

1. Introduction

This chapter first examines models for professional development (PD) for teachers in four different countries (the United States, Romania, Israel, and the Republic of Armenia) and then, using this comparison, considers the notion of professional development as a form of globalisation. Comparing and contrasting familiar PD models for established democracies (the US and Israel) and with less familiar PD activities (from countries exiting socialism/communism) helps us extend our understanding of the roots, strengths, and flaws of PD as we practice it. This comparison and contrast opens a window to cross-cultural aspects and invites examination of pivotal issues and challenges in conducting international PD. Based

B. Atweh et al. (eds.), Internationalisation and Globalisation in Mathematics and Science Education, 285–302.

on our experiences in conducting international PD, we reflect on our notion of globalisation and consider cross-cultural PD as a true form of globalisation.

2. Theoretical Referents: Learning Theory and Emancipation

The developmental and social aspects of what Yager (1991) referred to as the "constructivist learning model," and that Tobin, Tippins, & Gallard (1994), defined as social constructivism paint a portrait of learning as a social sense-making process of experiences based on extant knowledge, further experiences, and applications of the knowledge to new or unique circumstances. This epistemological standpoint for how we approach PD is consistent with our learning experiences as we applied our knowledge to new and unique circumstances. In any PD situation, learners of all ages need to perceive the experience as relevant and the results as useful. In effective PD settings, the "teacher-learners" are motivated by the value and applicability of the experience to their particular context – translating new learning into classroom instruction (Kelleher, 2003).

A second theoretical referent for our approach to PD comes from the literature on critical pedagogy and the development of teachers as transformative intellectuals (Giroux & Simon, 1989). This perspective is dramatically different from earlier positivism-dominated paradigms considering learners as empty vessels ready to be filled. In our practices, we acknowledge and build PDs on the wealth of contextual knowledge and in light of the local dynamics. Such teachers encourage students' challenges on both content and social norms levels. They understand that learning cannot be imposed, as individuals tend to resist (Anyon, 1997; Finn, 1999) and openly fight oppression (Freire, 1990). In a more recent work, Freire (1998) concentrates on the role of the teachers in our society as cultural workers and agents of change when he stresses

> …that those wanting to teach must be able to dare, that is, to have the predisposition to fight for justice and to be lucid in defence of the need to create conditions conducive to pedagogy in schools; though this may be a joyful task, it must also be intellectually rigorous
>
> (p.6)

In this statement Freire is advocating for teacher empowerment, creating professionals who are "critical, daring, and creative" (p. 8), willing to resist administrative attempts to "domesticate" them to the status quo.

Our work and this chapter are set in a milieu of growing autonomy, a sense of shared responsibility, and the development of collaborative learning communities akin to those richly described in the literature by authors such as Sergiovanni (1994), Senge P.M (1990), The fifth discipline New York, NY: Doubleday, Fullan (2003, 2005), and Martin-Kniep (2005). As often as possible, the participants are *required to become* partners in choices. These choices can range from the types of interactive students' experiences, collaborative methodologies, and determination of techniques for and content of emerging curricula. Particularly for teachers among the NIS ("Newly Independent States" from the old USSR) and Romania, this aspect of freedom or "freer thinking" as the Armenians refer to it, is *very significant* and gives a sense of otherwise never experienced emancipation within their teaching.

3. Essentials of Professional Development

Techniques for and studies of professional development continue to garner considerable interest among educators. With the publication of the National Science Education Standards (National Research Council, 1996) and its valuable follow-up on images of inquiry science in classrooms (National Research Council, 2000), science teacher educators realized that the science teachers are not prepared to take the role of the teachers as described in these documents. These documents also propose to the facilitators of professional development experiences to involve science teachers in scientific inquiries in an attempt to develop, define, and then refine the various dimensions of their pedagogical content knowledge (Shulman, 1986; Gess-Newsome & Lederman, 1999). Professional development experiences should strengthen the collaborative aspect of learning science in an environment that is supportive, collegial, and integrative (Hammer, 2000; Wallace & Louden, 2002; Armenian Ministry for Education and Science, 2004). The leader of such professional development endeavours is viewed as a facilitator, a consultant, a reflective practitioner, and a model. In all professional development settings, teachers consciously or unconsciously expect their "trainer" to not only know their subject, but to "walk the talk" reflecting the best practices that they are advocating.

Previous works in the area of professional development uncovered a variety of essential elements that ensure an effective experience that translates into the K-12 classroom. For example, Garet *et al.* (2001) used a national sample of 1,027 teachers who took part in professional development offered under Eisenhower grants. Based on teachers self-reported data, the authors concluded that sustained and intensive professional development activities that focus on academic subject matter (see also Crowther & Cannon, 2001; Govett & Hemler, 2001; Pyle, 2002; Moss & Lonning, 2001), provide teachers with "hands-on" experiences, and get integrated into the daily practices of the participant teachers have a higher chance of producing enhanced knowledge. Guskey (2002) goes one step further from self reported data and defines five levels for professional development evaluations: 1) participants' reaction 2) their learning (for reasoning skills see also Weld et al, 2002), 3) availability of organization support and change, 4) participants' use of new knowledge and skills, and 5) student learning outcomes. He sees professional development as a systematic effort that eventually results in improved learning opportunities for students. Gess-Newsome (2001), in her literature review focusing on the elements of successful PD experiences, supports this standpoint as well. Her recommendations for a successful PD include long term with follow-ups, sustained support, collective participation, clear and consistent intersections between content and pedagogy, and application of active learning principles in all the aspects of the PD. White, Russell, and Gunstone (2002) reemphasize the role of the classroom students in the translation of new pedagogical ideas into the classroom. Retraining of students experienced in traditional non-demanding curricula is necessary to help the participants rise to the challenges of transitioning to a curriculum requiring active participation by the students. The PD facilitator too must consider this as they work to enable their participants to become successful.

Other researchers have examined the correlation between enhancing the quality of professional development experiences and the composition of the PD leadership team. Their conclusion supports the idea that teams of content experts (i.e., scientists), science educators, and teachers (Thompson & Gummer, 2002; Hederson & Lederman, 2002; Gilmer *et al,* 2002; Caprio & Borgesen, 2001) who lead in a cohesive way. All these studies emphasized the role of the development of the community of learners, a cultural unit where the goal (learning) unifies professionals and de-emphasizes borders and roles across the group of leaders. The creation of networks—the natural extension of such models–become a powerful tool where teachers, as well as the leadership team equally share their knowledge, experiences, failures, and successes within their setting as well as within professional networks beyond the confines of their own educational institution (Varrella, G., Weld, J., Harris, R., Enger, S., Yager, R., 2005; Morris, Chrispeels, & Burke, Burry-Stock, J., 2003).

A common theme across all of the above studies and cited literature is that active learning—meaning that PD participants need to /manipulate/ what they learn and apply what they are learning in their own situations—is an absolutely critical element for successful PD.

4. Comparing the Familiar With the Unfamiliar—Models of PD

Over the last ten years, the authors of this chapter participated in PD situations in Romania, Israel, United States, and the Republic of Armenia and collected data. Papers highlighting different aspects of these PDs were presented at different professional conferences, including the International Association for Science Teacher Education and the National Association for Research in Science Teaching. Vignettes from the different PDs together with pictures provided the conference participants with tangible examples of how PDs are structured in the different countries that we studied. Two examples are provided below.

Romania, 2002
26 secondary science and mathematics teachers (teaching at the middle and high schools levels) from all Romania gathered in an exclusive hotel in the mountains (Sinaia Holiday Inn Resort) to learn about and develop correlated/transdisciplinary units that will involve science and mathematics concepts. All teachers were at the highest professional level (Level 1), meaning they had proved through testing, years of teaching experience, and professional development that they had achieved

professional excellence. Some served also as principals. After a short introduction, and speaking fluent Romanian, the international PD provider introduced the topic of international trends of interdisciplinary teaching (e.g., role and purpose of integration, levels of integration, group work as teachers and as students, integrity of addressing content standards and integration, etc...) for the morning session and after a short break involved participants organized in groups of 2–4 in activities that required expertise in science as well as in mathematics (e.g., calculation of reaction time, gravitational force, understanding the relationship and explanation of the raising water in the beaker turned upside down on lighted candles). Each participant contributed about $80 to the cost of the PD. Knowing that the salary of a Level 1 teacher (highest level) was about $130/month, the teachers were committed to learn by paying almost a monthly salary to attend.

It was interesting to see their commitment when participants reduced the morning break to less than 10 minutes (as they did not feel comfortable to leave for restroom breaks while the PD provider was talking), and decided to give up their lunch break in order to learn about more activities involving science and mathematics concepts and be able to ask questions regarding the preparation of their interdisciplinary/transdisciplinary units.

Israel, 2000

Every year, just before the summer, the list with the possible PDs comes out from the local branch of the Ministry of Education. As it was always the "first come, first serve" policy, teachers tend to register immediately as "thinking about it" might result with lack of spots. Registration is ensured through a check for a small amount. Graded and not graded PDs, can be stored up and translated into up to 8% of the salary. Most teachers have much beyond the 8% as courses taken during the sabbatical are considered as PD and accumulate over the years. About 24 teachers gather for an average of 3–5 days, most with follow-up days during the year, others with summer-to-summer continuation to deepen their understanding of science content concepts and consider pedagogical alternatives. A team of content and pedagogy experts from the University, invited guests with international reputation, and science specialists from the Ministry of Education are leading the participants. Time is allocated to analyzing specific teaching conditions, levels of content, and groups of students.

The examination of professional development models from these different countries: Romania, Israel, the Republic of Armenia, and United States allow for further definition of effective PD, helping to characterize elements of successful models particular in cases where the type of PD provided by international consultants comes in contrast to the traditional methods. For convenience, we selected a series of elements that cut across all four PD events allowing for a convenient comparison and contrast. These include (see Table 1):

- participants and funding sources,
- length, time, or duration of the PD (from one day to many years),

Table 1. Differences and similarities among professional development activities for teachers in Romania, Israel,Armenia, and U.S.

COUNTRY	PARTICIPANTS	FUNDING SOURCES	LENGTH	FREQUENCY	FOCUS	ADVANTAGES
Romania	Teachers and Administrators	Teachers +/- Districts and Grants	Days – Semester – Years	One Time, Once Every Week, Once Every Month	Content Only ⇒ Content and Pedagogy	University Credits/Points
Israel	Teachers and Administrators	Teachers +/- Districts and Grants	Days – Semester – Years	One Time, Once Every Week, Once Every Month	Content and Pedagogy	Salary Increase (up to 8% of salary)
Armenia	Teachers and Administrators	Teachers + Internationally-Sponsored Projects	1-3 Days (usually) at a time. For special state sponsored programs the total duration may be 72 hours or longer	One Time, Once Every Week, Once Every Month	Content and Pedagogy	The "certificates" awarded are highly valued by teachers
U.S.	Teachers	Districts +/- Grants	Days – Semester(s) – Years	One Time, Once Every Week, Once Every Month	Pedagogy ⇒ Content and Pedagogy	University Credits/Salary Increase

- focus of the PD and composition of the facilitating team (such as science educators, scientists, district representatives),
- length and intensity of the PD (number of hours/session, number of sessions, and session frequency), and
- advantages and level of commitment from participants i.e., "what's in it for me" (e.g., monetary, professional advancement, intrinsic motivation), autonomy for the participants, involvement or commitment (i.e., active participation during and/or after workshop)

Please note that it is impossible to feature all elements of PD from any one the individual countries used as case examples in this chapter. The authors have endeavoured to select widespread and often-used elements from each of the PD cultures, using their first hand experiences to create consistent comparisons, contrasts, and contextual examples.

4.1 Participants in PDs

In Romania, Israel, and Armenia, teachers and administrators participate in professional development activities together. In fact, many of the participants are both teachers and administrators, as Romania and Israel require administrators to teach as part of their job description. This is frequently the case, though not required in Armenia as well. In the United States, when administrators participate in the professional development for teachers, it is a "show of support," or just to check attendance, rather than a commitment to learn along side of teachers in a PD to make the notion of a learning community an actual event. This reality, particularly in Romania, binds together two of the critical stakeholder groups—i.e., teachers and administrators—more closely, which helps build community and represents part of the critical mass encouraging community building, change, and sustainability (Guskey, 2002; Fullan, 2005). Of course without similar investment by other critical stakeholders, (e.g., parents and elected or appointed government officials) deep change and sustainability are severely hampered, beyond the individual school site.

With the exception of the United States, participants in the PD are responsible for at least part of costs for the experience (even if just transportation fees) among the four countries. The PD application is a competitive process, and the monetary commitment from the participants ensures space in the PD. In the United States, PDs that are organized by the school districts are mandatory and teachers attend as part of their contractual time. In some instances in Armenia, which is in the midst of restructuring their educational system, attendance may be required as well. If the PD is supported through external funding (e.g., grants), not only are the teachers not required to pay to secure their space, they are paid by the grant to attend the PD. With the exception of the mandatory PDs performed during school day (on contractual time) in the US, in most, if not all the other instances, teachers have the power to choose the PD subject that best fits their needs and interests.

4.2 Length and Intensity of PDs

The length of the PD offerings varies from a number of days to a number of years. Their intensity also varies. The Romanian (Ms. Otilia Pacurari, verbal communication) and the Israeli (Dori, 2003) examples had the most "intense," i.e. demanding, PD models. PD participants meet for 3 years (Romania) and 5 years (Israel). In Romania, the PD—for the grant related work of the first author—was offered in a hotel in the mountains about two hours drive from Bucharest. Participants met collectively during the summer for one week (10–12 hours/day), developed materials, and based on the quality of their work, they were selected to participate for year two of the programme. An expected part of the teachers' commitment to the PD included follow-up visitations during the school year. In Israel, participants in the 5-year PD met over the summer for intensive collaborative training and for two-hours/week teacher meetings during school year. Training and discussion sessions were standard activities in the PDs included in this discussion. Experts in content and pedagogy conducted these sessions. In Armenia, the length varies, but most state sponsored workshops are a total of a standard 72 hours of "seat time." International projects in Armenia sponsor varied types of PD lasting from a week with follow-up training, to multiple years, depending on project goals and level of funding.

There is literature support suggesting that PDs that last longer and include numerous follow-up sessions are more successful in terms of translating ideas and techniques into the classrooms and improving pedagogical content knowledge (Gess-Newsome, 2001; Garet et al., 2001). As PD participants infuse their new learning into the curriculum, it is essential for them to have access to experts as well as peers.

4.3 Focus of the PDs

Focus of the PDs experiences in all countries except Armenia combined elements of content and pedagogy in a rich-resources environment (e.g., knowledgeable professionals, text, Internet). It is interesting to address various shifts that happened in the last years in the foci of the PDs in the different countries analyzed in this paper. In this instance Romania and the US, had the most dramatic shifts.

Until recently, the focus of professional development in Romania was content-based. Individuals would enrol and pass a university course in the content they were teaching. In the last decade, and as a response to the TIMSS results and our growing knowledge of how students learn (National Research Council, 2005), the focus shifted slightly and PDs experiences include some pedagogical elements in addition to the mandated content knowledge standards (Cucos, 2002). Romanian teachers know their teaching subject very well. Professional advancements are based on exams that are heavy on the content side and can be taken only after they had spent many years in the classroom including trial period of two years–for level 2, and five years after that–to reaching level 1, the highest level for a professional teacher. In terms of the new pedagogical ideas that were emphasized during the summer PDs and that slowly penetrate the university teacher preparation courses were teaching in cultural and ethnic pluralistic environments, and especially looking at teaching the Romanian gipsies and having students become active during the teaching and learning processes (Cozma, 2001; Cucos, 2002).

In the US, after a system that was heavy on the content, PDs shifted toward focus on practical pedagogies, methods, and use of technology. The requirement for content knowledge somehow was lost (Shulman, 1986), though recently we have witnessed a return to emphasizing teachers' content knowledge (pendulum once again swinging), as one cannot teach what one does not know. For example, in response to the lack of content knowledge of the population teaching at the elementary level, the California Council for Teacher Education requires that all elementary teaching credential candidates pass a content examination.

Armenia is attempting to create a balance between content and best practices. As is the case with other NIS and Romania, content dominated the attention of PD—teaching content, knowing content, etc.—however, there has been a shift toward discovering some type of balance between content, pedagogy, and the methods and skills appropriate to teacher specific content, i.e., the develop of PCK (Gess-Newsome & Lederman, 1999, Shulman, 1986). This shift in emphasis is incomplete, but is evident even among Armenian Ministry of Education and Science sponsored events that tend to emanate from a philosophy of central control and were entirely dominated by a philosophy of *training* and *knowledge acquisition.* The contrast with the old state is the slow emergence of an emphasis that also includes *reflection, sense-making,* and *advancement of methodological abilities.*

4.4 What's in it For Me?

Analyses of the various PD systems uncover an interesting picture. There is a fine line between intrinsic and extrinsic motivational factors that affect teachers' attendance and participation in the PDs. For example, teachers in Romania seem to be intrinsically motivated to attend PD events – learning new teaching techniques and content with a group of knowledgeable peers. If their PD is offered through the university, they receive university credit. In terms of extrinsic motivation, knowing more is helpful when they take the exams that allow them to advance (10 years after 2-year probation to move to level 2, and 5 years after that to move to level 1). Teachers in Romania also need to fulfil 40 hours of PD per year in order to maintain their credential.

In Israel, teachers are allowed a maximum of 8% salary increase based on PD experiences that include PDs that are graded and not graded (pass/fail). Most teachers continue to attend PDs even after the meeting the 8% goal. During the sabbatical (allowed after 6 years of teaching, teachers are paid 70% of their salary to attend academic and non-academic courses reaching the equivalent of one year of university study. The Ministry of Education pays the tuition for the sabbatical. During this period, teachers are not allowed to teach. Under these conditions, teachers in Israel reach the 8% maximum for salary increase rather quickly.

In most states in the US, teachers are required to enrol in courses or PD in order to renew their credential. The usual number is 150 hours of PD over a period of five years that also leads to an increase in teachers' salaries. Many of the PDs, especially in the past, were organized around meeting this requirement. It is a cruel reality that in many of these types of PD, the leaders pretended to teach while the teachers pretended to learn. Teachers did not have to become active in any way, just to be able to prove attendance. Excellent evaluations at the end of the PD ensured the sustainability of these courses. Because of the lack of translation of PD experiences into the classroom, and because of the extensive data on effective PDs, these experiences are shifting toward having teachers actively involved in the PD experience. From the teachers' point of view, the new PD requires time and translation of ideas into the classrooms. At the same time, agencies that allocate grant money for PD (e.g., the National Science Foundation) requires participant teachers to receive modest stipends for attending these professional experiences. School districts are often required to pay teachers—or conduct PD—during salaried periods of the day and year because of the local teacher unions. Unfortunately, these changes led many US teachers to come to the PD because of the money and/or the convenience of never having to step outside the standard workday to meet "continuing education requirements."

In recent years, the motivation to participate in PD in Armenia has shifted from primarily extrinsic, to a combination of intrinsic and extrinsic factors. Before, money and "certificates" (recognizing seat time) were the primary extrinsic motivators. The former was ordinarily provided through projects externally funded by foreign governments or international foundations (e.g., Soros and the Open Society Institute). The latter are important as a demonstration of "knowledge" and commitment to education, which can be a significant influence on retention, hiring, and firing events. Based on a temporary downward trend in population—due to emigration and a brain-drain—the number of teachers needed is being reduced in Armenia (known as "optimization"). A nation-wide reduction in the number of teachers in the classroom will result. Hence, the certificates take on more significance yet, given their value as noted above.

Some of the international projects—which provide stipends as well as certificates that may or may not be state sanctioned—are experiencing a dramatic rise in to the number of "volunteer" or non-paid participants (e.g., from 10–30% of the total number of participants to more than 60% of the participants). Although it is difficult to assess the specific motivator without conducting a specific study, evidence indicates that combinations of intrinsic and extrinsic factors are driving the shift. The extrinsic factors are the highly valued certificates, if they are recognized by the state, which some grant-based organizations have accomplished. The intrinsic factors are a combination of :

- a renewed hunger for learning among teachers educated under the old Soviet system,
- an interest and growing commitment to the "new" interactive techniques, and

- the growing reputations and regard—among Armenian teachers and teacher educators—for specific externally funded and long-running projects known for supportive environments and "progressive" and "efficient" techniques.

This represents a cautious, but consistent move toward a conceptual change (Posner, G. J., Strike, K. A., Hewson, P. W., & Gertzog, W. A., 1982; Freire, 1998), in which case once understanding, value, and results are demonstrated, the educators involved become genuinely motivated and choose *to change*. Once the latter *choice* has been made, the chances of sustainability improve.

4.5 The International Flavour

It is very important to understand the effect of the international exchange in the case of PD. In many ways, it can be described as cross-pollination where the international PD leader shares their skills and knowledge in the international setting and brings back elements from the international experience to become integrated into PDs offerings at home. The cross pollination is facilitated when the PD leader knows the language and the customs of the participants in the international PD experiences. This can be achieved by a native returning to work with teachers "at home" (first author) or through a long-term and intense commitment by a non-native who returns multiple times, establishing an identity and deeper understanding of the local culture (second author).

Teachers attending the PDs in Romania, Israel, and Armenia learned about the PD conditions in the US, while teachers in the US learned about the teaching and PD conditions in other countries. The result: US teachers realize that they are fortunate to receive stipends and work in a technology rich environment for their time in the PD, that they need more PDs that are content-related, and perhaps a sabbatical (Israel model) to recharge their batteries.

US teachers discovered common ground with their counterparts in other countries when they came to understand that in other countries teachers believe that they can change the curriculum and see the role of the PD in that reform process. As one Armenian teacher noted to the second author, "[I now believe that we can] change our approaches regarding teachers' responsibilities and rights and [learn to] participate in decision-making and policy creating processes related to education."

For the Romanian teachers the stipends and the intensive recruitment efforts common in the US were enigmatic. One teacher commented to the first author, "So, you are paying them to come and participate in something that professionally will advance them? I just don't get it!" Later, she questioned the grant developer about the possibility of receiving monetary recompense for her time like, in the US. "It's not in the grant," was the reply, since the granting agency–as compared to the National Science Foundation in the US—did not include teacher compensation for the Romanian teachers as a condition in the original RFP (request for proposals).

5. Do International Professional Development Activities Contribute to Globalisation?

Globalisation, as a term, crept into our vocabulary after Canadian futurist and educator Marshal McLuhan formulated the idea of the "global village" in the 1960s. Only in the last decade has the term been popularized. Now the term is commonly heard in a broad series of venues including from political stump during campaigns, has been paraphrased by a former first lady of the US…" "Taking a village to raise the child," and is frequently referred to on the major television networks. Wells, Carnochan, Slayton, Allen, and Ash (1998), argued that "global shifts" could shape the very meaning in the 21st century.

Attaching meaning to the notion of globalisation is a difficult task. Globalisation can be good, bad, threatening, or for many citizens of the global village, a term with murky meanings at best. Robert Hanvey, nearly 30 years ago (in 1976) offered a definition of "global education," thus providing a convenient starting point:

> Global education is learning about those issues that cut across national boundaries and about interconnectedness of systems, ecological, cultural, economic, political, and technological. Global education involves perspective taking, see things through the eyes, minds, and ears of others; and it means the realization while individual and groups view life differently, they also have common needs and wants.
>
> (p. 163)

In this definition, one sees part of the rationale for multicultural education and the ripples of the post-modern movement. However, it is the implications for business, industry, government, and culture—rather than educators—that have helped transform "globalisation" from this first notion of global education to that of a force of change.

Wells, et al. (1998) consider globalization a "…complicated set of economic, political , and cultural factors" (p. 323), citing a statement of Kumar from 1995, who notes "Political, economic, and cultural life is not strongly influenced by developments at the global level." (in Wells, et al., 1998, p 323). If this is indeed the case, whether in a formal or non-formal setting for these events to be reality, there is a continual and growing movement toward "global education."

The above typify the distinct, but overlapping interpretations presented by various globalists. These interpretations are worthy of debate, since the ideologies of globalisation will affect the direction and scope of educational reform. The role of globalisation and the distribution of material or natural resources and globalisation and the role of nation states are the crux of the matter, but are well beyond the modest discussion here, considering PD as a form of globalisation. For convenience, using the modern lenses on globalisation, we see the notion of "global education" residing with the bigger framework and find it an excellent frame-of-reference to consider the "globalisational" aspects of international PD work.

Beginning with a willingness to consider multiple perspectives is important and valuable for educators interested or invested in international PD. Though developed countries are inclined to fund globalisation efforts that effect stable economies and governments, and the development of civil society among developing nations such as in the Newly Independent State of Armenia, and post-socialist Romania, as providers of PD, our globalisation agendas must include the simple notion of global education as a two-way street. Hence, many of those involved in international PD are driven—or at least influenced—by political and economic engines of their nations, or by collaborative efforts among nations (e.g., World Bank). However, it does not diminish the value of our work as global educators nor limit our potential for success and encouragement of sustainable change beyond the scope of our personal work.

Our world has become more interconnected through immigration, world conflicts, television, and of course the Internet. In particular, the first and the last items in this list have a powerful effect on the peoples and PD conducted in Romania and in Armenia. School curricula are recognized as outdated, needing modification, and requiring new knowledge and skills first for the educators and then for the next generation of learners in these countries (Cucos, 2002). Vardumyan, Hoshannisyan, and Varrella (2003) suggest that to prepare students for our interconnected world, we must develop education systems that:

- offer a global perspective in all subjects at all levels, and
- is open to new ideas as they emerge.

They point out that instead of focusing on innumerable facts, a global education requires discerning the big ideas and applying pertinent facts, emphasizing the contemporary focus on developing thinking skills for general and in content-specific applications.

In particular, in this closing discussion, we wish to guide the reader to the issue of co-learning. Quite simply, the educator that sets about providing PD for developing and transitioning nations must live the notion that we learn from and with one another. For example, the meaning of "alternative assessment strategies" as assessments that are different from the ones used usually has different meaning in the US and countries that are coming out of socialism and communism. In the US testing is predominantly based on multiple-choice exams while in contrast Romania it is based on open-ended questions. What is the alternative assessment strategy in one country, is the commonly used assessment strategy in the other. However, global success on TIMSS and PISA requires knowing how to answer to multiple choice testing as well as to open ended question (PISA having more open-ended questions than TIMSS). Thus high performing nations (e.g., Finland), can provide examples and lessons for those that are further down in the rankings (e.g., the US and the UK).

Professional development situations can be transformative learning experiences that influence everyone present as well as the K-12 students (who are not usually present during professional developments). We sometimes take elements of our contemporary Western models of PD for granted, not considering best practice or ways that we can improve our home-based PD efforts. International perspectives on PD are extremely important for the science teacher education community—for that matter the entire teacher education community—as these allow us to study and understand the elements of these experiences in other contexts, broadening our horizons and enhancing our efficacy. Professional exchange and the development of the international professional development principles, considering the elements of professional development viewed through the eyes of others, are necessary to moving education forward globally, preparing a future generation that may ***think and act*** globally.

This discussion leaves us with questions, as well as conclusions. Among the conclusions are the parallel findings of our work and that reported by Garet *et al.* (2001) and by Gess-Newsome 2001 regarding PDs that produce knowledgeable and reflective teachers. The commonalities include sustained and intensive agendas, include inquiry-based and hands-on experiences that are integrated into the everyday teacher practices. These kinds of PD experiences also lead to the development and enhancement of pedagogical content knowledge, the knowledge that differentiates novice from experienced teacher (Shulman, 1986; Gess-Newsome & Lederman 1999). Having teams of knowledgeable professionals as PD leaders enhances the quality of the PD experience and adds to the will of the participants to learn and teach together. For example, in Armenia, the teams of teacher trainers train second waves of trainers that then work among the newest of the pilot school groups. The original teams continue to teach and train as well, sometimes in support of the new trainers and other times in new and emerging areas of priority (e.g., using cooperative learning for conflict resolution). As the leaders develop their understanding and PCK regarding the new topics, they collaboratively infuse their knowledge into the k-16 classrooms, sometimes directly in their own classrooms and other times through the second tier trainers and pilot school groups. These waves of teachers teaching teachers creates a sense of community, advances general understanding as well as personal PCK, and with each subsequent wave, the rate of mastery of the fundamentals is accelerated (Gilmer *et al,* 2002; Moscovici & McNulty, 2003, Varrella *et al*, 2003).

Our role as international PD providers, and the rights and responsibilities of the PD participants are equal priorities. In terms of our role, being able to offer PD in a different context allows us to view PD from the outsider's frame of represent. It is relatively easy to identify patterns, particularly the highly unfamiliar ones. Comparing PD patterns from the international experience with their parallels in the US allows us to develop a critical eye and continue to improve the PD offerings. PD participants and leaders from the international experience learn from our descriptions and observations during the international exchanges and, as a result, they modify their experiences to enhance the quality of their PD offerings and experiences. Perhaps the most influential method that has spread through formal instruction and modelling—from the US side—is that of cooperative learning. A second priority was that of questioning strategies designed for higher order thinking and inquiry-based activities in science and social studies. The impact was first on the international populations through modelling at workshops and conferences and later through publications such as the work by Hovhannisyan, Zohrabyan, Ohanova, Arnaudyan, & Varrella, (2002).

In terms of the rights and responsibilities of the PD participants, we feel strongly that the participants must share in identifying needed areas for growth and take responsibility for their learning. They become the cultural workers who dare teach (Freire, 1998). At the same time, the right to be able to choose/help develop the PD that satisfies one's needs comes with the commitment and responsibility to learn and use learned material in every-day practice. PD participants take an active role in learning and using learned material to improve students' learning.

Should teachers receive payment for attending PD as a form of low paid summer job with the sign-up sheets in place to make sure teachers do not cheat? Perhaps, as Giroux & Simon (1989), and Aronowitz and Giroux (1993) suggest, we need to rethink the meaning of the teacher as a transformative intellectual, someone who is confident with the content and pedagogy at the level that encourages students to become critical inquirers. Do we have the support from the system to create such teachers? Is the system interested in having such a teaching force, or is it more interested keeping education domesticated and "in-line" with the current bureaucratic agenda? Unfortunately, it is often the latter in the experiences of the authors. In our experience educational leaders—who might be "bully" advocates for reform—stop short when it comes to effecting their power on the financial bottom line. The former point is no surprise, but the latter cannot be ignored, since in most instances in the NIS and other developing nations, educators' salaries are dismally low.

5.1 For the International Participant

The first exchange between the PD provider or academic advisor (the "official titles" of the first and second author in order) initiates the process of globalisation. As we noted above, this is a two-way event; but it is more likely to be an unfamiliar event from those who have lived and been educated in relatively closed societies. The international PD provider can anticipate resistance and intense scepticism of "these new ideas." Certainly, this is not "news" for the experienced educator who has conducted PD; however, the resistance is deeper and culturally based in some international settings. Along with the intimidating experience of integration into the world economic system with its varied forms of relationships, there is a concern about "kinship" relationships and dilution of culture and treasured traditions.

In the case of the former, a less mobile society—for reasons of politics, finance, or culture—can promote strong family ties, which are the sustaining element of the culture and the cultural heritage spanning thousands of years. Hence, kinship both constrains and sustains the cultural identity of the society, providing stability as well as cultural inertia. When the PD specialists arrive, advocating "freer thinking" (an Armenian descriptor for active learning) among the pupils and creating circumstances where a new generation of idealistic educators are willing to think beyond the old rules, tensions result. These tensions are not isolated to those outside of the cadre of "progressive educators," in our experience these concerns exist within the groups of reform minded-educators as well. These educators become the true conduits for change and enhancement as they temper their kinship issues—that are culturally based—with the value and practicality of the new or alternative practical and theoretical elements of the PD experience. Thus, elements of family identity and cultural identity are examined in the context of the worthiness of alternative conceptions regarding teaching, learning, and by extension, the development of a democratic society. At this point the authors have recognized and are beginning to understand this issue with greater clarity, but have not adequately unpacked it to discuss it further. Hence, it is an area in need of further study and attention.

5.2 From you and I to us

We have found that one of the key indicators of progress is the blurring of the lines. As noted above, this only happens in sustained or unique circumstances (i.e., the PD provider has strong roots and language skills based on the society and culture of the participants). As the lines blur, *everyone becomes a teacher educator*, whether by example—i.e., modelling in their own schools—or position as a teacher educator or government provider of PD. In this same spirit, everyone becomes an international participant because time on task and cultural commonalities—whether by birth or emersion—provide a common base, a shared agenda, and the potential for true learning communities. The metaphor of the cell comes to mind. All cells have common elements such as membranes, a need for nutrients, and genetic material (to name a few). However, the cells are also unique, have specialized functions, shapes, and locations within the organism. The cells remain unique, but together the whole is greater than the sum or their parts. As the PD becomes contextualized to the circumstance, realities, and dreams of the populations served, one discovers "joint ownership" and the emergence of a learning community. Hence, the most viable solutions—owned by those in the developing culture—are discovered. A similar process can yield a re-examination of our own motivations and realities at home [for the authors] in the United States, thus shaping new and better solutions on this side of the Atlantic as well.

Finally, to answer the question used in the title above, yes, international PD does contribute to globalisation. However, there is one caveat—as the ones "who dare teach" (Freire, 1998)—our experience and a preponderance of evidence indicates that immersion and sustained contact is necessary for measurable progress, co-owned development agendas, and eventual diffusions of innovation.

References

Anyon, J. (1997). *Ghetto schooling: A political economy of urban educational reform*. New York: Teachers College Press.

Armenian Ministry for Education and Science. (2004). *National curriculum for general education*. Author.

Aronowitz, S., & Giroux, H. A. (1993). *Education still under Siege* (2nd ed.). Westport, CT: Bergin & Garvey.

Caprio, M. W., & Borgesen, D. S. (2001). Teaching to learn: Why should teachers have all the fun? *Journal of College Science Teaching*, *30*(6), 408–411.

Cozma, T. (2001). *A new challenge for education: Intercultural issues* (available only in Romanian). Jassy, Romania: Polirom.

Crowther, D. T., & Cannon, J. R. (2001). *Professional development models: A comparison of duration and effect.* Paper presented at the annual meeting of the International Association for the Education of Teachers of Science, Charlotte, NC, January 10–13.

Cucos, C. (2002). *Pedagogy* (5th ed.) (available only in Romanian). Jassy, Romania: Collegium Polirom.

Dori, Y. J. (2003). From nationwide standardized testing to school-based alternative embedded assessment in Israel: Students' performance in the Matriculation 2000 Project. *Journal of Research in Science Teaching*, *40*(1), 34–52.

Finn, P. J. (1999). *Literacy with an attitude: Educating working-class children in their own self-interest.* New York: State University of New York Press.

Freire, P. (1998). *Teachers as cultural workers: Letters to those who dare teach.* Boulder, CO: Westview Press.

Freire, P. (1990). *Pedagogy of the oppressed.* New York: Continuum.

Fullan, M. (2003). *Change forces with a vengeance.* New York: RoutledgeFalmer.

Fullan, M. (2005). *Leadership and sustainability – Systems thinkers in actions.* Thousand Oaks, CA: Corwin Press.

Garet, M. S., Porter, A. C., Desimone, L., Birman, B. F., & Yoon, K. S. (2001). What makes professional development effective? Results from a national sample of teachers. *American Educational Research Journal*, *38*(4), 915–945.

Gess-Newsome, J. (2001). The professional development of science teachers for science education reform: A review of literature. In J. Rhoton & P. Bowers (Eds.), *Professional development: Planning and design.* Arlingrton, VA: National Science Teachers Association Press.

Gess-Newsome, J., & Lederman, N. G. (Eds.). (1999). *Examining pedagogical content knowledge: The construct and its implications for science education.* Dordrecht, The Netherlands: Kluwer Academic Publishers.

Gilmer, P. J., Hahn, L. H., & Spaid, R. M. (Eds.). (2002). *Experiential learning for pre-service science and mathematics teachers: Applications to secondary classrooms.* Greensboro, NC: Southeast Eisenhower Regional Consortium for Mathematics and Science Education (SERVE).

Giroux, H. A., & Simon, R. (1989). Popular culture and critical pedagogy: Everyday life as a basis of curriculum knowledge. In H. A. Giroux & P. L. McLaren (Eds.), *Critical pedagogy, the state and cultural struggle*, (pp. 236–252). New York: State University of New York Press.

Govett, A. L., & Hemler, D. (2001). *Changing teachers' attitudes and perceptions of science and scientific research.* Paper presented at the annual meeting of the International Association for the Education of Teachers of Science, Charlotte, NC, January 10–13.

Guskey, T. R. (2002). Does it make a difference? Evaluating professional development. *Educational Leadership*, *59*(6), 45–51.

Hammer, D. (2000). Teacher inquiry. In J. Minstrell & E. H. van Zee (Eds.), *Inquiring into inquiry learning and teaching in science* (pp. 184–215). Washington, DC: American Association for the Advancement of Science (AAAS).

Hanvey, R. G. (1976). *An attainable global perspective.* Denver: Center for Global Perspectives, University of Denver.

Henderson, S., & Lederman, N. G. (2002). *Partnership between science teachers and laboratory-based scientists: Delineating best practices.* Paper presented at the annual meeting of the National Association for Research in Science Teaching, New Orleans, LA, April 7–10.

Hovhannisyan, G., Zohrabyan, A., Ohanova, B., Arnaudyan, A., & Varrella, G. F. (2002). *Constructivist teaching methods: A manual for teachers.* Yerevan, Armenia: International Research and Exchange Board.

Kelleher, J. (2003). A model for assessment-driven professional development. *Phi Delta Kappan, 84,* 751–756.

Martin-Kniep, G. (2005). *Developing learning communities through teacher experience.* Thousand Oaks, CA: Corwin Press.

Morris, M., Chrispeels, J., & Burke, P. (2003). The power of two: Linking external with internal teachers' professional development. *Phi Delta Kappan, 84,* 751–756.

Moscovici, H., & McNulty, B. (2003). *The effect of collaboration: Learning together toward the development of the mutualistic science educator of the future.* Paper presented as a part of a symposium at the annual meeting of the National Association for Research in Science Teaching, Philadelphia, March 23–26.

Moss, D. M., & Lonning, R. A. (2001). *Beyond "sit and get" workshops: Professional development for standards-based science education.* Paper presented at the annual meeting of the National Association for Research in Science Teaching, St. Louis, MO, March 25–29.

National Research Council. (1996). *National science education standards.* Washington, DC: National Academy Press.

National Research Council. (2000). *Inquiry and the national science education standards.* Washington, DC: National Academy Press.

National Research Council. (2005). *How students learn.* Washington, DC: The National Academies Press.

Posner, G. J., Strike, K. A., Hewson, P. W., & Gertzog, W. A. (1982). Accomodation of a scientific conception: Toward a theory of conceptual change. *Science Education, 66*(22), 211–227.

Pyle, E. J. (2002). *TIGERS of a different stripe: Two-way professional development exchanges between middle grades and science education.* Paper presented at the annual meeting of the International Association for the Education of Teachers of Science, Charlotte, NC, January 10–12.

Sergiovanni, T. J. (1994). *Building community in schools.* San Francisco, CA: Jossey-Bass.

Shulman, L. S. (1986). Those who understand: Knowledge growth in teaching. *Educational Researcher, 15*(2), 4–14.

Thompson, T., & Gummer, E. (2002). *Negotiating necessary science content in professional development exercises for teachers.* Paper presented at the annual meeting of the National Association for Research in Science Teaching, New Orleans, LA, April 7–10.

Tobin, K., Tippins, D. J., & Gallard, A. J. (1994). Research on instructional strategies for teaching science. In D. L. Gabel (Ed.), *Handbook of research on science teaching and learning* (pp. 45–93). New York: Macmillan Publishing Company.

Vardumyan, S., Hoshannisyan, A., & Varrella, G. F. (2003). *Modern pedagogical approaches: Theory, practice, partnerships, and assessment—a handbook for educators.* Yerevan, Armenia & Washington, DC: International Research and Exchange Board.

Varrella, G. F., Weld, J., Harris, R. L., Enger, S. K., Yager, R. E., & Burry-Stock, J. S. (in press). School based curriculum frameworks: Supporting local science education reform. *Electronic Journal of Science Education, 8*(1), 1–30.

Wallace, J., & Louden, W. (2002). *Dilemmas of science teaching.* New York, NY: RoutledgeFalmer.

Weld, J., Stier, M., & McNew, J. (2002). *Measuring scientific reasoning development under contrasting pedagogical styles: A novel approach.* Paper presented at the Annual Meeting of the International Association for the Education of Teachers of Science, Charlotte, NC, January 10–12.

Wells, A. S., Carnochan, S., Slayton, J., Allen, R. L., & Vasudeva, A. (1998). Globalization and educational change. In A. Hargreves, A. Lieberman, M. Fullan, & D. Hopkins (Eds.), *International handbook of educational change.* Dordrecht, Netherlands: Kluwer.

White, G., Russell, T., & Gunstone, R. F. (2002). Curriculum change. In J. Wallace & W. Louden (Eds.), *Dilemmas of science teaching: Perspectives on problems of practice* (pp. 231–244). New York, NY: RoutledgeFalmer.

Yager, R. E. (1991). The constructivist learning model: Towards real reform in science education. *The Science Teacher, 58*(6), 52–57.

17

DOING SURVEYS IN DIFFERENT CULTURES: DIFFICULTIES AND DIFFERENCES – A CASE FROM CHINA AND AUSTRALIA

Zhongjun Cao[1], Helen Forgasz [2] and Alan Bishop[2]
[1]*Postcompulsory Education Center, Victoria University, PO Box 14428 Melbourne City, VIC 8001, Australia;*
[2]*Faculty of Education, Monash University, Clayton, VIC 3800, Australia*

Abstract: International studies represent one important aspect of the phenomenon of the internationalization and globalization of mathematics education and have attracted the interest of many organizations and researchers. There are challenges and difficulties in conducting surveys in different cultural settings which have not been well documented in the mathematics education literature. This chapter accounts for some of the challenges and difficulties involved in selecting a survey topic, designing the survey, and administering it as occurred in a recently conducted study exploring students' attitudes towards mathematics in China and Australia. It is suggested that awareness of cultural differences is a key issue that researchers should pay attention to when conducting cross-cultural research in mathematics education

Keywords: international studies; cultural differences; survey; students' attitudes towards mathematics; sociocultural factors

1. Introduction

As one aspect of the internalization and globalization of mathematics education, international studies have drawn the interest of many organizations and researchers over the past decades. For instance, in the First International Studies on Mathematics conducted by IEA (International Association for Educational Assessment) (Husén, 1967), twelve countries participated; in the Second International Mathematics Study, twenty countries participated (Burstein, 1993; Robitaille & Garden, 1989; Travers & Westbury, 1989); there were over forty participating countries in the Third International Mathematics and Science Study (Beaton & Robitaille, 1999).

B. Atweh et al. (eds.), Internationalisation and Globalisation in Mathematics and Science Education, 303–320.

The current Programme for International Student Assessment (PISA) organized by the OCED (Organisation for Economic Co-operation and Development) is also attracting over forty countries' participation, and in each participating country, between 4,500 and 10,000 students are answering the tests (OCED, n.d.).

In addition to those large scale international studies, many smaller scale studies were also conducted. To name only a few, Mayer, Sims, & Tajika (1995) compared how textbooks teach problem-solving in the USA and Japan; Cai (1998) explored students' problem-solving and problem-posing in the USA and China; Ma (1999) investigated teachers' knowledge in China and the USA; Cao & Bishop (2001) compared students' attributions of success of failure in mathematics in China and Australia; Park & Leung (2002) studied the differences in the mathematics textbooks of China, England, Japan, Korea and the USA; and Neubrand (2002) analysed the features of mathematics lessons in Germany, Japan and the USA.

These large and small international studies have revealed huge differences in various aspects of mathematics education in different cultures: mathematics curriculum, teachers' knowledge, students' problem solving, and students' attitudes towards mathematics. It is believed, at least by some, that international studies have provided useful information for mathematics educators and policy makers, as maintained by Stigler & Perry (1988):

> Cross cultural comparison also leads researchers and educators to a more explicit understanding of their own implicit theories about how children learn mathematics. Without comparison, we tend not to question our own traditional teaching practices and we may not even be aware of choices we have made in constructing the educational process.
>
> (p.199)

However, some doubt the fairness and the validity of the large scale international studies in mathematics education. For instance, Keitel & Kilpatrick (1999) raised a range of issues concerning large scale international studies, ranging from who funded the projects, who designed the projects, who administered the projects, who made the criteria for data analysis, and who benefited from the projects.

Irrespective of whether researchers are or are not in favour of international studies, it is a fact that a vast literature has been created in international studies in mathematics education. However, within the literature reviewed, little has been found of studies with a focus on documenting the challenges and difficulties that researchers face in the process of designing and administering a survey, and how cultural factors can influence researchers' activities and research results, even though international studies evidently take place in different cultural settings, and cultural factors impact those survey activities.

Drawing on the experiences of the authors in carrying out a recent study exploring students' attitudes towards mathematics in China and Australia, this chapter tries to fill this gap. It documents the challenges in the survey design and the differences in the survey procedures and people's attitudes towards the survey between the two cultures.

2. The Study

The study took place during 2000–2004. The aims of the study were to investigate firstly students' attitudes towards mathematics in China and Australia, and secondly the influence of various external factors on the formation of those attitudes. The study was a PhD project conducted by the first author, and supervised by the other two authors in Australia. An instrument was developed to measure the attitudes of students, and the influences of the external factors. The construct of "attitudes towards mathematics" was based on a tri-component model (Affective-Behaviour-Cognitive) (Cook & Selltiz, 1964). The external factors included the influences of teachers, parents, peers, and mathematics tasks as perceived by the students. The instrument consisted of 56 items in total, which comprise five subscales: Attitudes scale, Teacher Influence scale, Parents Influence scale, Peers Influence scale, and Task Influence scale.

The study collected the answered questionnaires from 356 primary and secondary school students in China, and 406 primary and secondary school students in Australia. They were distributed at three year levels, Years 5, 7, and 9. The students in China were from three primary schools and three secondary schools in Kaifeng, a middle sized city in Henan Province. Of the three secondary schools, one was a "key-school" while the other two were non-key schools. Key schools in China were schools with better teaching facilities, higher quality teachers, and higher rates of graduated student enrolments in the tertiary institutions or the next level of schooling. The students in Australia were from six primary schools, and seven secondary schools in metropolitan Melbourne. The participating primary schools were a mixture of schools from suburbs with higher and lower socio-economic levels. The selection of secondary schools was based on VCE (Victoria Certificate of Education – a two-year programme for the final years of schooling) performance in the year 2000. Three of the schools' VCE performances were rated very good; the rest were rated average.

The challenges and difficulties took place when designing the survey and when administering the survey in the two different cultural settings, and the main sections of this chapter will document the details of these difficulties.

3. The Issues in Survey Design

The issues encountered in this study design included topic choosing, instrument designing, and translating of the instrument between English and Chinese. These are elaborated in the following subsections.

3.1 Choosing the Topic

The first issue in choosing the topic (students' attitudes towards mathematics in China and Australia) was the difficulty of justifying the equal importance of the topic to researchers in both countries.

Students' attitudes towards mathematics, as an important learning outcome, have clearly been documented in mathematics curricula both in China and in Australia (State Education Commission of the People's Republic of China, 1992; Ministry of Education of the People's Republic of China, 2003; Australian Educational Council, 1991). As a teacher with several years' experience in China, the first author thought it was important to conduct this study as he found from his own teaching experience in China and Australia that students' attitudes towards mathematics were always an issue in mathematics classrooms. There were always students in the mathematics classroom who appeared bored with mathematics, but the reasons for this were unclear. When talking with other mathematics teachers in China, he felt that many of them were also concerned with students' attitudes towards mathematics. He believed that through the investigation of students' attitudes towards mathematics in two different cultures, the forces that influence students' attitudes towards mathematics could be better understood, which could then benefit teachers' practice. Through an extensive search of the literature, the first author also found that there was not much literature exploring students' attitudes towards mathematics in China, nor on the differences in the attitudes towards mathematics of students in China and Australia.

However, when the first author originally talked to some fellow researchers in Australia about the intention of conducting this project, they all believed that this topic was not currently interesting, as they believed that many people had worked on this area before in Australia and in other Western cultural contexts. Thus they were concerned about whether the research could produce any new outcomes for the research community and for teaching practice. The first author convinced his supervisors and the relevant researchers that the significance of the study was to reveal how the students' attitudes towards mathematics changed across different year levels and how various external factors influence these attitudes in both countries. Nevertheless, when talking with other researchers in Australia later regarding this research topic, many still believed that this topic was a rather old one.

3.2 The Instrument Design

There were also some difficulties in designing the research instrument. These included choosing the right format of the instrument, covering adequate content, determining the number of choices in the response formats for the Likert items. These difficulties are discussed below.

3.2.1 The Format of the Instrument

The right format of the instrument is vital for a study. The format of the instrument is largely dependent on the purposes of a study, as well as on the research questions to be asked. However, cultural factors should also be considered regarding the appropriateness of the instrument's format. Difficulties can rise with both aspects.

Considering the purposes of the study, it was anticipated that the participating students in each country should number approximately 400, and it was felt that survey methods could achieve the purposes of the study better than other methods such as observation and interview. Survey methods suit Australian students as surveys are conducted often in Australian schools and students are supposed to know how to complete surveys. In contrast, school students in China are not familiar with surveys as surveys do not happen often in schools. They have more experience of dealing with examinations, in which students should try their best to give each question a correct answer. However, in this survey to measure their attitudes towards mathematics, students were asked to express their true attitudes towards mathematics. Did the students still treat the survey as an examination, in which they tried to give "correct answers"? Extra instructions regarding how to answer the survey were provided, however it is not known whether the instructions had any effect.

3.2.2 The Content of the Instrument

When doing surveys in different cultures researchers should make sure that the survey content adequately covers the important variables related to the target constructs in each culture. In fact, initially the authors proposed using a version of the attitude questionnaire developed in a previous study conducted in Australia (Forgasz , 1995), but this was not done as the first author noted that the previous questionnaires could only reflect the important features in mathematics education in Australia, and not those of importance in China. For instance, students in China are often asked by their teachers to concentrate on the teaching in the class, the students are told by their teachers that they must work hard in mathematics, parents pay extreme attention to their children's education, and teachers are felt to be particularly important in the eyes of the students. Thus the authors developed a new instrument for this study by using items from the earlier Australian questionnaire (Forgasz , 1995) and from other Chinese questionnaires, such as in a study in Hong Kong (Wong, Lam, Leung, Mok, & Wong, 1999), to cover the important aspects of mathematics education both in China and Australia. Some sample items from the instrument which the authors thought would reflect the important features of Chinese mathematics education influences included:

> Item 16. When the teachers teaches us in the mathematics classroom, it is important that I always concentrate my mind on her/his talking
>
> Item 25. My teacher reminds us frequently that a pupil must work hard in order to do well in mathematics
>
> Items 35. My mother expects me to be the best student in mathematics and other subjects
>
> (Cao,2004, p.301)

The developed instrument was presented to a mathematics education seminar at Monash University, Australia and was commented on by some fellow mathematics researchers, resulting in some small revisions being made. Although effort was made to make the instrument cover adequate content for both countries, subjectivity in designing the instrument may still have existed to a certain extent, as the authors could not find the opportunity to have the questionnaires commented on by any other researchers in mathematics education who were familiar with the Chinese context.

3.2.3 The Number of Choices in the Likert Scale

Likert scales are comprised of items or statements with different response choices; in attitude measurement, "strongly agree", "agree", "undecided" (or "unsure"), "disagree", and "strongly disagree" are frequently used. The number of choices in the response format of a Likert scale is usually five, however with variations from two to twenty (Nunnally, 1967). People have different opinions regarding whether an even number or an odd number of choices should be used (Cunningham, 1986). People who prefer the odd number of choices hold that respondents should be given the choice to express their neutral attitudes towards the attitude objects, by offering them a neutral mid-point.

Reliability measures are often associated with Likert scales to indicate the extent to which responses to the collection of items are consistent. Cronbach Alpha (α) is a commonly used measure of the reliability of a Likert scale. The validity of the scale, that is that it is measuring what is intended, is determined in a variety of ways, including, for example, its "construct validity" which might involve its consistency with definitions from the literature used in developing the scale items.

In the pilot study of this project, the authors first used a five-point Likert instrument, which is, for each statement, the respondents are asked to indicate whether they "strongly agree", "agree", "undecided", "disagree", and "strongly disagree". The instrument was tested among two classes of Year 7 students in China and Australia. It was found that the reliabilities (Cronbach Alphas) of three of the scales (Attitude scale, Teacher Influence scale, and Task influence scale) in the instrument did not reach a satisfactory level for the Chinese students, while all five scales did for the students in Australia (Regarding the reliabilities of the Likert scale, the value of Cronbach Alpha should be larger than 0.7 for a reliable scale, according to Nunnally, 1967).The reliability coefficients (Cronbach Alphas) of each of the five scales of the instrument for the students from the two countries are summarized in Table 1.

Table 1. Reliability coefficients (Cronbach Alphas) of the five-point Likert scales of the instrument in the pilot study in China and Australia

	China (N=20)	Australia (N=18)
Attitude scale	0.67	0.78
Teacher Influence scale	0.64	0.78
Parent Influence scale	0.75	0.84
Peers Influence scale	0.77	0.83
Task Influence scale	0.61	0.75

Then the reliabilities of two versions of the instrument were tested, with exactly the same items, but one with four-point response formats for the Likert scales (i.e., for each statement, there were four choices: "strongly agree" ,"agree", "disagree", and "strongly disagree"), and the other with five-point Likert scales. These were used with another class of 48 Year 7 students in China, and it was found that the four-point Likert scale instrument was more reliable than the five-point Likert scale instrument. The reliability coefficients (Cronbach Alphas) of each of the five scales of the two versions of instrument are listed in Table 2.

Table 2. Reliability coefficients (Cronbach Alphas) of four-point Likert scales and five-point Likert scales of the instrument in the pilot study

	Five-point version (N=24)	Four-point version (N=24)
Attitude scale	0.66	0.74
Teacher Influence scale	0.59	0.72
Parent Influence scale	0.76	0.87
Peers Influence scale	0.79	0.83
Task Influence scale	0.75	0.76

The authors decided to adopt a four-point Likert scale in the main study. Regretfully, due to time and other restrictions, the authors did not pilot test and compare the effectiveness of a five-point and a four-point Likert scale for the Australian students. Fortunately the analysis of the main study results indicated that the reliabilities of the scales in the four-point Likert instrument reached a satisfactory level among Australian students as well. The item analyses tended to suggest that cultural factors may play a role in the differences in reliabilities between the four-point Likert scale and the five-point Likert scale: some items might have been too sensitive for the Chinese culture, for example, "Item 20. My teacher likes maths very much" and "Item 21. My teacher's teaching of maths is very interesting", but we were unable to pursue this point any further.

3.3 The Issue of Translation of the Questionnaire

As the survey was conducted among students in two cultures, and the item pool comprised both English and Chinese items, there was also an issue of the exactness and appropriateness of translation from one language to another. Although considerable care was taken, with back-translation being used, there were some specific problems as well. For example, for the five-point Likert version questionnaire, the authors had a discussion with some other Chinese speaking researchers on how to translate the middle choice of "undecided" from English to Chinese. If it was translated into Chinese "bu ken ding", it might be understood by students as "do not know" for particular statements. In order not to confuse students, the word was translated as "ji bu zan cheng, ye bu fan dui" (neither favoured nor unfavoured).

Another issue emerged with respect to an item on the four-point Likert scale instrument. Item 54 was originally taken from Chinese questionnaires. It stated: "Most exercises in mathematics textbooks are (1) very difficult, (2) a little difficult, (3) a little easy, (4) very easy". The other authors suggested that it was not appropriate to directly translate it this way, as many primary schools in Australia do not use textbooks in their mathematics teaching. Therefore the item was changed to "Most exercises in Mathematics are (1) very difficult, (2) a little difficult, (3) a little easy, (4) very easy" in the final English version of the instrument. Similar changes were also made on other items which contained the phrase "mathematics textbooks".

4. The Survey Process

In order to understand what happened in the survey process, it is necessary to have a brief explanation of the cultural differences between the two countries.

4.1 Cultural Differences in China and Australia

China and Australia are two societies with different social and cultural backgrounds: one is a developing country with an Eastern cultural tradition, another is a developed country dominated by Western culture. There are huge differences between the two systems. Of importance to this chapter, two differences stand out.

One concerns the completeness of the rules, and the regulations. With the adoption in China of the "open-door" policy and the rapid development of the economy, in recent years, various laws have been made to protect the benefits of different groups of people (New Laws Take Effect, 2004). However, the authors' and many other people's observations are that much more needs to be done to set up a complete law/regulation system, and to protect the groups of weaker people in China (i.e., the people with disabilities, unemployed people, and children). In contrast, Australia is a developed country with a much more complete legal system. Various laws have been made to protect the benefits of different groups of people. The weaker groups of people, such as the unemployed, the senior citizens and children are well guarded by the regulations of the society.

The second difference between China and Australia relates to the relationships among people. Research (e.g., Hsu, 1981) has suggested that hierarchy is a strong feature of Eastern societies. People pay more attention to those who are superior to them in the hierarchy, and people usually obey the opinions of senior people and people with higher social status. For example, children are often instructed to listen to their parents' instructions, and students are often told to respect their teachers. In contrast, in Western societies, individuals are considered of extreme importance (Ling, Burman, & Cooper, 1998), and all individuals, no matter whether they are seniors or young people should have their rights fully respected.

The third difference reflects the way of contacting people and doing certain kinds of business. In China people are used to contacting people in an "informal way".

For example, it is usually effective to make contact in person without making an appointment. Written documents are usually not necessary except for very serious business which may result in economic conflict if no written agreement is made in advance. In Australia, more formal procedures are needed when doing certain types of business. Appointments and written documents are usually necessary to proceed with the business.

4.2 Different Procedures for Conducting Surveys in China and Australia

Because of the cultural differences in China and Australia, the procedure for conducting surveys in the two societies is quite different. One is simple, the other is complicated!

In China, if a researcher wants to conduct a survey among school students, the researcher can approach the school principals directly. With the permission of the principal, the researcher can approach the teachers, and ask the teachers to distribute the surveys to their students directly. Thus the students can answer the surveys without the need to get the permission of their parents. In some cases, if a researcher knows the teachers, the researcher can approach the teachers directly, and the teachers can then give the surveys to the students to answer.

In Australia, if a researcher plans to conduct surveys among school students, he/she must go through the following steps:

(1) *Get the approval from the university's Ethics Committee.* The researcher must provide a copy of the questionnaire to the Ethics Committee, and prepare to answer various questions raised by the committee. This process often takes a few weeks.

(2) *Get approval from the State Education Department.* The researcher must provide the approval letter from the Ethics Committee, the explanation letter about the research, and a copy of the questionnaire to the Department. This procedure also takes several weeks.

(3) *Get consent from the principals.* The researcher needs to approach intended participating schools' principals to get their consent for participation.

(4) *Get consent from the mathematics coordinators/teachers.* After getting consent from principals, the researcher then needs to approach mathematics coordinators/teachers to get their approval for their students' participation.

(5) *Get consent from parents/guardians.* The researcher needs to write to the parents/guardians of students, with letters of explanation, to seek their permission for their children to participate.

4.3 The Process of Conducting this Survey in China and Australia

There are striking differences between the two cultures in the approval procedures, the attitudes towards surveys of principals, teachers, and students; and the participating schools' interest in the survey results. This section mainly explains these differences as they related to this study.

4.3.1 The Approving Procedure in Australia

After completion of the literature review and the design of the Students' Attitudes towards Mathematics questionnaire, the first author submitted the application for conducting the survey among students in China and Australia to the university's Ethics Committee, with an explanation letter, a copy of the survey, and explanation letters to principals, teachers, parents, and students. The first submission was not approved. Six weeks after the submission of the application, the Ethics Committee asked the researchers to clarify many steps/procedures. For instance, how will the schools be selected? How can the students' rights to refuse to participate be guaranteed? How can the researcher make sure that the students' answers were not to be used by their principals to evaluate the teachers' teaching? What permissions are needed for conducting this survey in China?

The first author, in discussion with the other authors, sent a letter to the Ethics Committee, explaining and answering all the questions raised, and the application was approved four weeks after sending the letter.

The authors then wrote to the State (Victoria, Australia) Education Department to seek approval for the study, with copies of the explanation letters, and the approval letter from the Ethics Committee. It took approximately four weeks to receive approval notification from the Department.

4.3.2 The Participation Rate

Table 3 shows the participation data for the study of schools and students in China and Australia. It can be seen from Table 3 that in China, six out of the seven schools contacted agreed to participate, with a participation rate of 86%. In Australia, by contrast, only 13 out of the 25 schools contacted agreed to participate in the survey, the participation rate being only 52%. In China, 356 out of 380 distributed questionnaires were answered and returned; the return rate was 94%. In Australia, 408 out of the 600 questionnaires were answered and returned; the return rate was only 68%. The length of time for achieving the anticipated number of answered surveys in China was only 6 weeks, while the length for achieving the anticipated number of answered surveys in Australia was 26 weeks.

Table 3. Participation data of schools and students in China and Australia

	China	Australia
Number of schools contacted	7	25
Number of participating schools	6	13
Number of questionnaires distributed	380	600
Number of questionnaires returned	356	408
Time (weeks) from contacting schools to receiving questionnaires	6	26

4.3.3 Attitudes Towards this Survey of the Principals, Teachers, Parents and Students in China and Australia

Principals

In China, the (first) author usually went to see the intended participating schools' principals directly, without making any appointment in advance. The author introduced himself to the principals, the purpose of the survey, and the procedure for conducting the survey. To avoid the sense of competition (principals only asking students of high academic achievement to participate), the author purposely did not mention that the study was a comparative study between China and Australia, he only pointed out that this was a study to investigate students' attitudes towards mathematics and the factors that might impact on students' attitudes towards mathematics. Almost all the principals showed their willingness to help the author to conduct the study among their students after they heard the author was doing a PhD. Only one was reluctant to participate as the principal was concerned that the students would perform poorly in the survey, and the results could be exposed to the public. Those principals who agreed introduced the author to the relevant mathematics teachers.

In Australia, the author wrote to the primary and secondary school principals, with the supporting documents (approval letter from the Ethics committee, the State Department of Education, the supervisors' letter, and a copy of the survey) to apply for permission for conducting surveys in their schools. Interestingly, the primary school principals and secondary school principals demonstrated a different degree of interest in participating. Among the seven primary school principals contacted, only one principal refused to participate as there were only a few Year 5 students in his school; the other six primary school principals were very happy to have their students participate. Among the eighteen secondary school principals contacted by mail, none of them replied to the author's initial requests to participate. The author usually rang the principals two weeks after mailing the requests. Some principals said they would pass on the requests to the mathematics coordinators/teachers and suggested that the author contact the mathematics coordinators/teachers to discuss the matter, while some principals just refused to participate. The reasons for refusing to participate included there being too many surveys interrupting the normal teaching activities, and the possibility of the data being exposed to the public. Another interesting point was that it usually took several attempts to speak to the secondary school principals directly on the phone as receptionists usually took the author's calls. First the receptionists enquired about the author's identity and the reasons for speaking to the principal, and then they spoke to the principals. Most of the time, the answer from the receptionist was "the principal is in a meeting, can you call back some time later", or "the principal will call back to you later". However, the author always had to call back again as he never received any phone calls from the principals.

Mathematics Coordinators/Teachers

Mathematics coordinators/teachers in China were very supportive of the study, so long as their principals agreed; they also helped to instruct their students on how to answer the surveys. The students were given instructions with an emphasis that the survey was not an examination, with no right or wrong answers, and no names would appear on the questionnaires. They were to put the answered questionnaires into sealed envelopes and return them to their teachers.

In Australia, most mathematics coordinators were also supportive so long as their principals passed on the author's requests for participation, and they agreed to assist the author to administer the surveys among their students. However, the researcher got the feeling that generally their help was given because of their sympathy for the difficulty of a non-native English speaking PhD student conducting a survey and pursuing a PhD study in another culture, instead of the significance of the study itself. For instance, they cared little about how the study was designed, how the findings would inform further teaching, and only two of the secondary mathematics coordinators wanted to know any findings from the study.

Students

The students in China completed and returned their surveys swiftly – usually within one or two days after they were given the questionnaires. The analyses of students' answers in China showed that the majority of students tried to answer all the Likert scale items. However, for the non-compulsory open-ended question "Please give any comments on your mathematics learning if you would like", 97 out of the 356 students provided answers, accounting for 27% of the answered questionnaire. Only six students gave negative comments about their teachers. They tended to use mild words to comment on their teachers, for example, they expressed their wish for their teachers' teaching to be more interesting.

The students in Australia needed to get permission from their parents to participate in the survey (the procedure was that students were given the survey, explanation letters, and the consent forms to take to their parents, and the students were asked to bring back the consent forms with their parents' signature and the answered questionnaires). Around 70% of the distributed questionnaires were returned. Among the students who returned their questionnaires, most returned them within one week, and a small number of students returned their surveys within two weeks. Some students needed their teachers to remind them to bring back the answered questionnaires. For the open-ended question "Please give any comments on mathematics learning if you would like", 122 students provided answers, accounting for 30% of the 408 answered questionnaires. Seventeen of the students commented negatively about their teachers; some of the students (six) used very strong words to criticize their teachers. For instance, they said that they hated their mathematics teachers as their mathematics teachers never provided any help to their study of mathematics!

4.3.4 Participating Schools' Interest in the Survey Results

The school principals in the two countries demonstrated differences in the extent of their interest in the survey results (summarised in Table 4). In China among the six participating schools, only one secondary school principal said that she wanted

Table 4. Number of schools interested in the survey results in China and Australia

	China	Australia
Number of schools participated	6	11
Number of schools interested in the survey results	1	6

to know the survey results. In Australia among the 11 participating schools, four primary school principals said they would like to know the survey results, while two secondary mathematics coordinators said they wanted to know the results. It appears that the participating Australian school principals/mathematics coordinators showed a much higher interest in the survey results than the school principals/mathematics coordinators in China.

5. Conclusions and Discussion

To summarise, a few important issues emerged from the survey design process and the survey administration process:

(1) When designing a survey that was intended to apply to two cultural contexts, the difficulties encountered included: choosing a topic of equal importance in both countries, choosing the right format of the instrument, the appropriate number of choices in response formats of the Likert scales, the adequateness of the survey content, and the precision in the translation of the questionnaire.
(2) There were striking differences in the procedures and the relevant people's attitudes towards the survey, which include:

  (i) The procedure for conducting surveys in China is not as complicated as in Australia; conducting a survey among school students in China is less time-consuming than in Australia;
  (ii) The principals' attitudes towards surveys in the two countries were different: the majority of the principals in China showed their support for conducting the survey in their schools. In contrast, the primary school principals in Australia demonstrated considerably more interest in the survey than the secondary principals;
  (iii) Students' attitudes towards the survey were also different: a much higher percentage of the distributed surveys in China were answered and returned than in Australia.
  (iv) Participating principals / mathematics coordinators in China demonstrated very low interest in the survey results; the figure in Australia was also not very high among the secondary school principals/coordinators.

The difficulties in choosing an equally important topic, the right format of the instrument, adequate content, the appropriate number of choices in the response format for Likert scales, and the precision in the questionnaire's translation are, perhaps, all culturally rooted. For example, the reason that researchers from the two cultural settings thought differently about the importance of the research topic may result from the fact that more research on the topic has been conducted in one culture than in the other. The difficulty in justifying the appropriateness of the survey in both cultures may be due to the prevalence of surveys in schools differing between the two cultures. The content covered by surveys are related to the educational practices in a culture, and there are always differences in educational practices between different cultures. The issue of the number of choices in the response formats for Likert scales may be related to items with high cultural sensitivity, and the translation issue results from the differences between languages.

The differences in procedures and people's attitudes towards surveys in the two societies may also be attributed as a few sociocultural factors. For instance, the fact that doing surveys among school students is a less time-consuming and less complicated procedure in China than in Australia may be attributed to two reasons. First, as mentioned earlier, at the current stage of development in China, many laws/regulations have not yet been made. Perhaps there has been much attention paid to setting up laws in areas such as the economy, but no one has considered seriously whether research activities in social and educational areas could have negative effects on children. However in Australia, as a developed Western country, relatively complete regulations have been set up to protect individuals, particularly children. Second, there are differences in the ways of contacting people when doing certain types of business; in China, people can be contacted without making appointment in advance, while in Australia written documentation is needed.

The differences in principals' and students' attitudes to surveys in China and Australia are perhaps due to the differences in relationships between people in the two societies. In China, a PhD student is often considered to be a person of rich knowledge; he/she is respected by many people in the society, including school principals; therefore requests in relation to academic activities are easily satisfied by the principals. People are used to obeying senior peoples' opinions in China. Therefore, because the principals agree to participate in the study, the teachers also agree; and because the teachers agree to participate, the students agree to participate. However, in Australia, individuals are considered extremely important, individuals' rights are fully protected, and each individual can decide to participate or refuse to participate in activities with much greater freedom. Therefore, in Australia there were larger proportions of principals and students choosing not to participate. An interesting phenomenon which this study could not explain was the difference between the Australian primary and secondary school principals' attitudes towards the survey.

The fact that there was low interest in research and research results among the principals and teachers also suggests that there is a disconnection between educational research and educational practice in both countries. That is, the research communities have not really convinced the educational practitioners (principals/teachers) of the usefulness of their research, and the educational practitioners have not really understood the significance of educational research. It is true that the principals and teachers in China willingly assisted the researcher to conduct the survey. However, their help was perhaps in most part due to the respect they had towards a PhD student. In Australia, the fact that many principals showed no interest in conducting the survey in their schools suggests that they may not consider research to have much benefit for their educational practice.

With the trend towards the internationalization and globalization of mathematics education, a considerable number of cross-cultural studies have been undertaken, with more cross-cultural studies likely in future. This study provides some points for doing cross-cultural studies worth further consideration in our mathematics education research community.

5.1 Researchers' Awareness of Cultural Differences

Researchers should be fully aware of the differences in cultural and educational practices between different cultures when designing surveys for different cultural settings. Instruments used effectively in one cultural context are not necessarily suitable for another context, and researchers should take measures to minimise the errors caused by the instrument design through developing more relevant knowledge in different cultural settings. In order to facilitate the survey process, researchers should possess systematic knowledge about the customs, rules, and regulations in the societies in which they intend to conduct their studies, as researchers cannot undertake surveys without following these customs, rules and regulations.

5.2 Balance between the Efficiency and Privacy

It was easier to undertake surveys among school students in China than in Australia. However, in China, questions should be asked about how we can make sure that surveys do not interrupt the teaching in schools, that participating school students' privacy (i. e., their personal information, and their answers) is protected, and that students' participation is voluntary, not because of others' willingness to participate. In Australia, could the efficiency of the procedure for conducting a survey be improved? As researchers at a higher education institution in Australia, we often hear that many Masters level students have to give up their plans for undertaking surveys (and other types of research) in schools within their limited period of candidature time (one or two years) due to the complicated, time consuming permission procedures.

5.3 Difficulties in Probing "True" Attitudes

As surveys aim to measure students' attitudes towards mathematics, the researchers would like to know that the attitudes reflected are the "true attitudes" of students.

This means that instead of students choosing right or wrong answers, they are expected to make their choices based on their own experiences. However, as students in China are influenced by a culture of obedience, was it possible that they only chose the "right" answers, meaning the answers that they believed that their teachers or their society would expect them to choose? Did they fear that if they did not choose the expected answer, they would not be considered good students by their teachers, and accordingly their class results could be affected? If this were the case, it would obviously have affected the validity of the survey. If this were the case, how can that bias be overcome?

5.4 Connections between Research and Practice

How can researchers make our research more relevant to educational practice? It has been pointed out that there is a disconnection between mathematics educational research and practice (Sowder, 2001). Our experience in this research project, and other projects, also supports that conclusion. Why did this happen? Was it because of the choice of topics, or the design of the study? Should the communication between research communities and policy makers, principals, teachers be enhanced? Is it possible for researchers to distribute the findings to principals and teachers in plain language, instead of only publishing our findings in high level academic journals? Should research skills be introduced into teachers' professional development programs/principal training programs to develop teachers' and principals' interest in research?

To conclude, the conduct of surveys in different cultures faces many challenges and difficulties, as well as raising many issues. On the one hand, researchers should be aware of cultural differences when designing and administering research projects, take measures to minimise measurement biases, and get to know the regulations and customs in different societies to enable surveys to be conducted smoothly. On the other hand, researchers should also make efforts to enhance the relationships between researchers and educational practitioners, and to enable their research to produce better benefits for educational practice.

Acknowledgements

The preparation of this paper was supported by a *Postgraduate Publication Award* from Monash University.

References

Australian Educational Council. (1991). *A national statement on mathematics for Australian schools.* Melbourne: Australian Curriculum Corporation.

Beaton, A. E., & Robitaille, D. F. (1999). An overview of the third international mathematics and science study. In G. Kaiser, E. Luna, & I. Huntley (Eds.), *International comparisons in mathematics education* (pp. 19–29). London: Falmer Press.

Burstein, L. (Ed.). (1993). *The IEA study of mathematics III: Student growth and classroom processes.* Oxford: Pergamon.

Cai, J. (1998). An investigation of U.S. and Chinese students' mathematical problem posing and problem solving. *Mathematics Education Research Journal, 10*(1), 37–50.

Cao, Z., & Bishop, A. J. (2001). Students' attributions of success and failure in mathematics: Findings in China and Australia. In J. Bobis, B. Perry, & M. Mitchelmore (Eds.), *Proceedings of the Twenty-Fourth Annual Conferences of the Mathematics Education Research Group of Australasia. Numeracy and Beyond* (pp.139–148). Sydney: Mathematics Education Research Group of Australasia.

Cao, Z. (2004). *An Investigation of Students' Attitudes towards Mathematics in China and Australia.* Unpublished PhD thesis, Monash University, Melbourne, Australia.

Cook, S. W., & Selltiz, C. (1964). A multiple indicator approach to attitude measurement. *Psychological Bulletin, 62*, 36–55.

Cunningham, G. K. (1986). *Educational and psychological measurement.* New York: Macmillan

Forgasz, H. J. (1995). *Learning mathematics: Affect, gender, and classroom factors.* Unpublished PhD thesis, Monash University, Melbourne, Australia.

Hsu, F. L. K. (1981). *Americans and Chinese: Passage to differences.* Honolulu: University Press of Hawaii.

Husén, T. (Ed.). (1967). *International study of achievement in mathematics education* (Vols. I & II). Stockholm: Almqvist &Wiksell.

Keitel, C., & Kilpatrick, J. (1999). The rationality and irrationality of international comparative studies. In G. Kaiser, E. Luna, & I. Huntley (Eds.), *International comparisons in mathematics education* (pp. 241–256). London: Falmer.

Ling, L., Burman, E., & Cooper, M. (1998). The Australian study. In J. Stephenson, L. Ling, E. Burman, & M. Cooper (Eds.), *Values in education* (pp.35–60). London: Routledge.

Ma, L. (1999). Knowing and teaching elementary mathematics: Teachers' understanding of fundamental mathematics in China and the United States. Mahwah, NJ: Lawrence Erlbaum.

Mayer, R. E., Sims, V., & Tajika, H. (1995). A comparison of how textbooks teach mathematical problem solving in Japan and the United States. *American Educational Research Journal, 32*, 443–460.

Ministry of Education of the People's Republic of China. (2003). Mathematics curriculum standards (for trial) [in Chinese]. Beijing: People's Educational Publishing House.

Neubrand, J. (2002). Characteristic features of problems: Comparing German, Japanese and US mathematics lessons from the TIMSS-Video-Study. In F. Lopez-Real (Ed.), *Pre-conference proceedings of the ICMI comparative study conference 2002* (pp. 194–200). Hong Kong: University of Hong Kong.

New Laws Take Effect. (2004). Retrieved September 2, 2004 from http://english.people.com.cn/200409/eng20040902/html

Nunnally, J. O. (1967). *Psychometric theory.* New York: McGraw-Hill. OCED (n.d.). Retrieved January 8, 2005 from http://www.oced.org/pages/0,2966/html

Park, K., & Leung, F. K. S. (2002). A comparative study of the mathematics books of China, England, Japan, Korea, and the United States. In F. Lopez-Real (Ed.), *Pre-conference proceedings of the ICMI comparative study conference 2002* (pp. 225–234). Hong Kong: University of Hong Kong.

Robitaille, D. F., & Garden, R. A. (Eds.). (1989). *The IEA study of mathematics II: Contexts and outcomes of school mathematics.* Oxford: Pergamon Press.

Sowder, J. (2001). Connecting mathematics education research to practice. In J. Bobis, B. Perry, & M. Mitchelmore (Eds.), *Proceedings of the twenty-fourth annual conferences of the mathematics education research group of Australasia. Numeracy and beyond* (pp.1–8). Sydney: Mathematics Education Research Group of Australasia.

State Education Commission of the People's Republic of China. (1992). Mathematics curriculum for junior high schools [in Chinese]. Beijing: People's Educational Publishing House.

Stigler, J., & Perry, M. (1988). Cross-cultural studies of mathematics teaching and learning: Recent findings and new directions. In D. Grouws & T. Cooney (Eds.), *Effective mathematics teaching* (pp.194–223). Hillsdale, NJ: Lawrence Erlbaum.

Travers, K. J., & Westbury, I. (Eds.). (1989). The IEA study of mathematics I: Analysis of mathematics curricula. Oxford: Pergamon Press.

Wong, N., Lam, C., Leung, F. K., Mok, I. A., & Wong, K. (1999). *An analysis of the views of various sectors on mathematics curriculum.* Unpublished research report, Hong Kong Chinese University, Hong Kong.

18

THE BENEFITS AND CHALLENGES FOR SOCIAL JUSTICE IN INTERNATIONAL EXCHANGES IN MATHEMATICS AND SCIENCE EDUCATION

Catherine P. Vistro-Yu[1] **and Kathryn C. Irwin**[2]
[1]*Ateneo de Manila University, Philippines*
[2] *University of Auckland, New Zealand*

Abstract: International exchanges in higher education are influenced by globalised economies. Whether globalization is seen as a relabelling of capitalism or as the homogenization of cultures, social justice must not be an alien concept in academic endeavours because of the many inequalities that exist both within and between nations and peoples. This chapter presents an analysis of an exchange between two universities by examining certain aspects that impinge on issues in social justice such as fairness, equality, and equity. While the exchange benefited both universities in various ways, the authors who were the principal actors in the exchange admit that many lessons in social justice have been learned. Thus, the chapter lays out some reflections on how future academic exchanges might be conceived to ensure fairness and equity for all parties involved

Keywords: desert principle; difference principle; distributive justice; equity; fairness; higher education; international exchange; principle of redress; social justice

1. Introduction

The history of international aid, ranging from agreements between universities to those sponsored by large organizations like United Nations agencies, is littered with examples intended to further social justice that have not been realized because of factors that result from an imbalance of power. Atweh, Clarkson and Nebres argue that a globalised economy isolated from globalised social justice is problematic (2003, p. 191). From the point of view of social construction, all situations are inherently unequal and involve a power struggle. Globalization heightens these inequalities and in fact creates different forms of divide, e.g. economic or technological (Annan, 2000). However, social justice is difficult to achieve in an aid

B. Atweh et al. (eds.), Internationalisation and Globalisation in Mathematics and Science Education, 321–342.

program for the very reason that one party to such an agreement has more access to resources than the other, leading to unequal bargaining power.

International exchanges in higher education are influenced by globalised economies. Whether globalization is seen as a relabelling of capitalism or as the homogenization of cultures (Dash, 1998), social justice must not be an alien concept in academic endeavours. This chapter presents an example of an exchange between two universities that allows readers to explore relevant issues in social justice. The initial financial sponsor of this exchange specifically ruled out the possibility of this being considered an aid project by requiring the poorer nation to contribute an amount equivalent to that provided by the government of the richer nation. This chapter discusses the extent to which this and other factors impinged on issues of social justice throughout the exchange. It also seeks to explore sources of inequities and sources of convergences that are present in this inherently unequal relationship.

The second section describes and discusses key foundational concepts of social justice, fairness, equality, and equity. In this section, much reference will be made to Rawls' democratic interpretation of equality of opportunity, which makes up the bulk of the *Difference Principle* of social justice (Rawls, 1999; Blocker & Smith, 1980) and Miller's proposed theory of justice that takes into consideration human relationships (Miller, 1999). The third section presents some background of the exchange between the University of Auckland and the Ateneo de Manila University. In Sect. 4, the exchange is analyzed by discussing factors necessary for a possible mutually beneficial exchange and examining the different benefits and costs to each institution. The benefits include some material and professional gains as well as cultural and personal ones while the costs include a heavy financial burden for the poorer university. Sect. 5 discusses portions of the exchange with which issues of social justice may be critically examined. Implications will be drawn from various theorists on social justice. In the last section, Sect. 6, an outline and brief description of some of the lessons that have been learned from the exchange are offered in the hope of providing the seeds for developing an initial conceptual framework for implementing international academic exchange programs.

2. Social Justice: Fairness, Equality, Equity

Among the many social justice theories, Rawls' (1999) *Difference Principle*, which is part of distributive justice, is perhaps one of the theories most argued about (Swanton, 1981). Rawls states the *Difference Principle* as:

> Social and economic inequalities are to be arranged so that they are both (a) to the greatest benefit of the least advantaged and (b) attached to offices and positions open to all under conditions of fair equality of opportunity
> (Rawls, 1999, p. 72).

Social justice is often equated to equal distribution of goods, material wealth, or social values such as educational opportunities or advancement. This is a myopic view of social justice. Equality naturally implies equality in number or quantity. Thus, equality of opportunity, for instance, means giving each and every applicant for a job an equal chance of being accepted. This is manifested by giving each applicant a chance to go through every step of the application process regardless of the person's race, culture, religion and the educational institution from which the applicant finished. However, an applicant who completed studies from a recognised world-class university, as opposed to someone who completed studies from a less prestigious university, already has an advantage. Clearly, this situation is lacking in social justice.

Rawls' concept of democratic equality explains the situation just described. Schaar (1980) argues that Rawls wanted to revise the present understanding of equality of opportunity to a new meaning that takes into consideration the *difference principle* and the *principle of redress*. A new and revised version of the concept of equality is the democratic equality. If we are to genuinely treat all people equally and not allow benefits of social life to be determined by natural endowments or social fortune, then a different interpretation of equality is in order (Schaar, 1980, p. 170). The *difference principle* asserts that inequalities are justifiable only if they are to the advantage of the worst-off representative person or group (Schaar, 1980, p. 170). Thus, in the situation above, if it benefits the applicant from the less prestigious university to be given an extra opportunity to be evaluated further for the job through a trial work period, social justice has prevailed even though there is an inequality of opportunities between the two.

One important ingredient for democratic equality to be accepted is cooperation amongst the individuals or groups concerned. Everyone must agree to the notion that individuals deserve equal shares unless an alternative distribution, which may result in unequal shares, is more beneficial to everyone (Mithaug, 1996, p. 20). In fact, Rawls argues that

> rational persons who are free to construct principles of justice based on this conception of cooperation for mutual advantage would agree on two principles to guide their construction of fair and just social institutions – one principle guaranteeing equality in the assignment of rights and duties in the group and the other tolerating social and economic inequalities to the extent that they benefit all members of the group
>
> (Mithaug, 1996, p. 20).

Does Rawls' concept of social justice end here? Once a fair or democratic equality is achieved, has social justice been completely served? The second principle of democratic equality is the *Principle of Redress*. Without it, democratic equality would not be complete. If the *Difference Principle* intends to minimize undeserved advantages, then the *Principle of Redress* seeks to compensate for undeserved disadvantages. The *Principle of Redress* is meant to ensure that victims of personal, social, and economic disadvantages have the same prospects for pursuing a better life as others who are in better situations right at the beginning (Mithaug, 1996, p. 21).

> Thus … in order to treat all persons equally, to provide genuine equality of opportunity, society must give more attention to those with fewer native assets and to those born into the less favourable social positions. The idea is to redress the bias of contingencies in the direction of equality
>
> (Rawls, 1999, p. 86).

One interpretation of the *principle of redress* is the notion that persons deserve to receive goods according to their needs and not according to their abilities. Rawls suggests, for example, that in the area of education greater resources might be spent for less intelligent members of the group rather than for the more intelligent or accomplished members (Schaar, 1980, p. 171).

Rawls' concept of social justice has limited application because he assumes that principles of justice are meant to "regulate a well-ordered society" (Rawls, 1999, p. 8). In succeeding discussions of this assumption, it becomes clearer that Rawls actually confines his social justice theory to nations governed by well-defined institutions. Miller (1999, p. 17) disdains the fact that social justice theorists, Rawls being one of them, make the assumption that their theories are "to be applied within a self-contained political community" without justifying it. Miller feels this is no longer enough and suggests that the world must now "enlarge the universe of distribution to embrace transnational constituencies or even the world as a whole" (p. 7). Consequently, Miller argues that we must now think of global justice and not only justice within nations. However, he cautions that "we must continue to think of social justice as applying within national political communities, and understand global justice differently" (p. 19) since, for one there is no one governing body in the world that takes care of distributing fair amounts of goods and values among nations.

Young (1990 as cited by Atweh, 2004a) criticizes social justice theories within the distributive paradigm such as Rawls' because they focus more on "having" rather than "doing" (Atweh, 2004, p. 2a) or perhaps, more appropriately, on "acquiring". Miller (1999) elaborates that for Young, social justice requires the "elimination of institutionalized domination and oppression," and that distribution issues must be addressed from this perspective (p. 15).

As a possible compromise and a way to address inadequacies of existing social justice theories, Miller (1999) sketched a theory that begins with "modes of human relationship" (p. 25) rather than starting with social goods. He labels the three modes as *solidaristic community*, *instrumental association*, and *citizenship*. Miller argues that there is a need to first look at the kind of relationship we have with other people in order to understand which demands of social justice someone can make (Miller, 1999, p. 25). The exchange that is being analysed falls under the mode of *instrumental association*. In this mode, people relate to one another in a utilitarian way. Collaboration is a key ingredient to achieve the goals of each individual. The relevant social justice principle in this mode is *distribution according to desert* (Miller, 1999, p. 28). Under this principle, each person comes to the group with a set of talents that she or he uses to advance himself or his goals. Justice is served if he gets his fair share of reward based on his contribution.

Miller's principle of distribution according to desert has shades of similarity with Nozick's theory of entitlement. In Nozick's theory, freedom is a major aspect of fairness and justice. Autonomous individuals who freely interact and work independently to achieve their own goals comprise Nozick's society. They are not forced to share their fortune with the lowly members of the society but instead are free to do so voluntarily. Fairness, in Nozick's theory, is not the same as equality. Rather, fairness means respecting the rights of individuals to acquire and own what they have worked for and not coercing them to share their goods without their consent. Compared to Rawls' and Miller's concepts, Nozick's theory fails to consider responsibility as a crucial factor to achieve social justice. If nations want to move towards a just and fair society, individual freedom is no longer enough. More importantly, individuals must accept the responsibility of ensuring that the marginalised and disadvantaged are given the opportunity to overcome their state in life.

Entitlement, fairness, equity, mutuality, responsibility, desert, and redress are key ideas that one must keep in mind when engaging in programs that involve groups of people who come from differing economic, academic, and social backgrounds and status. These are important considerations for developing possible theories or frameworks for academic exchange programs.

3. Description of the Exchange in Mathematics and Science Education Between the University of Auckland and the Ateneo De Manila University (1998 – 2003)

The exchange that is the subject of this chapter originated from an advertisement by the Asia New Zealand Foundation (Asia: NZ, formerly Asia 2000 Foundation) for academic exchanges between New Zealand and the Philippines. The Asia: NZ Foundation sponsors a variety of activities that are seen as encouraging mutual understanding between Asian countries and New Zealand. One branch of this foundation is the Higher Education Exchange Programme (HEEP) which has the stated objectives to: "Promote mutually beneficial and ongoing exchange relationships between tertiary institutions in New Zealand and Asia; increase opportunities for New Zealand students, researchers and lecturers to undertake study and research in an Asian country; encourage the development of internationalization strategies at New Zealand educational institutions; and foster understanding and ongoing cooperation between academics in New Zealand and Asia" (Asia: NZ Foundation, HEEP, 26 April 2006). A desired goal of HEEP was "an increased number of tertiary students and academics across disciplines who have cultural, language and professional experience relevant to New Zealand's place in the Asia-Pacific region," (letter from P. Barton, 24 Feb 1999). The agency requires the Asian universities in their exchanges to be of top standing. The application form for the grant required a clearly thought out program and letters of support from both organisations.

The Asia-NZ Foundation is a quasi-governmental organisation partially funded by Asian and New Zealand industry and partially funded by New Zealand government departments such as the Ministry of Foreign Affairs and Trade. Its website states that, "Asia New Zealand Foundation is a non-profit, apolitical organization dedicated to promoting greater understanding between New Zealand and its Asian neighbours. The Foundation works to develop New Zealanders' knowledge and understanding of the countries and peoples of Asia, help New Zealanders acquire the right skills to work effectively with Asian counterparts, build New Zealand's links with Asia, and promote and assist New Zealanders' participation in regional activities." (http://www.asianz.org.nz/ 26 April 2006). This statement needs to be seen in the wider context of the fact that New Zealand is a small nation, dependent on trade. Its former major trade partner, the United Kingdom, stopped being that major partner when it joined the European Common Market. Now New Zealand's economy is dependent on good trade relations with many countries, including Asian countries.

The exchange was clearly to be a mutual one. Asia: NZ required equal funding from a prestigious university in the Philippines to match that provided by the New Zealand fund. Though not required to match dollar-for-dollar, the Philippine institution was required to provide airfares for their academics to New Zealand and board and room for New Zealand academics when in the Philippines. The New Zealand author, who proposed this exchange, originated a search for a Filipino partner university through email exchanges with several friends in the Philippines; she already had some knowledge of that country and its people. Ateneo de Manila University (Ateneo) was found to be the most appropriate university to form an exchange with, initially in the field of Mathematics Education but also including Science Education upon the recommendation of the president of Ateneo. This Philippine university has as its president a distinguished mathematics educator who was anxious to promote exchanges and, therefore, fund such an exchange. Five other prestigious Philippine universities were eligible but were not appropriate partners for this particular exchange. Other universities might have been interested in the exchange but did not have the required academic standing.

A shared language for communication was never an issue. Filipinos, in general, speak English very well.

The authors, who were actively involved in the exchange, always saw this as an activity in which they could learn from one another. If applied to Atweh's (2004b) schema for aid projects, the exchange would be seen as having multicultural attributes involving both recognition and affirmation of different educational systems. The motivation for the exchange was a desire to expand professional horizons and publishing opportunities of staff in both countries.

At the suggestion of Asia: NZ the initial exchange was an exploratory one. It enabled a science educator and a mathematics educator from the University of Auckland to travel to Ateneo de Manila University, meet the academic and administrative staff there, visit schools, and hold informal discussions as arranged by the Ateneo. It was intended to allow an equivalent party of Filipino academics to visit the University of Auckland. Because of other factors, only one Filipino mathematics educator participated in that initial exchange. One international publication came out of this first round of the exchange. This was a paper written for teachers that provided examples of best practices in one sphere of school mathematics education, with sample lessons from both countries (Ferrer, Hunter, Irwin, Sheldon, Thompson, & Vistro-Yu, 2001).

This initial exchange was followed by another grant. At the time of approval, the chairperson of Asia: NZ commended the authors on a truly mutual exchange (P. Barton, personal communication to K. Irwin). Much of the credit for this mutuality goes to the president of Ateneo who found money to fund the travel of his staff at a time of economic hardship in his country when funds for international travel were severely limited. These funds for airfares had to be in US currency. This use of US currency was prohibited to members of the state universities at that time but Ateneo, being a private university, was not bounded by such policy.

The second exchange enabled one science educator from the Philippines to travel to New Zealand and two mathematics educators from New Zealand to travel to the Philippines. Each visitor carried out activities with their counterparts in the other country. For example, the main goal of the science educator who came to New Zealand was to explore a wider range of journal publications in science education than were available in her country, with the guidance of a science educator in New Zealand. Likewise, the authors of this chapter set up a larger research study, now published, that explored young children's concepts and misconceptions in measurement in both countries, and compared these with the curriculum in the two countries (see Irwin, Vistro-Yu & Ell, 2004). One of the Filipino visitors decided to undertake his doctoral studies in New Zealand, thus increasing the knowledge that he will take back to the Philippines. The second mathematics educator to visit the Philippines was able to explore collaborative work in the area of technology in mathematics education, a field in which both partner universities excel.

Some remaining money from the Asia: NZ Foundation, put together with additional money from funds earned by the mathematics education community at the University of Auckland, and some funds from Ateneo, enabled an additional Filipino mathematics educator to come to New Zealand in order to discuss and complete the paper on children's concepts in measurement.

4. Analysis of this Exchange Program

It is critical for the partners who carry out an exchange to have the same interest, commitment, and finances to allow such an exchange to commence. In the case of Ateneo, the interest and enthusiasm came from the Philippine author who felt important enough to have been contacted by the New Zealand author, a well-published mathematics educator. As mentioned, the commitment of the university president to strengthen mathematics and science education in the Philippines contributed much to the exchange, but it was really the Philippine author and the New Zealand author who should be credited with the realization of the exchange. While Ateneo may have more money for such endeavours than other Philippine universities, without the enthusiasm and commitment of the major players, the finance needed would not have been released for this exchange. It is also important that the partners see tangible benefits. More importantly, the benefits must outweigh the costs for the exchange to thrive.

One key element in an exchange is respect for the partner country – respect for its manpower, financial and other resources, its pride as a sovereign nation capable of contributing to international scholarship, and its needs. For the Ateneo de Manila University, an exchange program that was not an aid program was novel. An aid program would have been beneficial although Ateneo was not averse to the idea of a mutual exchange program. In many cases, Ateneo is a financial provider for science and mathematics education activities in the Philippines through scholarships and low cost professional development seminars and workshops for teachers. However, Ateneo would not engage in a program where the aid comes in the form of foreign counterparts lending a hand to Ateneo's faculty as if its own people are lacking in talents and abilities. Ateneo takes pride in its own pool of scholars many of whom have been recognized nationally and internationally. Again, locally, Ateneo is usually very generous in lending the talents of its faculty to government agencies such as the Department of Science and Technology, the Commission on Higher Education and the Department of Education, whenever educators of their calibre are needed. Thus, a mutual exchange of skills and talents was more appropriate to Ateneo than an aid program as such. The mutual exchange with New Zealand considered Ateneo personnel as partners from whom New Zealand researchers could learn.

Another strength of the exchange with Ateneo was the interest and recognition of the needs of the partner country and of the partner institution. Ateneo's greatest needs in the area of mathematics and science education were very clear from the start – faculty development and exposure in the area of publications. New Zealand recognized this and was quite accommodating and responsive.

An exchange program will not materialize without a key person from the initiating country. In the case of the Asia: NZ exchange, the New Zealand author was the key person. A well-travelled, well-published and seasoned mathematics educator, she had studied at a university in the Philippines, had Filipino friends and was therefore enthusiastic to pursue an exchange with a qualified university in the Philippines. The initiator of an exchange should be open to the culture of the partner country and the peculiarities of that culture. The initiator should be experienced in the field and must be knowledgeable about collaborating with various sectors: government agencies, private foundations, school leaders, university colleagues, as well as ordinary teachers and students.

Equally important for an academic exchange to thrive is the contact person from the partner country. The contact person from the partner country should be as interested and committed as the initiating person. Whether the object of their interest and commitment are identical or not may not matter as long as they work in the same field and have the same vision for excellence and service. In this exchange, the contact person was initially the university president but the Philippine author bore the responsibility of maintaining the exchange in terms of needed personnel and its program of activities. It helped that the Philippine contact person was also as enthusiastic about the exchange and committed to ensuring that the activities did align with the nationalistic goals and professional goals of the institution.

The benefits and costs to each partner can be analysed more fully. They often intertwined.

4.1 Benefits to the Partners

The goals of the financers of the partnership, The Higher Educational Exchange Programme (HEEP), within the Asia: NZ Foundation, and the Ateneo de Manila University, as personified by its President, were clear and benefits can be measured against them. It is important to note that these goals were different, if overlapping. HEEP's first goal was to promote mutually beneficial exchange relationships between tertiary institutions in New Zealand and Asia. This sits within the wider goal of enabling New Zealanders to improve their knowledge and understanding of the peoples of Asia. This exchange enabled one academic to study in another country and five academics to travel to the partner country, one of them more than once. On each visit, academics spoke with many others in their field in the partner country. All of these contacts were seen as mutually beneficial. They increased both partners' knowledge of each other's place in the Asia-Pacific region. HEEP's stated goals appear to have been met. The President of Ateneo wanted to foster mathematics and science education and to encourage his staff to enrol for advanced degrees in a variety of countries. These goals were also met. Neither partner set out to redress imbalances in this exchange, but to satisfy the needs of its own country or university.

The exchange benefited mathematics educators from both countries in their professional careers. The New Zealand author was at a later stage in her career than the Philippine author was, and thus she could pass on some of her experience in having articles published in international journals. Both of the authors find mathematics education an exciting field and are willing to work hard to improve the mathematics education in their respective countries. They benefited from being able to work collaboratively, sharing their enthusiasm, and expanding their network of colleagues. The science educator from Ateneo learned about the kind of research conducted by New Zealand colleagues. This gave her insights into the possible areas of research to be conducted in the Philippines, which eventually led the science education faculty of Ateneo to engage in a collaborative textbook project.

Two publications in respectable international journals in mathematics education were a boost to the professional career of the Philippine author and to those of all three of the schoolteachers whose teaching is reported in the first publication (Ferrer, Hunter, Irwin, Sheldon, Thompson, & Vistro-Yu, 2001). These teachers are all now earning higher degrees and are likely to move into careers where publications are important. The publications also benefited the New Zealand author. It is not easy to publish in international journals. Moreover, the type of mathematics education research in the Philippines is of a different flavour and does not match the research paradigms that are currently favoured by many Western journals. This was consistent with what Atweh (2003) gathered from a focus group discussion among Filipino mathematics educators. The preferred research studies in the Philippines are those that seek to identify effective methods and formats that benefit the large population of Filipino students and teachers, and thus largely use quantitative methods for credibility and acceptability. The collaborative research studies born out of the exchange taught the Philippine educators what some international publications entail in terms of perspective, quality, and nature of the research as preferred by these Western journals.

The exchange program benefited the science and mathematics educators of the Ateneo in a unique way. For several years prior to the exchange, the Ateneo mathematics education faculty and science faculty who are active in science education endeavours had been trying to organize themselves into a mathematics and science education group, but had not been successful. The Asia: NZ exchange program provided them with a common activity within which they were provided the opportunity to formulate future goals and strategies for a more organized group and to develop more academics in science education. It would be after the exchange program had ended that the faculty began establishing a more formal organization for themselves within the Ateneo School of Science and Engineering, leading to the development of a graduate program in science education. The structure of the Education Unit in the Mathematics Department cooperating with the School of Education at the University of Auckland served as an inspiration to this move.

The University of Auckland and its staff also benefited from the joint publications resulting from this exchange. In addition to the two written by the authors of this chapter, the Filipino staff member who became a doctoral student has published papers jointly with his supervisor (see delos Santos & Thomas, 2001, 2002a, 2002b, 2003a, 2003b, 2005). Such publications strengthen this university's reputation as the pre-eminent research university in New Zealand. Universities in New Zealand are dependent on research publications for some of their funding from the government and staff are dependent on publications for promotion.

The Ateneo faculty member who is now a PhD student in the University of Auckland has been a gain to the Mathematics Education Unit in the Mathematics Department. His field, like that of one of the New Zealand visitors to Ateneo, is the use of technology in mathematics education. Both universities have a strong interest in this field and can share expertise.

The cultural benefits of the exchange for participants cannot be understated. Except for the initiator of the exchange, it was the first time for faculty from both institutions to have experiences with academics from the other country. For the participants from the Philippines it was their first opportunity to visit the other country, and one of the best aspects of the exchange. This was interesting because, while the Philippines is an importer of New Zealand products, very few Filipinos have actually had opportunities to interact with New Zealanders. Such interaction made the Filipino educators involved in the exchange become more aware of and more attuned to New Zealand. The Western country that most Filipinos know best is the United States of America. Of the Filipinos in this exchange, one had earned a doctorate in the USA, one had earned a doctorate in Germany, and one had earned a masters degree in Japan. Visiting New Zealand added to their worldly experiences.

New Zealanders would also put the experience of visiting the Philippines as a major benefit of the exchange, with one person considering taking his sabbatical there. New Zealand is a small country (population: 4 million); people have usually heard of the Philippines, which has a much larger population (almost 87 million). However, because of their history of being members of the British Commonwealth, New Zealanders are more likely to have travelled or worked in Commonwealth countries than in the Philippines. All of the New Zealand participants had spent time in other countries, but only the New Zealand author had been to the Philippines. Like the Filipino participants, the exchange added to their international academic experience.

In both countries, the visitors observed classroom teaching. New Zealanders could learn much from the ease with which teachers and students moved between English and their mother tongue in Philippine schools. They observed excellent teaching and resources in one primary school. They were impressed with the level of work in the prestigious secondary schools for science, for which there is no equivalent in New Zealand. They benefited from sharing both informal thinking and the focused thinking that has gone into the joint publications. Related outcomes have included discussing good Filipino pedagogy with New Zealand student teachers and pondering New Zealand's approach to teaching students who do not have English as a first language in the light of the Philippine example.

Filipino participants learned unexpected things. For example, the Philippine author learned that, in New Zealand, schools that teach children from lower economic backgrounds have more classroom materials and a lower ratio of students to teachers than do schools that educate children from middle or upper economic backgrounds. This could be considered a method to redress differences that can be related back to relative poverty (Schaar, 1980). In the eyes of the Philippine author, these well-equipped schools looked like private, fee-paying schools in the Philippines. She wondered how and when the Philippines would ever reach that point where their country could afford to fund schools in lower economic areas as well, in its educational system.

The exchange helped develop friendships among the participants. A most important feature of these friendships is an appreciation of the differences between the two cultures and traditions. Learning about these differences is a benefit in itself. Ateneo's long history of institutional collaborations with colleagues from Japan and Australia particularly in mathematics and science has shown that personal relationships are the key to sustaining these institutional partnerships (Atweh and Clarkson, 2001). This exchange bolstered that point. Certainly, the initial respect for each other's institution and personal reputation was most helpful. The visits, which were the most expensive part of the exchange, offered the most gain, personally and professionally, for the participants. Face-to-face encounters are irreplaceable despite the advances in technology and electronic mail, consequences of globalization. This has been particularly relevant when engaging in joint research and writing. The writing of one of the joint publications reached a near-impasse because qualitative features are hard to understand and discuss by e-mail, so the chance to arrange an additional visit was most beneficial.

4.2 Costs to each Partner: Difficulties Encountered

Overall, it is likely that the benefits outweighed the costs or difficulties. However, both partners experienced difficulties, foremost of which was the financial situation in the Philippines. As mentioned earlier, a financial crisis hit Asia at the time the initial exchange was to commence. Communication records and electronic mail exchanges indicate that the Ateneo had to shift around some of its money in order to begin the program. This was because the exchange began in the middle of the Philippine academic year and thus was not included into its budget for that year. The situation improved slightly in the succeeding years because expenses for the exchange were always included in the budget preparation. However, the continuous depreciation of the Philippine peso made it harder and harder for Ateneo to come up with the needed funding. The value of the peso against the US dollar at that time dropped from 26.4 pesos to a dollar in June of 1997 to 42.1 pesos to a dollar in June of 1998 (Bangko Sentral ng Pilipinas, 2004). New Zealand exchange rates did not go through such inflation during this period and there was little change in the price of airfares for that country. The Asia: NZ funding for this exchange in New Zealand highlighted the stark contrast between New Zealand and other relatively affluent Western countries and less affluent nations. The Asia: NZ Foundation acknowledged the problems caused by this financial crisis in Asia and permitted alterations to the number of people in the exchange travelling from the Philippines and in the timing of visits while not altering their basic requirement that this be an exchange rather than an aid program.

Across the five years, Ateneo paid for the roundtrip airfare of two faculty participants, contributed to the travel allowance of one faculty participant to New Zealand in US dollars, and covered the board and lodging costs of an equivalent of four New Zealand colleagues in pesos. Faculty at the University of Auckland paid for the travel expenses of one Filipino from other sources and the Asia: NZ Foundation paid for part of the travel of one Filipino and the board and lodging for three Filipinos. There was no agency equivalent to Asia: NZ in the Philippines that financed the academic exchange for Ateneo. In fact, it is unlikely that there was any agency that would have agreed to do that without imposing too many conditions. The thought of actually asking, for example, the Science Education Institute of the Philippine Department of Science and Technology to finance the exchange entered the minds of the Ateneo administrators. They suspected, however, that a government agency, though probably capable of financing the exchange, would most likely prefer to promote a state university as the Philippine counterpart for the exchange rather than a private university such as the Ateneo. Thus, Ateneo decided to simply finance the exchange from its own funds.

There were other costs. While Ateneo celebrated the fact that one of its mathematics education faculty members has been accepted into the doctoral program at the University of Auckland, it also felt the huge loss of his services. Even now, in spite of the service conditions imposed (Ateneo partially financed his studies), there is no assurance that he will return to the country to teach. Ironically, however, Ateneo's loss was Auckland's gain as mentioned earlier. The loss of scholars who go overseas to study is a major academic and economic problem for poorer countries (Atweh, 2003). Many Filipinos go overseas to study and do not return despite the needs of their country. One must not take this wrongly, however, because Ateneo has been encouraging its faculty in the science and mathematics departments who engage in mathematics and science education endeavours to pursue their graduate education in these fields, whether in the country or abroad. This is an ongoing problem in the Philippines and other relatively developing countries.

Another major difficulty that the Ateneo encountered was the lack of additional faculty who could contribute actively to the exchange. The difficult situation could be tied to the fact that there were hardly any science educators and only a couple of mathematics educators who were formally trained in the respective areas. Thus, these were the only ones who were comfortable in collaborating with the New Zealand counterparts. The lack of personnel to carry out the operations when the New Zealand colleagues arrived for a visit drained the energies of the very few who were committed. Although the Ateneo faculty who were involved tried to enable more Filipino mathematics and science educators to benefit from the exchange, still it did not seem enough.

Another cost was the time involved in setting up and running the exchange. Although the initial communication was between the New Zealand author and the president of the Ateneo, it was the Philippine author who prepared the program content of the exchange for the Ateneo and took charge of all the details. That was considerable work for one person. What assisted in the situation was the continuous dialogue among the Ateneo participants. Group meetings among the person in charge, the mathematics and science educators, the university president, and academic vice-president held periodically as well as individual meetings sought between the person in charge and the academic vice-president became the venues for dialogues and discussion of ideas on how to further the exchange. Similarly, establishing the exchange took time for the New Zealand author who wrote the application, found a willing partner in Ateneo, arranged air travel, booked housing for the visiting Filipinos, arranged speaking engagements, and arranged visits and some entertainment. She was on leave and out of the country when it was necessary to make the second application, so she needed the cooperation of a colleague in making that application.

There were periods when there were no visits and one would wonder whether the exchange was still alive. Ateneo university officials did express doubt at one time or another about whether the exchange program was worth the money they were spending because it looked like the program did not benefit as many people as the Ateneo had hoped. However, due to the high degree of mutual trust between and among the university officials and faculty involved, and their belief that the exchange would help the faculty expand their linkages and not remain tied to American ideas, the exchange program and it's funding from the Philippine side continued.

At the time that the exchange program began, Ateneo's president was more interested in pursuing collaboration in science and mathematics education with Asian neighbours. Also, at that time, Ateneo was very involved in domestic mathematics and science education projects, particularly those that directly benefit the public schools. Although New Zealand was quite a distance from the Philippines, the commitment by the administration to pursue programs that benefited mathematics and science education in the country was most important during this time, so Ateneo continued their financial commitment.

There was some concern from the Philippines side about whether the New Zealand colleagues were satisfied with the exchange program and whether they learned from the Filipino educators as much as the Filipino educators did from them. This worry is uppermost in the mind of the Filipino science educator because she and the New Zealand colleagues failed to identify common research interests. In fact, while science education was a growing field at Ateneo it was a decreasing one at the University of Auckland, as one science staff member lost his position and the other was not in good health. Such unexpected and unspoken factors can be harmful for an exchange.

While both partners in the exchange benefited, it was hard to judge if the exchange was equally beneficial to both parties. One factor in this was that the colleagues from the University of Auckland were more experienced and often older. Equity of age is certainly not a requirement in an exchange, however, from an Asian cultural perspective differences in age and experience may cause respect that actually gets in the way of a free exchange of views. Mathematics education and science education as academic disciplines rather than areas for outreach services were just getting started at Ateneo. These fields were about 15 years old in New Zealand, a difference that may have been more important to the Filipinos than to the New Zealanders.

Timetabling of visits was difficult for both parties because of different academic years in the two countries and the fact that all participants had teaching and research responsibilities that meant that they could not be away from their own university for long periods. No one managed a two-week visit as envisaged initially. Among the five mathematics educators at the University of Auckland, two were unable to travel because of family commitments. The science educator who lost his position would have been interested in visiting the Philippines, but could not do so in his new position.

While a goal of the exchange was ongoing cooperation, this is dependent on stability of staffing in both institutions. Two of the Filipino academics are still in the same positions but only one of the New Zealand participants is still employed. Two older participants have retired. If the on-going benefit is based on personal relationships, retirement of some parties becomes a cost to the on-going nature of the exchange. If the Philippine member who is enrolled in a PhD in New Zealand returns to Ateneo, the relationship might continue through his collaboration with his New Zealand colleagues.

Another ongoing issue is the possibility of misunderstanding with people from a different culture. There have been unintended differences in research approach, when both the authors thought they were doing the same thing. If the Filipino participants wondered if New Zealand participants were satisfied, those at the New Zealand end were not accustomed to the respectful manner in which the younger Filipinos treated them. New Zealand academics are usually informal with one another, and are used to having students call them by their first name. Many Asian students find this informality disrespectful. Cultural differences and their effects continue to come to light in writing this chapter jointly. Growing understanding of these differences is one of the benefits of the exchange.

5. Issues in Fairness, Equality, Equity, and Social Justice

Upon reflection, the international exchange between the University of Auckland and the Ateneo de Manila University was not free from issues in social justice. First and foremost was the issue of the process of selecting the partner institution. New Zealand's requirement for a top quality university immediately disqualified more than 90% of the tertiary institutions in the Philippines. Again, "the rich get richer and the poor get poorer". If international activities contribute to an institution's prestige, how can an institution that needs to build up its reputation internationally do so? This is consistent with Rawls' justification for the need to shift the traditional concept of social justice to democratic equality. Moreover, although language was not an issue in the case of Philippine partner institutions, it is so for non-English speaking countries such as Colombia (Atweh & Clarkson, 2004). English, being the universal language in education, certainly limits the potential involvement of the non-English speaking countries in international collaborations (Atweh 2004a).

The authors wondered if there had been an infraction of social justice by asking the Philippine university that was economically disadvantaged to contribute equal funding into the exchange as the New Zealand university did. Was it fair? Was it just? Ironically, the fact that the exchange required equal funding from both institutions for the exchange to be approved took it out of the category of an aid program with the inequities implied in that category, but increased the financial inequity between the two universities. Spending for the board and lodging of foreign visitors is a standard expense to the Ateneo. Foreign scholars who wish to collaborate with Ateneo fund their own travel but usually request Ateneo's contribution through free board and lodging at its housing facility or at other housing units in the neighbourhood. The financial requirements of Asia: NZ were heavy because the airfares needed to be paid in US dollars. Ateneo paid for the travel of only two of its faculty when the intent of the program was to match the number of personnel sent by New Zealand, which was four. Part of the reason for sending only half of the expected number was financial; the other part was workload. A roundtrip airfare to New Zealand costs about US$1000. That was a significant amount of money for the Ateneo, whose policy is to give only $300 to one faculty member for any significant scholarly activity abroad. Indeed, the poorer institution, financially, was at the losing end. It might have been more equitable if Asia: NZ had required financial contributions based on the institution's capacity to provide the counterpart funding needed.

Rawls' theory of democratic equality could certainly have been applied in this situation. Ateneo could have been allowed to put in less money than the University of Auckland and both parties could still have had the exchange with no major difference in the way it would be run. Unequal contributions from the two major players would not have been an injustice because one university was more financially capable than the other. The funding foundation permitted the Philippine university to send fewer people but did not allow the funding of additional Filipino participants.

Admittedly, the goal of New Zealand to ensure that the exchange was not an aid programme was noble and gratifying. It clearly acknowledged that the Philippine university was capable of contributing what was expected to the exchange. This indicates that there was mutual respect for each other and high regard for the Philippine partner university in spite of the disparate economic status of the two countries. According to Miller (1999), there was social justice evident in the project, although not in the financial sense. Since social justice does not only pertain to how material goods are distributed or acquired but also includes mutual recognition of each other's value in the society or community, social justice prevailed in this aspect of the exchange. Recall Miller's (1999) theory that takes into consideration the mode of human relationship. Recall the mode *instrumental association* under which this exchange belongs. The appropriate theory is *justice according to desert*. With academics from both universities contributing their fair share of talents and skills, they obtained corresponding rewards and benefits of various forms; foremost is, perhaps, increased national and personal pride.

Nozick's, R. (1974) theory of fairness, however, appears to dominate the exchange in the financial aspect. New Zealand has greater financial security than the Philippines. It is the funding agency's prerogative to allocate its resources. Yet, if this agency had chosen to absorb some of the financial burden that would have lessened the financial responsibility of the poorer university.

While acknowledging the agency's principles on use of their funds, we wonder who or what would convince them to recognize and practice a "fair inequity" or, in the words of Rawls, "democratic equality". Who or what will persuade a rich country to relieve the poorer partner of the financial burden? This is a pervading issue in social justice. As seen globally the gap between rich and poor economies grows wider and the poorer countries have limited means of addressing this. This is injustice as described by Young. According to Young (1990, as cited in Atweh, 2004a, p. 5), basic forms of injustice in "doing" are based on manifestations of oppression: exploitation, marginalisation, powerlessness, cultural imperialism and violence. What is manifested here in some way is powerlessness on the part of the Philippine partner university. In fact, whichever university in the Philippines was chosen would have been unlikely to influence changes in the financial provisions of the exchange. Because such an exchange is almost always desirable, the poorer country would acquiescently agree to the provisions despite the financial demands.

Such is a dilemma that poorer countries find themselves in. Poorer countries need financial assistance from richer countries in order to provide opportunities such as higher education for their nationals. However, when a country is always at the receiving end, it can lose a huge part of its national pride and dignity. Worse, it can become overly dependent on aid projects, making it difficult to get out of the "aid mode". On the other hand, if the poor and rich countries that collaborate do so on equal terms, including equal share in the financial obligations, the poor country, though gaining much national pride, may end up in a financially worse situation. This, again, is powerlessness as described by Young (1990, as cited in Atweh, 2004a).

Relative powerlessness is similarly manifested in other academic situations involving mathematics and science education. Writing for high-ranking international publication houses whose books or journals that authors from non-affluent countries or their institutions could not even afford to purchase or subscribe to presents a "no choice" situation for them. Adjusting the sale prices to fit the buying capacity of the poorer institutions could rectify such a situation. The Philippine author, for example, has had the distinct honour to co-write for an international book a chapter that is meant to address a global concern in mathematics education (see Gates and Vistro-Yu, 2003). However, except for a few private higher learning institutions including her own, other universities from her country are unlikely to be able to afford the book and, thus, lose the opportunity to benefit from it. The opportunity to write for the book might have been a once-in-a-lifetime experience and she would have been foolish to refuse the invitation as a form of protest or in order to influence the international publisher to ease the burden of the non-affluent countries. The demand for international recognition in mathematics education is great and therefore international experiences are very much desired. This could also be a case of cultural imperialism. In mathematics education, one does not just publish in any international publication; the publisher must be world renowned with an excellent record in maintaining the quality of the writings. However, who decides which publications are excellent?

To New Zealand's credit, the principle of redress was upheld in the latter part of the exchange. With the savings and additional funding from the mathematics education unit of University of Auckland, the third academic from the Philippine university was able to visit New Zealand and carry out a collaborative research study. The decision may be viewed as the New Zealand University's way of compensating for the undeserved financial position that the Philippine university was in.

In both situations described – cost of airfares and publications – it appears that a country with few financial resources has very little chance to engage collaboratively or benefit from high-level scholarly activities with other countries unless the costs of such collaboration are adjusted to suit the financial capability of the poor country. This is consistent with one perspective on the relationship between mathematics education and democracy that of dissonance. Dissonance suggests that mathematics education has established a systematic access denial on the grounds of people's gender, race, language, and class or socio-economic status (Skovsmose & Valero, 2001). Skovsmose and Valero (2001) highlighted the critical relationship between mathematics education and democracy, indicating that mathematics education activities could either support or deny democracy. With the high cost of international collaboration and the special importance attached to mathematics education activities, poorer countries are caught in the middle. The very nature and demands of the highly valued international mathematics education activities challenge democratic ideals and social justice. How can international academic exchange programs and other international collaborations work to support social justice and democracy for all countries involved?

6. Some Lessons Learned: What are the Necessary Ingredients for a Fruitful Exchange While Ensuring Social Justice?

The advantages and disadvantages of this New Zealand-Philippines exchange program have led the authors to theorize about the characteristics necessary for social justice in such exchanges. They considered what ingredients are necessary for a fruitful exchange and what factors contribute to or deny the development of social justice in a mutual academic exchange program. These are the points of convergence that can be further explored when analysing international academic exchange activities. Following this, the authors give this exchange a report card in relation to these factors.

The main factors for social justice in an exchange are:

1. Compatibility of goals between the financers of the exchange.
2. Compatibility of goals of the actual participants in the exchange.
3. Willingness of the participants to learn from each other – a 2-way benefit not 1-way.
4. Differential offerings of each country, so there is something to learn from each other.
5. Equity of real costs to the partners in the exchange.
6. Ability of the participants to take full advantage of the exchange. This includes both personnel and time.
7. Equity of outcomes for the participants in the exchange.
8. Exchange of academic ideas without overwhelming the needs or wishes of one country or participant with the needs or wishes of the other.
9. Collaborative opportunity to determine the social equity of an exchange.

Assessing the presented exchange on these qualities gives a mixed report.

1. Compatibility of goals of the financers of the exchange. The authors grade this as, "Cannot tell". If the goals of the New Zealand financers were as stated, "an increased number of tertiary students and academics across disciplines who

have cultural, language and professional experience relevant to New Zealand's place in the Asia-Pacific region," this was achieved (http://www.asianz.org.nz/about/education-heep.php#top, 7 December 2004). However, there are other goals of this organization, including ongoing relationships between tertiary institutions, that cannot be judged. The president of Ateneo financed his side of the exchange partly because he wanted to enhance the status of mathematics and science education in his university. If the Philippine faculty member does not return to Ateneo, this would be a loss to Ateneo. For both financers these goals could be seen as part of larger ambitions. New Zealand has a real need to know and be known by its Asian neighbours. The president of Ateneo no doubt wants to increase the international status of his university. Both of these are long-term goals and impossible to evaluate in relation to a short-term project.

2. Compatibility of goals of the participants in the exchange. The authors grade this as, "Yes." All of the participants in this exchange were interested in increasing their knowledge and publications in mathematics and science education. All participants benefited from spending time in each other's country.

3. Willingness to learn from one another. The authors also grade this as, "Yes." All participants were willing to learn from their counterparts in the other university and gained their respect as well.

4. The participants being different enough that both parties could learn from one another. The authors grade this as "Yes." This outcome is closely related to willingness to learn from one another, but goes further. The New Zealand author believes that this exchange provided her with more challenges to her thinking than did the sabbatical leaves she had taken in the USA, although those also had been valuable. Visiting schools and talking with teachers in other countries enabled visitors to learn about things that they were not even aware that they did not know. This added to perception of issues in one's own country.

5. Equity of costs. This must be graded as "No". Interestingly, the reason that Asia: NZ asked for the financial commitment of the partner university was that they did not want this to be viewed as aid. Yet their requirement increased the financial inequity.

6. Ability of both partners to take full advantage of the exchange. This factor also earned "No" for a grade. Alterations allowed by the Asia: NZ foundation that allowed fewer academics to travel from the Philippines meant that fewer Filipinos benefited. The co-authors and other participants had too many responsibilities in their home universities to take full advantage of the exchange by staying for longer periods.

7. Equity of outcomes for the participants. The grade for this would probably be "Yes" but this is somewhat difficult to determine. Both authors, as the main participants, were also engaged in other research and joint writing as well as the joint publications mentioned here.

8. Exchange of ideas without overwhelming the needs of one party with the needs of the other party. The authors grade this as "Danger – watch this space". This is the factor that comes closest to globalization that does not equate to social justice.

Thomas (2001) refers to this as cultural imperialism in mathematics education. By taking time to write articles for international journals, most of which are situated in western countries and have editors and reviewers who share western ideas, the authors have not spent time on activities that would have helped teachers working with students who are most in need. Both universities place a higher value on publications in internationally refereed journals than on works that would be read by teachers. The reviewers of international journals become the arbitrators of what is valuable in educational research. The publications from this exchange met the requirements of a select few western academics who may not know what the poorer nations have to offer in education unless it is couched in western terms. If the education or research of the poorer country is considered to be not of the flavour of these international publications, then it may be more likely to be rejected (Atweh & Clarkson, 2001). Unlike this book, these reviewers rarely see social justice as the main object of research. In the final analysis, the authors do not know if the exchange or the publications have helped anyone, anywhere to teach or learn mathematics or science subjects better. All they know is that the participants benefited from the exchange.

9. Equal opportunity to determine the social equity of an exchange. This must be rated "No". The poorer country had no opportunity to recommend variations in funding nor, for that matter, did the academic from the richer country who initiated the exchange. That decision had been taken by the funding agency. However, this issue could have been addressed during the exploratory visits. It might have been possible to make alterations to the exchange, for example by tying it to other sources of funding from New Zealand. This discussion did not take place. This could have been due to politeness, visitor-host relationships, age differences, or to the two authors' naïveté. The authors bear some responsibility for not addressing issues of social justice. We recommend that other parties in an exchange explore this issue.

7. Final Remarks

Despite the inequalities discussed, various forms of social justice (Rawls' democratic inequality, Miller's justice with desert principle) prevailed in the end. New Zealand did end up contributing more, financially, to the exchange, although that was not the initial intention. Asia: NZ paid the airfare for three New Zealand academics, one of whom travelled twice, and the University of Auckland paid the airfare for one Filipino academic. These organisations hosted all Filipino visitors. Ateneo sent only two of its faculty to New Zealand, paid part of the living expenses of a third, and hosted all visiting New Zealanders. This inequity of visitors was necessary for financial reasons, although not intended initially. It could be argued that New Zealand gained more from the exchange and that Ateneo did not gain as much. As in many financial situations, the more money one invests, the more benefits and rewards one gets and the less one invests, the fewer the benefits that one expects to reap – Miller (1999) theory that is interpreted differently.

Atweh, Clarkson and Nebres (2003) pose a very important challenge: How can communities with different political, cultural, and social conditions make ways to learn from each other more productively? Can nations across the globe practice social justice with one another the way people within one nation do? This chapter showed that through exchange programs, genuine learning from communities and, eventually, social justice could prevail. Participants in such programs need to be critical of each step in the exchange for a mutual collaboration to occur. Otherwise, as has been shown, social justice could be sacrificed, as could the real goals of the exchange.

Acknowledgement

The authors are grateful to the Asia: NZ Foundation and the Ateneo de Manila University who made this exchange possible. They also wish to thank Peter T. Howard of Australian Catholic University, Raul Lejano of University of California, and Alma Salvador of Ateneo de Manila University for their invaluable insights and comments on this chapter.

References

Annan, K. (2000). The politics of globalization. In O'Meara, Howard et al (Eds.), *Globalization and the challenges of a new century* (pp. 125–131). Indiana: Indiana University Press.

Asia: NZ Foundation website. (2006). http://www.asianz.org.nz/ Retrieved 26 April 2006.

Asia: NZ web page for HEEP. (2006). http://www.asianz.org.nz/education/education-heep.php/ Retrieved 7 December 2004, 26 April 2006.

Atweh, B. (2003). International aid activities in mathematics education in developing countries: A call for further research. Paper presented at the International Conference of the *Australian Association for Research in Education*, Auckland.

Atweh, B. (2004a). Injustice and international academic activities. Paper presented at the *Australian Association for Research in Education Conference*, Melbourne.

Atweh, B. (2004b). Towards a model of social justices in mathematics education and its application to critique of international collaboration. In I. Putt, R. Faragher, & M. McLean (Eds.), *Mathematics education for the third millennium: Towards 2010. Proceedings of the 27th annual conference of the mathematics education research group of Australasia* (pp. 47–54). Sydney: MERGA.

Atweh, B., & Clarkson, P. (2001). Internationalisation and globalization of mathematics education: Toward an agenda for research/action. In B. Atweh, H. Forgasz, & B. Nebres (Eds.), *Sociocultural research in mathematics education an international perspective* (pp. 37–94). New Jersey: Lawrence Erlbaum Associates, Publishers.

Atweh, B., & Clarkson, P. (2004). Some problematics in international collaboration in mathematics education. In B. Barton, K. Irwin, M. Pfannkuch, & M. Thomas (Eds.), *Mathematics education in the South Pacific: Proceedings of the 25th annual conference of the mathematics education research group of Australasia*. MERGA: University of Auckland.

Atweh, B., Clarkson, P., & Nebres, B. (2003). Mathematics education in international and global contexts. In A. J. Bishop, M. A. Clements, C. Keitel, J. Kilpatrick, & F. K. S. Leung (Eds.), *Second international handbook of mathematics education* (pp. 185–229). Dordrecht: Kluwer Academic Publishers.

Bangko Sentral ng Pilipinas. (2004). <http://www.bsp.gov.ph/> Retrieved on 15 December 2004.

Blocker, H. G., & Smith, E. H. (Eds.), (1980). John*Rawl's theory of social justice*. Athens: Ohio University Press.

Dash, R. C. (1998). Globalization: For whom and for what. *Latin American perspectives, 25*(6), 52–54.
delos Santos, A. G., & Thomas, M. O. J. (2001). Representational fluency and symbolisation of derivative, *Proceedings of the Sixth Asian Technology Conference in Mathematics,* Melbourne, 282–291.
delos Santos, A. G., & Thomas, M. O. J. (2002a). Teaching derivative with graphic calculators: The role of a representational perspective, In W-C. Yang, S-C. Chu, T. de Alwis, & F. M. Bhatti (Eds.), *Proceedings of the 7th Asian technology conference in mathematics 2002* (pp. 349–358). Malaysia: Melaka.
delos Santos, A. G., & Thomas, M. O. J. (2002b). Teacher perspectives on derivative, In B. Barton, K. Irwin, M. Pfannkuch, & M. O. J. Thomas (Eds.), *Mathematics education in the South Pacific: Proceedings of the 25th mathematics education research group of Australasia conference* (pp. 211–218). Sydney: MERGA.
delos Santos, A. G., & Thomas, M. O. J. (2003a). Representational ability and understanding of derivative. In N. A. Pateman, B. J. Dougherty, & J. Zilliox (Eds.), *Proceedings of the 27th conference of the international group for the psychology of mathematics* (pp. 325–332).
delos Santos, A. G., & Thomas, M. O. J. (2003b). Perspective on the teaching of derivative with graphic calculators. *Australian Senior Mathematics Journal, 17*(1), 40–58.
delos Santos, A. G., & Thomas, M. O. J. (2005). The growth of schematic thinking about derivative. In P. Clarkson, A. Downton, D. Gronn, M. Horne, A. McDonough, R. Pierce, & A. Roche (Eds.), *Building connections: Research, theory and practice. Proceedings of the 28th annual conference of the mathematics education group of Australasia* (pp. 377–384). Sydney: MERGA
Ferrer, B., Hunter, B., Irwin, K., Sheldon, M., Thompson, C., & Vistro-Yu, C. (2001). Which is the best way to pay, by the meter or the square meter? *Mathematics in the Middle School, 7*(3), 132–139.
Gates, P., & Vistro-Yu, C. (2003). Is mathematics for all? In A. J. Bishop, M. A. Clements, C. Keitel, J. Kilpatrick, & F. K. S. Leung (Eds.), *Second international handbook of mathematics education* (pp. 31–73). Dordrecht: Kluwer Academic Publishers.
Irwin, K. C., Vistro-Yu, C. P., & Ell, F. R. (2004). Understanding linear measurement: A comparison of Filipino and New Zealand children. *Mathematics Education Research Journal, 16*(2), 3–24.
Miller, D. (1999). *Principles of social justice*. Cambridge, MA: Harvard University Press.
Mithaug, D. E. (1996). *Equal opportunity theory*. Thousand Oaks: Sage Publications.
Nozick, R. (1974). *Anarchy, state and utopia*. New York: Basic Books
Rawls, J. (1971, 1999). *A theory of justice*. Cambridge, MA: Belknap Press of the Harvard University Press.
Schaar, J. H. (1980). Equality of opportunity and the just society. In H. G. Blocker & E. H. Smith (Eds.), *John Rawls' theory of social justice* (pp. 162–184). Athens: Ohio University Press.
Skovsmose, O., & Valero, P. (2001). Breaking political neutrality: The critical engagement of mathematics education with democracy. In B. Atweh, H. Forgasz, & B. Nebres (Eds.), *Sociocultural research in mathematics education an international perspective* (pp. 37–55). New Jersey: Lawrence Erlbaum Associates, Publishers.
Swanton, C. (1981). Is the difference principle a principle of justice? *Mind, New Series, 90*(359), 415–421.
Thomas, J. (2001). Globalization and the politics of mathematics education. In B. Atweh, H. Forgasz, & B. Nebres (Eds.), *Sociocultural research in mathematics education an international perspective* (pp. 95–112). New Jersey: Lawrence Erlbaum Associates, Publishers.

19

GLOBALISATION, TECHNOLOGY, AND THE ADULT LEARNER OF MATHEMATICS

Gail E. FitzSimons
Monash University

Abstract: The focus of this chapter is on the implications of globalisation and internationalisation of education for adults returning to study mathematics at all educational levels in the post-compulsory years. These inter-related phenomena bring tensions and contradictions for mathematics educators arising from competing needs for institutions to give greater flexibility to students – who are tending to form an increasingly diverse cohort – while controlling or even reducing tuition costs: Dangers arise from the commodification of education and particularly from the assumption that mathematics, and indeed new learning technologies, are culture- and value-free. The work of Basil Bernstein forms the basis of an analytical distinction between mathematics and numeracy as different discourses with different forms of pedagogy with consequences for program design and delivery. Sociocultural activity theory is employed as a theoretical foundation for analysis of new learning technologies in order to assist developers and users in the design and appropriation of pedagogical resources. The chapter concludes with a discussion of the tensions in mathematics/numeracy education in relation to internationalisation, together with some recommendations for program developers utilising new learning technologies

Keywords: adult learners, globalisation, internationalisation, mathematics, numeracy

1. Introduction

The phenomenon of globalisation has arisen over recent decades, supported in part by the rapid development of new information and communication technologies. The practice of internationalisation in education has been in use for considerably longer (Cambridge & Thompson, 2004), but is taking on new meanings as educational institutions seek to expand their markets (Schapper & Mayson, 2004). One outcome of this expansion is the utilisation of new learning technologies. The focus of this chapter is on adults returning to study mathematics at all educational levels

B. Atweh et al. (eds.), Internationalisation and Globalisation in Mathematics and Science Education, 343–361.

(including adult numeracy), with possible implications for other students in the post-compulsory years.

The chapter opens with a general discussion of the inter-related phenomena of globalisation and internationalisation of education, highlighting the tensions and contradictions faced by educators, particularly at the post-compulsory level. It then draws on Bernstein's language of description to distinguish analytically between mathematics and numeracy – the latter term gaining prominence in adult education in many countries around the world. Regarding them as different discourses with different forms of pedagogy has consequences for program design and delivery, whether programs be of short or long duration, face-to-face (proximal) or distal modes, on- or off-campus. Increasingly new learning technologies are coming to play an important role as vectors for the transmission of program materials as well as tools for understanding, manipulation, automation, and communication. Here, sociocultural activity theory is employed as a theoretical foundation for analysis in order to assist developers and users in the design and appropriation of pedagogical resources. Clarity of purpose, along with cultural sensitivity, is especially important in the case of adults returning to study at any level, and particularly in response to pressures of globalisation and internationalisation: simplistic notions of the universality of mathematics education cannot be allowed to prevail. In a similar vein, new learning technologies cannot be assumed to be value-free, and the chapter offers some critique from the relevant literature. The chapter concludes with a discussion of the tensions in mathematics/numeracy education in relation to internationalisation, together with some recommendations for program developers utilising new learning technologies.

2. Globalisation and Internationalisation

Burbules and Torres (2000) identify a range of characteristics of globalisation that most affect education. These include, in the economic sphere, new forms of work organisation and removal of barriers to trade and investment. In the political sphere they note the erosion of national autonomy and a corresponding weakening of the notion of citizen. In the cultural sphere they note the tensions arising between trends towards standardisation and greater homogeneity simultaneously with heterogeneity arising from fragmentation and locally oriented movements. (See also Jones, 1998, for discussion of these three patterns of globalisation in relation to world education.)

The rise of new information and communication technologies is seen as both contributing to globalisation and arising as a consequence of it. In particular, new learning technologies are emerging in adult, vocational, and higher education sectors, supported ostensibly by economic arguments for doing more with less, and for offering greater flexibility to learners in greater choice of time and place of study. Some educational decision-makers (e.g., administrators, bureaucrats, and politicians), faced with the ever-increasing costs of labour, naively presume that new learning technologies will offer a means to reduce teaching costs and increase profits for their institutions through the widespread dissemination of educational materials to fee-paying students, locally and even around the world (Clegg, Hudson, & Steel, 2003). Schapper and Mayson (2004) note that in the context of falling levels of government support for higher education (in Australia at least) and the need to maintain competitive advantage in global education markets,

> universities face contradictory tendencies: they must market and deliver their educational services across the globe while simultaneously accommodating the diverse and localised and decentred needs of specific student groups. … Education becomes a commodity … delivered to "customers" in rationalised and economic ways, with only lip service paid to learning outcomes or educational objectives of diverse student groups.
>
> (p. 192)

In other words, the economic impetus of expanding educational markets sees at one and the same time the commodification of education with tendencies toward homogeneity in large-scale operations, while the student population becomes increasingly diverse. In the case of mathematics, one underlying assumption is that mathematics is culture- and value-free (Ellerton & Clements, 1989). This is a convenient but dangerous assumption by technicist-oriented decision makers in policy and resource production, and runs contrary to the mathematics education research literature of the last two decades (e.g.,Bishop et al., 2003; D'Ambrosio, 1985).

In the adult and vocational education sectors in Australia, for example, there is evidence of the commodification of education generally, and specifically in the arrangements made for learners to take up workplace numeracy via face-to-face or online modules, such as *Calculations*. These modules are often based on curricula drawn from the early school years rather than actual workplace activities, reflecting a lack of respect for workers' integrity and their lifetime of experience (FitzSimons, 2000). Having said this, it is also true that in the higher education sector in Australia there are exemplary materials for adult learners, designed to meet the local and contextualised needs of a specific cohort of students, and which recognise that teaching and learning are socially, culturally, and historically situated activities (e.g., Taylor, 2001).

Exploring the tensions between globalisation and internationalisation, Cambridge and Thompson (2004) note that:

> The pragmatic 'globalist' current of international education may be identified with the processes of economic and cultural globalization, expressed in terms of satisfying the increasing demands for educational qualifications that are portable between schools and transferable between education systems, and the spread of global quality standards through quality assurance processes such as accreditation.
>
> (p. 164)

By contrast, they propose that:

> international education, as currently practised, is the reconciliation of a dilemma between ideological and pragmatic interests. The ideological 'internationalist' current of international education may be identified with a progressive view of education that is concerned with the moral development of the individual by attempting to influence the formation of positive attitudes towards peace, international understanding and responsible world citizenship. (p. 164)

However, international education may also assist in the maintenance of the privileged position of the transnational capitalist class, locally and internationally. Cambridge and Thompson (2004) state that international education may be identified with:

> - a transplanted national system serving expatriate clients of that country located in another country;
> - a transplanted national system serving clients from another country;
> - a simulacrum of a transplanted national system, for example the programmes of the IBO [International Baccalaureate Organisation], serving expatriate clients and/or host country nationals; and
> - an ideology of international understanding and peace, responsible world citizenship and service (Cambridge & Thompson, 2001).
>
> (p. 172)

It is apparent that mathematics education is associated with the first three at least. I believe that it can also contribute to the last, particularly in the case of adult and post-compulsory education. Cambridge and Thompson (2004) provide examples of the ambiguous and contradictory nature of international education, such as the celebration of cultural diversity whilst tending towards the development of a monoculture, or the encouragement of positive attitudes towards global citizenship and community service whilst it enhances positional competition and personal economic advancement. They conclude that "the globalist approach to international education is influenced by and contributes to the global diffusion of the values of free market economics" (pp. 172–173).

Schapper and Mayson (2004) assert that internationalisation in the university context results in the standardisation of subject content, and requirements of up to 100% commonality, with detailed and prescriptive weekly learning objectives, limiting academic flexibility in development of materials. Even the texts and references, which may be chosen to suit local circumstances, are often reduced to those which reflect the dominant culture – in their case, of North America for business studies. In addition, staff may be required to place lecture notes in the form of (hegemonic) *PowerPoint* presentations on *WebCT*, one of a number of products that use the internet as a delivery platform for program management. I argue that both technologies have inbuilt constraints on pedagogy, but they also offer possibilities for international collaboration and co-operation. In order to highlight and possibly resolve some of the tensions and ambiguities arising from internationalisation, and to inform the development of new learning technologies in mathematics education, I turn to recommendations offered by various authors.

3. Principles for Internationalisation

According to Vulliamy (2004, p. 270), comparative ethnographic studies "suggest that cultural values are more important for learning than pedagogical styles because underlying educational values give meaning to styles of pedagogy." Face-to-face teaching styles will inevitably undergo some transformation due to the prevailing culture of the national context where the interaction takes place. But, what does this mean for learners undertaking fully imported programs (e.g., standardised online programs or CD-ROMs) where the pedagogy cannot be modified to attend to the local culture in any way?

Schoorman (2000) defines internationalisation in the following way:

> Internationalization is an ongoing, counter-hegemonic educational process that occurs in an international context of knowledge and practice where societies are viewed as subsystems of a larger, inclusive world. The process of internationalization at an educational institution entails a comprehensive, multifaceted program of action that is integrated into all aspects of education. (p. 5)

She identifies five facets of this definition as: (a) counter-hegemonic, (b) ongoing, (c) comprehensive, (d) multifaceted and (e) integrated. In this context, a counter-hegemonic approach advocates moving beyond narrowly defined economic and instrumental conditions "to facilitate a knowledge base, skills and attitudes among students that prepare them for work and leadership in the context of global interdependence. To this end, a counter-hegemonic approach makes explicit the content, the instruction, and the rationale desired of the internationalization process" (Schoorman, 2000, p. 5).

In terms of content, Schoorman is critical of the privileging of a limited cultural base with respect to content (i.e., from the USA and/or Western Europe). She recommends that curricula represent truly global or international perspectives throughout. In terms of instruction, she advocates a critical pedagogy, where students are regarded as independent thinkers and decision makers. As well as providing students with the knowledge base about diverse cultures, it also helps in the development of "appropriate interpersonal skills and attitudes when working with persons with diverse world views and perspectives" (p. 6). She recommends drawing on the knowledge base of students, particularly when international students are present, because they are frequently ignored as a valuable educational resource. (This phenomenon of ignoring the contributions of immigrant children in school was noted in the Australian context by Wotley, 2001.) There should be a critical examination of the multiple perspectives through engaged dialogue within the whole group. In terms of the rationale, she suggests that a counter-hegemonic perspective on internationalisation requires working towards "the establishment of global democracy and international social justice" (p. 7). Further, these goals should be articulated. She also recommends the use of case studies and/or "an examination of the implications or applicability of diverse theories, practices or issues among diverse nations" (pp. 8–9). Notably, Schoorman observes that internationalisation has not been seen as relevant to the sciences, resulting in a limited effort in these areas.

Beside acknowledging the work of Schoorman, Schapper and Mayson (2004) also endorse guiding principles from Whalley (1997, cited in Schapper & Mayson, 2004). These are to: "encourage student participation, independent learning, self-reflection, critical thinking, problem solving and the examination of international and intercultural perspectives" (p. 210). Whalley also recommends making explicit the cultural origins of the course content, as well as incorporating the techniques designed to support cultural diversity listed above.

In summary, globalisation and internationalisation offer contradictory tendencies for mathematics education at the post-compulsory level (i.e., university and adult education). Economic imperatives demand doing more with less and tendencies toward homogeneity with the increasing scale of operations. At the same time, access to education via new learning technologies is made possible for greater numbers of students – at least for those with the necessary physical and financial resources – resulting in greater cultural diversity of the (potential) learner population. Although improved access to mathematics education could further privilege the already privileged, it could also serve to disempower or marginalise groups whose characteristics do not conform to the mainstream culture of the program's origin. From a different perspective, however, internationalisation could serve to further the aims and practices of international understanding and peace, working towards global democracy and international social justice.

So far, I have discussed the topic of mathematics education for adults unproblematically. In fact, the term *adult numeracy* is becoming increasingly widespread, particularly for adults returning to study with limited formal educational backgrounds. The work of Basil Bernstein provides an ideal language of description (Daniels, 2004), especially in relation to issues of power and control, taken at the structural level as well as the interactional level.

4. Distinctions Between Mathematics and Adult Numeracy

In recent times in different countries and in different sectors of education, the term 'numeracy' has come to be used interchangeably with, or as a replacement for, the term 'mathematics', both for school children and for adult learners. In FitzSimons (2004) I drew upon the work of Bernstein (2000) in relation to *vertical discourses* and *horizontal discourses* to distinguish these analytically. In the educational field they are frequently used in oppositional rather than complementary senses as school(ed) vs. everyday common-sense knowledge, or as 'official' vs. 'local' knowledge. According to Bernstein, commonsense knowledge is likely to be: "oral, local, context dependent and specific, tacit, multi-layered, and contradictory across but not within contexts" (p. 157). He asserts that vertical discourses and horizontal discourses will result in different forms of knowledge construction. Bernstein describes vertical discourses as having a coherent, explicit, and systematically principled structure: Mathematics is defined as a vertical discourse. Bernstein defines horizontal discourse as entailing a "set of strategies which are local, segmentally organised, context specific and dependent, for maximising encounters with persons and habitats." (p. 157). Thus, the knowledges of horizontal discourses are "embedded in on-going practices, usually with strong affective loading, and directed towards specific, immediate goals, highly relevant to the acquirer in the context of his/her life" (p. 159). I argue that the construct of *numeracy* is an example of a horizontal discourse, in contrast to the vertical discourse of mathematics. This has pedagogical implications.

According to Bernstein (2000), the pedagogy of horizontal discourses is usually carried out in face-to-face interactions. It may be tacitly transmitted by modelling, by showing or by other explicit means. If the pedagogy is not sufficient in the context of its enactment it is repeated until the particular competence is acquired. Bernstein continues that: "From the point of view of any one individual operating within *Horizontal discourse*, there is not necessarily one and only one correct strategy relevant to a particular context" (p. 160). He concludes that horizontal discourse "facilitates the development of a *repertoire* of strategies of operational 'knowledges' activated in contexts whose reading is unproblematic" (p. 160) [italics in the original]. By contrast, vertical discourses such as mathematics consist of specialised symbolic structures of explicit knowledge. Procedures are linked hierarchically, and the official pedagogy is an on-going process. Whereas the discipline of mathematics is able to be abstracted into general principles, numeracy is necessarily integrated with its context. In practical situations, such as the need for efficient and timely solutions to (ever-evolving) workplace problems, high level mathematical abstractions alone are insufficient and may even prove counter-productive. Rather, contextualised and often shared practices of numeracy within the particular workplace (or other) community are called for.

Mathematical knowledge is developed, from earliest infancy, informally through socialisation in the home and, later, in the broader community and the workplace. This knowledge might be termed 'everyday' or 'commonsense' mathematics; even 'ethnomathematics' or numeracy. However, adult numeracy in practice draws on, or is enhanced by, formal mathematical knowledges and skills which are dialectically related to other generic competencies such as communicating, problem solving, and using technology (e.g., FitzSimons, 2002; Kent et al., 2004; Wake & Williams, 2001). Formal mathematics education is available in school and higher educational institutions for most young people, and also in the workplace, community setting, and/or home for adults who have the desire and/or need to further develop their existing mathematical knowledges and skills. (For further discussion of informal learning in the workplace, see FitzSimons & Mlcek, 2004.) Although much has been written about the (potential) clash between these two forms of knowledge (formal and informal) for certain social and cultural groups, these arguments will not be rehearsed here; however, implications for the development of new learning technologies in mathematics/numeracy will be considered below.

5. Adult Learners of Mathematics/Numeracy

Adults return to study mathematics for a complex array of reasons, often not articulated, and which may be in tension themselves. Increasingly, there may be some degree of coercion from employers or government authorities. They are concerned with maintaining a positive self-concept and, although they may be highly motivated to learn, they may lack confidence in their own ability and require continuing encouragement. One significant characteristic of adult learners is that they bring a rich and diverse range of experiences with them. For example, they generally have a substantial experience of formal education with often unsatisfactory histories of learning mathematics, from cognitive and/or affective domain perspectives. On the other hand, they have a wealth of experience accumulated over a lifetime of managing their affairs at work, at home, and in the broader community. Over recent decades, with the rise of planned immigration, together with the flow of refugees arriving in the relatively more advantaged nations, it has become more common to have a diverse cultural mix in 'basic' mathematics or numeracy classes for adults; likewise in tertiary preparatory programs. This is in addition to the trends toward internationalisation discussed above. The different kinds of experiences brought by the learners – educational, practical and multicultural – can provide important resources for contextualising their mathematics education, provided that the learners are willing to share them and feel confident to do so. (See FitzSimons & Godden, 2000, for a more comprehensive review of research on adults returning to study mathematics.)

Distance education has played a prominent role for many learners at post-compulsory level since the second half of the 20th century, mostly reliant on print-based materials, possibly supported by telephone conversations, audio- and video-tapes. My own experience of this form of teaching witnessed a very high attrition rate and, given the complexity of the symbolisation and the abstractness of the mathematics, together with the relatively weak educational histories and relative isolation of the students, this is not really surprising. However, new learning technologies in the form of CD-ROMs and online programs, supported by email, discussion groups, electronic blackboards, the worldwide web, and so forth, have the potential to ameliorate the difficulties faced by learners in the past. At the same time, providers such as universities and state education authorities are keen to optimise the use of these new technologies for the economic and political reasons discussed above – even for students who are able to meet regularly on-campus. I believe that it is critical that adults returning to study mathematics or numeracy are not further disadvantaged by the inappropriate use of technology mediated education, given the financial and personal investments they are making, with the concomitant risks of failure (yet again) and opportunity costs of other activities foregone.

To summarise, one aspect of the phenomenon of globalisation in recent years has been the rise to prominence of new learning technologies as vectors for mathematics (and other) education for adult learners and other post-compulsory students, whether by proximal or distal modes. Another aspect has been, for a variety of reasons outlined above, the increased tendency for people to enrol in programs offered by different countries at all levels, from adult basic education to higher degrees – or even just to use learning resources developed elsewhere. At the same time, a burgeoning corpus of research into how mathematics is used in the workplace and elsewhere, together with strategic political pressure in some countries, has resulted in a growing interest in adult numeracy which, as I have briefly asserted above, is a qualitatively different kind of discourse from mathematics. Each of these have implications for curriculum and pedagogy, not least the requirement for internationalisation.

In order to link these discourses of globalisation, internationalisation, new learning technologies, and mathematics/numeracy, I now turn to sociocultural activity theory. This theory allows a systemic view of the complex processes of teaching and learning, as it dialectically links the psychological focus on the learner or subject, the tools or mediating artefacts, and the object or outcome of learning, with the broader sociocultural setting in which the learning takes place. Unlike the work of Bernstein, briefly discussed above, it does not address issues of power and control (Daniels, 2004).

6. Sociocultural Activity Theory, Adult Learners, and New Learning Technologies

Engeström (1999) describes Activity Theory as providing a worthy unit of analysis for enabling a theoretical account of the constitutive elements of an object-oriented, collective, and culturally mediated activity system in all its complex interactions and relationships. The minimum elements of this system include the object, subject, mediating artifacts (signs and tools), rules, community, and division of labour (Figure 1). Engeström adds that the internal tensions and contradictions of activity systems are the motive force of change and development.

According to Kuutti (1996):

> A *tool* can be anything used in the transformation process, including both material tools and tools for thinking. *Rules* cover both explicit and implicit norms, conventions, and social relations within a community. *Division of labor* refers to the explicit and implicit organization of a community as related to the transformation process of the object into the outcome.
>
> (p. 28)

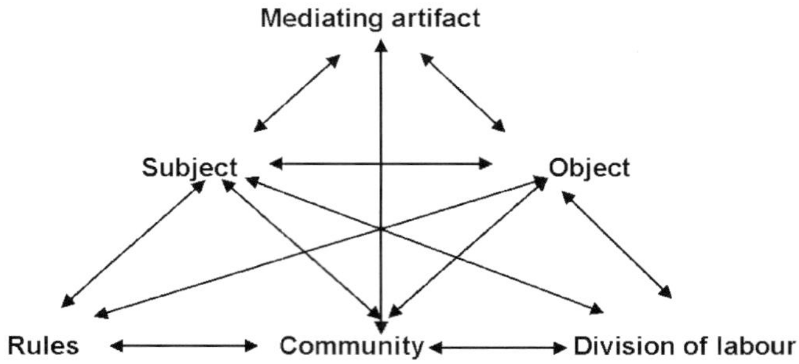

Figure 1. The Basic Mediational Triangle Expanded (after Engeström, 1987)

Following sociocultural activity theory in relation to the theme of this chapter:

- the *Subject* is the adult learner of mathematics/numeracy;
- the *Object* is the development mathematical knowledges & skills, leading towards numerate practices;
- the *Mediating artefacts* are technology as carrier of content and technology as tool for understanding, together with traditional mathematical tools and texts;
- the *Rules* encompass both pedagogical and mathematical constraints;
- the *Community* includes learner/s and possibly teacher/s or other support persons; and
- *the Division of labour* is shared among the learner/s, teacher/s, curriculum/IT developer/s.

  Following Matsushita (1999, p. 219), tensions in the activity system could be as follows:
- *Subject*: grade maker vs. sense maker. What are the motives for adult students to take on learning mathematics/numeracy via new learning technologies, wholly or in part? Is it to achieve a credential? And for what reasons – coercion from (potential or current) employer or government? desire for status? or some of each? Is it to achieve a new or deeper understanding? And for what reasons – personal development/satisfaction of unmet educational needs? To help family members in learning mathematics or in running a business, for example?
- *Object*: dead object vs. live object. Is there an anticipation of developing a deep understanding or of merely rote learning? Is there an anticipation of being able to use what is learned in a practical sense?
- *Mediating artefacts*: tools for the teacher's illustration vs. tools for students' inquiry. What pedagogical approaches will be adopted, especially in relation to new learning technologies? Who decides? Whose cultural background/s will be considered?
- *Rules*: accuracy, speed vs. creativeness, consistency. What philosophies of mathematics and of mathematics education will be assumed?

- *Community*: class of separate individuals vs. team of inquiry. Is the program designed for single users (even in a community setting) or for collaborative use?
- *Division of labour*: isolation and competition vs. cooperation. What levels of interaction if any, between participants are assumed? Are the learners able to contribute to the planning, implementation, and assessment processes? How?

Kuutti (1996) observes that:

> The tool is at the same time both enabling and limiting: it empowers the subject in the transformation process with the historically collected experience and skill "crystalized" into it, but it also restricts the interaction to be from the perspective of that particular tool or instrument only; other potential features of an object remain "invisible" to the subject.
>
> (p. 27)

He sees in the relationship with activities that information technology is able to: (a) automate and substitute human operations, (b) serve as a tool in manipulative and transformative actions, and (c) help in actions directed towards sense making. Drawing on the work of Leont'ev, Kuutti (see Figure 2) identifies three hierarchical levels of development – operations, actions, and activity – and applies these to analyse the uses of technology in education from the perspective of the six vertices of Engeström's model (Figure 1). These distinctions enable to developers of new learning technologies to ascertain which kind of support they are intending to provide. This is critical in light of the above discussion about globalisation, internationalisation, and mathematics/numeracy.

For example, whereas the operational level is intended to hone automatised responses and embed rules, the action-level of support is intended to help learners understand both the mathematical content and the operation of the mathematical tools and procedures themselves. It is here that an awareness of cultural diversity among potential learners is most critical. On the other hand, the activity level is intended to support creativity and innovation, as well as critical reflection, and this is where the importance of the community of learners really comes to the fore. This is the level which most closely approximates the attributes of numerate practice discussed above, whether at the semi-skilled or the professional levels, or somewhere in between. And it is here that new learning technologies are ideally placed to support information seeking and communicative practices on a global scale.

However, the transfer of knowledge from school to workplace contexts is complex. Eraut (2004) identifies five interrelated stages:

1. the extraction of potentially relevant knowledge from the context(s) of its acquisition and previous use;
2. understanding the new situation—a process that often depends on informal social learning;

Operation-level support	Action-level support	Activity-level support
Tool, instrument		
Automating routines	Supporting transformative and manipulative actions Making tools and procedures visible and comprehensible	Enabling the automation of a new routine or construction of a new tool
Object		
Providing data about an object	Making an object manipulable	Enabling something to become a common object
Actor		
Triggering predetermined responses	Supporting sense-making actions within an activity	Supportinglearningand reflection with respect to the whole object and activity
Rules		
Embedding and imposing a certain set of rules	Making the set of rules visible and comprehensible	Enabling the negotiation of new rules
Community		
Creating an implicit community by linking work tasks of several people together	Supporting communicative actions Making the network of actors visible	Enabling the formation of a new community
Division of labor		
Embedding and imposing a certain division of labor	Making the work organization visible and comprehensible	Enabling the reorganisation of the division of labor

Figure 2. A Classification of Potential Ways of Supporting Activities by Information Technology (Kuutti, 1996, p. 35)

3. recognizing what knowledge and skills are relevant;
4. transforming them to fit the new situation;
5. integrating them with other knowledge and skills in order to think/act/communicate in the new situation.

(p. 256)

Although, as noted by Eraut, points 2, 4, and 5 tend to be ignored in formal institutional education and workplace training processes, using the powerful resources offered by electronic resources such as the world wide web and CD-ROMs it is possible to simulate numerate practice in its social, cultural, and historic contexts – well beyond the usual 'text book' applications – even though such simulations can never address all of the complexities associated with actual, contingent workplace pressures.

6.1 Mediating Artefacts in Mathematics Teaching and Learning.

Mathematics teaching and learning is supported by a range of mediating artefacts. These include, as tools for thinking, texts, written and spoken, rulers, compasses, and so forth. Electronic forms include calculators (scientific and graphic), computer software packages (e.g., spreadsheets, databases, computer algebra systems, as well as statistics and geometry workbooks, including simulations) to aid in the automation or calculation processes, as well as the manipulative or transformational and sense-making processes, mentioned above. However, the construction of meaning via the use of such artefacts is far from unproblematic.

Referring to CAS instruction, Trouche (2004) distinguishes between an artefact as a given object, and an instrument as a psychological construct.

> the instrument does not exist in itself, it becomes an instrument when the subject has been able to appropriate it for himself and has integrated it within his activity. ... The construction of this [functional] organ, named *instrumental genesis*, is a complex process, *needing time*, and linked to the artifact characteristics (its *potentialities* and its *constraints*) and to the subject's activity, his/her knowledge and former method of working.
>
> (pp. 285–286)

From this it follows that that a tool or an artefact is not transparent. It is also the case that the knowledge acquired by the learner about the tool and the ways to use it may differ from what the tool was intended to do and the ways it was intended to be used. This is where the teacher, tutor, or other instructor has a vital role to play.

FitzSimons (2005) notes that in his discussion of instrumental orchestration Trouche (2004) exemplifies the three levels of activity theory as analysed by Kuutti (1996).

> At the operations level, the focus is on the ordinary utilisation of the artefact, and students may be helped learn the mechanics of computing and to understand the limits of functions on symbolic calculators. At the second level, artefacts as instruments correspond to both representations and action modes, in order to encourage debates and to make procedures explicit, or as a means to reintegrate remedial or weak students into the class. At the third (or 'meta') level the artefacts offer reflective methods of self analysis of the activity, both individually and collectively – in the case of symbolic calculators students are left with observable traces of their own-instrumented activity.
>
> (p. 772)

However, many authors make the point that technology use in education is non-neutral.

7. Critical Perspectives on New Learning Technologies

Bryson and De Castell (1998) identify three kinds of meta-narrative accounts concerning new learning technologies in education. The first, a technicist/modernist viewpoint, characterises the educational use of computers as a powerful information processing technology, portraying computers as value-neutral tools, to aid the autonomous, rational learner in developing higher-order thinking skills which enable the processing and interpretation of data in problem-solving exercises, and ultimately the accumulation of cultural capital (Bourdieu, 1991).

The second, critical discursive perspective, identifies the problematic nature of the first – especially the value-neutral role accorded to technology. In particular, Bowers (1988, cited in Bryson & De Castell, 1998) is critical of an ideological view of progress which includes "(a) increased control over access to and manipulation of information, (b) abstract rationality as the most effective form of human thinking, and (c) individualism and entrepreneurship as constituting the most effective models for human commerce" (p. 73). Bowers is concerned with the prevalence of the Cartesian view and the obscuration of human authorship of knowledge; also the dominance of digital thinking over metaphorical thinking, including the lack of consideration given to historically and ethically informed decision-making in the adoption of a masculine model of knowing and its distancing gaze – all of which contribute to the personal alienation of so many, especially women.

Discussing the third, postmodernist, account, Bryson and De Castell (1998) turn to a 1987 study by Griffin and Cole to conclude that "technology provides a means for reconstructing the division of labor in classroom tasks and for restructuring power relations between participants in educational contexts who typically occupy very unevenly positioned discursive roles with respect to power" (p. 81). (See, for example, Goos, Galbraith, Renshaw & Geiger, 2000; Motaung, Henning, & van der Westhuizen, 2004) Ultimately, Bryson and De Castell claim that their focus is on possibilities for agency and equity, eschewing the critical perspective which they claim plays into the hands of the conservative and already privileged members of the educational community.

Like Kuutti (1996), Kaptelinin (1996) underlines the importance of tool mediation in transmitting cultural knowledge, shaping external activities, influencing internal mental processes, and even shaping the goals of those who use the tools, as these are implicitly built into the tools by their developers. Discussing technology and didactics, Nordkvelle (2004) draws on the 1998 work of Wallin who called for:

> a 'reflective approach to educational technology.' He [Wallin] elaborated this as a technology that questions the relations between ends and means and illuminates the contextual and situated nature of teaching. This 'new' technology of teaching should be able to operate according to the rules for action within a strategic, not a rule-directed, framework. He specified that a 'reflective educational technology' should explicitly declare its value premises and its relation to the social context, demonstrate consistency within the knowledge base of the technology—implying the justification of ends and means—and help learners analyse the relationship between the values and knowledge of the technology and the consciousness of the learner.
>
> (p. 438)

Nordkvelle concludes that "didactics is a technology, in a broader politicized sense, a technology that should be used in ethically and socially responsible ways. …

new didactical artefacts, such as ICT in education, are not automatically tested and proven" (p. 440). Both critical and creative didactic approaches are necessary.

8. Tensions and Recommendations in Relation to Internationalisation

Experiential learning is popular in school and in some parts of post-compulsory education (especially adult education), yet Edwards and Usher (1998) warn that, although it is a valuable pedagogical strategy, the tendency to manage that experience can have a disempowering effect. Given that experience is unique to each person with a range of possible, sometimes conflicting and contradictory meanings, the normalising tendency of education, as a technology of power, is to privilege a particular, univocal, reading of that experience, to privilege the universal knower, thereby eliminating the differences which might be expected to surface under a postmodern perspective. In mathematics education the use of new learning technologies as a form of experiential learning is becoming common, and undoubtedly an asset to skills development and conceptual understanding, but the question remains: Whose experience counts? In the burgeoning accounts of the successes and potentials for new learning technologies in mathematics education, whose experiences are privileged? Whose life experiences are valued and whose are ignored or devalued in the quest for uniformity made increasingly necessary and 'desirable' when there are local, national, and international high-stakes examinations dominating curriculum and assessment practices throughout the developed world and increasingly in developing countries – possibly influenced, intentionally or unintentionally, by aid donors? Perhaps the broader question is: Are new learning technologies value-free? This question has been long discussed in the case of mathematics and the enculturation/acculturation of learners of all ages into its practices. Are new learning technologies more than just powerful information processing tools to aid the autonomous, rational learner in developing higher-order thinking skills which enable the processing and interpretation of data in problem-solving exercises? It is essential that the issues raised above by Bowers, Bryson and De Castell, and others, are not dismissed in the rush for technical competence, and that actual life experiences of all participants, internationally, are taken into account – in a much more serious manner than so-called 'real-life' applications which actually bear little resemblance to the lived experience of the diversity of students around the world and across the whole life span.

What is the language of instruction? English is hegemonic, but its use is not surprising if the program is developed in an English-speaking country, primarily for local students. There are many variants to English language spelling: UK, USA, Australian, etc.: for example, the spelling of centre/center. Similarly for the use of vocabulary, idioms such as petrol/gasoline. Audio contributions will probably reflect the local accent of the country of origin, and perhaps even strong regional accents – audio is a useful complement in mathematics education through distance education, but may add to difficulties of English as a Second or Other Language [ESOL] students.

Mathematics education contexts vary widely internationally. This is reflected in the special edition of *Journal for Intercultural Studies*, 23(2), 2002, which addresses intercultural aspects of mathematics education. In terms of broad pedagogy: does theory precedes real examples and/or case studies, or vice versa? This is an issue raised by an Australian business management educator working in the People's Republic of China (Wrathall & Liu, 2004).

Contexts for students can vary widely, not only in terms of spatial distance, but also in terms of social and cultural distance. What is appropriate for urban students in an industrialised country may be completely inappropriate for indigenous students living in remote areas, even within the same country. However, it may be that programs which recognise the contribution of Indigenous cultures and which place importance on developing a holistic philosophical framework, embracing "ethical obligations towards the entire web of life" in an information society (Deer & Håkansson, 2005, p. 77), placing value on the importance of narrative, networking and community, may actually be a useful model for those adults (especially women) alienated by previous mathematics education experiences.

For adults returning to study, programs in 'Basic' Numeracy generally focus on rational numbers and measurement. These clearly need to be contextualised but, especially with internationalisation, the use of local currency can be problematic, both in terms of units and of realistic values (witness the recent and longer-term volatility of crude oil prices). One solution would be to ask learners, local and international, to identify or to collect real information about the prices of items such as petrol/gasoline, regular postage stamps, fruit and other supermarket or hardware items of interest to adults. Should decimal fractions present an initial difficulty, learners can be shown by example, via a hotlink, how to round off to nearest whole number of the relevant unit.

Secondary school and higher level mathematics texts, traditionally, have followed the format of demonstrating a new topic or procedure, and requiring students to solve a series of practice examples, often including so-called 'word problems' in order to maintain motivation. These take the form of what I call pseudo-contextualisations (FitzSimons, 2002), because they are often spurious and the solution process varies little from question to question. Part of the explanation is that real-life contexts are messy, and that simplification is necessary in order to keep the students on track. A possible solution here is to present one or more real examples and ask the students to collect experimental data and, where helpful or necessary, to use simulations grounded in students' practical experience or video clips.

At all levels, it is advisable to recognise that the discipline of mathematics is socially constructed, and that it has diverse cultural origins. Accordingly, the history of mathematics should play an integral part in exposition and/or student research. (See, for example, Fauvel & van Maanen, 2000, as well as other publications associated with the ICMI History & Pedagogy of Mathematics study group, http://www.clab.edc.uoc.gr/hpm/, as well as other historically- and ethnomathematically-oriented support materials, in print and via CD-ROMs or the internet). Another recommendation is to respectfully recognise and examine the multiple perspectives via the different contributions made by students of diverse cultures drawing on their own backgrounds, through engaged dialogue within the whole group. In line with the recommendations under *Principles for Internationalisation*, these strategies could prove very powerful for increasing international understanding.

As noted above, internationalisation requires working towards the establishment of global democracy and international social justice, with clear articulation of these goals. Although this may again present a challenge for mathematics education in particular, the use of case studies and/or an examination of the implications or applicability of diverse theories, practices or issues among diverse nations is also recommended.

9. Conclusion

Clearly there are correspondences between principles for internationalisation and for mathematics/numeracy educational activities and actions utilising new learning technologies. The challenge is to avoid the narrow packaging of parochial content and pedagogy for export; also to avoid students adopting hegemonic perspectives in their interactions with international students at home and abroad, in class and online, with respect to their diverse and rich histories and cultures. Especially in relation to mathematics, instructional designers need the capacity to incorporate, where appropriate, aspects of the history and cultural aspects of mathematics without being patronising or trivialising. This is becoming easier with the availability of hotlinks to rich and colourful websites. Internationalisation of curricula and pedagogy presents an ideal opportunity for supporting local and international dialogue between learners, enhancing understanding not only of mathematics and associated adult numeracy, but also of humanity.

In neoliberal economies, where there is a shift towards placing the burden of responsibility and costs for lifelong learning onto the learner (FitzSimons, 2002), it is imperative that adults who, for whatever reason, find themselves in need of developing further mathematics or numeracy knowledges and skills are not further disadvantaged by the technologisation of education. On the other hand, new learning technologies can offer a range of innovative means to enhance understanding of mathematical concepts and to promote new forms of communication between people. This chapter has addressed the complex issues associated with the introduction of new learning technologies for adult learners of mathematics at all levels in a globalised world.

Acknowledgement

This chapter is based upon work funded by a post-doctoral fellowship grant: Australian Research Council Discovery Project, DP0345726.

References

Bernstein, B. (2000). *Pedagogy, symbolic control and identity: Theory, research, critique* (Rev. ed.). Lanham, MD: Rowman & Littlefield.

Bishop, A. J., Clements, K., Keitel, C., Kilpatrick, J., & Leung, F. (Eds.). (2003). *Second international handbook of mathematics education*. Dordrecht: Kluwer Academic Publishers.

Bourdieu, P. (1991). *Language and symbolic power* (G. Raymond & M. Adamson, Trans.). Cambridge, UK: Polity.

Bryson, M., & De Castell, S. (1998). Telling tales out of school: Modernist, critical, and postmodern "true stories" about educational computing. In H. Bromley & M. W. Apple (Eds.), *Education/technology/power: Educational computing as a social practice* (pp. 65–84). New York: State University of New York Press.

Burbules, N. C., & Torres, C. A. (2000). Globalisation and education: An introduction. In N. Burbules & C. Torres (Eds.), *Globalisation and education: Critical perspectives* (pp. 1–26). New York: Routledge.

Cambridge, J. & Thompson, J. (2004). Internationalism and globalization as contexts for international education. *Compare, 34*(2), 161–175.

Clegg, S., Hudson, A., & Steel, J. (2003). The emperor's new clothes: Globalisation and e-learning in higher education. *British Journal of Sociology of Education, 24*(1), 39–53.

D'Ambrosio, U. (1985). *Socio-cultural bases for mathematics education*. Campinas: UNICAMP.

Daniels, H. (2004). Activity theory, discourse and Bernstein. *Educational Review, 56*(2), 121–132.

Deer, K., & Håkansson, A.-K. (2005). Towards an indigenous vision for the information society. In T. J. van Weert (Ed.), *Education and the knowledge society: Information technology supporting human development* (pp. 67–80). Boston/Dordrecht/London: Kluwer Academic Publishers.

Edwards, R., & Usher, R. (1998). "Moving" experiences: Globalisation, pedagogy and experiential learning. *Studies in Continuing Education, 20*(2), 159–174.

Ellerton, N. F., & Clements, M. A. (1989). Tea*ching post-secondary mathematics at a distance: A report to the commonwealth secretariat*. Geelong, VIC: Deakin University.

Engeström, Y. (1987). *Learning by expanding: An activity-theoretical approach to developmental research*. Helsinki: Orienta-Konsultit.

Engeström, Y. (1999). Activity theory and individual and social transformation. In Y. Engeström, R. Miettinen, & R.-L. Punamäki (Eds.), *Perspectives on activity theory* (pp. 1–16). Cambridge: Cambridge University Press.

Eraut, M. (2004). Informal learning in the workplace. *Studies in Continuing Education, 26*(2), 247–273.

Fauvel, J., & van Maanen, J. (2000). *What engine of wit: The ICMI study on history in mathematics education*. Dordrecht: Kluwer Academic Publishers.

FitzSimons, G. E. (2002). *What counts as mathematics? Technologies of power in adult and vocational education*. Dordrecht: Kluwer Academic Publishers.

FitzSimons, G. E. (2004). *An overview of adult and lifelong mathematics education*. Keynote presentation at Topic Study Group 6, 10th International Congress on Mathematics Education [ICME–10]. [Available from the World Wide Web: http://www.icme-10.dk]

FitzSimons, G. E. (2005). Technology mediated post-compulsory mathematics: An activity theory approach. *International Journal of Mathematical Education for Science and Technology, 36*(7), 769–777.

FitzSimons, G. E., & Godden, G. L. (2000). Review of research on adults learning mathematics. In D. Coben, J. O'Donoghue, & G. E., FitzSimons (Eds.), *Perspectives on adults learning mathematics: research and practice* (pp. 13–45). Dordrecht: Kluwer Academic Publishers.

FitzSimons, G. E., & Mlcek, S. (2004). Doing, thinking, teaching, and learning numeracy on the job: An activity approach to research into chemical spraying and handling. In J. Searle, C. McKavanagh, & D. Roebuck (Eds.), *Doing thinking activity learning: Proceedings of the 12th annual international conference on post-compulsory education and training* (Vol. 1 pp. 149–156). Brisbane: Australian Academic Press.

Goos, M., Galbraith, P., Renshaw, P., & Geiger, V. (2000). Reshaping teacher and student roles in technology-enriched classrooms. *Mathematics Education Research Journal, 12*(3), 303–320.

Jones, P. W. (1998). Globalisation and internationalisation: Democratic prospects for world education. *Comparative Education, 34*(2), 143–155.

Kaptelinin, V. (1996). Computer-mediated activity: Functional organs in social and developmental contexts. In B. A. Nardi (Ed.), *Context and consciousness: Activity theory and human-computer interaction* (pp. 45–68). Cambridge, MA: The MIT Press.

Kent, P., Hoyles, C., Noss, R., & Guile, D. (2004). *Techno-mathematical literacies in workplace activity*. Paper presented at International Seminar on Learning and Technology at Work, Institute of London, March, 2004. Retrieved November 24, 2004, from the World Wide Web: http://www.ioe.ac.uk/tlrp/technomaths

Kuutti, K. (1996). Activity theory as a potential framework for human-computer *interaction* research. In B. A. Nardi (Ed.), *Context and consciousness: Activity theory and human-computer interaction* (pp. 17–44). Cambridge, MA: The MIT Press.

Matsushita, K. (1999). From monologic to dialogic learning: A case study of Japanese mathematics classrooms. In M. Hedegaard (Ed.), *Learning in classrooms: A cultural-historical approach* (pp. 211–225). Aarhus: Aarhus University Press.

Motaung, E., Henning, E., & van der Westhuizen, D. (2004). Investigating the role of ICT in the transformation of teaching practices: A rural South African perspective. In *Proceedings of the international society for cultural and activity research [ISCAR] regional conference, Wollongong* (pp. 185–198). [Available on CD-ROM]

Schapper, J. M., & Mayson, S. E. (2004). Internationalisation of curricula: An alternative to the Taylorisation of academic work. *Journal of Higher Education and Management, 26*(2), 189–205.

Schoorman, D. (2000). What really do we mean by 'internationalization?' *Contemporary Education, 71*(4), 5–11.

Taylor, J. A. (2001). Affective research and the mathematics curriculum for distance and online education. In M. J. Schmitt & K. Safford-Ramus (Eds.), *A conversation between researchers and practitioners. Proceedings of the seventh international conference of adults learning mathematics – a research forum* (pp. 50–53). Cambridge, MA: NCSALL, Harvard University, in association with ALM.

Trouche, L. (2004). Managing the complexity of human/machine interactions in computerized learning environments: Guiding students' command process through intrumental orchestrations. *International Journal of Computers for Mathematics Learning, 9*(3), 281–307.

Vulliamy, G. (2004). The impact of globalisation on qualitative research in comparative and international education. *Compare, 34*(3), 261–284.

Wake, G., & Williams, J. (2001). *Using college mathematics in understanding workplace practice: Summative report of research project funded by the Leverhulme trust*. Manchester: The University of Manchester.

Wotley, S. E. (2001). *Immigration and mathematics education over five decades: Responses of Australian mathematics educators to the ethnically diverse classroom*. Unpublished doctoral dissertation, Monash University, Victoria.

Wrathall, J. P., & Liu, J. (2004, November). *Issues associated with the provision of management training and education designed in the west for delivery in the people's republic of China*. Paper presented at the Crossing New Learning Frontiers Showcase, Monash University, Melbourne, Victoria, 16th November.

SECTION 3

PERSPECTIVES FROM DIFFERENT COUNTRIES

20

BALANCING GLOBALISATION AND LOCAL IDENTITY IN THE REFORM OF EDUCATION IN ROMANIA

Mihaela Singer
Institute for Educational Sciences, Bucharest, Romania

Abstract: This chapter reports on the mechanisms activated by the reform of education at the confluence of economic, political and sociocultural factors in a system in transition to democracy and a functional market economy. The paper analyses the fluctuating balance between ideological and professional positions in the decision making process based on a case study of the curriculum reform in Romania, which has been developed within the Education Reform Project co-financed by the Romanian government and the World Bank. The tensions among customs, traditions, mentalities, the will to integrate, and the desire to keep a cultural specificity in a global world, are not yet resolved. Such tensions bring new issues into the contemporary debates regarding globalisation and the new polarization of power. In this context, the reform in mathematics and science education is approached from the perspective of the relationship between knowledge and power during the industrial époque in Eastern Europe, stressing some of its implications for the post-industrial era and sketching some future developments in the framework of the knowledge society.

Keywords: Globalisation, Educational Reform, Curriculum, Competence

1. Introduction: The School in the Equation "Knowledge is Power"

The tension between national interests and global imperatives manifested as tension between adopting global norms and adhering to national practices is frequently addressed in education (e.g. Ahonen, 2002; Kosma & Polonyi, 2004; Slaus, Slaus-Kotovic & Morovic, 2004). The history of the last century shows that this tension involves frequent confrontations at the local level, with effects on what is taught and learned in school. In Eastern Europe, the expansion of communism in the second half of the 20th century is a good demonstration of how imported politics transform

B. Atweh et al. (eds.), Internationalisation and Globalisation in Mathematics and Science Education, 365–382.

the domains of the local social and cultural life. The balance between knowledge and power becomes therefore a key issue in understanding the contemporary educational reforms in this part of the world. From this perspective, the present chapter approaches the reform in mathematics and science education through the relationship between knowledge and power during the industrial époque, stressing some of its implications for the post-industrial era and sketching future developments in the framework of a knowledge society.

If "knowledge is power", as Francis Bacon stated four centuries ago, and if school is the main institution for transmission/reproduction of knowledge, and hence for power distribution in society, then the school holds a fundamental position in any society. Further, the extent of the development of a society should be measured not only by aggregate indicators of the economy and quality of life, but also by those indicators concerning education, i.e. concerning the way of organizing knowledge and learning, the types of expected outcomes, the quality of the educational delivery, and the ways of distributing education within the society (e.g. Singer, 1999; Vlasceanu, 2003).

The school as it exists today has its roots in the industrial époque. It seems that the idea of knowledge as power was specifically understood in the industrial era in the West as well as in the East. As more knowledge means more power, the industrial era developed mass schooling as a necessary condition of its emerging powers. The industrial worker needed basic literacy and numeracy to follow the instructions for simple automatic technological processes used in the arising factories. The need for a larger number of skilled workers implied a movement away from an *elite school* for few people to a *mass school* for all. This reform, marked by its efficiency and effectiveness, may be regarded as the highest success ever recorded by the education system as a whole.

The mass school of the industrial era tends to self-reproduce today – and this inertia seems to be a global feature. Thus, in terms of educational outcomes, many of the schools are still preparing graduates for an industrial economy, bearing an industrial way of thinking, in spite of the fact that the socioeconomic conditions have drastically changed. The inadequacy of the contemporary school to meet the current and future needs of its students is recorded by many international studies. For example, a UNICEF report found that countries like Canada, New Zealand, Great Britain and the United States have more than 42 per cent of children aged 14 or 15 "unable to apply basic math knowledge" and an adult illiteracy rate of 10 per cent (Garner, 2002).

The large mass of workers in factories is replaced nowadays by a large mass of people working in services within a global market. While the industrial worker required basic instruction to develop the ability to handle clear, specific driven tasks in a large driven mechanism, today's employee needs the ability to communicate, to think fluently, and to adapt him/herself to a variable environment, including coping with changing the profession many times throughout a lifetime. However, the schools have been ineffective in meeting these needs, as demonstrated by the fact that most of today's graduates show poor ability to transfer their skills from one job to another, and many are inadequately prepared even for employment in one specific job. Perhaps as a result of its failure, school learning is less valued by the society as a whole. While a "drill and practice" technique assured the success of the economy in the industrial era, this way of learning and approaching problems is largely inappropriate today. In addition, the various informational sources external to school (for example, through internet or TV) contribute to erode the school's credibility. The contemporary school has to face challenges such as: How to deal with the amount of information doubling every one or two years (Gardner, 2000)? How to deal with the rhythm of changes, exponentially increasing? What type of knowledge will be significantly useful in the next 20–30 years when the generations in schools today will be productive adults?

Questions about the role of schools and especially the role of mathematics, science and technology education arise from the perspective of the emerging knowledge society (Rubenson & Schuetze, 2000; Berryman, 2000). In the past 40 years, the workers in factories, in mines, in transportation, which were by the mid-1950s an actual majority of the working population in the UK, West Germany and Japan, and at least two-fifths in the US, have declined rapidly. By the end of the 20th century, in the U.S. they were down to a fraction of what they were before World War I. As Drucker (1994) emphasized, it is the *knowledge workers* who are the group which is fast becoming the centre of gravity of the working population.

What roles does the knowledge society expect of schools? From a holistic perspective, these roles can be characterized by two words: *dispersion* and *extension*. School as a knowledge institution needs to be more fluid (i.e. the borders between formal, informal and non-formal learning tend to be diffused) and more extensive (i.e. the duration of school learning extends beyond the ages it used to cover, giving rise to concepts such as lifelong learning). Coming back to Drucker's words, "access to the acquisition of knowledge will no longer be dependent on obtaining a prescribed education at any given age. Learning will become the tool of the individual available to him or her at any age if only because so much of skill and knowledge can be acquired by means of the new learning technologies" (Drucker, 1994, p. 4).

The proliferation of school reforms throughout the world is a strong evidence of the fact that the school as a knowledge-transmission institution is in crisis. The "new wave" of educational reforms that started in the 1990s shows that education throughout the world is confronting difficult problems and many systems manifest "a growing sense of urgency about the need for large-scale reform and more appreciation of the complexity of achieving it" (Fullan, 2000). Further, as the world is interconnected, changes within a small subsystem could have wide effects even on remote systems (Bar-Yam, 1997). In an interconnected world, the network propagation of effects influences the economic, political, cultural and social dimensions of each country. The international assessments and comparisons in education are an expression of this trend.

2. A Particular Context

Within this context of an interconnected world, I will focus on a particular case: the Romanian education system. From the perspective of a globalised knowledge society, the way of producing and sharing knowledge, the new mechanisms activated in the transitional processes need to be instanced in many particular cases in order to extrapolate significant experiences (e.g. Rubenson & Schuetze, 2000). This particular case study is of special interest for several reasons. First, the history of the communist regimes is an eloquent expression of the relationship between knowledge and power. Second, the industrial époque has had specific traits in the region, generated by the superposition of the economic endeavour with ideological and political constraints. Third, recent and rapid reform of education in post communist Romania, with its successes and failures, could illustrate significant lessons about the complexity of educational change in the new times.

Romania is a South Eastern European country with an area of 237 500 sq. km and a population of 22 million (estimated in 2000) with a rapid decline of births, which is only 63.5% of the 1989 level. Romania is a country of considerable potential: rich agricultural lands; diverse energy sources (coal, oil, natural gas, hydro, and nuclear); a substantial, though aging, industrial base encompassing almost the full range of manufacturing activities; an educated work force; and opportunities for expanded development in tourism on the Black Sea and in its high mountains. The economy of Romania is still based to a large extent on agriculture and industry. The GDP per capita was estimated in 1999 at $6,000 rising to $7,700 in 2004. The private sector is rapidly expanding, but its contribution to the GDP (61.5% in 1999) is larger than to employment. By contrast, agriculture's share in total employment has increased although its share in GDP has remained relatively unchanged (OECD, 1999; OECD, 2001). The composition of the GDP by sector in 2004 was the following: agriculture: 13.1%, industry: 33.7%, services: 53.2%. In 2004, the composition of the workforce by sector was: agriculture: 31.6%, industry: 30.7%, and services: 37.7%.

Since 1989, successive governments have sought to build a Western-style market economy. The pace of restructuring has been slow, but by 1994, the legal basis for a market economy was largely in place. Despite delays in privatizing certain companies, the overall balance of the economy has shifted decisively. By 2004, Romania's private sector employed over 72% of the country's total workforce. Unemployment was officially 6.2% of the active labour force at the end of December 2004 and 5.7% at the end of April 2005. In the 1990s, inflation was one of Romania's most serious economic problems. The retail price inflation, averaging 12.1% monthly in 1993 (the equivalent of 256% annually) declined to 28% annually in 1995 and to 7% in 2005. The post communist political evolution is mostly dominated by an economic oligarchy forced to adopt a liberal policy but still holding communist mentalities. This political evolution has directly affected the educational reform.

In order to better understand the current state of change in Romania, it is helpful to consider a short history of the effect of the communist years on education.

After World War II, the distribution of powers in the new emerging European socialist countries followed a quite similar pattern: to gain and keep power, the new leaders needed a new distribution of knowledge in society. The communist regime has required a new social force on which to rely; this new social force was the working class, which replaced the elite that had run the country until that moment. However, this class was mostly unschooled. To achieve the goal of changing the roles in the society – that is the power distribution – many of the intellectual elite of the '50s were imprisoned (or at least removed from official positions), while new "intellectuals" were created through short-term courses. At the same time, the new socialist economy needed large masses of workers and peasants possessing basic literacy and numeracy in order to develop an industry able not only to compete with the West, but also to sustain the socialist ideology. This generated in the '50s and the '60s massive endeavours in educational development. The accomplishment of universal education was one of the triumphs of communism. Romania was among the first countries in Europe to make education free and compulsory for all children. In Romania, as in other former communist countries of Central and Eastern Europe, access to education was traditionally high by international standards. Due to the political and economic priorities of the socialist state, adult literacy was generally universal; participation and completion rates for children and youths of both genders were high at all levels of education; and level repetition and dropout rates were low (OECD, 1999; 2001).

By the '70s, as a sign of its independence from the Soviet Union, Romania had paid off all its international debts and had imported the leading-edge technology of the moment for many of its local industries. Unfortunately, this independence quickly turned, under the totalitarian regime, into total isolation. Further, science and technology were supposed to be developed merely inside the borders, with no international exchange of ideas, problems, solutions, or paradigms. Consequently, research got rapidly stuck in this closed shop thinking. This influenced the quality of knowledge development throughout the education system. In particular, this led to a huge contradiction between the level of abstraction of the courses the students were taught (for example, all the graduates of the "industrial" high schools were supposed to learn the algebraic structures of group, ring, and field, and advanced differential and integral calculus) and the low level jobs in narrow specialized areas in which the graduates were supposed to work (lathe operator or caster, for instance).

The few students able to cope with the highly competitive university entrance have had to follow a quite similar pattern: gain a highly abstract level of theoretical knowledge, graduate a narrow specialization, and get a compulsory job for at least three years in the same place, decided upon graduation, at national level. The stage of three years was compulsory even for the most advanced students, who might have been ready to do research in mathematics or science. They were obliged to teach in a school, usually a peripheral one or far away from their living place, before moving into the research area. In this way, the teaching profession became a second-rate job, as students were trained to become researchers and might have entered teaching without a proper preparation for it.

In 1989, at the time of transition, 90% of the youth were still trained in the so called "industrial high schools" in which they were supposed to learn mathematics, physics and chemistry at a very high theoretical level, while they received a narrow qualification in a typically low level industrial job. The inertia of the system was so high that these schools continued their training almost a decade after the fall of communism, even if the new social and economic conditions have claimed for people trained for the post industrial and information society.

3. General Reforms to Specific Contexts

Specific cases need specific solutions. This truism is, however, forgotten by the policies and reforms supported by many transnational bodies, such as the World Bank. As a factor of spreading global economic trends, the World Bank promotes universal strategies of development. After the fall of communism in Central and Eastern parts of Europe, and in Central Asia – Europe and Central Asia (ECA region in the World Bank terms) – the World Bank extended to the education sector its practice of giving loans to the new emerging democratic states to assist in the implementation of particular reforms. As a global economic entity, the World Bank has targeted to build a global education market (World Bank, 1996; 1994–2001 (2001)). Its goals for education in ECA countries can be summarized as follows (Berryman, 2000):

- Realign education systems with market economies and open societies.
- Combat poverty by increasing educational fairness.
- Finance for sustainability, quality, and fairness.
- Spend resources more efficiently.
- Reinvent the sector's governance, management, and accountability.

The Reform of Education in post-communist Romania started in the framework of the Project RO-3724, co-financed by the Romanian Government and the World Bank. It was contracted after four exploratory and debating years (1991–1994), and it was one of the three largest projects (Serrant & McClure, 2000) of this kind in Eastern and Central Europe in terms of its scope and budget (approximately 100 million USD).

After more than a decade since the project started, some lessons could be drawn from its achievements and failures. A starting question to address from the perspective of the local- global interaction is if this project was necessary for the country. Developed in isolation, the Romanian post-war education system lived within the myth of absolute performance: the Romanian children are often the winners of the international Olympiads in Mathematics and Science and the children of Romanian emigrants usually perform well everywhere in their adopted countries throughout the world. For many, this raised the question: why change anything? The trust in the superiority of the system was – and still is – a major blockage in explaining to the main actors involved in education why (and whether) structural and conceptual changes were necessary. The reform appeared to the general public as a political demand, and not connected to education. A popular feeling was that, after years of isolation, Romania should go for a western integration (European community, NATO) and the educational reform is just a pre-condition of this. Within the process of globalisation, the reform of education in Romania did have an assumed dimension of integration into the occidental world, from which it had been abruptly excluded after World War II. At the same time, however, it was not understood how deep the changes should be in a society in which mentalities were still influenced by the communist ideology and the infrastructure was linked to an over-centralised system.

At the beginning of the '90s, all levels of management were in the hands of people with strong political backgrounds (as former members of the communist party in superior hierarchical positions) but frequently with low levels of professional expertise. If, for the general public, the question was "Why should we change?" for the professionals in education the question was "Could the necessary changes be introduced and managed within the local human resources distribution and management?" Being encapsulated in political and ideological constraints, the education system was unable to self-develop and self-change. A retrospective view shows that an intervention from outside the system was needed to get freedom to the internal forces and to initiate a redistribution of powers and social roles. In this respect, the World Bank project was a necessity.

Further, the reform of education was conceived on a large scale: all the components of the education system (management, finance, curriculum, assessment, teacher training) were simultaneously targeted. The discussion on whether this kind of extended approach is an efficient one is still subject to numerous debates – even at the World Bank level (e.g. World Bank, 2001 and, from a different view, Ilon, 2002). Concerning the working methodology, after setting the terms of reference for each component together with the local counterparts, the World Bank team has visited Romania twice a year. During these appraisal missions, the World Bank team has discussed the outcomes of the project during the past period and has made recommendations.

Among the components of the reform project, curriculum development recorded the most visible results. This visibility was given by the development of the National Curriculum, which was published between 1998 and 2001. It has 27 volumes totalling over 3000 pages, which contains the curriculum framework plans and the subject curricula. In addition, to facilitate teachers' access to the new philosophy of education, teachers' guides for implementing the new curriculum were developed. It is about 31 methodological teachers' guides dealing with various subject matters and students' ages, totalling 3100 pages. About 1,000 teachers were involved in developing these materials, as members of the working groups in charge with curriculum development.

In general, the model being proposed in Romania retains a centralized system. A dilemma to be solved was "How can a centrally developed curriculum be reconciled to the idea of free enterprise?" A possible model of compromise was identified: central development and distribution of subject curricula and a free market for producing textbooks and supplementary materials.

The technical assistance for math and science education during the project was not a straightforward one. An international consultant appointed by the World Bank came to look at the system before the project started. He visited various schools and institutions connected with education and delivered a report containing a description of the mathematics and science taught in schools and a list of recommendations. Taking into account that this report was never officially translated and widely distributed, and hence it was not accessible to many of the individuals involved in the reform process, we can consider so far that the influence of World Bank experts in this area was very limited.

In general, the recommendations made periodically by the World Bank team were taken into account merely regarding quantitative financial aspects; qualitative aspects followed the local powers distribution. For example, instead of generating co-operation among the coordinators of the reform components, the steering committee at the ministry level remained a formal entity not functioning as a working team. Each member has reported directly to the secretary of state about the activities undertaken by his/her component with no discussion or debate by other members. Thus, the efforts in curriculum development, teacher training, and assessment did not align with each other, with severe consequences to the project's sustainability. Ten years later, we might see that the reform failed in its most important goals because the *tacit aspects* of a post communist culture were neglected. These refer especially to the real-time management, which used to be in the recent past only hierarchical, not collaborative, and to the communication among institutions, which used to follow only a hierarchical path (vertically, not horizontally). Both management and communication were also affected by a hidden disparity between thought and action, between intention and report, habituated in the old hierarchical pyramid. Beyond the financial arrangements, the education reform was designed with an implicit presupposition that a post-communist system will be able to behave like a learning organization. If this were the case, there would be no need for external consultation; the internal resources would be able to find the critical issues and to act appropriately.

Lacking a coherent program agreed to by all political forces and subsequently implemented, the reform evolved chaotically, even though it was acclaimed and celebrated by the successive administrations of the Ministry of Education (an institution that changed its name and main targets almost every time a new minister was appointed, and there were eleven ministers in fifteen years). To be effective, the World Bank project should have considered the customs, values, beliefs, some material aspects of culture, and various kinds of tacit knowledge pertaining to the local intangible culture. Such factors make each context of reform unique. These differences are very visible today in Eastern Europe, where observers of the educational decision-making processes find a broad range of policies and practices, although these countries apparently started at the beginning of the '90s from similar economic and political backgrounds (e.g. Kitschelt, 2001).

4. A Better Understanding of the Local Context

A widespread concept throughout the region is that national education is demonstrably of high quality as long as teams of well-coached students win international academic competitions. Thus, since 1959 – the first year of the International Mathematical Olympiad – the Romanian team won for 5 times the first prize, being ranked 27 times among the first five. The Romanian teams obtained similar results in international competitions for physics, chemistry, biology, and informatics. In contrast, out of the 41 countries participating in TIMSS in 1995 (Vári, 1997; Martin et al., 1999), Romania was the 34th in Mathematics and 31st in Science. A slight progress was recorded in the next two international assessments in mathematics and science, in 1999 and 2003. Clearly, while the Romanian system continued to do well for its high-ability pupils, the performance of children in the system as a whole was less than satisfactory.

A short examination of some of the TIMSS results (Table 1) might give a clearer image of this phenomenon. The following two items selected from TIMSS 1995 show the discrepancy in students' abilities.

Item R8. The graph shows the distance travelled before coming to a stop after the brakes are applied for a typical car travelling at different speeds. A car is travelling 80 km per hour. About how far will the car travel after the brakes are applied?

A. 60 m
B. 70 m
C. 85 m
D. 100 m.

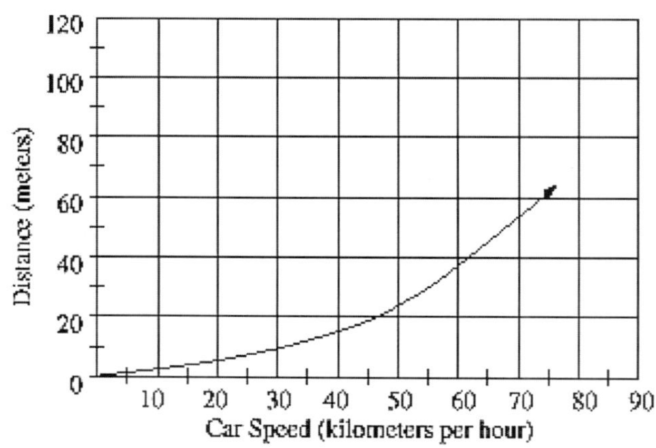

Item J11. A quadrilateral MUST be a parallelogram if it has:

A. one pair of adjacent sides equal
B. one pair of parallel sides
C. a diagonal as axis of symmetry
D. two adjacent congruent angles
E. two pairs of parallel sides.

The item R8 involves interpretation of a graphic representation and data analysis in order to solve a problem set in a practical context. Only 28.5% of the Romanian students answered this item correctly, compared with the international average of 49%. Moreover, the 51.4% concentration of the answers on the distracting factor A suggests that many students selected the most familiar solution in the problems involving speeds, 60, without other particular reasoning. By contrast, the Romanian average on item J11 (that involves the use of theoretical properties in recognizing a parallelogram) was well above the international average. This type of question is more in line with the theoretical approach typically used to prepare Romanian teachers and the way which they, in their turn, follow to prepare their students.

This atypical concentration can be explained by the traditional way of content development. In the '70s and '80s, the Romanian curricula for elementary and secondary schools contained lists of concepts to be acquired, organized by chapters in a fixed succession, unequivocally associated to a temporal distribution. As a rule, these lists were actually the contents of the single textbook for each subject and grade. The textbooks, especially the ones for mathematics and sciences, looked like collections of theoretical facts, being used by the teachers as a basis for step-by-step preparation of discursive lessons and by the students to do their tedious homework, consisting in solving exercises and problems from the books. As only the teacher had the knowledge to access the textbook – distributed free of charge to the students – he/she was maintaining an unsurpassable ascendancy over the learners.

Table 1. Comparative results for two TIMSS items

Item	Content Category	Performance expectation	International average - Percent of students responding correctly	Percent of Romanian students responding correctly	International Difficulty
R8	Data representation & probability	Solving Problems	49%	28.5%	565
J11	Geometry	Knowing	49%	67%	573

Source: Singer & Voica, 2004.

The relatively poor performance of Romania in the TIMSS for 13-year olds in 1995 raised questions about the myth of the "Olympiads". However, these questions remained mezzo-voice and TIMSS (1995) and TIMSS-R (1999) reports were completely ignored by the decision-makers. Moreover, the reports remained unknown to the public until 2001, when they were published under a neutral title, *Learning mathematics and sciences*, within a series of Methodological teachers' guides; the fact that it was a TIMSS study appeared only as a discreet subtitle. Even published, they continued to be ignored at the decision making level.

Regarding the content, while the mathematics curriculum has demonstrated the characteristics that has made it part of a global curriculum (Atweh & Clarkson, 2001), the science curriculum was far behind. In the last decade of the 20th century, science taught in Romania was similar to that taught in the '50s in Western Europe and the U.S. Physics, chemistry, biology were descriptive and focused on an analytic presentation of disconnected topics. For example, the focus in biology was on the tissue and organ level. This is where research and development in the life sciences were at the turn of the 19th century and which prevailed in the school curricula of the U.S. into the early 1960s, when modern textbooks, which reflected current research and development efforts in biology, emerged (Dawson, 1992).

This analytical approach continues to dominate even today the teaching of sciences in Romania. Changing a paradigm within a scientific community is a complex process. "The process of paradigm change is closely tied to the nature of perceptual (conceptual) change in an individual. Novelty emerges only with difficulty, manifested by resistance, against a background provided by expectation" (Kuhn, 1996, p. 64). It becomes even more complicated when the scientific beliefs are influenced by political ideology. The history of developing a new curriculum in Romania offers many examples for the social understanding of Francis Bacon's equation as it was stated in the first section. Thus, in search for a student centred way to teach biology, the working group (WG) in charge of developing a new curriculum proposed an integrated model focused on biological processes and principles. After many critical debates, pressure groups composed of people from the Academy of Science, universities, and trade unions, which were afraid to lose the positions of "their" disciplines in the school curriculum, succeeded in their efforts to constrain the WG to return to the old paradigm of treating botany, zoology, anatomy, and ecology as separate, non-connected, subject matters. A plethora of actions were used to achieve this final decision, from contesting WG members' competencies to threatening and dismissing key persons. The result was that a more "collaborative" WG continued a "half tone" reform.

The same fate faced the trial to introduce a discipline called "Science" in primary and lower secondary education. The main reason for promoting this idea was that the child lives in an integrated world and therefore an integrated presentation is the most appropriate way to introduce young students to science. This trial also failed under the pressure of centres of power who decided to keep the discontinuity of the traditional model, in which biology starts in the 5th grade by botany, physics in the 6th grade and chemistry in the 7th grade, as distinct disciplines.

These two examples of rejecting an integrated approach in teaching science in primary and lower secondary grades are meant to show the strength of conservative positions that might neglect universal achievements of the educational community. Moreover, only in an open context, a system can take knowledgeable decisions concerning what does fit with its needs, and becomes able to oppose the global trends that are not suitable. Otherwise, arbitrary interferences, eventually in conjunction with inappropriate local management, can make the loss bigger when striving to preserve the local specificity.

5. Balancing Opposite Tendencies

The last decade of the 20th century was marked by deep-rooted changes in almost all European education systems. In search of a better balance of power in curriculum decision-making, countries with a centralised education system (as in France and most Eastern countries) have tried to decentralise, while countries having a decentralised education system (as in Great Britain and the Nordic countries) have tried to centralise. These trends are reflected in most of the conceptual curriculum frameworks developed in the 90s (e.g. *Department of Education,* 1989; 1995; *Ministry of Education,* 1993a; *Ministry of Education,* 1993b; *National Board of Education,* 1994; *Ministry of Education and science,* 1998; *OECD,* 1999; *Department of Education and science,* 2001; *Department for Education and skills,* 2002). The fluctuating balance between centralising and decentralising within the education reform is interesting in its complexity and some issues need to be further analysed in relation to the targets of this essay.

In the first half of the '90s, the Romanian education system was still reflecting the values and characteristics of a culture dominated by economic industrialism and sociopolitical authoritarianism. Thus, the curriculum was based on centralized administrative and bureaucratic political thought. In particular, the driving documents – the instructional plans containing the compulsory subjects and the number of hours per week for each subject – were developed from the perspective of academic disciplines and teachers' workloads, rather than from the perspective of the students and their formative needs. The instructional plans, which remained unchanged until 1998 as regulatory documents, did not allow any variation in lesson numbers from one school to another, nor flexibility to adapt teaching to the specificity of certain communities or to students' needs. The instructional plans were drafted in a centralized manner, so that all Romanian schools of the same level had to go through the same syllabus, with timetabling being the only exception which was established by each school.

At first sight, the new National Curriculum for England and Wales, implemented in a directive manner starting from 1989 (Department of Education, 1989), could be perceived as similar to this description. However, there are some important differences in focus and purpose between the two curricula. While the English curriculum was objective-centred, being driven by attainment targets, the old Romanian curriculum was content-based and content-centred. When a curriculum is focused on objectives, it has a student-oriented dimension. But when it is focused on contents, it is implicitly teacher-oriented. An important aspect of the science and mathematics curriculum which illustrates this difference is problem solving. In a student-centred approach, a problem in science and mathematics education is understood as a kind of a unique puzzle about the natural world, which needs to be solved. In contrast, usually in Romania a problem solving activity has been a problem requiring the application of a memorized rule (or of a complex succession of theoretical rules). Western problem solving in modern curricula would begin with a question such as "What would happen if . . .?" or "Why are these data different from the data collected from the other site?" (Dawson, 1992). In Romania, an example of a problem might be "If the solvent moved at 0.8 mm per second, how long would it take to get to point X?" While in the first example it is about exploration and multiple pathways to get to the solution, the second example involves only the application of a known algorithm to get the single possible answer.

Within the new curriculum implemented in 1998 in Romania, more emphasis was put on approaching problem solving through exploration and investigation. Within this paradigm, mathematics might be seen as a dynamic discipline, closely related to the society by its relevance in everyday life, in natural sciences, in technology and social sciences (Singer, 2003, 2007). The overall idea behind the new mathematics and science curriculum was to produce some major shifts:

- regarding **the content**: from a theoretical approach of scientific facts to a variety of contexts that generate scientific thinking;
- regarding to what is **expected from students**: from merely applying algorithms to using problem solving strategies in various problem solving situations;
- regarding **learning**: from memorization and repetition to exploration and investigation;
- regarding **teacher's role**: from an information provider to an organizer of a wide variety of learning activities for all children, adapted to individual levels of attainment and rhythms of development;
- regarding **assessment**: from the rigidity of marks meant to label students to self-assessment and progress assessment as parts of learning.

These changes demanded some major shifts in the way teachers think about their classroom activities. In a society trying to redefine itself, there was a need to emphasize the values and attitudes to be promoted through the new curriculum. The values and attitudes listed in the curriculum were supposed to help textbook designers and teachers to focus not only on acquiring knowledge, but equally, on developing specific personal qualities in students. To be effective, they were formulated for each subject matter. As an example, the following values and attitudes were to be developed within Mathematics:

- Demonstrate curiosity and imagination in creating and solving problems
- Manifest tenacity, perseverance and the capacity of focusing on a specific problem

- Manifest a spirit of objectivity and impartiality
- Manifest independence in thought and action
- Demonstrate initiative and interest in approaching various tasks
- Manifest confidence in using technology
- Develop an aesthetic and critical sense, appreciation of rigor, order and elegance in the architecture of problem solving or theory building
- Develop the habit of making use of mathematical concepts and methods in approaching everyday situations or in solving practical matters
- Develop motivation for studying mathematics, as a relevant field for the social and professional life.

The development of values and attitudes was considered important because knowledge not accompanied by ethics and sensibility, with positive effects on personal life, leads to personal failure and to degradation of the social life, especially in countries in transition, searching for their identity (e.g. Duthilleul, 2001; Kosma & Polonyi, 2004).

The new philosophy of education was presented in detail in more than 6000 pages that have been published between 1998 and 2001. The real implementation of these changes, however, was less successful. What was missing was the knowledge on how to disseminate this new philosophy to the actors involved in education. Under the effect of the totalitarian mentalities, accepting change would involve a movement from one political ideology to another. The justification of the rationale was not clearly explained to convince teachers. It seems that freeing conceptual domains from politics is a difficult process, requiring time, analyses and subtle influences, carried out within a coherent projection of changing mentalities.

Unfortunately, the termination of the project co-financed by the Romanian government and the World Bank coincided with a political change: after a liberal oriented government, the new elections were won at the end of 2000 by the social-democrats, with many former communist leaders regaining the decision-making positions. Thus, the evolution of the educational reform has continued to interfere with the political level: the new leaders in the Ministry of Education took the stand to change most of the decisions taken during the reform project. As the new philosophy has not yet pervaded profoundly the system, it was relatively easy to cancel most of the reform achievements.

Returning to the comparison between the UK and the Romanian experience in curriculum development, some other aspects can be paralleled. The new curriculum for England and Wales was developed at a central level by teams of high professionals and implemented as compulsory regulations for all schools starting with 1989. While having a strong international impact, the feedback from "the bottom" determined successive revisions resulting in a curriculum that is more suitable to the habits and traditions of the schools, in order to best meet the needs of their pupils and local community (e.g. Department of Education, 1995; Department for Education and Skills, 2002). The Romanian new curriculum, appreciated in many reports as very conceptually advanced (e.g. OECD 2001) was successively revised after the implementation. If the English curriculum was revised in order to give more freedom at the local level (school and district), the Romanian curriculum was revised exactly in the opposite direction. Large parts of the school based curriculum initially recommended in the regulatory documents were replaced by a centrally directed curriculum. It seems that in both cases, and in different ways, the ensemble of the education system was not enough prepared for the top-down suggested reform and went back to the previous habituated models. This illustrates that curriculum reform, whether stemming from national or global initiatives cannot be free from local constraints and risks, which sometimes can be anticipated, but hardly can be prevented.

6. Some Conclusions

Both the large scale aspects of change and the individual subjective perception of change have to be monitored for understanding the difficulties and pitfalls likely to occur. Otherwise, and especially in a system in transition, the change remains declared, possibly accepted, but not accomplished. Being in its essence a conservative one, the education system preserves its old customs, values and attitudes, which generate a wider anti-reform reaction, usually passive, but even more efficient. Formally, science teachers teach today a new curriculum in Romania, but what happens in the classroom tells the same old story of a teacher-centred approach.

In a period of deep structural reforms, a centralised education system has the advantage of a relatively fast propagation of change from above. On the other hand, when the policy is designed in a top-down manner, the imbalance may be compensated for by autonomous professional bodies that can provide "from below", in real-time, a feedback on the reform movement. If the global tendencies are understood, it is possible to use them as a lever to avoid the powerful effects of arbitrary decisions. A competitive market for teacher training programs; schools that undertake specific missions, related to their context and local populations; local programs focused on preserving and valuing some aspects of traditional intangible culture – all these are important factors in balancing the central power with an array of social forces animated by the desire of preserving and integrating the traditions within coherent programs for better education. It is about reproducing at the national level the globalisation tendencies from above and below (Atweh & Clarkson, 2002). The Romanian experience of a decade of educational reform shows that in order to cope with arbitrary top-down interventions, a necessary condition (but not a sufficient one) is therefore the existence of professional competitive bodies and of social decentralisation mechanisms able to provide an ethos for transparency, communication and moral values.

In its role of knowledge transmission (knowledge accumulated over centuries by the previous generations), the school is meant to be conservative. However, this role is in conflict with the global knowledge society and the induced accelerated rhythm of change. In addition, within the context of globalisation, school has the mission to preserve the national values. That means that school is supposed to become even more conservative, opposing the internationalisation tendencies. Another inconvenience is the fact that, because its outcomes are visible after many years, the process of transforming the school is – and has to be – a long-term projection. However, to project on a long term is quite impossible when the initial conditions (new generations in school, new aspects of the social-cultural life) are significantly changing during the reform process. Nevertheless, the school has to prepare the generations better for a changing future, beyond all these conflicting contexts. The large scale educational reforms of today are answers to this challenge but, in order to make these answers more suitable, the fragile balance between knowledge and power has to be considered, with all its implications.

The lack of sustainability of the Romanian education reform after the termination of the project constrains many of the efforts of the actors involved and dissipates the achievements. What is most needed now is training the teachers for promoting and sustaining the spirit and the general lines of the reform started more than a decade ago. Although the reform was meant to accelerate the pace of change, recent history shows that shortcuts are not always a successful solution. Between national and global, mathematics and science education in Romania tries to find its own way to progress. Beyond this, a lesson to be learned is that the international projects should carefully look into the traditions and local context, and contribute to create *local* viable professional structures able to implement change within the local ethos. These professional structures need to be independent from the political ones, in order to build their power based on their own knowledge. Otherwise, looking at a historical scale, the result of a huge financial and human investment remains just a spike in an unpredictable non-linear evolution.

References

Ahonen, S. (2002). From an industrial to a post-industrial society: Changing conceptions of equality in education. *Educational review, 54*(2), 173–181.

Atweh, B., & Clarkson, P. (2001). Internationalisation and globalisation of mathematics education: Towards an agenda for research/action. In B. Atweh, H. Forgasz, & B. Nebres (Eds.), *Sociocultural research on mathematics education: An international perspective* (pp. 167–184). New York: Erlbaum.

Atweh, B., & Clarkson, P. (2002). *Globalisation and mathematics education: From above and below.* Paper presented at annual conference of the Australian Association of Research in Education, Brisbane, available at http://www.aare.edu.au/02pap/cla02360.htm

Bar-Yam, Y. (1997). *Dynamics of complex systems.* Addison Wesley Longman, Inc.

Berryman, S. E. (2000). Hidden Challenges to Education Systems in Transitional Economies. Education Sector Strategy Paper for ECA Region. Washington, DC: World Bank.

Dawson, G. (1992). *Report on mathematics and science teaching in Romania.* World Bank Working Document.

Department for Education and Skills. (2002). *Extending opportunities, raising standards* [Green. Cm.3390]. London: Stationery Office, Department for Education and Skills [DfES].

Department of Education. (1989). *The national curriculum of England and Wales.* London: Department of Education.

Department of Education. (1995). *The national curriculum of England and Wales.* London: Department of Education.

Department of Education and Science. (2001). *The new national curriculum framework. Preschool, primary and secondary education* (Discussion white paper). Pristina, UNMIK: Department of Education and Science.

Drucker P. F. (1994). *Knowledge work and knowledge society. The social transformations of this century* The 1994 Edwin L. Godkin Lecture, available at www.ksg.harvard.edu/ifactory/ksgpress/www/ksg_news/transcripts/drucklec.htm

Duthilleul, Y. (2001). *Education and the learning economy.* Background paper for Romania Country Economic Memorandum, ECSHD.

Fullan, M. (2000). The return of Large-scale Reform in *Journal of Educational Change, 1*:5–28, Kluwer: Academic Publishers.

Gardner, H. (2000). *The disciplined mind.* New York: Simon & Schuster.

Garner, R. (2002). *Four in ten British pupils cannot do sums, says UNICEF.* Independent.co.uk, Nov. 26 available at http://education.independent.co.uk/news/story.jsp?story=355923

Ilon, L. (2002). Agent of global markets or agent of the poor? The world bank education sector strategy paper. *International Journal of Educational Development, 22*(5): 475–482.

Kitschelt, H. P. (2001). Accounting for post-communist regime diversity: What counts as a good cause? In R. Markowski, (Ed.), *Transformative Politics in Central Europe.* Warsaw: IP-PAN Publishers.

Kosma, T., & Polonyi, T. (2004). Understanding education in Europe-East. Frames of interpretation and comparison. *International Journal of Educational Development, 24*, 467–477.

Kuhn, T. S. (1996). *The structure of scientific revolutions.* Chicago: University of Chicago Press.

Martin, M. O., Mullis, I. V. S., Gonzales, E. J., Kelly, D. L., & Smith, T. A. (1999). *School contexts for learning and instruction: IEA's third international mathematics and science study center.* Boston: College.

Ministry of Education. (1993a). *National curriculum.* Budapest: Ministry of Education.

Ministry of Education. (1993b). *The New Zealand curriculum framework.* Australia, Wellington: Ministry of Education.

Ministry of Education and Science. (1998). *Curriculum for the compulsory school, the pre-school class and the after school centre.* Stockholm: Ministry of Education and Science in Sweden, Regeringskansliet.

National Board of Education. (1994). *Framework curriculum for the comprehensive school.* Helsinki: National Board of Education.

OECD. (1999). *Review of national policies for education. Romania.* Draft Examiners report.

OECD Centre for Co-operation with the Economies in Transition. (2001). *Education at a Glance: OECD indicators.* Paris: OECD.

Rubenson, K., & Schuetze, H. G. (Eds). (2000). *Transition to the knowledge society. Policies and strategies for individual participation and learning.* Institute for European Studies, University of British Columbia.

Serrant, T. D., & McClure, M. W. (2000). Secondary education reform: Policy briefing paper: The World Bank [On-line]. Available: http://www.seryp.org/review/wb.html

Singer, F. M. (coord.). (1999). *The new national curriculum* (a synthesis in English). Bucharest: Prognosis Publishing House.

Singer, F. M. (2003). From cognitive science to school practice: building the bridge. In N. A. Pateman, B. J. Dougherty, & J. Zilliox (Eds.), *27th PME*, CRDG. 4, pp. 207–214.

Singer, F. M. (2007). Modelling both complexity and abstraction: a paradox? In W. Blum, P. Galbraith, H. W. Henn, & N. Mogens (Eds.), *Applications and modelling in mathematics education*, New York: Springer.

Singer, F. M., & Voica, C. (2004). *Challenging the future: Mathematics education in Romania between ideals and reality.* Baia Mare: Cub, ICME-10.

Slaus, I., Slaus-Kotovic, A., & Morovic, J. (2004). Education in countries in transition facing globalization – a case study Croatia, in *International journal of educational development, 24*, 479–494.

Vári, P. (ed.). (1997). Are we similar in maths and science? A Study of Grade 8 in Nine Central and Eastern European Countries [incl. Romania]. Budapest, IEA and TIMSS.

Vlasceanu, L. (coord.). (2003). *The school at crossroads – Change and continuity in the curriculum for compulsory education. The impact study concerning the implementation of the national curriculum.* Bucharest: Polirom P.H.

World Bank. (1994–2001). *Staff appraisal report: Romania education reform project.* Washington: The World Bank, 1994 (and subsequent mission reports, 1994–2001).

World Bank. (1996). *World development report 1996: From plan to market.* Washington, DC: The World Bank.

21

VOICES FROM THE SOUTH: DIALOGICAL RELATIONSHIPS AND COLLABORATION IN MATHEMATICS EDUCATION

Mónica E. Villarreal[1], **Marcelo C. Borba**[2] **and Cristina B. Esteley**[3]
[1] *Universidad Nacional de Córdoba-CONICET, Argentina, mvilla@famaf.unc.edu.ar*
[2] *Universidade Estadual Paulista, Rio Claro, Brazil*
[3] *Universidad Nacional de Villa María, Argentina*

Abstract: This chapter presents a collaborative experience between two neighbouring countries from South America: Argentina and Brazil. Our purpose is to share a model of international collaboration that we consider to be an alternative to the classical movement of early mathematical and scientific knowledge between East and West and between North and South. We start our chapter with a general discussion about the phenomenon of globalization considering some local examples. We characterize our collaboration exploring the tensions and difficulties we faced along our own professional development at the local as well as the international level. We describe the development of our prior collaborative work that established the foundation for our international collaboration portraying the local mathematics education communities. We refer to some balances that were created among our relationships, the expansion of our collaborative network, and how this particular collaboration allows us to contribute to the regional field and inform the international one. We discuss the way that the search for balance and symmetry, or at least a complementary asymmetry in our collaborative relationships, has led us to generate a genuine and equitable collaboration

Keywords: Globalization, collaboration, mathematics education, North – south, South – south

1. Introduction

Globalization refers to a set of phenomena associated with technological, economic, political and cultural exchanges. It is also a term used to describe how human beings are becoming increasingly intertwined with each other around the world economically, politically, and culturally. Although globalization can be analyzed from these various perspectives, we find that most of the discussion is associated

B. Atweh et al. (eds.), Internationalisation and Globalisation in Mathematics and Science Education, 383–402.

with the discourse of democracy, in which democracy will be achieved through economic growth and opportunity for all. For example, when the main proponents of “free trade” push the globalization agenda, they apparently frame their discussion within this notion of democracy, arguing against any restrictions on commerce. They equate freedom with democracy, as they claim that everyone should have freedom to trade, and claim that a democratic world is one with “free trade”. In this line of thinking, as proposed by many voices in the State Department of the United States (U.S.), globalization would bring freedom and development to the world, but in effect they are obscuring an underlying agenda of opening up new markets for U.S. products. While the U.S. State Department works hard to promote the democratic agenda, their discourse rapidly changes the moment U.S. farm interests become threatened by more efficient agricultural production in other countries such as Brazil, for example. This situation illustrates the ambiguity of the discourse of democracy as it relates to globalization, especially when other less economically powerful countries clamour to open up U.S. markets to their own products. What is claimed to be democratic is actually a form of neocolonialism in which trade is far from free, to say nothing of fair. The barriers the developed countries want to bring down are very selective, and of course they want more and more barriers to stop the free movement of people: more visa requirements, new walls along the border between Mexico and the U.S., new restrictions in French immigration laws, etc.

This ambiguity can also be seen in the economic policies adopted by some Latin American countries during the nineties which were influenced by the Washington Consensus.[1] Employing the discourse of democracy, the Washington Consensus promoted economic reforms designed to make local economies similar to those of the First World countries. The result, however, was a disbandment of the state, consolidation of oligopolistic economies, and decreased spending on health, education and social programs in order to pay the international debt (loans from the IMF and World Bank). The discourse of democracy associated with globalization led the former Argentinean president, Menem (1989–1999), to promote the illusion of being part of the First World, as shown by the following excerpt from a speech given in 1996 during a ceremony on the first day of class in a school in an impoverished village in northern Argentina (Tartagal – Salta): *“Shortly, a system of space flights will be tendered for a contract, and from a platform, that will possibly be installed in Córdoba, those spacecrafts will leave the atmosphere, will rise above the stratosphere, and from there they will choose the place where they want to go, in such a way that, in an hour and a half, we will be able to be in Japan, Korea or in any place in the world...”*.[2] This is just an example of the ridiculous kind

[1] The Washington Consensus is a set of policies promulgated by many neoliberal economists as a formula for promoting economic growth in many parts of Latin America and other parts of the world (http://es.wikipedia.org/wiki/Consenso_de_Washington).

[2] *“Dentro de poco tiempo se va a licitar un sistema de vuelos espaciales mediante el cual desde una plataforma, que quizá se instale en Córdoba, esas naves van a salir de la atmósfera, se van a remontar a la estratosfera, y desde ahí elegirán el lugar donde quieran ir, de tal forma que en una hora y media*

of globalization that the Argentinean political class proclaimed in the nineties in a country where benefits for the few led to poverty rates that almost exceeded 50% of the population in 1989, or where the government provided computers to schools where electricity was not yet available. During the Menem government, Argentina implemented most of the Washington Consensus policies which are today the target of sharp criticism. The result was the Argentinean economic crisis of 1999–2002, which is often cited by critics as the case in point as to why the Washington Consensus policies are flawed.

In the global relationship between nations, when some Third World countries receive international financial support from First World countries to develop their scientific or technological fields, the criteria and decisions for distributing those resources are usually determined by those external funding agencies. For example, the implementation in 1994 of a local program of economic incentives for Argentinean teacher-researchers from public universities, known as the *Program of Incentive for Teachers-Researchers,*[3] was based on neoliberal educational policies promoted by the World Bank. This program benefits only those academics who are classified as "researchers" and who are participating in actual research projects. It is worth noting that the extra funding was not paid as financial support for the research project, but rather as an increment to the researcher's salary.[4] In part due to the precarious salaries at universities,[5] a multitude of educational research projects were submitted. Teachers who had never done research became researchers. From 1994 until 2001, the total number of teacher-researchers included in the Program of Incentives grew almost 75%.[6] In the mid-1990s, a kind of over-production of publications and projects related to mathematics education could be observed in Argentina. While many of these projects were classified by their authors as research, a study conducted by Villarreal and Esteley (2002) suggested that only 8.4% of the presentations in local mathematics education conferences held annually (Reunión de Educación Matemática – REM[7]) could be characterized as research. Clearly, this program of incentives did not benefit research activity in the field, and it represents an example of the patronizing facet of global neoliberal policies. The effects of the program were ambiguous. On the one hand, increases in teachers' salaries can be seen as positive. On the other hand, this increase in salary was not accompanied by

podremos estar en Japón, Corea o en cualquier parte del mundo…" (http://es.wikiquote.org/wiki/Carlos_Menem).

[3] Programa de Incentivos a Docentes-Investigadores: this programme was created through decree number 2427, dated 11/19/1993, of the Secretary of University Politics of the Ministry of Culture and Education, during the government of President Menem.

[4] Such increment depends on the researcher category and the time dedicated to research and represents approximately 28% of the annual salary.

[5] As of march 2007, the current monthly net salaries range from US$88 (or €66) up to US$1377 (or €1025).

[6] Source: Argentinean Ministry of Education, Science and Technology – Secretary of University Politics – Programme of Incentives to Teachers-Researchers (MECyT – SPU – Programa de Incentivos a Docentes-Investigadores).

[7] This annual meeting is organized by the Argentinean Mathematics Union (Unión Matemática Argentina) since 1977.

the worker benefits guaranteed by the constitution, nor did it represent a permanent increase in salary, or ease the effect of low government spending on education. Moreover, it fostered confusion regarding the concept of research in an emergent research community.

Critiques of the above ambiguous, illusory and patronizing aspects of globalization have been developed by many authors who propose "fair trade instead of free trade". This slogan attempts to give new meaning to globalization, suggesting that doing away with boundaries may have its positive side. The veiled neocolonialist discourse of globalization – which preaches about democracy and freedom but in fact attempts to open up new markets for the North – is opposed to a view in which globalization can be seen in a positive light. For instance, it can be associated with the creation of the World Social Forum,[8] which tries to coordinate responses of social movements throughout the world. The Internet, as an icon of globalization, can be seen as a means of increasing profits on Wall Street, but also as a means of organizing the Chiapas resistance. In a similar fashion, mathematics education gained momentum during the Cold War, when the soviets launched a rocket before the United States, generating a new concern about students' mathematical capabilities in the U.S. But it can also be seen as a movement that helps build bridges between countries that transcend national governments.

Many collaborative efforts in the field of mathematics education began when the main capitalist countries provided assistance to Third World countries to "develop their mathematics". In the beginning, this was part of the Cold War, in which among other actions, rich capitalist countries developed programs in poor countries in an attempt to deter socialist revolutions. Later, this "development of mathematics" was perhaps motivated by economic and political considerations. In this case, there is collaboration in which there are hidden agendas, such as selling textbooks, or imposing a particular view of mathematics education upon a given country. It is easy to see that some inequity still persists in global mathematics education; consider, for example, the much higher number of citations of publications authored by First World than Third World researchers. Of course there are many types of collaborations, including some discussed in this book, that do not follow this "neo-colonialist model of collaboration". But these are exceptions that confirm the rule.

From a macro-social perspective, Martínez Miranda (2004), points out that globalization can be seen as an entity with its own life concretized through social actors, *"as a logic of capitalism that imposes a set of consequences in the different settings of the social reality"*, as a phenomenon that involves *"homogenising the economical, political and cultural expressions of the world"*. However, Martínez Miranda argues that, in spite of the homogenizing force of globalization, local forces are always present, and that social actors are needed to act as analytical mediators of the globalizing processes as expressed at the local level. In this sense, the analysis by the actors in their local contexts is vital to open other comprehensions of the phenomena

---

[8] For more details about the World Social Forum you can access http://www.forumsocialmundial.org.br.

of internationalization and globalization. According to Martinez Miranda, "*...To impregnate the analysis of globalization with the action of the actors may help to divest of idealism the analysis of the construction of the contemporary society*" and to break down the illusion of homogeneity frequently associated with the process of globalization. In this same line of thought, Atweh and Clarkson (2005) acknowledge that *"the different experiences of globalisation are necessarily functions of the particular historical, social, economic, language and political contexts in which it is experienced"*.

The process of globalization and the vertiginous changes provoked by it have resulted in the emergence of asymmetries in the relationships between countries, mainly between developed countries and Third World countries. The search for new models of collaboration, alternatives to the neoliberal models of international aid, becomes paramount to achieving equitable and genuine collaborations.

Issues related to international collaboration in mathematics education in light of the globalization process were considered during ICME10 (Denmark, 2004) in the Discussion Group 5: *International Cooperation in Mathematics Education* (Atweh, Boero, Jurdak, Nebres, & Valero, 2004) and in articles that analysed and characterized different modes of possible collaboration among academics from different cultures (Atweh, 2004; Atweh & Clarkson, 2005). Atweh and Clarkson (2002) proposed some requirements for genuine critical collaboration between developed and developing countries. They argue that collaboration should empower each country to be self-reliant and should be based on mutual respect and trust in the ability of the different partners.

The coordinators of the discussion group mentioned above posed the following question: *What other forms of international cooperation in mathematics education exist at the international scene?* (Atweh et al., 2004). In part, this chapter presents one possible response to this question, showing a case of South-South collaboration. Our interest in emphasizing South-South relationships and collaboration has been spurred by the similarities of the educational challenges we face, the relative ease of communicating given the similarity of our languages (Spanish and Portuguese), and our certainty regarding the need for regional alliances. We aim to present dialogical and "horizontal" relationships in the way discussed by Freire (1992). It is in this sense that our relationship between researchers from Brazil and Argentina has developed, over approximately ten years, an experience in collaboration in which we try to avoid the traps that have marred this internationalization-globalization process in mathematics education.

We try to share a model or mode of international collaboration that we consider to be an alternative to the classical movement of early mathematical and scientific knowledge between East and West and between North and South. We intend to characterize the model through an exploration of the tensions and difficulties we faced along the course of our own professional development and different types of collaboration. We believe that our case meets the criteria that Atweh and Clarkson (2002) proposed for a genuine critical collaboration.

2. A Case of a Dialogical Relationship Among Researchers from Argentina and Brazil

The fruitful collaboration that has been ongoing between Mónica and Marcelo, and with this chapter includes Cristina, is strongly impregnated by our personal development as researchers and teachers in our local contexts as well as the history of the development of the discipline in our countries. Our present collaboration is a synthesis of several years of interactions and shared concerns, and constitutes segments of our individual paths. In this section, we would like to walk through our individual paths, which led us from local to international contexts and from those international landscapes back to our local realities. Our travels from one context to another provoked tensions that we had to deal with. In using the word tension, we are referring to a specific situation that leads to the need to decide or opt between different alternatives which are in some way antagonistic to each other, contradictory, or incompatible for the person who is facing the situation. Our notion of tension is subjective, since what may be considered a tension for a person in a given circumstance may not be for another, or for the same person in a different circumstance.

The three of us had the opportunity to do postgraduate study abroad and then returned to our native countries. We were all exposed to influences from the host countries with respect to the choice of our research problems, the methodological options, and the way of conducting research according to the dominant research paradigms of the postgraduate programs of studies and/or restrictions imposed by supervisors. At the same time, we had our own concerns related to mathematics education, which may or may not have been compatible with those of our supervisors, but which were never given up while staying abroad. Each of us had to cope with an *"option tension"*: the option between our local interests and concerns, and the interests and concerns of postgraduate programs or supervisors.

In this section, we present a summary of our individual journeys and discuss aspects of the development of mathematics education in our countries to contextualize the tensions and difficulties encountered on our paths.

2.1 Argentina: Sowing the Seeds for a Future International Collaboration

Here we describe the development of collaboration between two Argentinean colleagues, Cristina and Mónica, which preceded the international collaboration, emphasizing some obstacles encountered at the local as well as at the international level.

Cristina received her Masters degree in mathematics education at the City University of New York at the end of 1985. Her research topic was anxiety towards mathematics, mainly among women (Esteley, 1985). The main motivation for choosing this topic was her feminist concern about the role that mathematics could potentially play in excluding women in an increasingly technological society.

When she returned to Argentina, she started working at the Faculty of Mathematics of the University of Córdoba, and became a member of the *Science and Technology Education Group.*[9] The main focus of the group was on training of secondary teachers. In accordance with a national trend at that time in Argentina (Villarreal & Esteley, 2002), the group concentrated on "what to teach" and "how to teach", with a strong focus on the structure of academic mathematics and the development of teaching techniques.

In that local scene, Cristina's prior research interest regarding mathematics anxiety among women seemed unrelated to the local problematic.[10] She decided to choose another option: she started preparing material for the teaching of statistics at the secondary level and conducted research related to applications of content. The option tension was there again: to continue researching issues related to mathematics anxiety, or to start researching a new topic adapted to the local interests. Even though her concerns about mathematics anxiety were strong, the possibility of access to updated references on this topic was almost nonexistent in Argentina.

At that time, the libraries at the Argentinean universities did not have enough books or journals devoted to research in mathematics education. Obtaining a copy of an article sometimes took months. The lack of access to international references in Argentina contrasted with the wide access to references while living abroad. This difficulty could be interpreted as a lack of recognition of the activity of research in the field of mathematics education by the local academic community, mainly by some mathematicians.

It was in this local scenario, with little interest in mathematics education as a research field and limited access to references, that Cristina and Mónica began working together in 1988. Mónica was teaching calculus in the Agronomy Faculty of the University of Córdoba. The main concern that motivated and guided their studies from the beginning was the role of exclusion[11] played by mathematics and its teaching among students at different educational levels. Their first collaborative experience was the implementation of new strategies in mathematics courses for Agronomy majors, and research developed to monitor the application of those strategies. The aim of this experience was to seek answers to some questions arising from the teaching practice with the objective of going beyond proposals that were based only on the practitioners' common sense. This kind of activity was in accordance with the concerns of the growing mathematics education community in Argentina: the "deficient" mathematical knowledge of students entering the university, and the generation of proposals to overcome such deficiencies. Their research methodology was compatible with the positivist research paradigm that dominated the international mathematics education scene in the seventies.

---

[9] Grupo de Enseñanza de la Ciencia y la Tecnología.

[10] Problematic: Set of problems that are recognized by an individual or a group. The situations as far as the things are not working (Young, 1993, p. 158).

[11] We can see the exclusionary role played by mathematics reflected in the high rates of students failing in their basic mathematics courses at the university level, the influence this has on the student's decision to give up their careers, etc.

Based on this experience, they decided to focus on the problematic of the students' errors while solving mathematical exercises or problems (Esteley & Villarreal, 1996). In a sense, the study of errors is also related to the role of exclusion of mathematics. This new concern, and a desire to explore and understand the question of students' errors more deeply than was possible with the quantitative methods used up to that point, provoked changes in their methodological options, leading to a shift from the positivist to the interpretative paradigm.

During this first period of interaction, they wrote papers, presented talks at mathematics education national conferences, and participated in mathematics education research projects. Such projects were always chaired by mathematicians from the Faculty of Mathematics. Financial support for research projects received from state agencies allowed the acquisition of books and journals to start a mathematics education library at the Faculty of Mathematics. The research activity in the field was growing and gaining recognition by mathematicians.

Meanwhile, the Argentinean community of mathematics educators started growing, and people interested in mathematics education as a research domain started forming study groups. Even though those groups were isolated with little interaction between different institutions, personal and informal interactions were increasing. Visits by foreign specialists began to take place, most of whom were French academics working under the *Didactique* tradition. Such international contacts represented asymmetric interactions between local and international research. International authors, mainly from the North, through their presence in international journals or books, were influencing the local community, but the local research was shared only within the local community of mathematics educators.

In the context of this local scene, in 1994, Mónica decided to begin her doctoral studies. Graduate programs in mathematics education were practically nonexistent[12] in Argentina at the beginning of the nineties. The idea of studying abroad to "broaden her mind" was strong, and at that time she faced an option tension: to study in a recognized university in the North with all the possibilities that it could offer, or to look for regional alternatives closer to her local reality and political and social ideology. The options regarding graduate programs in mathematics education in South America are limited, and the Graduate Program in Mathematics Education (GPME) at the State University of São Paulo at Rio Claro is a regional option with a recognized international trajectory. She finally decided to study in Brazil: a neighbouring Latin American country with social and educational problems compatible with those in Argentina, the land of Paulo Freire, whose ideas that education could be liberating had great impact in Latin America and around the world. Moving to Brazil seemed to be the right option. The South-South collaboration among the authors was about to start.

---

[12] In 1994 a Masters Programme in Didactics of Mathematics was starting in the University of Río Cuarto (Córdoba). The supremacy of the French *Didactique* was remarkable in the contents of the courses and the origin of the foreign teachers invited to the programme.

2.2 *Brazil: the Beginning of an International Collaboration*

By the time that Mónica and Cristina started their collaborative work in 1988, the Brazilian Society of Mathematics Education was created, promoting the status of the field in the country. According to Fiorentini (1996), in addition to the concerns related to what and how to teach mathematics, new dimensions of research and teaching were emerging: epistemological, philosophical, historical, socio-cultural and political. The great variety of dimensions could be explained by the presence of supervisors coming from diverse disciplines (mathematics, philosophy, education), not specifically from mathematics education. The contributions of researchers such as Ubiratan D'Ambrosio, María Bicudo, Roberto Baldino, among others, played an important role in the constitution of the Brazilian community of mathematics educators and gave a fundamental impulse to the creation of the GPME at the State University of São Paulo (UNESP) at Rio Claro. Although there were other short-term, graduate level courses in Latin America,[13] it can be said that this program was one of the first graduate programs in mathematics education in Latin America.

Marcelo was one of the first students to obtain a Masters degree in the program in 1987, conducting research on ethnomathematics (Borba, 1987). His concerns about culture, social justice, and political dimensions of education were in accordance with the changes in methodological approaches and new research problems arising in Brazil. Then, Marcelo faced the *option tension* when he begun his doctoral studies at Cornell University, U.S., becoming a member of the Mathematics Education Research Group, led by Jere Confrey, which focused on the role of computers in mathematics education. Ethnomathematics and computers were seen as very different areas of research and writing. However, over the years, he has come to terms with this tension, tying these seemingly disparate themes together and with others, such as modelling. Since then, he has reflected on computers, their presence in education, and related epistemological issues. His earlier research interests related to ethnomathematics and mathematical modelling gained new momentum and began to interact with some of the epistemological issues related to computers that he had developed during his PhD research (Borba, 1993).

Since 1990, a scientific community of researchers in mathematics education has emerged and continues to grow in Brazil. This period was characterized by the return to Brazil of mathematics educators with doctoral degrees from foreign universities, the constitution and consolidation of local and national research groups, and the creation of new graduate programs at several universities. In 1992, the Doctoral Program at UNESP – Rio Claro was created. In that context, Marcelo returned to his native Brazil in 1993, where he started working at the Department of Mathematics at UNESP. As an undergraduate teacher, he started teaching a pre-calculus/calculus course for biology majors. In his classes, diverse kinds of activities took place, from traditional solving of problems in the book, to others which are intertwined with modelling as a pedagogical approach. The Brazilian

[13] For instance, in 1975 D'Ambrosio was involved in creating a Masters programme in the teaching of sciences and mathematics at the State University of Campinas (UNICAMP).

authors D'Ambrosio (1978, 1985) and Bassanezi (1994) proposed throughout the 1980s the notion of modelling as a pedagogy characterized mainly by the use of problems from the real world, choice of the problem by students, the possibility of interdisciplinarity, and resemblance with mathematical modelling. This pedagogical perspective was enriched by the use of technologies, such as graphing calculators.

As a researcher and postgraduate teacher, Marcelo joined the faculty of the GPME and founded the research group GPIMEM[14] which he has coordinated since then. GPIMEM has been investigating information and communication technologies associated with mathematics education. The group is composed of professors of the Graduate Program (UNESP – Rio Claro), a technician, undergraduate students in mathematics, and graduate students in mathematics education. The group emerged at a time when educational research in Brazil was characterized by the influence of so-called qualitative research (Lincoln & Guba, 1985) as a predominant paradigm. The development of this type of research became relevant in the story of our international collaboration as both Mónica and Cristina began to see research in a different way as they started interacting with Marcelo and other Brazilian researchers in mathematics education. Later on, already influenced by his interactions with his Argentinean colleagues, Marcelo discussed the quality of different qualitative research seen in the light of the critical mathematics education movement (Skovsmose & Borba, 2004).

It was in July, 1994, that Mónica met Marcelo for the first time in the II Iberian-American Congress of Mathematics Education,[15] organized by the Brazilian Society of Mathematics Education and the Regional University of Blumenau. During their first meeting, the possibility of Monica studying in Rio Claro was discussed. Finally, Mónica moved to Rio Claro at the beginning of 1995. As her interest in studying the conjunction of technology and mathematics education was compatible with the research project that Marcelo was developing in Brazil, he became her supervisor.

As a doctoral student, Mónica studied processes such as visualization and experimentation, usually considered as "second class citizens" inside mathematics, but which are intrinsically related to mathematics education and technologies. She conducted a detailed study to characterize the thinking processes of university students while working mathematical questions related to differential calculus in a computational environment (Villarreal, 1999).

The arrival of Mónica to the Graduate Program introduced her into the particular context of the GPIMEM and into the distinctive culture of the Graduate Program at Rio Claro. In this new landscape, two things caught her attention: the passionate way in which some people defended their points of view in the *Seminar of Mathematics and Mathematics Education* (SMME) held weekly in the graduate program, and the collective way in which the supervision sessions were conducted.

In the SMME, foreign or local specialists coming from diverse traditions in mathematics education, as well as teachers and students in the program, presented

---

[14] *Grupo de Pesquisa em Informática, outras Mídias e Educação Matemática* (Technology, other media and Mathematics Education Research Group). In Borba & Villarreal (2005), a complete report about the trajectory of the group is presented.

[15] II Congresso Ibero-Americano de Educação Matemática.

talks and research seminars, discussing issues such as methodological procedures, results, and research questions. This particular context for sharing and discussion was characteristic of the culture of the Graduate Program in Rio Claro and it was not a common practice in Argentina. The SMME was a window to see the international landscape of mathematics education.

The supervising sessions of GPIMEM were forums to analyze and criticize the research projects of the members of the group, present and discuss theoretical frameworks, or prepare presentations for meetings. In this context, Marcelo introduced the ideas of Tikhomirov's (1981) and Lévy's (1993). Lévy's (1993) concepts of thinking collective and cognitive ecology, and his particular perspective which considers orality, the written word, and information technology as all being technologies of intelligence, were compatible with Tikhomirov's (1981) theory of reorganization that discusses the role of the computer as a mediator that reorganizes human activity. Marcelo started developing research in his classroom which attempted to make sense of these ideas in the context of mathematics education. Other members of the group kept bringing new inputs from empirical research and suggestions for reading that challenged the idea that knowledge is constructed solely by "a human" or more than one, regardless of media. In most epistemologies that emerged from constructivist or social historical perspectives, the basic unit that produces knowledge is a "lonely knower", or more than one human. The role of technology is, at the most, peripheral and humans and technology are seen as disjoint sets. Marcelo's research, however, suggested that knowledge is constructed by collectives of humans-with-media.

It was in this context that Marcelo, and later many members of GPIMEM, started using and developing the notion that knowledge is produced by a collective of humans-with-media, which emphasizes the role of historical and cultural tools in knowledge production. One of the most important products of our South-South collaboration to date was the publication of a book in which this notion is elaborated (Borba & Villarreal, 2005).

Constant dialogues, exchanges, informal conversations, and the contact with new research perspectives and new authors strongly influenced Monica's perspective regarding what it means to teach and learn mathematics and to conduct research in mathematics education. While Mónica was influenced by new pedagogical approaches such as modelling, research methodology issues regarding an in-depth view of qualitative research and the collective dynamic of the research group, GPIMEM, she has also, at the same time, influenced Marcelo. She helped him to balance the emphasis he gave to modelling, where the students have great importance in choosing the theme to be investigated, and the experimental approach, in which the students are encouraged to generate conjectures. Mónica always reminded him of the importance of keeping those approaches intertwined with formal aspects of academic mathematics. This influence also helped Marcelo, and other members of GPIMEM, to stretch to the limits the difference between mathematics in textbooks and the mathematics which may be emerging as new information and communication technology interfaces become important actors in the classrooms.

The interaction of Mónica, as a doctoral student, with Marcelo as her supervisor, led them not only to share their research studies with their local communities, but also to present them in international meetings of mathematics educators (Borba & Villarreal, 1998; Villarreal & Borba, 1996,1998). They utilized international contributions for their local research, and then informed the international field with their local results.

Even though the student-supervisor relationship is, in fact, asymmetric, the fruitful work that Mónica and Marcelo carried out during that relationship let them set the stage for further collaborative work between Brazil and Argentina. The next stage of collaboration will be more symmetrical and horizontal.

2.3 The Collaboration Grows

In the previous sections, we described our personal paths in different contexts, the tensions and difficulties we experienced, and some of our prior collaborative work. Now, we will describe new collaboration, some balances that were created among our relationships, the expansion of our collaborative network, and how this particular collaboration allows us to contribute to the regional field and inform the international field within the particularities imposed by the local contexts.

2.3.1 The Collaboration and Scenario in Argentina: Shifts and Changes

When Mónica returned to Argentina in 1999, she resumed her activities as a teacher and researcher at the University of Córdoba. She started working again with Cristina, but now both had experienced the tensions of being abroad and the tensions of returning to Argentina. The interaction between those tensions led to a growing need to rethink themselves as local members of an international community of mathematics educators; the need to seek an identity. This situation led them to initiate their first research project in Argentina: a study about the state of the art of mathematics education in Argentina (Villarreal & Esteley, 2002). That inquiry was carried out with the aim of shedding light on reflections about the local field, and understanding the constitution of the local community of mathematics educators. This study led them to see the field as a local state of affairs and to consider it from an international point of view. Their oscillations between the local and the international perspectives continued to develop. Their interaction with other colleagues from inside and outside Argentina also grew.

In 2000, Cristina and Mónica decided to conduct further in-depth studies regarding their previous theme: errors in exercises and problem solving. They determined to research a phenomenon that had attracted their attention for a long time: the overgeneralization of linear models in non-linear contexts (Esteley, Villarreal & Alagia, 2004; Villarreal, Esteley & Alagia, 2005a, 2005b). This phenomenon is widespread among university students, and is characterized by being very persistent. If we look at this set of studies about overgeneralization, and the previous ones about errors, it is possible to find some important shifts in patterns, due to the growing impact of the international scene and the first step in our international collaboration. First, there was a methodological shift, from factorial analysis to inductive/constructivist data analysis (Lincoln & Guba, 1985). Second, there was a participation shift, from local to international meetings or publications. Third, there was an epistemological shift, from looking at the students' errors to recognizing students' mathematical production beyond the dichotomy of right/wrong.

By the time that Mónica and Cristina resumed their collaborative work, a new phase in the development of mathematics education had begun in Argentina. The Argentinean Society of Mathematics Education was created in 1998, and research activity related to mathematics education as a field of research was growing in the country. For example: new national congresses in mathematics education[16] have been organized since 1999, with the participation of Latin American specialists; many people have obtained their Masters degrees in national universities, a few have earned Doctoral degrees in foreign universities, and fewer still have returned to Argentina. In this particular context, Cristina and Mónica have initiated collaborative work with researchers in another Argentinean public university, evaluating papers to be presented at local and international mathematics education meetings or submitted to journals, and also co-supervising Masters students.

In the international landscape, new phases in collaboration also started. Mónica initiated a new stage in her relationship with GPIMEM as an *associated* researcher. Cristina, in turn, participated in the first virtual distance course offered by Marcelo and members of the GPIMEM.

2.3.2 *New Phases of the International Collaboration and Two Faces of Globalization*

Being an associated researcher of GPIMEM allowed Mónica to keep in contact and work with Brazilian mathematics educators and to deepen the study of the problematic of information and communication technologies in mathematics education.

In Brazil, Marcelo had continued conducting research on technologies, mathematics education and modelling as a pedagogical approach and developing the notion of *humans-with-media*. Although Lévy continued to be the main reference for this work, the writings of other authors like Kerckhove (1997) and Castells (1999, 2003) helped to transform some of the main ideas. He was also carrying out a research project on modelling as a pedagogical approach in a mathematics course for biology majors. It was time to systematize and intertwine the theoretical ideas with examples from research; it was time to write a book that shares what has been learned and developed by this particular research group. Marcelo invited Mónica to help him with this endeavour, and in the process she became a co-author of the book. She returned twice to Brazil to work on the book, in 2002 and 2003,

[16] The Argentinean Conference of Mathematics Education (Conferencia Argentina de Educación Matemática) organized by the Argentinean Society of Mathematics Education since 1999; and the Symposium in Mathematics Education (Simposio de Educación Matemática) organized since 1999.

sponsored by FAPESP, a funding agency of the State of São Paulo. It is worth noting that, without this financial support, it would be difficult, if not impossible, to solve a *"financial tension"* that Third World researchers frequently have to cope with.[17]

In 2005, Springer published *Humans-with-media and the reorganization of mathematical thinking*. This theoretical construct provides a view of knowledge that permeates the studies carried out by GPIMEM. This view values understanding, not results, and affirms that knowledge is produced by collectives composed of human and non-human actors. The relevance of this book can be seen from many different perspectives. For the purpose of this chapter, it should be emphasized that it was the first book published in the "Mathematics Education Library" whose authors were from south of the equator and whose native language was not English. Moreover, it was the result of an international collaboration of two neighbouring countries in South America, which usually does not play a major role in international mathematics education. It is still too early for an analysis regarding the importance of the book in epistemological or educational terms, but the idea was to present to the international community of mathematics educators a fresh approach to research on information and communication technology, with an emphasis on a relativistic perspective. In this view, ethnomathematics helps set the stage for seeing the production of mathematics as being bounded by different technologies of intelligence, which can be seen as artifacts which are historically and culturally bounded. As will be illustrated in the next example of collaboration, the Internet, one of our themes of study, not only changes the nature of knowledge, as argued in the book by Marcelo and Mónica (Borba & Villarreal, 2005), but it also makes new collaboration possible.

As mentioned before, in 2000 Cristina participated in the first virtual distance course offered by Marcelo and members of the GPIMEM. This course was offered over the Internet and provided material for the development of research on distance education. Since it was offered only for teachers, it had much more of a collaborative environment, in which experiences were shared and different actors played the role of discussion leader. The course was focused on some international trends in mathematics education, including Critical Math Education, Ethnomathematics, Modelling and Information Technology in mathematics education, Philosophy of Mathematics Education, and Teacher Education. Through the Internet, the door was opened for Cristina to interact with Brazilian colleagues in a very unique way. In one of the virtual classes, she was not only introduced to Ethnomathematics, but also had the opportunity to interact with one of its founding fathers, Ubiratan D'Ambrosio. In another, she not only read about and discussed the work of Arthur Powell, but also participated in a virtual conversation with him. While Ubiratan was in São Paulo, and Marcelo and Arthur were in a city in the U.S., other

[17] This financial tension appears in many situations: participation in international meetings, face-to-face collaborative work between researchers from different locations, access to technological devices, etc. For instance, you have to opt between: spending most of the research project grant just for paying a conference fee like PME or ICME or buying materials for the research project.

colleagues were in different cities in Brazil. Meanwhile, Cristina was at home reorganizing her thinking with others in a virtual space created by the chat. The global Internet space provided a platform for sharing, for creating meanings, for producing knowledge, or developing collaborative work. A learning process was taking place through a singular global process of interaction. This illustrates one of the faces of globalization: the inclusion, the *being-part-of*. But, while Cristina was part of that empowering global process, many other teachers were unable to participate. Some of them did not have a computer at home; others had computers but did not have access to the Internet. This situation showed the other face of globalization: the exclusion, the *not-being-part-of*. The difficulty of access presents itself again, creating a sort of paradoxical situation: technology and the global Internet provide possibilities for collaboration and diminish distances, but at the same time, they create a gap between those who have access to them and those who do not. Fortunately, in our case, we are able to experience the positive, inclusive face of globalization.

With this virtual contact, a new phase in collaboration was initiated. This distance course, and the interaction associated with it, led Cristina to start designing a collaborative project with secondary school mathematics teachers in Argentina. The project focused on the professional development of these mathematics teachers with modelling as a pedagogical strategy. In 2004, the project was implemented in three secondary schools, and some of the results were presented at a Latin-American meeting (*19 Reunión Latinoamericana de Matemática Educativa*, in Montevideo-Uruguay.)

Currently, Cristina and one of the teachers who participated in the project are participating in a new virtual distance course, offered by Marcelo and members of the GPIMEM, focusing on modelling as a pedagogical approach. Their experiences in the classroom have enriched the course and contributed to practical aspects of the discussion.

As a mathematics education teacher for pre-service mathematics teachers at the University of Villa María (Córdoba-Argentina), all the courses and the collaborative work with the secondary school teachers influenced Cristina's decisions about the organization and contents of her own classes. In this way, the Brazilian authors become virtually present in a face-to-face course for pre-service teachers taught by Cristina in Argentina. While the evidence on the outcomes of this interaction is slim due to its early stages, it can be said that the notion of humans-with-media has strongly influenced Cristina. By the same token, Cristina has influenced both distance courses in strong ways as she has brought examples from a different country to an online course taken mostly by Brazilians.

Interestingly, the Internet, an icon of globalization, has been used in this case as a means to exchange ideas and build respectful collaboration between researchers who have met face to face only a few times. This illustrates the inclusive face of globalization. Also, it should be emphasized that, in this case, the option tension was quite different, if not nonexistent. The option tension, as well as the financial tension, were minimized. Cristina did not have to move to a new location, and she did not have to opt between her activities in Argentina and the course in Brazil. There were no financial tensions involved in her decision to take the course because the costs were minimal – especially compared to the cost of taking a traditional face-to-face course abroad. In this case, she was able to continue working on Argentinean educational issues at the same time she was debating and developing ideas in a course led by a Brazilian mathematics educator. Also, unlike a face-to-face Masters program, the teacher has much less power over the students in distance courses such as this. So in this case, one of the symbols of globalization, the Internet, can help mathematics educators engage in discussion with people from different communities without loosing touch with their own local community.

Our movement from local contexts to international landscapes, and vice versa, always shows us two faces of globalization: the inclusion and the exclusion, the *being-part-of* and the *not-being-part-of*. While our participation in international congresses and publications in recognized international journals or books in the field, usually in English, allow us to share our research with the international community of mathematics educators, many of the members of our local communities do not have the same opportunity to share their experiences with the international community, due to financial reasons, or because they cannot read our own publications published in a foreign language, or speak and write in English in order to present their own work in international conferences such as ICME. In a sense, we live a contradiction: whenever we get closer to the international community, we move away from our local communities. This situation can be exemplified by a comment made by some Argentinean teacher-researchers who are reading, with great effort (due to the English), the book *Humans-with-media and the reorganization of mathematical thinking*. They said: *It seems like the book was written for us … the only problem is the language. Write the next one in Spanish or Portuguese*. The book allowed us to get closer to the international community of mathematics educators, but the English, considered to be the global language of scientific communication, moved us away from our local community.

3. Final Remarks

Castells (1999, 2003) has analyzed how globalization has created new opportunities for business, and argues that public funding is needed to increase access to the Internet among the poor in areas where big business has no interest in financing the construction of Internet backbones. Examples are shown of the double-edged sword of the globalization processes, but emphasis is given to perverse processes that generate "Fourth Worlds" throughout most of the big cities. Different authors, including Castells, have shown the asymmetries of globalization. These imbalanced processes also permeate mathematics education, where there have even been working groups in conferences that call for new forms of international cooperation (Atweh et al., 2004).

Throughout this chapter, we have mentioned the appearance of different *tensions* in diverse modes of collaboration in local and international scenarios: the *option tension* and the *financial tension* were discussed. Our narratives describe tensions we have faced and the paths followed in order to cope with them. We have described many manifestations of the option tension: the choice between different places to study, the election of research problems that differ from our prior interests. We have presented situations in which we had to cope with financial tension. We also referred to the emergence of difficulties regarding access to information and communication technology in different circumstances. The ways that we have dealt with the tensions and difficulties depend on local realities and possibilities as well as on international influences. In our present international collaboration the tensions and difficulties were attenuated. The search for balance and symmetry, or at least a complementary asymmetry in collaborative relationships, has led us to generate a genuine and equitable collaboration. The mutual influence, respect and recognition of the other's work are the solid foundation on which our fruitful collaboration is based.

Brazil and Argentina have similarities and differences in terms of their history of mathematics education. It is still interesting to note the difference in participation of the two countries in international conferences such as ICME and PME, for instance. This difference can be attributed to different national policies in terms of financing research, and how strongly they embraced the neoliberal policies at the end of the twentieth century, as discussed previously. Both countries have received specialists from countries north of the equator who have influenced the growth of mathematics education. In Argentina, we can observe a prevalence of influence from the French *Didactique* in many of the studies presented in national or international meetings, and in the orientation of the graduate programs. In this sense, mathematics education in Argentina became mostly "monolithic". In Brazil, the sources of the main influences, as well as cooperation, were from different countries: France, England, U.S.A, Germany and Denmark. For instance, since 1994, for almost ten years, Ole Skovsmose and Marcelo Borba have been cooperating in research endeavours which have led to joint publications with references from both South and North. Of course, there are many other examples of North-South cooperation that have maintained the neo-colonialist model already discussed.

Although we stress the role of a South-South relationship, we do not want to claim that North-South interactions are necessarily wrong or evil. As a matter of fact, we ourselves have had positive interactions with educators from the North, and we plan to have more. We do hope, however, that by stressing the way these relationships take place, we can show the positive possibilities that globalization can bring. Maybe we can even inspire different North-South relations, including more attention to the research being conducted in the South.

In this chapter, we want to claim that globalization has also generated a new kind of collaboration that we have called South-South. We illustrated how this relationship can be horizontal, dialogical in the Freirean sense that there is a constant listening from both sides. Although there were power relations involved (the second author was the PhD supervisor of the first, and the teacher of the third for a period of time), this did not prevent independent work from being done by all the actors involved, nor a fruitful, respectful collaboration among them. Mónica and Cristina have published together, without Borba, developing their own ideas, contributing to Argentinean mathematics education, and also transforming it. On the other hand, in a book that summarized research conducted by both authors, Borba and Villarreal (2005) present new theoretical ideas that bring an ethnomathematical perspective to the use of information and communication technology in mathematics education. For the purpose of this chapter, this means that our collaboration resulted in a South-South publication that could possibly influence international mathematics education. This "subversion" will, of course, only be complete if the book is not overlooked by researchers who are not accustomed to learning from researchers based in Third World countries, and cannot see possibilities for horizontal academic exchanges with colleagues from the South. This does not mean that others, from the North or the South, will have to agree with the ideas expressed in references from the South, but it does mean that we, as a community, have to admit that there are social relationships in the "quotation game", something that has been studied already in the sociology of the "hard sciences".

Our continuing collaboration includes occasional visits to each others' countries, encounters at international conferences, intense use of electronic communication, and new projects. Some day our South-South collaboration may grow to include neighbouring countries. It is our hope that the South-South collaboration described here may serve as a feasible, attractive model for researchers struggling against financial and language barriers that impede them from taking full advantage of the positive, inclusive side of globalization. We believe that in our case free trade means fair trade.

Acknowledgements

Although not responsible for the contents of this article, we would like to thank the two anonymous reviewers for their suggestions and comments made on earlier versions of this article. Specially, we would like to thank Anne Kepple and Bill Atweh, for their insightful comments and editing of our "foreign English".

References

Atweh, B. (2004). Towards a model of social justice in mathematics education and its application to critique of international collaborations. In I. Putt, R. Faragher, & M. McLean (Eds.), *Mathematics education for the third millennium: Towards 2010. Proceedings of the Annual Conference of the Mathematics Education Research Group of Australasia 2004 (47–54).* MERGA, James Cook University: Townsville, Australia.

Atweh, B., Boero, P., Jurdak, M., Nebres, B., & Valero, P. (2004). International cooperation in mathematics education: A discussion paper. Proceedings International Congress of Mathematical Education 10. Denmark.

Atweh, B., & Clarkson, P. (2002). Some problematics in international collaboration in mathematics education. In B. Barton, K. Irwin, M. Pfannkuch, & M. Thomas (Eds.), *Mathematics education in the South Pacific* (pp. 100–106). Auckland: Mathematics Education Research Group of Autralasia.

Atweh, B., & Clarkson, P. (2005). Conceptions tensions in globalisation and its effects by mathematics educators. In P. Clarkson, A. Downton, D. Gronn, M. Horne, A. McDonough, R. Pierce, & A. Roche (Eds.), *Building connections: Theory, research and practice*. Sydney: Mathematics Education Research Group of Australasia.

Bassanezi, R. (1994). Modelling as a teaching-learning strategy. *For the learning of mathematics, 14*(2), 31–35.

Borba, M. (1987/1994). *Um Estudo de Etnomatemática: sua Incorporação na Elaboração de uma Proposta Pedagógica para o "Núcleo-Escola" da Favela da Vila Nogueira-São Quirino*. Master thesis, Universidade Estadual Paulista, Rio Claro, Brazil. Published by Associação de Professores de Matemática (APM). Portugal, 1994.

Borba, M. (1993/1994). *Students' understanding of transformations of functions using multi-representational software*. Doctoral dissertation, Cornell University. Published by Associação de Professores de Matemática (APM). Portugal, 1994.

Borba, M., & Villarreal, M. (1998). Graphing calculators and reorganization of thinking: the transition from functions to derivative. In A. Oliver & K. Newstead (Eds.), *Proceedings of the 22nd Conference of the International Group for the Psychology of Mathematics Education* (Vol. 2, pp. 135–143). Stellenbosch, South Africa.

Borba, M., & Villarreal, M. (2005). *Humans-with-media and the reorganization of mathematical thinking: Information and communication technologies, modelling, experimentation and visualization*. Mathematics Education Library, V. 39. Springer.

Castells, M. (1999). *A sociedade em rede*. São Paulo: Editora Paz e Terra.

Castells, M. (2003). *A galáxia da internet: reflexões sobre a internet, os negócios e a sociedade*, Jorge Zahar Ed, Rio de Janeiro. Translated by Borges, M.L. Translation of: Reflections on Internet, business and society.

D'Ambrosio, U. (1978, April). Relationship of integrated science to other subjects in the curriculum. A paper presented at WG W-8, International Conference on Integrated Science Education Worldwide, Nijmegen, The Netherlands.

D'Ambrosio, U. (1985). Ethnomathematics and its place in the history and pedagogy of mathematics. *For the Learning of Mathematics, 5*(1), 41–48.

Esteley, C. (1985). *Mathematics anxiety in collage students as a function of gender and effectiveness of math tutorial sessions in reducing math anxiety*. Master Research Project, School of Education, The City College, City University of New York. United States of America.

Esteley, C., & Villarreal, M. (1996). Análisis y Categorización de Errores en Matemática. *Revista de Educación Matemática*. Córdoba, *11*(1): 16–33.

Esteley, C., Villarreal, M., & Alagia, H. (2004). Extending linear models to non-linear contexts: an in-deph study about two university students' mathematical productions. In M. Høines & A. Fuglestad (Eds.), *Proceedings of the 28th Conference of the International Group for the Psychology of Mathematics Education* (Vol. 2, pp. 343–350). Bergen, Norway.

Fiorentini, D. (1996). Um estudo histórico da Educação Matemática Brasileira enquanto campo de investigação. In *Atas de História e Educação Matemática* (pp. 214–221). Portugal: Braga.

Freire, P. (1992). *Pedagogia da Esperança: um Reencontro com a Pedagogia do Oprimido*. Rio de Janeiro: Paz e Terra.

Kerckhove, D. (1997). *A Pele da Cultura. (Uma investigação sobre a nova realidade eletrônica)*, Relógio D'Água Editores, Lisboa. Translated by Soares, L., & Carvalho, C. Translation of: The skin of culture (Investigating the new electronic reality).

Lévy, P. (1993). *As tecnologias da inteligência. O futuro do pensamento na era da informática*, Editora 34, São Paulo. Translated by Costa, C. Translation of: Les technologies de l'intelligence.

Lincoln, Y., & Guba, E. (1985). *Naturalistic inquiry*. Newbury, California: SAGE Publication.

Martínez Miranda, R. (2004). La internacionalización de la educación, la ciencia y la tecnología. Revista Electrónica de Psicología "La Misión". Facultad de Psicología de la Universidad Autónoma de Querétaro. México. (May 22, 2006); http://www.uaq.mx/psicologia/lamision/p_sociales4.html

Skovsmose, O., & Borba, M. (2004). Research methodology and critical mathematics education. In R. Zevenbergen & P. Valero (Eds.), *Researching the socio-political dimensions of mathematics education: Issues of power in theory and methodology* (pp. 207–226). Dordrecht: Kluwer Academic Publishers.

Tikhomirov, O. K. (1981). The psychological consequences of computarization. In J. V. Wertsch (Ed.), *The concept of activity in Soviet psychology* (pp. 256–278). New York: M.E. Sharpe Inc.

Villarreal, M. (1999). O Pensamento Matemático de Estudantes Universitários de Cálculo e Tecnologías Informáticas (Doctoral dissertation, Universidade Estatual Paulista). Rio Claro, Brazil.

Villarreal, M., & Borba, M. (1996). Computers and calculus: Visualization and experimentation to characterize extremes of functions. In Luengo, R. (Ed.), *Book of abstracts of short presentations international congress on mathematical education 8* (p. 408). Sevilla.

Villarreal, M., & Borba, M. (1998). Conceptions and graphical interpretation about derivative. In A. Oliver & K. Newstead (Eds.), *Proceedings of the 22nd Conference of the International Group for the Psychology of Mathematics Education* (Vol. 4, p. 316). Sevilla.

Villarreal, M., & Esteley, C. (2002). Una caracterización de la Educación Matemática en Argentina. *Revista de Educación Matemática*. Córdoba, *17*(2): 18–43.

Villarreal, M., Esteley, C., & Alagia, H. (2005a). As produções matemáticas de estudantes universitários ao estender modelos lineares a contextos não-lineares. *BOLEMA - Boletim de Educação Matemática*. Year 18, n. 23, pp. 23–40, Brazil.

Villarreal, M., Esteley, C., & Alagia, H. (2005b). Sobregeneralización de modelos lineales: estrategias de resolución de estudiantes de profesorado en Matemática. *Resúmenes de la 19ª Reunión Latinoamericana de Matemática Educativa* (RELME 19) (p. 101). Montevideo, Uruguay.

Young, R. (1993). Teoría Crítica de la Educación y Discurso en el Aula. Barcelona: España. Editorial Paidós.

22

GLOBALIZATION AND ITS EFFECTS IN MATHEMATICS AND SCIENCE EDUCATION IN MEXICO: IMPLICATIONS AND CHALLENGES FOR DIVERSE POPULATIONS

Edith J. Cisneros-Cohernour, Juan Carlos Mijangos Noh, and María Elena Barrera Bustillos
Universidad Autónoma de Yucatán, México

Abstract: In this chapter we examine the effects of global and international trends in mathematics and science education in the Southeast of Mexico, in particular the implications and challenges that these bring for the education of diverse populations

Keywords: Globalisation, Mexican Educational System, Mathematics and Science education

1. Introduction

As Castro, Carnoy and Wolf (2000) argue, the growing interdependence of economic markets and the escalating intellectual knowledge production require a workforce with solid mathematics, language and communication knowledge and skills. Hence, students' achievement in mathematics, and science we may add, is necessarily related to the nation's capacity for competing in a global market.

Globalisation effects on economic markets have been the subject of discussion by different scholars around the world. As the UNESCO Position Paper on Higher Education in a Globalised Society states, "globalization affects each country in a different way due to the individual history, traditions, cultures, resources, and priorities." (UNESCO 2003, p. 4). Here we argue that more attention needs to be given to its implications for education.

In this chapter, globalisation is understood both as a contemporary and historical phenomenon. As a contemporary phenomenon, it mainly has economical and political characteristics, associated with the development of communication,

B. Atweh et al. (eds.), Internationalisation and Globalisation in Mathematics and Science Education, 403–419.

information and transport media with an increasingly broad world coverage and effect on many nation-states and in the people and cultures in them. However, as an historical phenomenon, we argue that its current and future manifestations are, necessarily, determined by the interaction between the diversity of economic, political and cultural forces existing in the global, national and local contexts. In other words, we argue that globalisation, in its current state, is a step in the process of reconfiguration of the economic, political and cultural forces of the world society which, in the specific context of each nation-state, acquires particular characteristics and offers specific opportunities for ways to overcome the problems and difficulties now posed by capitalism.

Contrary to what some scholars of the globalisation theory have argued, such as Ulrich Beck (1997, p. 16) for instance, we do not consider the forces of the transnational financial powers to have no opposition in the world political arena. In defense of the German sociologist, we must say that when he published his book about globalisation, the massive and frequent public protests organized by the alter-worldist movement against the neo-liberal policy of globalisation, as represented by the World Trade Organization (WTO), World Bank (WB) and the International Monetary Fund (IMF), had not taken place yet. Beginning with the protests in Seattle in 1999, then Sydney in 2002, and Mexico in 2003, the neo-liberal policy of globalisation has been questioned by a growing, organized and proactive alternative movement.

Moreover, we do not think that the nation-states, even those more prone to the orthodox application of the neo-liberal formula, as is the Mexican case, can disregard adjustments in their policies to meet the demands of students in a global world. They, and the Mexican government in particular, must attend to the effects caused by the neo-liberal policies because, as we examine in the following paragraphs, even assessments such as the OECD's Program for International Students Assessment (PISA) show the failure in education of the neo-liberal policies followed by the government of Mexico.

Such characterisation of globalisation discussed here implies the following observations. Firstly, there is not only a single path for globalisation; the teleological objectives of the different possible ways are not homogeneous, and the destiny of any single proposal for globalisation is mutable. Secondly, globalisation in a specific national context, in this case Mexican, must be understood within the framework of globalisation as an evolving and contradictory process. Of course this is valid also for the educational sphere in general and, in particular, for the mathematics and sciences educational policy and practice which are the focus of this book.

What does this theoretical perspective imply for the analysis of policy and practice of mathematics and sciences education, which, at first glance, might appear to be not linked to politics? This theoretical approach impels us to say: the Mexican neo-liberal policies in education are questionable and reversible, and moreover must be subject to critical analysis in search of solutions for the obvious and urgent problems generated by such policies. This chapter provides evidence about the current Mexican situation in education with emphasis on the areas of mathematics and sciences policy and practice. The statistically poor results are interpreted as a consequence of the neo-liberal policies of the Mexican government, and we encourage a deep review of those policies. In addition, we criticize state management in education that leads to the standardization of current pedagogical practices in basic and elementary education. We also question the effectiveness of the Mexican neo-liberal policies in preserving human resources, academics who are under- or even un-employed and professionals who immigrate to developed countries, thus leaving Mexico with fewer working professionals to overcome its national problems. Here we argue that all of that must be associated with the consistent neo-liberal dogma of reducing the nation-state structure to the minimum. These policies, and in general the neo-liberal understanding of globalisation, has been the subject of sharp criticism, not only from the left wing politics area, but also from some scholars of main stream academics, such as Paul Krugman (2005), Dani Rodrik (2002, 2002a), and Ha-Joong Chang (2002).

Prior to examining the effects and implications of the neo-liberal policies in mathematics and science education, it is important to provide some background on the Mexican context and the characteristics of the Mexican educational system.

1.1 A Short Introduction to the Mexican Context

The United Mexican States, better known as Mexico, is located in the southern part of North America. To the north, Mexico is bounded by the United States; to the south by Guatemala and Belize (former British Honduras); to the east, by the Gulf of Mexico and the Caribbean Sea; and, to the west by the Pacific Ocean and the Gulf of California (OECD, 1997). In 2000, the country had just over 97 million inhabitants (INEGI, 2001). Now Mexico is the world's largest and most populated Spanish-speaking country with 104 million inhabitants and an Economically Active Population estimated at 36,580 (OIT, 2005). It is also one of the nations of the Americas in which a mingling of races took place. During colonial times, seven main ethnic groups comprised Mexico's population: Españoles (whites of Spanish descent born in Spain); Criollos (Whites of Spanish ancestry born in Mexico); Mestizos[1] (mingling of Indian and Spanish); Mulattos (People of both Spanish and black ancestry); Negros (blacks); Zambos (people of both black and Indian ancestry), and Indigenous[2] (originally 270 groups, now only 87).

Over the years, new immigrants from China, Korea, Lebanon, France, Italy, Germany and Spain began to settle in various parts of the country. These immigrants and the refugees of Central America contributed to the cultural and ethnic mosaic that constitutes the Mexican people of today, of whom 30% are Indians, 10% Whites, and 60% of mixed ancestry. As Collins & Jones, (1997) observe: "To the foreigner, it takes little effort to realize that the Mexicans are not homogeneous

---

[1] Over the years the word "Mestizo" has changed its meaning. Now it is used to refer to people of Mixed ancestry not necessarily of White and Indian background.

[2] Native people of the Americas.

people, they resemble more a collection of peoples and types formed by traditions and a turbulent history" (p. 21).

Although the Mexican government recognizes the status of the indigenous culture, this is not reflected on the segregation of Indians in the rural areas, and their isolation from the mainstream Mexican society. The marginalization of the Indian people is reflected in the small number of students of indigenous ancestry in middle and high school, and particularly in higher education (Blat-Gimeno, 1983; Prawda, 1989; Cisneros-Cohernour, 1998). The Indian students usually attend rural schools, many of which provide bilingual instruction because different indigenous dialects are still in use, including "more than 20 different Maya and six Nahuatl languages" (Von Bleyleben, Von Bleyleben, Hassenpflug, & Klug, 1993, p. 23).

Seventy-three percent of Mexico's population lives in urban centres where most of the hospitals, schools, banks and social services are concentrated. Rural areas are isolated and lack essential services and work opportunities. Although in recent years the federal government has expressed an interest in decentralizing the educational system and other social services, the country remains heavily centralized. Men occupy most leadership positions and, with a few exceptions, dominate the political processes of the country. Since the government of Miguel de la Madrid Hurtado in 1985 to the present, Mexican rulers have been following neo-liberal economic policies, which has affected all aspects of national life, and particularly education.

1.2 Characteristics of the Mexican Educational System

The Mexican educational system is divided into three educational levels: basic education, high school and higher education (Mexican Department of Education, 1998). The duration of each level is illustrated in Figure 1.

Basic education serves the needs of children between 3 and 15 years of age. It includes kindergarten, *educación primaria* (elementary education), and *educación secundaria* (middle school). Kindergarten education serves the needs of children who are 3–5 years old. Primary education serves the needs of children aged 6 to 12. Secondary education (middle school) became compulsory in 1993 and includes three levels and serves the needs of children who are between 12 and 15 years of age.

High school education (in Spanish, *educación media superior* or *bachillerato*) and higher education (*educación superior*) are not compulsory. High school education serves the needs of students whose ages are between 15 and 18. There are four types of high schools: general (or preparatory), technical (combined preparatory plus vocational preparation), open general high school and open technical high school. The last two are addressed to the needs of the adult population. In former years, students selected their majors during the third year of high school. Currently they study the same general curriculum. As a result of these changes, some subjects like calculus were eliminated from the third year of the high school curriculum for students interested in studying a career in Mathematics or Science. High schools graduates are called *bachilleres* (Chavels, 1980).

Basic Education			High School	Higher Education[3]	
Kindergarten	Elementary	Middle school		Undergraduate/Graduate	
3 years. (1 required)	6 years (all required)	3 years (all required)	3 years (non required)	4[4]–6 years plus 1 year of field experience (baccalaureate)	Specialization Master and Doctoral degrees (variable duration)

Figure 1. Educational Levels

Higher education includes the baccalaureate degree and graduate education. There are at least three different types of bachelor degrees. The first type is the one offered by universities. The program of study at this level of education involves mainly five or six levels and is usually addressed to individuals between the ages of 18 and 24. Mexican universities can be classified as autonomous (self-governed), state universities (organized by the state), and private universities (see Villa-Lever, 1988). The second type of bachelor degrees is offered by the normal schools.[5] It involves the preparation of kindergarten, primary and secondary teachers. The third kind is the technical bachelor degrees, offered by polytechnic institutes, which are responsible for preparing technicians for the industrial, farming and stockbreeding, and fishing sectors. Colleges and universities do not have a general education curriculum. College education focuses only on the discipline of study. Graduates who complete their bachelor degree are called "*licenciados*".

The Mexican government spends approximately 6.3% of its Gross National Product on education (roughly around 37 million U.S.A. dollars at current exchange rates). This percentage has been increasing over time, in spite of the economic crises experienced by the country, but the investment per student is still low (OECD, 2005). Furthermore, the following statement of the OECD reveals the deeplinks between the Mexican economic policies and the investment in education: "México is increasing its monetary expense in education, much more than its wealth, as a measure of the Gross National Product. However, the country has not yet reached an economic level which allows more discretion in the monetary expense

[3] Length of study varies.
[4] The duration of most bachelor degrees is five plus one year of field experience. This is a graduation requirement.
[5] Normal schools now offer bachelors in kindergarten, primary, and secondary education as well as special and physical education. They also offer graduate studies at the Master and Doctoral levels. Before 1993, normal schools did not require the high school diploma as a prerequisite for a teaching career.

assignment" (OCDE, 2005). For us, this is the recognition of the main cause of failure of the neo-liberal policies in education.

Ninety-nine percent of the Mexican population has completed elementary school, 53% secondary school, and only 15% higher education (Oxford Encyclopedic World Atlas, 1997). The school year for k-12 education consists of 200 working days and it operates from the last week of August to the first week of July. Some schools are only open for the morning session or the afternoon session, while others are open for both. In this way, schools can address the needs of at least three groups of students: in the morning, students who work in the afternoon and adult students in the evening. The school day for the schools varies in length but is generally six hours long for each session.

The Mexican educational system has been strongly influenced by France and recently by reforms from the United States. The influence of the French model is reflected in the different ability tracks or educational levels. Higher education curriculum with emphasis on the preparation of professionals is also a reflection of the influence of the French system. The influence of the American model is reflected on the organization of grades by chronological age as well as on the new curricular reform of 1993 and the current standardized assessment reform.

2. Globalisation Effects in Mexico's Context and Education

Globalization has influenced several changes in Mexican society. Prior to 1990, a small number of transnational companies, mainly from the United States, were located mostly in the capital, Mexico City, with other provincial states having their own local economies and business. This situation has changed over a period of 14 years. Nowadays, in most provincial states companies such as McDonalds, Kentucky Fried Chicken, Pizza Hut, Baskin & Robbins, SAM's Club, JC Penny and Wallmart compete with other national and local companies for customers. According to Castellanos (1999), thanks to North America Free Trade Agreement (NAFTA), foreign companies' participation grew from 1.9% in 1994, to 2.5% in 1998, to 10.8% in 1995, to 12.7% in 1998.

In addition to the increased number of transnational companies, there is a growing rate of immigration of Mexican families to the United States and Canada. Although Mexicans from rural areas traditionally migrated to the US, the economic crisis of the country in 1995 began to influence migration of working class citizens from urban settings, and in some cases return migration when the immigrants go back and forth from Mexico to the foreign countries. According to Adams (2003), "For a handful of labour-exporting countries, international migration does cause brain drain. For example, for the five Latin American countries (Dominican Republic, El Salvador, Guatemala, Jamaica and Mexico) located closest to the United States, migration takes a large share of the best educated" (p. 13). Recent government reports indicate that the income provided from the immigrants to their families that remain in Mexico is resulting in the creation of new services and benefits for the country that equals and sometimes is higher than the government expenditure on social services (Instituto de los Mexicanos en el Exterior, 2005).

Moreover, further changes have taken place in education. In 1990 after a long process of assessment and elaboration, Mexico began to create a new national curriculum for elementary and secondary schools, highly influenced by educational reform in the US. Once the reform plan was approved, the Mexican Department of Education initiated a process of evaluation review that led to the creation of detailed curricular programs and textbooks. The new curriculum reform was expected to accomplish various goals such as the improvement of educational quality, an increase in the educational attainment and skills of Mexican children, and the creation of national standards (Colosio, 1990; Beltran-Vera, 1990; Carranza, 1990). Additionally, the reform agenda was expected to accomplish both modernization and decentralization of the educational system, and a stronger participation of Mexican society in the decision-making process for improving the educational system (Gonzalez-Cantu, 1990; Pescador-Osuna, 1990; González-Torres, 1990). A strong effort was also put on increasing the number of mathematics and science students attending college, and to improve the educational skills of these students, as well as on the use of technology in teaching. Indeed, the Mexican Department of Education has strongly supported the acquisition and use of technology in schools, as well as teacher training and preparation on the use of educational technologies.[6] In secondary education, the reform focused the teaching of mathematics and sciences on problem solving, the increased knowledge of science, and the development of attitudes towards the protection of the environment (SEP, 1993, p. 11).

The reform is characterized by promoting a more equitable access to basic education, modifying curriculum contents and plans, increasing the required level of education (from K-6 to K-9); reassigning responsibilities to the states and municipalities for improving the quality of education at the schools, and improving teacher salaries and creating teaching awards for elementary and middle school teachers. According to Buenfil (1997), the reform, influenced by the neoliberal movement from North America, "implies a subordination of education to the productive system, conception of education as an investment (human capital theory), liberation of offer and demand, use of the educational system as a limitation for the demands of employment" (p. 6). As Ornellas asserts, the neoliberal influence penetrated stronger in Mexico, even though reform documents do not make reference to the original authors.

Since its implementation the new curriculum has been subject to controversy, both because of the decision to initiate decentralization by creating the national curriculum at the central level, and specifically by the rewriting of history books that changed the traditional perspective of Mexico's history (De Palma, 1995).

[6] As a result of the reform, teacher preparation increased. Prior to the reform teachers started their preparation in the Normal schools after completing middle school; since the reform they must complete high school before starting their teaching training.

The interest of Mexican authorities for making these changes, particularly in mathematics and science education, was also influenced by the findings from the Third International Mathematics and Science Study (TIMSS) in 1993. In this study, conducted by the International Association for Student Assessment, Mexican students obtained the lower scores when compared with their peers from 40 other countries. On average, Mexican students were 100 points below the world mean. Students at the secondary level obtained the lowest scores (US Department of Education, 1996).

In 1994, the year of the implementation of the curriculum changes, a new assessment reform was initiated by the Mexican government following the international trend for standardized assessment. This reform resulted in the creation of entrance examinations for universities and the development of minimum competencies for different disciplines and professions (ANUIES, 1993). To implement these changes a new organization was created, the Centro Nacional para la Evaluación de la Educación Superior (CENEVAL). Additional efforts cantered on the development of indicators for institutional and program quality by the Consejo Nacional de Ciencia y Tecnología (CONACYT).

In 2000, after international comparisons showed discouraging results for the country (Vidal & Díaz, 2004), the Mexican government established a National Institute of Educational Evaluation (INEE) with the purposes of developing quality indicators for assessing the educational system, test development, as well as school evaluation (INEE, 2002). This new institute is using standardized assessment to determine the quality of achievement of elementary, middle and high school students. In addition, current policy documents, such as the Programa Nacional de Educación (2001–2006) and local policy documents and plans, are stressing a growing trend for accountability, assessment and accreditation for all educational processes, schools, personnel and results. This trend is reflected in the strong effort devoted to the adaptation of international standards for accountability purposes.

Despite those attempts, the general landscape of education in Mexico is still dark. Of the population 15 years and older, 53.1% do not finish middle school (INEGI, 2004). That means the government and its neo-liberal policies have not yet accomplished the mandate of the Mexican General Law of Education (Ley General de Educación, 2005) which states: "All the inhabitants of the country must graduate kindergarten, elementary and middle school."

Indigenous students have a special- and not exactly better- situation in relation to the Mexican educational landscape under neo-liberal policies. The following statistics describe the situation of indigenous students in the elementary school:

- The national drop-out rate is 1.3%; while among indigenous students it is 3.05%.
- The national average of failure is 5%; among indigenous students it is 9.81%.
- In Mexico, the rate of completion of elementary school is 89%, but among indigenous students it is only 81.44% (SEP, 2005).

After elementary school, indigenous communities receive no education in their own languages. Moreover, none of the mathematics and science textbooks are available in the native languages.

In Math and sciences the results of the 2000 and 2003 PISA surveys show evidence of the continued failure of the neo-liberal policies in Mexico. The statistics published by the INEE, show that Mexican students have an average of 387 points in mathematics in the PISA survey of 2000, far from the OECD average of 500, ranking Mexico in 35th place out of 40 countries. In 2003 the results became even worse, dropping to an average of 385 points, placing Mexico 37th out of 40 countries. Results in the sciences are no better. The PISA survey of 2000 shows Mexico in the 33rd place of 40 countries, with an average of 422 points. In 2003, the country was in the 37th position with a student average of 405 points. Both surveys set Mexico among the countries with a notably lower average among the OECD countries surveyed in mathematics and science (Vidal & Díaz, 2004).

As new standards for assessing graduate programs were adopted by CONACYT, Mexican colleges and universities began to increase their effort for recruiting personnel with doctoral degrees, encouraging their faculty to obtain higher degrees, and creating new doctoral programs. At the same time, a number of private institutions devoted primarily to the preparation of secondary students began to offer undergraduate and graduate programs that began to compete for students with the major public universities.

The support provided by CONACYT, and other funding agencies, for study abroad has contributed to the increased number of students who complete their graduate education abroad. Although a great majority of these students return to their home country, a significant number remain in the host countries seeking better job opportunities. Among the returning graduates, some have readjusted to their country's educational system, but many have faced problems after being exposed to different value systems and experiences abroad. A significant number of graduates do not remain in their country for long, mostly due to lack of academic opportunities for growth and development. A survey conducted in a listserv developed for students from CONACYT in 1999 indicated that the majority of the students who graduate from overseas universities wish to go back to the foreign countries due to problems in the recognition of their graduate studies by the Mexican accreditation systems, the lack of permanent job opportunities, or problems adapting to the highly bureaucratic and politicised context of Mexican higher education institutions.

Research about the extent of the brain drain in Mexico is limited. A study conducted by Carrington & Detragiache, 1999 shows the following evidence about the immigration from the Western Hemisphere to the United States: "Mexico is by far the largest sending country (2.7 million), with the large majority of its migrants (2.0 million) having a secondary education and fewer than 13 percent having a tertiary education". That means, in absolute numbers, 351,000 professionals, which represents around 2% of the higher educated Mexican population. For a developing country, this is a significant loss of a long-term and expensive investment. More research is needed in this area because, according to Vitela (2002), while 12% of

Mexico's workers live in the United States, 30% of Mexicans with PhDs live in that country, and 79% of the science students Mexico funds to study abroad never return to work in Mexico.

In addition to the number of professionals who actually leave the country, a large group of professionals remaining in Mexico just find jobs below their schooling skills. They are under-employed. Although there is no data about the rate of professionals in this category, in a country with 26.1% under-employed workers, the rate can be expected to be high (INEGI, 2005). Further, according to the International Labour Organization (OIT 2005a), whilst there were 7,492,102 Mexican professionals in 2004, 271,700 of them were unemployed.

This situation is similar to other developing countries, as Williams (2000, p. 2) states: "just as international mobility of skills may deplete a nation's productive capacity, so may the malfunctioning of internal labour markets and weaknesses in personnel management at the level of employing organisations lead to ineffective deployment of human capacities within countries".

Lastly, English has become the second most important language in Mexican education. This is evident in the inclusion of English content in the curriculum of middle and high school, as well as the increasing number of private centers for teaching English. Since most scientific journals and textbooks are written and published in English, this language is also the predominant language in the academic arena. The increasing pressures for accountability and the limited number of research publications in Spanish is making Mexican scholars face the challenge of publishing in journals with international coverage and recognition that require them to write in a second language and to follow unfamiliar publication guidelines (Altbach, 2000).

2.1 Implications and Challenges in Mathematics and Science Education

To identify the implications and challenges from global and international trends in mathematics and science education in Mexico, we used data from three research studies. The first was a study on the evaluation of curriculum reform in Mexico (Cisneros-Cohernour, Merchant & Moreno, 1999). The research involved the study of four secondary schools in the southeast of Mexico. The issues addressed by the research were: (a) What are the needs being served by the Mexican curriculum reform of 1993? (b) Were the needs of the different ethnic groups being considered? Was the curriculum satisfying some needs while excluding others? (c) Do different stakeholders have similar perceptions and expectations about what the educational system is promoting and about what the curriculum should be promoting? and (d) What advantages and disadvantages occur when choosing a national, instead of a local or regional approach in the creation of a curriculum?

The second study was on the influence of national and local culture on the role of school principals (Cisneros-Cohernour, & Merchant, 2005). This case study research used Hofstede's framework (1994, 1995), to examine the influence of the sociocultural context in the lives of five high school principals in the southeast of Mexico. The research was part of an international study on the lives of high school principals in seven nations. The third study focused on the conditions of mathematics and science education in a state in the southeast of Mexico (Cisneros-Cohernour, López, Canto, & Alonzo, 2004). This research, conducted in 27 secondary schools in the southeast of Mexico, focused on examining: (a) What are the aims of science education and the way in which they were interpreted by administrators, officials at the State Department of Education, teachers and others involved in the implementation and support of the curriculum? (b) What were the conditions that affect science education in relation to the curriculum teaching, student assessment and professional development of teachers? (c) What are the challenges that schools and teacher preparation centers are confronting, and what solutions are proposed to confront these challenges?

Findings from these research studies show that some benefits have resulted from the adoption of a new curricular reform. According to different stakeholders, the new reform and international comparisons have resulted in more support from the Mexican government to improve the conditions for the teaching of mathematics. Since mathematics is one of the subjects where student performance has been very poor in recent international comparisons, curriculum reformers increased the length of mathematics teaching within the secondary curriculum. The mathematics instructors interviewed were very pleased with these changes because they allow them to devote enough time to the teaching of their subject. Science teachers, however, have a different perspective about the reform. They struggle trying to cover the increased content coverage of their courses with a reduction in time devoted for science. Biology teachers also face problems with the curriculum sequence in the first and second years of middle school. According to different stakeholders, the lack of coordination between elementary and secondary education is still a problem that seriously affects the teaching and learning of the two subjects in schools. The lack of articulation between educational levels needs to be improved because students not only lack prerequisite knowledge from one level of education to the next one; they also need to improve their critical thinking skills and develop their reading, writing and mathematics skills as well as their ability to use library resources.

The emphasis put by the Department of Education on the use of technology in the teaching of mathematics and science has led to the provision of professional development workshops for secondary teachers as well as the acquisition of computers at the school level. Some of these workshops train teachers on the use of the software Mathematica, E-Math, and other kinds of software. In the first level of secondary education, students have also been asked to buy scientific calculators for their mathematics courses, even though some come from families that cannot afford this expense. There are also serious problems when adopting standardized practices in Mexico due to the significant differences between the characteristics and resources of private and public schools, and especially between urban and rural schools. This can be illustrated with the following excerpt from one of the interviews with a principal in a rural school from the Cisneros-Cohernour and Merchant' study (2005):

> The Department of Education wants all schools to use computers. So, they sent us a computer, but we can't use it because the electric capacity of the school doesn't allow for it. We also lack teachers who know how to use the computer. I asked if we could get books instead because they can be more useful for our students, but the people at the Department of Education said that we can't do that.

This example illustrates how the lack of consideration given to the context and available resources of the schools can result in the waste of resources, primarily in schools that are most in need of improvement. This could bring even more negative consequences to rural and inner city schools because of the limited resources of this kind of school and the low socioeconomic status of the students that attend them.

Similarly, the assessment reform that was implemented in 1994 has also resulted in some benefits and problems for students and their families. As with the curriculum reform, the transfer and mobility of students from one state to another has improved as a consequence of the standardization. In addition, the assessment reform is providing for a validation of the teacher-based assessment conducted at the schools. However, the cost of the standardized assessment examinations is almost impossible to afford for most families in rural communities. For these families that are already struggling to support their children to continue with their education after elementary school, the cost of the standardized assessment has become a heavy burden that could further diminish the educational opportunities for these children.

In addition, the changes in the grading system at the school level were the sources of both benefits and problems. On one side, the standardized assessment reform is beneficial, since having an external assessment can help to identify problems in student learning that are not detected, or remained uncovered, by the internal assessment procedures at the school level. However, the change has also resulted in other problems. As a result of the curriculum reform, there was a change in the rating scale for student achievement at the school level. Prior to 1993, students were graded using a scale from 0–10. With the reform, the scale changed to 5–10. In other words, in the past the minimum grade that a student could have obtained in a test or classroom assignment was zero, but now he or she will obtain a minimum of five points for the same test results or assignment. According to teachers and school administrators, a negative consequence of the change is that students are paying more attention to how to obtain a minimum passing grade than to improving their learning. In addition, although the Education Department states that there has been an improvement in student achievement scores, this statement needs to be revised because the scale change provides a false impression of improvement. The demands for increasing student achievement scores and retention have also increased in the schools. The overwhelming majority of teachers interviewed in the Cisneros-Cohernour et al. study (2004) expressed concern that they felt themselves to be under pressure to approve students so the school could show that there was an increase in student retention, even though they themselves are convinced in some cases that students are not ready to pass to the next level.

There are also some problems related to the decentralization efforts that deserve attention. The decision to implement the reform within the former bureaucratic structure is negatively affecting the quality of teaching and learning in mathematics and sciences. Teachers were especially concerned about the strong role of the federal government in the decentralization process. According to the majority of the instructors interviewed, the decentralization has resulted in more bureaucracy limiting even more their academic freedom and motivation for introducing changes in teaching.

Moreover, although the reform increased the requirements for teacher preparation and provided a new award called "Carrera Magisterial" for elementary and middle school teachers, the instructors and administrators who participated in the Cisneros-Cohernour et al study (2004) indicated that they need better preparation in the subject matter, as well as on developing teaching skills under a constructivist approach, and also on the teaching of adolescents. An assessment conducted with a sample of middle school science teachers in the southeast of Mexico during the fall of 2005 provided evidence of lack of teaching preparation in these areas (Cisneros-Cohernour, Lopez, Barrera, Baas, & Domínguez, 2005). In spite of the efforts made by the State Department of Education and its centers to provide continuing education, most of the science teachers interviewed stated that they do not participate in professional development activities because they are hired on an hourly basis which demands that they work in three or four schools. Since most professional development courses are designed for half-time and full-time teachers, who are also eligible to receive a teaching award called "carrera magisterial," hourly instructors are left with less possibilities of professional development. This situation is serious, since the overwhelming majority of mathematics and science instructors who participated in the study in the southeast of Mexico were hired on an hourly basis.

When asked if the needs of the different ethnic groups had been addressed by the new secondary school curriculum, different participants provided different responses. Most principals eluded the question and only re-iterated the official rationale for initiating the reform. Teachers and one of the principals interviewed, however, indicated that although the Mexican Government organized several forums to identify strengths and weaknesses of the former secondary school curriculum, and organized groups of educators and scientists over the country to prepare recommendations for the development of the new curriculum, the recommendations made by these stakeholders were not taken into consideration when designing the new curriculum. As one teacher indicated, "Every six years we get a new educational reform with the new president elected. I think the reform was inspired by the US educational system because our last two presidents were educated there". Another teacher added, "I do not believe they (Department of Education) looked at the information they collected through a national consultation. There was not enough time to look at the results of the forums and have a new design on time for the implementation date already announced." A colleague of this teacher added: "If they (Department of Education) looked at our feedback, how can you explain that they ignored our comments. You can see that they just copied something that was already prepared by the time they were organizing the national consultation."

Officials at the Department of Education shared the perceptions that the curriculum was not generally responding to the needs and contexts of the schools, especially those attended by indigenous children. Teachers shared this perception that although the reform stressed the importance of linking the new program to student lives, it disregarded student cultural context. Moreover, rural schools face additional problems that were not taken into consideration in the implementation of the reform. Some of these problems include the limited economic resources, inadequate facilities, poor quality teachers, most of whom do not attend the school during all days of the week, and a high number of students who need to work full time to help support their families. The lack of supervision in the rural areas by the Department of Education makes it easy for teachers who live in the capital of the state to hold a position while living in the city. But having teachers who do not live in the communities where they teach and leaving them with limited supervision, makes it easy for teachers to skip days or hours from their work at the schools. In addition, the new curriculum emphasizes new expectations from student roles that may not be consistent with their cultural context. In some schools with high numbers of indigenous students, the change in curriculum orientation created value conflicts between home and school. The school requires students to be more assertive, but this behaviour has not been fully accepted by the parents who see this change as a threat to their authority at home. In general, the different stakeholders stressed that the emphasis on homogeneously implementing national standards without consideration of local contexts has created a situation where the tension between national standards and local needs could affect the quality of students' education.

3. Conclusions

Globalisation, in its neo-liberal agenda, has influenced changes in Mexican society, which is reflected in the presence of transnational companies, migration and return migration to the US and Canada, as well as a growing interest in quality, accountability and evaluation in terms of the OECD. In addition, international comparisons of student achievement such as the TIMSS, and PISA studies have led to government decisions for establishing standardized assessment policies for k-12 and higher education as well as curriculum changes in basic education (k-9), but always under the economical limitations of investment in education established by the general neo-liberal policies of the Mexican government.

Global vs. local is a significant topic, not only for Mexico but also for other nations, as is the issue of diverse ethnic groups and the extent to which students will be well served by national and international standards. The changes influenced by international and global trends have provided both benefits and problems to the teaching of mathematics and science. As a result of the changes, there is a growing concern among politicians and education authorities for teacher training and professional development, the use of technology, and assessment of educational outcomes, particularly in mathematics and science. Other efforts have led to the creation of teaching awards for elementary and middle school instructors.

However, the adoption of reforms influenced by other countries and cultures into a different educational system presents problems, given the differences in resources and the bureaucratic structure of Mexican schools. As the findings of the research studies used for examining the implications of educational and assessment reforms in the schools indicate, the emphasis on homogeneously implementing national standards without consideration of local contexts has created a situation where the tension between national standards and local needs could affect the quality of students' education. The educational possibilities for children of indigenous ancestry are particularly at risk, not only because of the economic burden of new standardized testing, but also because of the disregard for the intuitive way of learning maths and the sciences which is demanded in the new curriculum.

The situation, however, can be reversed, because there is not only one path for globalisation. Mexico needs to adopt its own path while being sensitive to its specific national context and multicultural diversity. Finding a different path requires a deep reassessment of the status quo prior to the implementation of new policies addressing globalisation in a way that responds to Mexico's needs. These policies should allow for the country to define what would be its commitment to globalisation, what the country will not compromise in order to preserve and protect the development of its multicultural diversity and how this will be accomplished, Due to the complex situation outlined above, diverse and multiple kinds of actions and changes will be needed in policy, politics, economy, teacher education, schools management, curriculum, as well an increasing support for education. This requires enormous amounts of talent, work and commitment from Mexican scholars, politicians, teachers and parents.

References

Adams, R. K. (2003). International migration, remittances, and the brain drain: A study of 24 labor exporting countries. World Bank Policy Research Paper WPS3069. Available online: http://www-wds.worldbank.org/servlet/WDSContentServer/WDSP/IB/2003/07/08/000094946_03062104301450/Rendered/PDF/multi0page.pdf

Altbach, P. (2004). Globalization and the university: Myths and realities of an unequal world. *Tertiary Education and Management, 1,* 1–20. Available online: http://www.bc.edu/bc_org/avp/soe/cihe/publications/pub_pdf/Globalization.pdf

ANUIES. (1993). Propuesta del Cuerpo Consultivo sobre el Examen Nacional Indicativo previo a la Licenciatura y el Examen General de Calidad Profesional. México. *Tertiary Education and Management, 1,* 1–20. Available online: http://www.bc.edu/bc_org/avp/soe/cihe/publications/pub_pdf/Globalization.pdf

Beck, U. (1997). ¿Qué es la globalización? Falacias del globalismo, respuestas a la globalización. Barcelona, España: Paidós.

Beltrán-Vera, J. M. (1990). Propuesta de educación básica. In L. D. Colosio (Ed.), *Modernidad Educativa y Desafíos Technológicos*. México: PRI (Partido Revolucionario Institucional).

Blat-Gimeno, J. (1983). Education in Latin America and the Caribbean: Trends and prospects, 1970–2000. United Kingdom: United Nations Educational, Scientific and Cultural Organization.

Buenfil, R. N. (1997). Globalización: Significante nodal en la Modernización Educativa. Colección Pedagógica Universitaria, 27–28, Universidad Veracruzana, Instituto de Investigaciones en Educación. Available in http://www.uv.mx/iie/Coleccion/N_2728/pagina_n9.htm

Carranza, J. A. (1990). Visión global del sistema educativo. In Modernidad Educativa y Desafíos tecnológicos. Edited by PRI (Institutional Revolutionary Party). Mexico.

Carrington, W. J., & Detragiache, E. (1999). How extensive is the brain drain? *Finance and Development, 36*(2). Available in: http://www.imf.org/external/pubs/ft/fandd/1999/06/carringt.htm

Castellanos, C. (1999). Foreign interest. Bridge News' Mexico Bureau. Available online: http://www.mexconnect.com/mex_/travel/bzm/bzmtop50-99.html

Castro, C de Moura, Carnoy, M., & Wolff, L. (2000) *Secondary schools and the transition to work in Latin America and the Caribbean*. Sustainable Development Department Technical Papers Series. (ERIC No. ED474307)

Chang, H. J. (2002). Kicking away the Ladder. Development strategy in historical perspective. London, England: Anthem Press.

Chavels, S. G. (1980). *A study of the educational system of Mexico and guide to the academic placement of students from Mexico in United States educational institutions* (Rev. ed.). Mexico: World Education Series.

Cisneros-Cohernour, E. J. (1998). *National versus Regional: Challenges for the new Secondary School curriculum of Mexico while serving the needs of diverse populations at the local level.* Unpublished report for the Tinker Foundation and the Center for Latin American and Caribbean Studies at the University of Illinois.

Cisneros-Cohernour, E. J., Merchant, B. (2005). The Mexican high school principal: The impact of the national and local culture. *Journal of School Leadership, 15*(2) 215–231.

Cisneros-Cohernour, E. J., Merchant, B., & Moreno, R. P. (1999). *Evaluating curriculum reform in Mexico: Challenges addressing a diverse population. US Government.* (ERIC Publication No. ED 436474)

Cisneros-Cohernour, E. J., López, M. T., Canto, P., & Alonzo, C. (2004). *La enseñanza de las ciencias en Yucatán: Cuestiones críticas para el mejoramiento de la enseñanza y la evaluación del aprendizaje.* Unpublished research report. México: Universidad Autónoma de Yucatán.

Cisneros-Cohernour, E. J., Lopez, M. T., Barrera, M. E., Baas, M., & Domínguez, G. (2005). Fortalecimiento de Recursos Humanos para la Innovación de la Enseñanza de las Ciencias en Yucatán. Research Project financed by CONACYT, Fondos Sectoriales SEP-SEByN, Project proposal. Universidad Autonoma de Yucatán.

Colosio, L. D. (Ed.). (1990). Modernidad Educativa y Desafíos tecnológicos. Mexico: PRI (Institutional Revolutionary Party).

Collins, J., & Jones D. M. (1997). *Mexico: Blue guide.* New York: W.W. Norton & Company.

DePalma-Anthony. (1993). New battles flare over Mexico's past (controversy over the textbooks). New York Times (Late New York Edition). Aug. 30 '93, p. A9.

Estudios sobre los beneficis que aportan las comunidades mexicanas en Estados Unidos. Available in http://www.ime.gob.mx/investigaciones/aportaciones.htm

Gonzalez-Cantu, R. (1990). Evaluación sectorial. In *Modernidad Educativa y Desafíos Tecnológicos.* Edited by PRI (Institutional Revolutionary Party). México.

González-Torres, E. (1990). Educación de valores. In L. D. Colosio (Ed.), *Modernidad Educativa y Desafíos Tecnológicos.* México: PRI (Partido Revolucionario Institucional).

Hofstede, G. (1994). Hofstede's cultural dimensions: An independent validation using Rokeach's value survey. *Journal of Cross-cultural Psychology, 15*(4), 417–433.

Hofstede, G. (1995). Managerial values. The business of international business is culture. *Cross-Cultural Management, 5*, 27–48.

INEE. (2002). ¿ Qué es el INEE?. Available online: http://www.inee.edu.mx/images/stories/documentos_pdf/que_es_el_inee.pdf

INEGI. (2001). XII censo de población y vivienda, 2000. Aguascalientes, México: Instituto Nacional de Estadística Geografía e Informática. Available in: http://www.inegi.gob.mx/inegi/default.asp

INEGI. (2004). El rezago educativo en la población mexicana. Aguascalientes, México: Instituto Nacional de Estadística Geografía e Informática. Available in http://www.inegi.gob.mx /prod_serv/contenidos/espanol/biblioteca/default. asp?accion=&upc=702825497538&seccionB=bd

INEGI. (2005). Porcentaje de trabajadores con menos de 35 horas trabajadas a la semana. Instituto Nacional de Estadística Geografía e Informática. Available in: http://www.inegi.gob.mx/est/ contenidos/espanol/rutinas/ept.asp?t=mtra21&c=3612

Instituto de los Mexicanos en el Exterior. (2005). Estudios sobre los beneficios que aportan las comunidades mexicanas en Estados Unidos. Avialable in: http://www.ime.gob.mx/investigaciones/aportaciones.htm

Krugman, P. (2005, March 18). The ugly American Bank. *New York Times*, (p. 21).

Mexican Department of Education. (1998). El Sistema Educativo Mexicano. *Estructura y Organización*. México, DF: Secretaria de Educación Pública.

OECD. (1997). Reviews of national policies for education. Mexico: Higher education. Paris, France: Organization for Economic Co-operation and Development.

OECD. (2005). Panorama de la educación 2005. Breve nota sobre México. México: Organización para la Cooperación y el Desarrollo Económico, Centro México. Available in: http://www.oecd.org/dataoecd/28/22/35354433.pdf

OIT. (2005). Total and economically active population by age group. México: International Labour Organization Available in: http://laborsta.ilo.org/cgi-bin/brokerv8.exe

OIT. (2005a). Unemployment by level of education. Mexico: International Labour Organization. Available in: http://laborsta.ilo.org/cgi-bin/brokerv8.exe

Oxford Encyclopedic World Atlas. 1997. (4th ed.). New York: Oxford University Press.

Pescador-Osuna, J. A. (1990). Descripción de los Servicios Educativos. In L. D. Colosio (Ed.). Modernidad Educativa y Desafíos Tecnológicos. México: PRI (Partido Revolucionario Institucional).

Prawda, J. (1989). Logros, inequidades y retos del futuro sistema educativo mexicano. Colección Pedagógica, Grijalbo.

Rodrik, D. (2002). Globalization for whom? Published in Harvard Magazine. Available in: http://ksghome. harvard.edu/~ drodrik/shortpieces.html

Rodrik, D. (2002a). After neoliberalism, what? Available in: http://ksghome.harvard.edu/~drodrik/ shortpieces.html

Secretaría de Educación Pública. (2001). Programa Nacional de Educación 2001–2006. México, DF: Author.

SEP. (1993). Programa para la Modernización Educativa 1989–1994, Secretaría de Educación Pública, México.

SEP. (2005). Lay General de Education (2005). México: secretaría de Educación Pública. Available in http://www.cddhcu.gob.mx/leyinfo/pdf/137.pdf

SEP. (2005b). Estadística de Educación Inicial y Básica Indígena. México: Secretaría de Educación Pública, autor.

UNESCO. (2003). Higher education in a globalized society. Private Linkages, Public trust.

U.S. Department of Education. (1996). Pursuing excellence: A study of U.S. eighth-grade mathematics and science teaching, learning, curriculum, and achievement in international context. Initial findings from the Third International Mathematics and Science Study (TIMSS). Washington, DC: National Center for Educational Statistics. http://nces.ed.gov/pubs97/97198.pdf

Vidal, R., & Díaz, M. A. (2004). Resultados de las pruebas PISA 2000 y 2003 en México. *Habilidades para la vida en estudiantes de 15 años*. México: INEE.

Villa-Lever, L. (1988). *Los libros de texto gratuitos: La disputa por la educación en México*. Guadalajara, Jalisco, Mx.: Universidad de Guadalajara.

Vitela, N. (2002). Brain Drain, cited by Hector Carreón, The Great Mexico Brain Drain: The other face of Mexican immigration. Available online: http://www.aztlan.net/mexico_brain_drain.htm

Von Bleyleben, K. A., Von Bleyleben, A., Hassenpflug, W., & Klug, F. J. (1993). Mexico: Baedeker, NJ: Prentice Hall Inc., Englewood Cliffs.

Williams, P. (2000). Brain drain. Federation for American Immigration Reform. Wasington, DC: Author. Available online: http://www.fairus.org/site/PageServer?pagename=iic_immigrationissuecenterse514

23

IN BETWEEN THE GLOBAL AND THE LOCAL: THE POLITICS OF MATHEMATICS EDUCATION REFORM IN A GLOBALIZED SOCIETY

Paola Valero
Aalborg University, Denmark, paola@learning.aau.dk

Abstract: Understanding recent globalization processes demands grasping the relationship between the social practices in micro-contexts and macro-contexts, in search of the mechanisms that have worsened the distribution of material, human and knowledge resources in the world. If mathematics education research and practices are to be committed to social equity and justice in the midst of globalization, then they need to address the ways in which they are implicated in the production of a particular social order. In order to do so, mathematics education research needs to open its scope and allow linking the classroom with other spheres of social action. In this way it is possible to gain a broader understanding of the multiple forces that constitute mathematics education, and particularly its reform. Based on a case of a Colombian school, the paper illustrates the way in which mathematics education is being constituted in between macro, global and micro, local contexts in a time of reform

Keywords: Mathematics education reform, socio-political research in mathematics education, macro-micro analysis in mathematics education research, mathematics education as social practices

A first glance at Esperanza Secondary School

I got ready for my first visit to the school. As I drove from my home towards the west of the city, the landscape changed and green residential areas were replaced by many small businesses: bakery shops, beauty parlors, groceries stores, and plenty of garages. Public transportation became denser as buses stopped to collect the crowds of people on their way to work. Uniformed school children also tried to catch public transportation, bearing the burden of big school bags on their shoulders. I reached the school gate at around 6:30 A.M. The place was deserted, and I could only see two street children

B. Atweh et al. (eds.), Internationalisation and Globalisation in Mathematics and Science Education, 421–439.

approaching slowly in their desperate search for food in the piles of garbage. The school redbrick building was notorious for its high walls and for its height that exceeded most of the two to three stories houses in the neighbourhood. The many broken windows and the thin grid-bars behind them were also a typical symbol of a District school building. I waited outside and observed the large, empty square in front of the school. This sand-dusty area was decorated with scattered rubbish, the natural result of having many school kids around. Ten minutes later, teenagers and teachers started arriving; some jumped out of a public bus, others arrived walking, and some of the teachers came by car and parked inside the walls of the school. I decided to go in as well.

The gatekeeper let me in and told me to wait in the roughly paved schoolyard. Viviana, my contact, welcomed me warmly. She introduced me to all teachers who crossed our way to the second floor, where the academic coordinator's office was situated. Irma, the academic coordinator, expressed their willingness to share their everyday with me, and their expectation of feedback. Viviana gathered the group of mathematics teachers and suddenly five women in their thirties and forties were standing by the door, ready to meet me and to go to their classrooms to start the school day. Laura, Julia, Mercedes, Ana and Viviana introduced themselves cheerfully. We shook hands and they ran to class when they heard Juan, the discipline coordinator, approaching to hurry them up into their respective rooms. I was also introduced to Juan whose cordiality and toughness helped in keeping the school under control.

Irma showed me the school building, the administration offices, the staff room, the cafeteria, the library, the computer room and the science lab. When the school bell rang, well-dressed teachers with aprons switched classrooms. There were more women than men among the 40 teachers in the school. The 910 students were adolescents between 11 and 19, and were distributed in six grades of three or four groups per grade. Each class had in average 41 students. Students wore uniforms and their appearance ranged from being very clean and neat to being dishevelled and sometimes dirty. The building, classrooms, desks, offices, walls and toilets were worn out by the continuous use and abuse that students in three different school shifts make of them during the ten months of each school year.

The school day was over. Many things struck me that day: a desk graveyard at the end of a corridor, the ill smelling toilets, the bars behind the many broken windows, and the highly risky building with its slight inclination and a severe crack. The square in front of the school was now alive: Ice-cream vendors, pirate CD and cassette retailers, fast-food sellers, hand towels and clothes bargainers spread their little informal shops all over the place. Students stormed out of the school. Teachers left in a hurry to make it on time to their next job, and I left back to my protected, privileged home, at the opposite extreme of the city.

Esperanza Secondary School, as many schools in the world at the end of the 1990s, was undergoing change. This was also evident in the teaching and learning of mathematics. The story illustrates my impressions when visiting this typical, low-class state school in Bogotá, Colombia. The story captures observations that do not seem to be related with the teaching and learning of mathematics. However, they highlight the environment in which those practices take place. While carrying out my research, I realized that it is not possible to disassociate mathematics education practices from their context, and that my research should find ways of making that context part of its gaze.

It is my contention that mathematics education research needs to open its scope in order to gain a broader understanding of the multiple levels of social action implicated in mathematics education reform. I also argue that such an opening in scope is a demand posed by globalization processes which, on the one hand, make more evident the constitutive relationship between social practices in micro-contexts and social practices in macro-contexts, and, on the other hand, have worsened the distribution of material, human and knowledge resources in the world. If mathematics education research and practices are to be committed to social equity and justice, then they need to address the ways in which they are implicated in the production of a particular social order.

In order to unfold my arguments, I start the chapter with some considerations about the notion of context and its inclusion as part of the research focus in mathematics education when adopting a perspective interested in current processes of globalization and internationalization. Then, based on the case of the Colombian educational reform in mathematics education during the 1990s, I present an analysis that highlights the connection between global trends and particular events in Esperanza Secondary School.[1] The examination of the case does not only intend to provide insight into a particular context, the Colombian context (about which international mathematics education research has little information), but also and primarily to illustrate that change in mathematics education is at the crossroad of global and local contradictory forces. Finally, I present some concluding reflections about the challenges that globalization poses to mathematics education research with a concern for social equity at local and global levels.

1. Addressing Context in Mathematics Education Research

Let me start with some remarks about the notion of context and its significance for mathematics education research in the light of current processes of globalization. According to the Websters' Encyclopaedic Unabridged Dictionary (1996,

---

[1] Although it is not pertinent to enter into the methodological details here, I want to point that the study followed a variety of qualitative methods that made possible to gather information about teachers and students in the classroom, teachers as a collectivity in the school, the role of school leaders in mathematics education related decision-making, and about the process of policy making at national and local level concerning mathematics in secondary schools (Valero, 2002).

p. 439), the term *context* (what goes together with the "text") refers to "the set of circumstances or facts that surround a particular event, situation, etc." In research, the definition of the problem and focus of a study brings with it a definition of the context of what is being researched. In mathematics education research, one can always point to the circumstances that surround the particular event or phenomenon under study. As Vithal and Valero (2003, pp. 552–554) and Valero (2002, pp. 107–112) argue, different research trends have dealt differently with the notion of context according to how theoretical assumptions help shaping the research focus of particular studies. In studies with an emphasis on mathematical learning as cognitive processes context is often related to the mathematical problems, which motivate the learning of a given content. In studies with an emphasis on the significance of interpersonal interaction for learning, context is often conceived as the social, interpersonal relation among learners and/or learner and teacher, which trigger individual mathematical thinking processes. In studies with an emphasis on the socio-cultural situation in which learning takes place, it is a socially structured situation either in the form of a classroom community or an out-of-school group, what constitutes the context of individual participation in mathematical practices and, therefore, learning. In studies with a sociological perspective, context is often constructed as the large social, political and economic structures within which meanings, practices and discourses about mathematics learning and education are generated. In the four research trends mentioned above, it is clear that the interest of the researcher is the "text" or the focus, and not so much the "context". The context of a research object may sometimes be described or referred to, but it is very seldom taken into consideration as part of the study. Somehow the context is seen as a "recipient" which contains the substance of the study, but which hardly makes a significant alteration of that substance, and, therefore, does not need to be directly addressed.

Let us recall Esperanza Secondary School. As mentioned above, it was clear to me that the surroundings of the school with street children and the daily blooming of an informal economy at the front gates permeate the school. The following episode showed me its significance for mathematics education: A pair of students in a 10th grade class seemed not to be interested in maths and adopted a quite irritating attitude towards their surroundings. Despite of my attempts to convince them about the advantages of being good in school and at maths, one of them expressed his concern:

> José: The only class I would like to pay attention to is English because I want to get out of this fucking place and go to the USA. Though, I don't even manage to say 'Hello, good morning'.
>
> (Valero, 2004a, p. 38)

How can one interpret José's concern without linking his intentions to learn (mathematics) with the fact that, given a deep economic crisis, migrating was the only possibility for many Colombians to have a life? How can one consider policy-makers', school leaders' and teachers' intentions to transform mathematics education practices without considering the intricacies of the larger education reform at the moment? In other words, how could I engage in the study of mathematics education practices in that school without making a clear effort to look at their larger context and making it part of my research interest? Contrary to what seems to be the case in most research in mathematics education, I tried to expand my methodology and my analysis to be able to find ways of connecting the larger social, political and economic context with the practices of teachers, leaders and students in the school and, in this way, venturing in an understanding of the sociological complexity of the social practices of mathematics education.

Out of my analysis and based on the socio-political approach to research that I had adopted (see Valero, 2004b), there emerged a notion of context that refers to the series of historical and structural macro conditions interpenetrating the micro conditions and organization of the practices of mathematics teaching and learning in schools. This definition points to the constitutive and dialectical relationship between the "text" and its "context". It implies that research intending to grasp practices at a sociological micro level – the level of individual agency of teachers, students and school leaders in relation to mathematics education practices – has to find ways of linking them to a sociological macro level of social action – the level of structures – where meanings, discourses and systems of reason are historically constructed.[2]

This view of context and its significance for my research had implications for how to conceive the scenario of globalization and internationalization at the turn of the 20th century. Drawing on sociological studies, Skovsmose and Valero (2002) propose to link mathematics education practices and research with some of the characteristics of the contemporary world order. This order is characterized by the consolidation of an informational society (Castells, 1999) in which technological expansion has changed the source of value from material capital to knowledge and learning capital. It is also characterized by globalization, the process due to which our environment – in political, sociological, economic, or ecological terms – is permanently reconstructed through inputs from all corners of the world. Internationalization, understood as the increase in the exchange of social and political actors across national boundaries, is also part of the mechanisms for the constitution of a "world-village". These processes go hand in hand with the expansion of particular economic and political models, namely late capitalism, neoliberal regimes and representative democracy. In this sense, globalization also relates to the expansion of homogenizing discourses, based on dominant Western, post-industrial culture, that install the belief on the desirability of a given social order and on the universal commitment to the achievement of certain political ideals.[3] More often than not, globalization processes

---

[2] Behind this formulation there is the classical debate of the "macro-micro link" in the social sciences. For details about the implications of this debate in recent educational studies see Martín Izquierdo and Moreno Mínguez (2003).

[3] For a critical discussion of globalization and neoliberalism see for example Ocampo (2003), Raplay (2004) and Rupert (2000).

are effected at structural levels and impact people in their everyday lives. The meeting of processes happening in macro levels with social processes at micro level is not smooth but rather filled with contradictions and dilemmas that are sorted out in the multiple levels of social action through which globalization operates.

Such a conflictive nature has been captured in the formulation of the two salient paradoxes of the informational society, which are of relevance for the role of mathematics education in the current order (Skovsmose & Valero, 2002, pp. 384–387). The *paradox of inclusion* designates the contradiction between the discourse of the current neoliberal, globalised model of social organization, which emphasizes democracy, universal access and inclusion as a stated principle, and the deep disempowerment and exclusion that certain social sectors actually experience as a result of the practices associated to that discourse. The *paradox of citizenship* refers to the contradiction between the role of education intending to prepare for active, autonomous, critical citizenship, but at the same time ensuring adaptation of the individual to the given social order. This paradox emerges from the fact that the learning society, claiming the need of relevant, meaningful education for current social challenges, reduces learning to a matter of necessity for adapting the individual to social demands. The presence of these two paradoxes in the current global order is problematic since it challenges very basic principles of a social, radical democracy with a concern for equity and social justice. If mathematics education is seen as key practices in the current order, it is in the critical position of either contributing to the installation of the two paradoxes or of challenging the production and reproduction of the imbalances of globalization at both local and global levels.

Mathematics education research can also be implicated in the consolidation of the paradoxes of the informational society (Skovsmose & Valero, 2002). One of the ways of doing so is precisely the elimination of a serious consideration of the context of teaching and learning practices. As I have argued previously, the opening of the scope of research to see micro levels in relation to macro levels of practice is a way of placing mathematics education in the complexity of systems of action and meaning that give a role to mathematics education in our current contemporary society. Nowadays it is unavoidable to consider that many phenomena of social, political and cultural nature influence the practices and discourses of mathematics education in schools. In what follows I present an analysis of the educational reform process in Colombia and of the associated changes in mathematics education. I concentrate on three levels, namely, the general national and international frame, at the end of the 1990s, in which the educational reform took place; the policy frame for the mathematics curriculum; and the series of initiatives taken by the staff of Esperanza Secondary School in their implementation of intended educational and curricular changes. I attempt to make evident the connection between these levels in a search for possible clues about how different actors in these levels respond to the challenges posed by the paradoxes of the informational society.

2. Educational Reform in Colombia in the 1990s

Colombia is a diverse country. The diversity of peoples and life experiences generated through our collective and individual history make part of both the context of Esperanza Secondary School and of its very same constitution. The teaching happening there, the experiences of leaders, teachers and students – as much of my own experience as a researcher in the school – has to be discussed in relation to the structural factors that interpenetrate the world of the school. Therefore, a starting point to examine the interplay between the context of the school and the teaching and learning of mathematics in it is considering the historical framing of events happening there during the 1990s.

Three events had a strong influence in the educational development during the 1990s in Colombia, and provided a general context for the emergence of the new Currricular Guidelines in School Mathematics (MEN, 1998). These events are: the introduction of neoliberalism as a "necessary" model for economic and political governance for the new times of globalization and internationalization; the proclamation of a new Political Constitution that in 1991 emerged as the first inclusive and participatory political agreement in the democratic history of the country; and the promulgation of a new Law of Education that intended to adjust the national educational system to both internal and external challenges.

The decade of the 1980s represented a time of change in most Latin American countries. In political terms, it was a time of democratic transition after long dictatorships. Although this was not the case of Colombia, having the most stable representative democracy in Latin America, the country entered a process of democratic progression intended to renew the old frames of political participation. Political changes went together with economic changes. The protectionist economic management that characterized the region entered a crisis when international demands for globalization pushed the adoption of neoliberal models. In economic terms, neoliberalism stands on the assumption that the internal logic of a free market can regulate all exchange relationships and, as a consequence, shape social and political relationships among rational individuals. Opposed to a centralist, paternalist and protectionist organization, neoliberalism proposes a strong decentralization of administration, a reduction of the State including its welfare functions, and transference of the provision of different services into the realm of the private (Ocampo, 2003). Together with this, the advance of the informational society (Castells, 1999) has also positioned knowledge and information as sources of value, and has altered time and space boundaries dissolving the limits of national States into broader boundaries for the exchange of traditional and virtual goods. In particular, science, technology and constant learning capacities are part of the conditions that a country must have as sources for a powerful positioning in the global, virtual village. This political and economic agenda was implemented in different ways in each Latin American country.

In Colombia, the neoliberal presidency of César Gaviria (1990–1994) echoed the international scene and introduced dramatic changes, among them a reform in the administration of education. This reform of decentralization was seen as an effective way of tackling an endemic education crisis, made evident in the inefficient, bureaucratic, central administration; in the low levels of student achievement; in the high levels of student drop-outs; and in the lack of acknowledgement of local differences in the student population, among others (Londoño, 1998). The diagnosis was compatible not only with similar reports from other Latin American countries, and the reform was in line with the recommendations that emerged in international scenarios such as the UNESCO "World Conference on Education for All" (UNESCO, 1992). Neoliberalisation in education meant in Colombia the adoption of a universal, global agenda where "achieving education with equity and quality, decentralizing curricular policies, educating democratic subjects, adopting the scientific-technological revolution, and determining a core curriculum" (Londoño, p. 54) are priority issues. These tenets certainly do not disagree with the development of educational systems in richer nations, as Apple (1996, 2000) has discussed extensively.

The Political Constitution of 1991 also contributed significantly on shaping the political and educational landscape in Colombia. The 1990s were a critical moment of national reconciliation that led to a constitutional reform. The New Political Constitution of 1991 was the first inclusive attempt of bringing together a historically fragmented society. A popularly elected Constitutional Assembly formulated a *carta magna* that marked the end of a constitutionally closed democracy and the start of an open participatory democracy (Murillo & Valero, 1996). The Constitution declared education as a key element in the democratic reconstruction of the country and, therefore, part of the social rights to which all children should have access. The Colombian State must not only regulate and inspect the provision of education, but also protect and make effective the right of education in its fundamental function of building a democratic society (Londoño, 1998, p. 62).

Within this broader frame, the General Law of Education (MEN, 1995) and its corresponding administrative act established a new educational organization in the country. This general law embodied the vision of education and national development that underlay the whole educational reform, according to the Constitutional intentions. In order to build a Colombian society for future challenges, the individual, in all her potential, has to be put in the center of a society that needs to learn how to be socially democratic. Knowing is at the service of not only more knowledge production, but also the maintenance of the environment, the boosting of production and technology, and the consolidation of a nation that can also participate as a part of a global world. In order to achieve this vision in the formal education system, the general law set up the general aims for the whole formal education system and for each level of schooling, including the compulsory school subjects. The law also established different mechanisms of reform implementation and institutional development, created new regulations for teacher education, set up the frames for a whole new assessment system based on outcomes, and regulated the local, regional and national administration of the educational service. Shortly, the law set a framework that intended to bring the transformation of the Colombian educational service in many aspects such as policy, administration, funding, curriculum, teachers' professional development, assessment, school organization, etc. The broad educational change was conceived as a general national transformation for the challenges of both the national demands and the internationalized world.

The previous account points out the complexity of the broad macro-context in which mathematics education reform was introduced. This macro-context is far from being unproblematic. Some points of critique have been raised to the neoliberal project and the advance of democracy in general, and to the implications of this discourse in the formulation of particular educational policies in Colombia. First of all, it is not a secret that the theoretical assumptions of neoliberalism have not worked so well in practice. After at least a decade of the expansion of these ideas in many developed nations, and their enforcement into almost all developing nations, it is clear that freer markets and freer individuals do not necessarily lead into stronger economies and democratic political relationships (McLaren, 1999). On the contrary, the gap between rich and poor, nationally and internationally, has increased due to the adjustment of national political and economic structures to the new international parameters. Problems of equity and social justice have also increased, putting at stake the democratic discourse associated to all these adjustments. In education there are significant examples of these contradictions. In the cases of the USA, England and Australia, Apple discusses how the assumption of the improvement of quality in the educational service given the introduction of competition into it does not hold in reality. He shows that the atomization of decision-making in highly stratified societies creates the fallacy of equal opportunities and participation. Apple concludes that "neoliberal policies involving market 'solutions' may actually serve to reproduce – not subvert – traditional hierarchies of class and race" (Apple, 2000, p. 247).

Second, globalization in education has been associated with discourses that emphasises the need to form flexible, ready-for-change individuals, whose main competency is their ability to "learn to learn". Masschelein (2001) argues that this type of discourse about the learning society reduces the whole educational enterprise to a mechanism of *zoological* survival. This view opposes a conception of life as *human* existence, where unique subjects search for meaning, in an attempt to initiate events that contribute to secure a sustainable, durable common world. As a richer conception of education is suppressed, education becomes a mechanism that strengthens individualization and the selection of the most adaptable beings. Education – under a façade of accessibility–promotes stratification. As Flecha (1999, p. 67) points, "the knowledge prioritized by the new forms of life is distributed unevenly among individuals, according to social group, gender, ethnic group, and age. At the same time, the knowledge possessed by marginalized groups is dismissed, even if it is richer and more complex than prioritized knowledge. More is therefore given to those who have more and less to those who have less, forming a closed circle of cultural inequality".

Third, Londoño (1998), in her analysis of the Colombian General Law of Education as a national project of school autonomy and democracy, discusses how the grounding of the Law in neoliberal ideology poses two main difficulties. On the one hand, neoliberalism has set homogenizing forms of thinking and feeling, by means of an "unlimited expansion of the rational domain" (p. 58) and the institutionalization of techno-scientific processes for the production of the social world. Educational autonomy and the discourse of educational democratization can only have an instrumental connotation since they refer to the "efficient and effective execution of assigned tasks according to parameters previously established by extra-social organisms" (p. 59), but not to an authentic possibility for participants in educational processes to set their agenda. In this sense the contradictions of a restricted autonomy are installed in the educational arena. In such a landscape it is very easy for the most apt – the quickest to grasp the official discourse and the strongest to bring it into practice – to "survive" and be able to put forward their interests. A growing gap between people and their actual capacity to act constitutes a democratic challenge. On the other hand, the extreme emphasis on the individual and the conception of liberal democracy through representation in decision-making goes against views of radical, collective, deliberative democracy in which groups of people act together for the transformation of their life conditions (Valero, 1999). These two obstacles point towards the inconsistency of achieving a democratic social reconstruction, such as the one intended by the Political Constitution of 1991, through education, as proposed by the law of 1994, in a dominantly neoliberal economic and political scenario. As Londoño (1998, p. 60) concludes it is questionable the extent to which it is possible to build a new social imagination, that can represent a change from old, traditional democratic and educational projects.

The problems and contradictions in the macro context may pose serious challenges to mathematics education in countries such as Colombia and in schools such as Esperanza. It is clear that the Colombian process of educational reform is formatted by global trends in ways that evidence how easily the paradoxes of the informational society can also become paradoxes at a national level. In what follows I get closer to mathematics education in Esperanza Secondary School through the presentation of the policy frame in which mathematics education in the country was supposed to be carried out and the way it meets the mathematics education practices in the school. My hope with such a presentation is to provide elements to analyze whether, at least in a policy level, there is a possibility for mathematics education to engage in responding to the paradoxes of the informational society.

3. Mathematics Education Policy in Colombia during the 1990s

The General Law of Education established a broad aim for mathematics education in both basic secondary and middle school: mathematics education should develop "reasoning capacities through the mastering of numerical, geometrical, metric, logic, analytic, sets, and operations and relations systems; and through their use in the interpretation and solution of scientific, technological and everyday problems" (MEN, 1995, Amend. 22). This aim was the basis for the formulation of curricular guidelines in mathematics. Led by the mathematics education research team in the Ministry of Education, a national group of researchers, teacher educators and teachers produced the Curricular Guidelines in School Mathematics (MEN, 1998). This document was meant to inspire teachers to discuss fundamental issues about the teaching and learning of mathematics. In this way, it could help teachers making informed decisions in their task of designing and implementing curricular programs.

The Curricular Guidelines intended both to overcome the limitations of the previous centralized, national curriculum of the 1980s and early 1990s, and to build on its achievements. The old curriculum, based on "Systems Theory" in mathematics,[4] integrated the positive aspects of the structuralist program of the "New Mathematics" with a Piagetian constructivist learning theory and its didactical implications (MEN, 1991). This old curricular proposal did not have the expected results in improving student's conceptual understanding due to its internal contradiction between contents, teaching methods and aims for the teaching and learning of mathematics (Agudelo, 1996); teachers' rejection of its structuralist and didactic-technological nature (García, 1996); its detailed prescription of syllabi together with national high stakes tests in mathematics which led to procedural teaching and rote learning (Valero, 1997); and the lack of opportunities for teachers to question traditional teaching practices in favour of the adoption of the new proposals (Perry, Valero, Castro, Gómez, & Agudelo, 1998). Despite the pitfalls, the curriculum of the 1980s represented advances for mathematics education in the country because it was the first systematic attempt to spread strong ideas about the didactics of mathematics among mathematics teachers and it institutionalized discussions about mathematics education, especially around the idea of "systems" (García, 1996).

The new Curricular Guidelines of 1998 (MEN, 1998) kept the idea of school mathematics as knowledge systems, but introduced the advances of international mathematics education research concerning constructivist, problem-solving oriented teaching and learning processes as fundamental ideas about how teachers and students could interact in a classroom. In contrast with the previous guidelines – and in agreement with the discourse of autonomy and decentralization in the Political Constitution and the Law of Education – these guidelines do not intend to be a centralized and detailed prescription of teachers' work, but an open guide for reflection among teachers in their role as curriculum designers and implementers. Mathematics teachers have the responsibility of choosing contents and methodologies that are appropriate to their particular students, and which are in agreement with their school's educational project. Nevertheless, the guidelines set the ultimate goals and outcomes of all mathematics education, which are "to improve students' conceptualization capacities, to promote their understanding of their possibilities, and to develop competencies for tackling the complexity of life and work, dealing with the resolution of conflicts, managing uncertainty, and strengthening the culture for a healthy, holistic life" (MEN, 1998, p. 17, my translation).

[4] The curriculum was based mainly on the ideas of Carlos Vasco who defines a mathematical system as a set of objects and their relations and operations (MEN, 1991).

The Guidelines put forward five topics that teachers need to address when designing their own curriculum. First, teachers need to reflect about different positions in the philosophy of mathematics and the implications that adopting a particular view has on mathematics education in the school. Second, teachers need to consider the reconceptualisation of mathematics education based on research results. In particular the guidelines highlight the work of Ernest (1991) about the connection between philosophy of mathematics and mathematics education, and Brousseau's (1986) theory of didactic situations.[5] Third, teachers need to consider views of school mathematics as a "powerful intellectual tool whose mastery provides intellectual privileges and advantages" (MEN, 1998, p. 29, my translation), and as knowledge embedded in a social and cultural context. Therefore, the learning of such type of knowledge demands taking into consideration students' interests, feelings and culture. Fourth, teachers have to discuss curricular models. A three-dimensional tool for curriculum organization includes a reflection about the general learning processes at stake, the basic knowledge and specific processes linked to a particular mathematical system, and the students' contexts which give meaning to the mathematics they learn and which are captured in problem situations – of mathematics, of daily life, of other sciences – that teachers present to students in order to learn mathematics. These three dimensions can be combined in different ways in order to produce different curricular models. The guidelines invite teachers to develop their own. Fifth, teachers need to consider assessment. The guidelines discuss the central ideas for mathematics education related to broader changes in the assessment system where a variety of assessment forms are put together in a qualitative system in which general outcomes, described in terms of minimum levels of achievement, are operationalised as observable indicators of behaviour. The latter allow teachers to make a judgment of students' achievement in relation to stated outcomes.

The Curricular Guidelines for School Mathematics are at the crossroad of the global international transformation of education and the particular national and educational change in Colombia, during the decade of the 1990s. On the one hand, these policy recommendations about the direction of mathematics education in Colombian schools is permeated by global political agreements about the role of education in current societies, as expressed in UNESCO (1992), and are implemented within the administration frames of neoliberal regimes. It is interesting to notice that many of the changes that have been described for the Colombian case are similar to processes in other countries in the world. Despite national particularities, educational changes have followed similar patterns in, for example, the marketisation, corporatisation, commercialisation and privatisation of educational services.[6] On the other hand, the intended changes in mathematics education are influenced by a growing political interest in the strengthening of mathematical

[5] For a discussion of the effects of the internationalization of mathematics education research and national curricula in developing countries see Vithal and Valero (2003).

[6] See for example Mok and Welch (2003) for the case the Asia Pacific region and Berryman (2000) for some European and Central Asian countries.

competencies for improved economic productivity and social democratization (see Skovsmose & Valero, 2001) and by the expansion of an international research field which provides a foundation for discussions about the different components of the mathematics curriculum.

In relation to the latter, an analysis of the discourse of the Colombian Curricular Guidelines (Valero, 2002, pp. 131–133) shows that it privileges a view of mathematics as a powerful thinking tool, and focuses on conceptualization and abstraction as aims of the curriculum. The maintenance of the view of school mathematics as systems (in contrast to other possible organizations of school mathematics) keep a link to structuralist views of school mathematics which have been associated with the abstract, procedural character of school mathematics in the previous curricular guidelines. The choice of particular research literature (Brousseau's theory of didactical situations) for the understanding and reconceptualising school mathematics reinforces an internalistic view of the discipline and of the practice of teaching and learning mathematics (Skovsmose & Valero, 2001, pp. 40–41).

The depth of the explanation given to these ideas contrasts with the loose and tentative statements provided in relation to other curricular dimensions such as the psychological, cultural and sociological (Skovsmose & Valero, 2002). Concerning the psychological orientation of the guidelines, there is not much that will give an indication of the concern for the inclusion of psychologically meaningful ideas in the curriculum, except for the mention of constructivism as a recommended epistemological position to adopt when thinking about mathematical learning. Although the guidelines mention the issue "paying attention to students' culture" (MEN, 1998, p. 30), it is not clear whether the interpretation of culture adopted transcends a simple recognition of all people making part of a human group – with particular features and particular preconditions for learning. The mention and discussion of culture and its role in mathematics education is not as broad as the description and emphasis on the mathematical content. Besides, considerations of cultural context seem to be reduced to the necessity of providing a task-context in problem situations. Finally, there is not a clear indication in the document of a critical, sociological interpretation in relation to the role of mathematics and mathematics education itself in social and technological action. Whereas other curricular guidelines (e.g., NCTM, 2000) clearly justify the teaching of mathematics in society (what may open a sociological and political space in the curriculum), the Colombian guidelines do not provide a contextualization for the teaching and learning of mathematics within a broader social and political theoretical justification.

The Curricular Guidelines for mathematics in Colombia, as a frame of reference for teachers' curricular design in mathematics, offers an incomplete frame for tackling the challenges raised to mathematics education by the current global, informational society. Mathematics education, intending to face the paradoxes of inclusion and citizenship, needs to adopt logical, psychological, cultural and sociological perspectives of mathematics and its curriculum as essential and complementary. The dominance in a curriculum of one of those interpretations may reproduce imbalances that have been associated with the exclusion of many students from mathematics education practices (Skovsmose & Valero, 2002). Still, as a guide for curriculum design and implementation, there could be possibilities for teachers to integrate the missing perspectives.

4. Mathematics Education Change in Esperanza School

In 1999 Esperanza Secondary School was in constant transformation. The slow bureaucracy in the local administration had left the school without a formally appointed rector. Confrontations between the Teacher Union and the national government, which started plans of privatization of state schools, led to frequent long strikes, which forced teachers to reschedule activities during weekends to catch up with missed lessons. Parents supported school activities but also caused troubles. Extraordinary teachers' meetings took place to solve conflicts with parents. The change in the National System of Evaluation of Educational Quality forced the school to participate in a standardized test for 7th and 9th graders. This implied not only students' preparation, but also discussion of the effects of such an evaluation on teachers' career. In short, mathematics education at Esperanza was taking place in an environment of constant perturbation. Mathematics teachers had to respond not only to mathematics-related challenges, but also to the larger challenges that the local, national and global contexts were imposing on them.

Mathematics teachers have sailed through the changes that the educational reform of 1994 introduced. They had organized the teaching of the approximately 910 students in the school, from 6th to 11th grade, all of whom had mathematics as a compulsory subject. Julia, Mercedes and Viviana had been involved in a one-year professional development program run by a team of mathematics education researchers in a private university. This program, part of a teacher professional development initiative of the local educational authorities, gave them practical and conceptual tools to actively engage in curriculum design, implementation and inquiry in their school. Ideas such as higher order mathematical thinking, social constructivism and use of graphing calculators and educational software such as Cabri Geometry in teaching had got their interest. These three teachers have also managed to "infect" Ana and Laura and the school leaders with the bug of innovation. The group was moving forwards, but still struggling with the many demands of change.

Although the reform provided a frame for teachers to design their own curriculum – in agreement with the school's institutional educational project and taking into consideration the Ministerial Curricular Guidelines – the team of teachers had not been able to dedicate much time to this activity given the multiple new demands that divided their attention. One of the most challenging transformations was the creation of a new assessment system and the demand of constructing a qualitative outcome-based assessment system that suited the school and teachers' collective formulation of outcomes. Teachers perceived this as a task that could not wait, not only because of its implications for everyday assessment, but also because of the general change in the national high stakes examination at the end of Grade 11. Teachers had to make sure that students had a chance of coping with qualitative assessments both in the daily life and at the end of their schooling. Thus, teachers continued to use the previous compulsory curriculum (MEN, 1991) as a reference, but made advantage of a diversity of new textbooks to guide their teaching. Teachers individually and as a group tried to respond as well as they could to the process of change.

The team of mathematics teachers and the physics teacher designed a project in order to tackle the problems that they had identified in their subject. The project starts from the recognition of the connections between the general current demands to students from the Colombian society – as formulated in the Political Constitution and the General Law of Education; the commitment of the school with the improvement of students' life conditions – as formulated in the school's institutional educational project; and the potential contribution of mathematics education to the education of democratic citizens – as formulated in the Colombian Curricular Guidelines and international trends of mathematics education reform. The justification for the project and its theoretical foundations make clear that teachers interpret the demands of mathematics education reform in relation to the large social order in which their practice is embedded. The project intended to formulate a set of actions that teachers could use and develop in their teaching.

It is possible to find relations between the project and diverse ideas that make part of the cultural frames that, at the end of the 1990s, dominated in Colombia. The project was conceived as a long-term strategy with the intent of developing seven interconnected actions: (1) The use of portfolios as means of consolidating students' learning and supplementing assessment. (2) The realisation of special workshops in which students will have the chance of using diverse concrete materials in playful activities that contribute to a more meaningful mathematical learning. (3) The development of real-life projects which allow students establishing connections between mathematics and society, and developing a critical stance towards the role of mathematics in society. (4) The introduction of graphic calculators as a technological resource in mathematics teaching and learning in order to promote richer mathematical experiences for students. (5) The realization of school and inter-schools "Mathematics Olympiad" in mathematics and physics, with the intent of promoting problem-solving capacities and developing a sense of healthy competition among students. (6) The participation of teachers in diverse professional activities, inside and outside the school, in order to implement the previous five actions, reflect on their practice, and communicate to other colleagues the achievements reached. (7) The engagement of teachers in a systematic enquiry of student's learning as a means of tracking the impact of the previous six actions on mathematics education in the school, and of providing feed-back for the further development of the strategy in the long-term (Esperanza, 1999, pp. 3–4).

Even though these actions emerged from the teachers' perceptions of their needs and problems in the teaching of mathematics to the students of Esperanza School, they keep a connection with different discourses that, from the macro-context, pose demands to teachers' work. These discourses penetrate practice and become reified in the micro-context of mathematics education in the school. It is possible to identify the following ideas in operation. First, the need for accountability of social processes illustrated by means of educational assessment has been particularly emphasized by the technocratic rationality of political parties with neoliberal and neoconservative agendas around the world and in Colombia, as well. The discourse of evaluation and assessment is recontextualised by mathematics teachers and, in the case of Esperanza, is expressed in a concern for constructing functional assessment devices. Second, the need of individual engagement in learning through motivation and creation of affective bounds is one of the central points of individualization processes. For teachers in Esperanza attention to the individual goes hand in hand with providing sources of meaning for mathematical instruction. Third, the need of connecting school with other arenas of practice, such as everyday-practice and work, are demands being posed by the role of education as a tool of governmentality. The idea of education as preparation for active participation in society and in the labor market is transformed by the group of teachers to a concern for making mathematics learning useful in students' lives. Fourth, the need of getting involved in the technological development of society and in the consumption of technology is constantly highlighted as part of the discourse of the informational society. For teachers this is transformed in the concern for the involvement of IT tools (computers and calculators) to support mathematical learning (in an environment with precarious resources). Fifth, the need of creating standards of competition which are an important element in the discourse of internationalization, globalization and the open markets economy is transformed by teachers in a concern for allowing students to have positive competition experiences in mathematics, among themselves and with other schools. Finally, the need for the professionalisation though innovation and research is a paramount feature of the "learning society" discourse. Teachers have participated in professional development strategies and have translated these ideas and demands into a commitment to an active learning from a systematic, collective examination of their own practice.

These discourses and their reification in different spheres of practice constitute frames of action for mathematics teachers in Esperanza. In the every day of school life, and in each of the mathematics lessons teachers are finding ways of coping with and responding to the multiplicity of challenges that fall upon their shoulders. With this variety of strategies they struggle to provide a mathematics education that may contribute to students' lives. Whether their efforts actually represent a reaction to the paradoxes of inclusion and citizenship is uncertain, but it is at least the best they can offer to the students of Esperanza Secondary School.

5. Linking the Local and the Global in Mathematics Education Research

There is no doubt that school mathematics change, alongside general school change, is a complex phenomenon and an understanding of its intricacies and contradictions has been a central challenge for mathematics education research in the last two decades. A great amount of research in many countries has been associated with the development of improvement initiatives and with the evaluation of reform projects, and many results have illuminated the advances and pitfalls of change. Kilpatrick (1997) reminds us that one of the most important lessons to be learned from the attempts of reform in mathematics education since the time of the "Sputnik Shock" is that, despite being often construed as a technical problem, "changing how and what mathematics is taught to our children is not a technical problem. It is a human problem that demands an understanding and appreciation of how people work together in classrooms to learn and teach and do mathematics" (p. 6). I would add to this realization saying that the human problem of mathematics education reform, especially at the turn of the 21st century, demands an understanding that goes beyond classrooms. In fact a great deal of research literature during the last decade has provided more insight in the operations of reform in classrooms (for example, Cobb, Yackel, & McClain, 2000). Such an understanding needs to embrace the organization of mathematics education in schools as a whole (Perry et al., 1998) and the organization and construction of mathematics education discourses and practices in larger fields of social action (Valero, 2002), including international and global spaces.

In my analysis of change in Esperanza Secondary School I illustrated how the global political, economic and educational trends are brought together in an historical moment, which determines the paths of educational transformation in Colombia. The larger scenario of change is reconstituted in particular initiatives and discourses about the mathematics curriculum. These discourses and ideas enter schools, where teachers in their everyday initiatives for improving mathematical instruction recontextualise and accommodate discourses and practices according to their possibilities. In all this process there are not only internationalized ideas about what educational development should be (such as the outcomes education agenda that dominated in the 1990s) but also about what the very same teaching and learning of mathematics could be (thanks to the expansion and internationalization of research findings in mathematics education). Through the case of Esperanza Secondary School I showed how the diversity of layers of context inter-connects and how the examination of their connections allow constructing a fuller picture of the complexity of mathematics education reform. The intermeshing of the global and the local allows raising questionings to the way in which mathematics education change may be contributing to the installation or the eradication of the paradoxes of the informational society.

Seeing the world from a perspective of globalization and internationalization does certainly imply widening the lenses of research. Research on mathematics education using lenses adjusted to see the intricacies of micro-levels of practice risks ignoring the challenges posed by current social processes operating far away from classrooms, but having a definite impact on them. The systems of reason that provide meaning to educational practices are partly produced in macro-structures. Mathematics education research needs to grasp the way in which macro and micro levels of practice are constantly intermeshing in the constitution of mathematics education. I hope to have illustrated that facing the paradoxes of the informational society is a task not only for practitioners but also for researchers. A globalised world also puts at stake the constructions of mathematics education research.

Acknowledgements

This paper is connected with the project "Learning from diversity", funded by The Danish Research Council for Humanities and Aalborg University. I thank Diana Stentoft, Troels Lange and Ole Skovsmose for comments to previous versions of this paper.

References

Agudelo, C. (1996). Improving mathematics education in Colombian schools: 'Mathematics for all'. *International Journal of Educational Development, 16*(1), 15–26.

Apple, M. (1996). *Cultural politics and education*. New York: Teachers College Press.

Apple, M. (2000). *Official knowledge. Democratic education in a conservative age*. New York: Routledge.

Berryman, S. (2000). *Hidden challenges to education systems in transtition economies*. Washington: The World Bank.

Brousseau, G. (1986). Fondements et méthodes de la didactique des mathématiques. *Recherches en Didactique des Mathématiques, 7*(2), 33–115.

Castells, M. (1999). Flows, networks, and identities: A critical theory of the informational society. In M. Castells, R. Flecha, P. Freire, H. Giroux, D. Macedo, & P. Wilis (Eds.), *Critical education in the new information age* (pp. 37–64). Lanham, USA: Rowman and Littefield.

Cobb, P., Yackel, E., & McClain, K. (Eds.). (2000). *Symbolizing and communicating in mathematics classrooms*. Mahwah, USA: Lawrence Erlbaum Associates.

Ernest, P. (1991). *The philosophy of mathematics education*. London: Falmer.

Esperanza School. (1999). *Mathematics project: "Sleeping with the Ghost"*. Bogotá: Author, unpublished project proposal.

Flecha, R. (1999). New educational inequalities. In M. Castells, R. Flecha, P. Freire, H. Giroux, D. Macedo, & P. Wilis (Eds.), *Critical education in the new information age* (pp. 65–82). Lanham, USA: Rowman and Littefield Publishers.

García, G. (1996). Reformas en la enseñanza de las matemáticas escolares: perspectivas para su desarrollo. *Revista EMA, 1*(3), 195–206.

Kilpatrick, J. (1997). *Five lessons from the new math era*. Retrieved October, 2004, from http://www.nas.edu/sputnik/kilpatin.htm

Londoño, M. (1998). *El proyecto de Autonomía y la escuela. Reflexiones para el Cambio en la Institución de la Sociedad Colombiana*. (Masters dissertation, Universidad de los Andes). Bogotá (Colombia).

Martín Izquierdo, H., & Moreno Mínguez, A. (2003). Sociological theory of education in the dialectical perspective. In C. A. Torres & A. Antikainen (Eds.), *The international handbook on the sociology of education. An international assessment of new research and theory* (pp. 21–41). New York: Rowman & Littlefield.

Masschelein, J. (2001). The discourse of the learning society and the loss of childhood. *Journal of Philosophy of Education, 35*(1), 1–20.

McLaren, P. (1999). Trumatizing capital: Oppositional pedagogies in the age of consent. In M. Castells, R. Flecha, P. Freire, H. Giroux, D. Macedo, & P. Wilis (Eds.), *Critical education in the new information age* (pp. 1–36). Lanham, USA: Rowman and Littefield Publishers.

Ministerio de Educación Nacional de Colombia (MEN). (1991). *Marco general de matemáticas. propuesta de programa curricular para noveno grado*. Bogotá: MEN - Dirección General de Capacitación y Perfeccionamiento Docente.

Ministerio de Educación Nacional de Colombia (MEN). (1995). *Ley general de educación. El salto educativo. Serie documentos especiales*. Bogotá: Empresa Editorial Universidad Nacional.

Ministerio de Educación Nacional de Colombia (MEN). (1998). *Lineamientos curriculares en matemáticas*. Bogotá: MEN - Dirección General de Capacitación y Perfeccionamiento Docente.

Mok, K., & Welch, A. (Eds.). (2003). *Globalization and educational restructuring in the Asia Pacific region*. New York: Palgrave Macmillan.

Murillo, G., & Valero, P. (1996). De una democracia restringida hacia una participativa: el peligro de la contra-reforma y la regresión en Colombia. In E. Diniz (Ed.), *O desafio da Democracia na América Latina* (pp. 493–511). Rio de Janeiro: IUPERJ.

NCTM. (2000). *Principles and standards for school mathematics*. http://standards.nctm.org/protoFINAL/cover.html

Ocampo, J. A. (2003). Globalization and development: A Latin American and Caribbean perspective. Palo Alto, USA: Stanford University Press.

Perry, P., Valero, P., Castro, M., Gómez, P., & Agudelo, C. (1998). *La calidad de las matemáticas en secundaria. Actores y procesos en la institución educativa*. Bogotá: una empresa docente.

Raplay, J. (2004). *Globalization and inequality: Neoliberalism's downward spiral*. Boulder, USA: L. Rienner.

Rupert, M. (2000). *Ideologies of globalization: Contending visions of a new world order*. London: Routledge.

Skovsmose, O., & Valero, P. (2001). Breaking political neutrality. The critical engagement of mathematics education with democracy. In B. Atweh, H. Forgasz, & B. Nebres (Eds.), *Sociocultural research on mathematics education: An international perspective* (pp. 37–56). Mahwah, USA: Lawrence Erlbaum Associates.

Skovsmose, O., & Valero, P. (2002). Democratic access to powerful mathematical ideas. In L. D. English (Ed.), *Handbook of international research in mathematics education: Directions for the 21st century* (pp. 383–407). Mahwah, USA: Lawrence Erlbaum Associates.

UNESCO. (1992). *Education for all*. Paris: Author.

Valero, P. (1997). A day to be true. Mathematics education for democracy in Colombia. *Chreods, 11*, 49–61. http://s13a.math.aca.mmu.ac.uk/Chreods/Issue 11/PaolaValero.html

Valero, P. (1999). Deliberative mathematics education for social democratization in Latin America. *Zentralblatt für Didaktik der Mathematik, 98*(6), 20–26.

Valero, P. (2002). Reform, democracy and mathematics education. Towards a socio-political frame for understanding change in the organization of secondary school mathematics. (Ph.D. dissertation) Copenhagen: Danish University of Education.

Valero, P. (2004a). Postmodernism as an attitude of critique to dominant mathematics education research. In M. Walshaw (Ed.), *Mathematics education within the postmodern* (pp. 35–54). Greenwich (USA): Information Age.

Valero, P. (2004b). Socio-political perspectives on mathematics education. In P. Valero & R. Zevenbergen (Eds.), *Researching the socio-political dimensions of mathematics education: Issues of power in theory and methodology* (pp. 5–24). Dordrecht: Kluwer.

Vithal, R., & Valero, P. (2003). Researching mathematics education in situations of social and political conflict. In A. Bishop et al. (Eds.), *Second international handbook of mathematics education* (pp. 545–592). Dordrecht: Kluwer.

24

SINGAPORE AND BRUNEI DARUSSALAM: INTERNATIONALISATION AND GLOBALISATION THROUGH PRACTICES AND A BILATERAL MATHEMATICS STUDY

Khoon Yoong Wong[1], **Berinderjeet Kaur**[2], **Phong Lee Koay**[3] **and Jamilah binti Hj Mohd Yusof**[4]

[1,2,3]*Mathematics and Mathematics Education Academic Group, National Institute of Education, Nanyang Technological University, 1 Nanyang Walk, Singapore 637616;* [4]*Department of Science and Mathematics Educations, Sultan Hassanal Bolkiah Institute of Education, Universiti Brunei Darussalam, Tungku Link BE 1410, Negara Brunei Darussalam*

Abstract: Internationalisation and globalisation involve many aspects. The first section of this chapter illustrates these aspects through practices of mathematics education in Singapore and Brunei Darussalam, including adapting global ideas about mathematics curriculum to local contexts, upgrading local publications to international level, and use of English as the main medium of mathematics instruction. On the other hand, comparative studies are often the medium through which both internationalisation and globalisation can take place. Many comparative studies such as TIMSS and PISA are multi-national and focus on students' performance on mathematics items designed by panels of mathematics educators from different countries. In contrast, the second section describes a bilateral study between Singapore and Brunei Darussalam about pupils' perceptions of mathematics learning. This study was a unique form of internationalisation because it was situated within a broader research agenda, with the noteworthy feature that mathematics educators met with non-mathematics educators to discuss education issues common to both countries. Despite wide differences in racial, cultural, and economic makeup, mathematics education in Singapore and Brunei Darussalam shares many similar practices. This chapter concludes with reflections about future issues of the internationalisation and globalisation of mathematics education

Keywords: bilateralisation, Brunei Darussalam, comparative study, culture, curriculum, homework, Islam, model drawing, problem solving, Singapore, teacher-centred, teacher quality, textbook

B. Atweh et al. (eds.), Internationalisation and Globalisation in Mathematics and Science Education, 441–463.

1. Introduction

The editors of this book have raised several interesting aspects related to internationalisation and globalisation: training of graduate students and movement of researchers from one country to another; publications with local to international standards; multi-national comparative studies; dissemination of curriculum documents and textbooks across countries; the role of English as an international language for communicating about educational research and practices. This chapter will discuss some of these aspects by examining practices of mathematics education in Singapore and Brunei Darussalam and using a bilateral primary mathematics study as a concrete example of how internationalisation can take place within an inter-disciplinary setting.

2. Mathematics Education in Singapore and Brunei Darussalam

The following descriptions covering practices in the past two decades are necessarily brief and general. An important observation is that despite vast differences in the make-up of the population and geography of these two countries and their education system, the mathematics curriculum, instruction in the classroom, and public examinations are very similar. This lends some support to the notion of a "global mathematics curriculum" at least at the systemic level and "global mathematics pedagogy", entrenched in the traditional method. Details now follow.

2.1 Mathematics Education in Singapore

Singapore is a small island-nation (699 sq. km) with a local population of 3.49 million, comprising mainly Chinese (77%), with other ethnic groups (14% Malay, 8% Indian, and 1% other)[1], and a sizeable expatriate population (0.75 million). Its multi-racial people have many different religious and cultural practices, so that inter-racial understanding and harmony is one of the most important factors for national survival. This message is strongly promoted in schools through the infusing of National Education in the school subjects (Wong, 2003; 2005). The main economic activities are in business and commerce, high technology products and services, construction, tourism, and transport. These activities require its people to be highly educated in mathematics, science, and technology as well as proficient in English. The per capita GDP is high (US$21,800), but the working environment is very competitive. The government often reminds its citizens that Singapore does not have any natural resources and its survival depends on the hard work and creativity of its people.

Singapore has a centrally controlled education system. The Ministry of Education[2] sets the mission (*Moulding the future of our nation*) and structure of the system,

---

[1] http://www.singstat.gov.sg, retrieved 25 July 2005.
[2] http://www.moe.gov.sg/, retrieved 25 July 2005.

develops the curriculum, organises public examinations (at ages 11, 15, 17), approves the textbooks used, employs the teachers, and carries out many other functions. In recent years, more decision-making functions, such as budget, teacher professional development, and niche school-based programmes such as Mathematics Olympiads, have been devolved to the schools.

Pupils are placed into various academic streams from age 10 years (Primary 5) onward. At the end of primary schooling (age 11 years), pupils sit for the public examination, called PSLE (Primary School Leaving Examination). The bilateral primary mathematics study reported in this chapter involved a small sample of these pupils. On the basis of the PSLE results, pupils enter secondary schooling in one of the four streams: Special, Express, Normal (Academic), and Normal (Technical). At the end of secondary schooling, pupils take the Singapore-Cambridge N- or O-level examinations. After this examination, they continue with further study or enter the workforce. Mathematics is a compulsory subject in primary and secondary schools.

English is the main language of instruction from primary to tertiary levels because of its international status and a neutral language for all the ethnic groups in Singapore (the three main ethnic languages, Chinese, Malay, and Tamil, are also official languages). However, pupils have to learn their mother tongue, and this is seen as a crucial way to help them retain their cultural heritage against some undesirable elements of Westernisation. This is the Singapore approach to address the tension between internationalisation (especially Westernisation) and local cultural transmission and preservation. Mathematics, together with other school subjects, can play a role in transmitting cultural values and citizenship to the pupils (Wong, 2003; 2005), even though this is not a commonly acknowledged goal of mathematics instruction. Indeed, values-based mathematics can become part of a "global curriculum", only if mathematics educators elsewhere will pay more attention to the integration of cultures and values in mathematics education (Winter, 2001). This will be a fruitful area for future effort in globalisation.

The Singapore mathematics curriculum is succinctly summarised by the so-called "Pentagon Framework" shown in Figure 1, first developed in 1990 and retained in subsequent curriculum reviews (Ministry of Education, Singapore, 2000). Its main focus is mathematical problem solving, where "problem" is defined to include both routine and unfamiliar problems as well as open-ended investigations. This definition is slightly different from the one commonly used in the mathematics education literature, where "problems" refer to only novel, non-routine, or unfamiliar ones. The framework underscores the belief that the ability to solve mathematical problems depends on five inter-related factors: concepts, skills, processes, metacognition, and attitude. When the framework was developed in 1990, the committee was influenced by three major publications: *Mathematics Counts* (Cockcroft, 1982), which introduced the idea of investigation; *An Agenda for Action* (National Council of Teachers of Mathematics, 1980), which emphasised the importance of problem solving; and *How to Solve It* (Polya, 1957), that explained the idea of heuristics. Although the committee had sought input from mathematics teachers at the review stage, there was a lack of local curriculum research that it

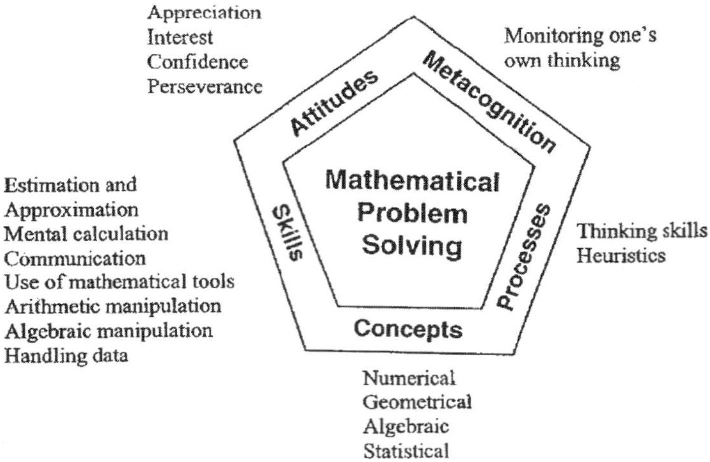

Figure 1. Singapore Mathematics Curriculum Framework

could use. This eclectic approach was very much in line with what Clements (2002) recommended that, given the complexity of curriculum work, curriculum developers should consider "all relevant theories and empirical work" (p. 622) rather than keep to one particular theory such as constructivism, because "overzealous applications … can limit practical effectiveness" (ibid.). This illustrates how important ideas in the international mathematics education community are adapted and modified to suit local priorities, values, and conditions.

The mathematics topics in the curriculum are arranged in a spiral fashion from primary to secondary levels. The pace and scope of coverage is differentiated for pupils from different streams. The American Institutes for Research (2005) describes this framework as "a balanced set of mathematical priorities centred on problem solving" (p. xi). It has some semblance with the five strands of mathematical proficiency, namely, *procedural fluency, conceptual understanding, strategic competence, adaptive reasoning,* and *productive disposition*, discussed by Kilpatrick, Swafford, and Findell (2001), but the Singapore framework uses terms that school teachers can easily identify with, which is an important consideration for effective curriculum implementation.

Since the early 1980s, Singapore primary pupils have been taught the locally-developed "model drawing" approach to solve challenging word problems without the use of algebra. In recent years, this method has become known internationally (Ferrucci, Yeap & Carter, 2003). Singapore mathematics textbooks are written by local authors and must be approved by the Ministry of Education before they are used in schools. Some of these textbooks are being analysed in relation to other mathematics programmes and standards (Adams, Tung, Warfield, Knaub, Mudavanhu & Yong, 2000) and used outside Singapore, for example, those used by the Massachusetts district of the United States[3] and Israel. Singapore educators have been invited to deliver addresses about Singapore mathematics education at international conferences (Kaur, 2003; Lee, 2004; Lee, 2005). This recent interest by the international education community in Singapore mathematics education and textbooks may be stimulated by the top performance of its fourth and eighth graders in the three TIMSS studies in 1995, 1999 and 2003 (Kaur, Edge & Yeap, 2003). Indeed, these activities illustrate fruitful outcomes from the globalisation of the Singapore mathematics education practices.

The predominant mathematics teaching style in Singapore still follows the traditional sequence, exposition → class practice → feedback → homework, coupled with worksheets consisting of practice exercises and questions for guided discovery. Within this standard pedagogical flow, many innovative teaching strategies have been trialled in the schools, including problem posing, journal writing, maths trails, use of Geometer's Sketchpad and other mathematics software in the classroom. The Ministry expects that each school subject should include about 30% of curriculum time using Information and Communications Technology (ICT). Singapore school teachers have prepared ICT-based mathematics lessons, some of which are shared on the ministry's edu.MALL website.[4] This push for ICT use in mathematics education is in line with the global trends, which have developed along different directions (Yang, Butler, Cnop, Isoda, Lee, Stacey, & Wong, 2003). Singapore teachers are entitled to 100 hours per year of professional development. Many of them take in-service modules on mathematics content and mathematics pedagogy conducted by the National Institute of Education (NIE), which is currently the sole teacher training institute in Singapore.

Finally, for internationalisation through publications, *The Mathematics Educator* began in 1996 as a local publication of the Association of Mathematics Educators. Since then, it has evolved into an internationally refereed journal (ISSN 0218–9100), included in the *Mathematics Didactics Database* (MATHDI)[5], with an editorial board consisting of mathematics educators from Australia, United Kingdom, United States, and Singapore. This development being a long process is made possible no doubt by the hard work of its past and present editors, but support from overseas authors through personal and professional contacts is an important success factor.

2.2 *Mathematics Education in Brunei Darussalam*

Brunei Darussalam, the "Abode of Peace", is very different from cosmopolitan Singapore. Its area is about 10 times (5,765 sq. km) that of Singapore, but its population (358,000) is about one tenth that of Singapore. The Brunei population composes mostly Malay (66%) with 11% Chinese and the rest from many minority groups[6]. It also has a high per capita GDP (US$18,600) due to its substantial oil

---

[3] http://www.smartbrief.com/latestIssue.jsp?i=14322&l=247746, retrieved 25 July 2005.

[4] http://www.moe.gov.sg/edsoftware/ir/maths.htm, retrieved 25 July 2005.

[5] http://www.emis.de/MATH/DI.html, retrieved 25 July 2005.

[6] http://www.brunei.gov.bn/index.htm, retrieved 25 July 2005.

and gas output. Besides oil and gas, the main economic activities are agriculture, fisheries, commerce, and banking. The life-style is fairly relaxed with little public entertainment.

The Brunei national philosophy is the Malay Islamic Monarchy, which places strong emphasis on the teaching of the Holy Quran. The education system is also centrally controlled with the vision "to mould a well-rounded (*syumul*) Bruneian race, for a flawless and blissful life on earth and hereafter (*akhirat*)". Islamic practices and values are evident at all levels of schooling, for example, Islamic attire, frequent recitals of prayers, and respect for teachers. The medium of instruction of the first three years of primary education is Malay, and from Primary 4 onwards, most subjects including mathematics are taught in English. This switch in the medium of instruction of mathematics from Malay to English at Primary 4 is a major factor that contributes to relatively poor performance in mathematics in upper primary level. After six years of primary schooling, pupils sit for the PCE (Primary Certificate in Education) examination. The bilateral primary mathematics study to be discussed later involved a sample of these PCE pupils.

Secondary schooling is divided into 3 years of lower secondary and 2 or 3 years of upper secondary level, depending on the ability streams of the pupils. At the end of lower secondary level, pupils take the Lower Secondary Assessment examination, and on the basis of their performance, will be channelled into the Science, Arts, or Technical stream. The pupils will then take the Brunei-Cambridge GCE O- or N-level examination, similar to the Singapore system. After this examination, pupils either continue with pre-university study or join the workforce.

Mathematics is a core subject in Brunei schools. Its primary mathematics curriculum has just been revised to include a new conceptual framework as shown in Figure 2 (Norjum & Veloo, 2004). Norjum and Veloo explain that the motivation for the change includes local factors (such as declining mathematics performance, need to move away from teacher-centred instruction) and global trends (such as use of ICT in education, increased need for mathematics literacy in a technological workplace, and world trends in mathematics education). This framework stresses the interweaving of processes and values into five specific content areas.

Brunei Primary 1 to 3 pupils use locally produced texts written in Malay. From Primary 4 to the end of secondary schooling, they use Singapore mathematics textbooks and workbooks because the pupils from both countries take similar public examinations at the end of secondary schooling. This is an interesting example of "bilateralisation" due to their similar colonial past (Wong, Zaitun, & Veloo, 2001).

Mathematics teaching in Brunei Darussalam is also predominantly teacher-centred and focuses on skill mastery. Unlike Singapore, there are few school-based programmes to stimulate the interest of pupils in mathematics. Since Brunei Darussalam has not taken part in any international comparative studies in mathematics, the mathematics performance of its pupils cannot be assessed against international benchmarks. Indeed, few educators outside Brunei know much about its mathematics education.

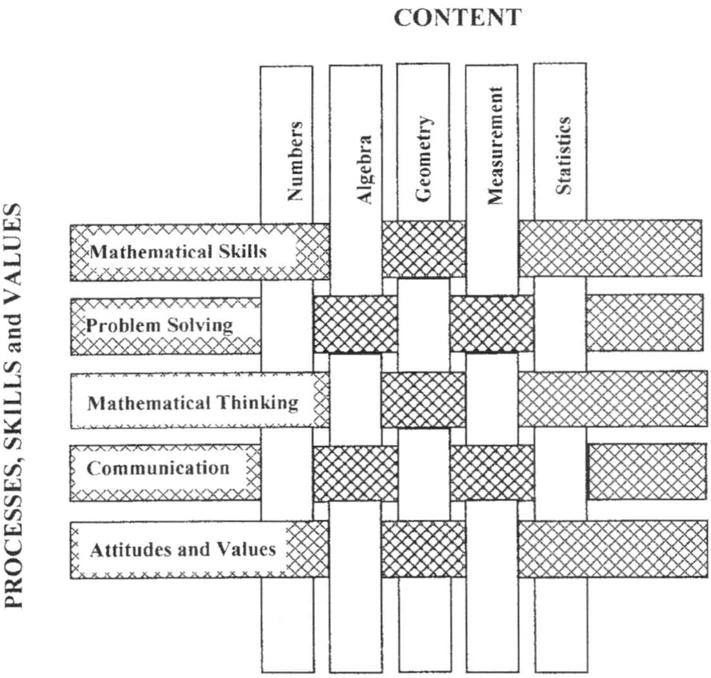

Figure 2. Mathematics Curriculum Framework of Brunei Darussalam

The Department of Science and Mathematics Education at the Sultan Hassanal Bolkiah Institute of Education (SHBIE), Universiti Brunei Darussalam (UBD), has undertaken two main activities to promote mathematics education locally with international participation. The first activity was to launch the journal, *Science and Mathematics Education* (renamed *Science, Mathematics and Technical Education* in 1998), in conjunction with the Fifth Southeast Asian Conference on Mathematics Education held at Brunei Darussalam in 1990. This journal published work by local mathematics and science educators, with overseas contributions. It was given free to all Brunei schools, and had not been published since 2003.

The second activity is the series of annual international conference on science, mathematics, and technical education (SMTE), beginning in 1996 to celebrate the 10th anniversary of UBD. This annual series is ongoing, with strong participation from well-known international educators. The conference themes over the past 10 years give a strong indication that globalisation is a serious issue discussed at these conferences. This international forum contributes to internationalisation at Brunei Darussalam through face-to-face exchanges, networking, and the publication of conference proceedings.

- 1996: Assessment and Evaluation in Science and Mathematics Education
- 1997: Innovations in Science and Mathematics Curricula
- 1998: SMTE for National Development
- 1999: Cultural and Language Aspects of SMTE
- 2000: SMTE in the 20th and 21st Centuries
- 2001: Energising SMTE for All
- 2002: Realities in SMTE
- 2003: Responsibilities in SMTE
- 2004: Globalisation Trends in SMTE
- 2005: Future Directions in SMTE
- 2006: Shaping the Future of SMTE

A noteworthy point about these conference themes is that SMTE appears in all the latest nine conferences. These conferences provide a useful forum for inter-disciplinary interactions among educators in SMTE. This adds an important dimension to internationalisation, similar to the effort made by this book and similar publications (e.g., Groves, Jane, Robottom & Tytler, 1998; Kelly & Lesh, 2000). This theme is taken up again in the next section.

3. The Bilateral Primary Mathematics Study

3.1 Multi-National Versus Bilateral Comparative Studies

We begin this section with some comments about multi-national and bilateral comparative studies. International collaboration in educational research has been conducted in the past thirty years, initially in a rather low-key fashion. Only in recent years have stakeholders outside the education research community begun to pay attention to international league tables of student achievement, in particular after the much publicised poor results of advanced nations such as the United States against the strong performance of Asian pupils in the *Third (Trends in) International Mathematics and Science Study* (TIMSS). The benefits of participating in these studies include: exposure of researchers to advances in research methodologies; identification of the strengths and weaknesses of schools and curriculum in each country based on multi-national benchmark; sensitisation of policy-makers and politicians to the need for school reforms, often leading to more stringent accountability through student testing (which may or may not be desirable); increased output in publications; networking among researchers; and the sale of research services such as data entry and data analysis, which can be quite expensive for the less developed countries. Most of these studies focus on students' performance and the social, curriculum, pedagogy, and school factors (especially the notion of "opportunity to learn") that might account for differences in performance. However, the validity of these comparative studies is being debated (e.g., Rowan, 2000) and their impacts on classroom teaching and learning remain unclear or questionable.

Bilateral comparative studies are easier to manage and cheaper to finance compared to multi-national ones, but few bilateral studies are reported in the literature. The two teams involved in such a study can discuss subtle differences in interpretations to greater depth, and it is easier to build friendship and partnership with fewer organisation and technical problems to contend with. The bilateral primary mathematics study to be discussed here differed from TIMSS by focussing on student, home, and teacher factors rather than on student performance. This study also has two other features not found in other multi-national studies. Firstly, it was one of 20 collaborative projects under a larger research agenda between the two institutes, NIE (Singapore) and SHBIE (Brunei Darussalam), so that the mathematics teams also participated in discussion about teacher education, language instruction, educational psychology, and teaching practice, which were of interest to both countries. This arrangement enriched the outlook and experience of all the researchers in the mathematics study as well as the parent project. This is different from comparative studies that are subject specific, such as mathematics, science, or English, and where the researchers do not work with those from another discipline. Secondly, the research meetings were held alternatively in the two institutes, and this allowed the team members to deepen mutual understanding through playing hosts and guests in different years. This change in roles is missing from the usual practice of multi-national studies, where the research teams from different countries usually meet at a "central" site. Now we turn to a description of the main research agenda.

3.2 The Main Research Agenda: Inter-institute Dialogue on Educational Advances (IDEA)

Informal collaboration in education has existed between Singapore and Brunei Darussalam for several years prior to this bilateral study. In September 1997, the then-vice-chancellor of UBD proposed an annual dialogue between its education faculty (SHBIE) and NIE. This dialogue, code named IDEA by Sim (2000), the then-Dean of SHBIE, was to serve as a platform for exchange of experience and collaborative research into teacher education in the two countries. An implicit goal was to strengthen the research culture in SHBIE by having its researchers "learn" from their Singapore counterparts, who have wider experience in educational research.

The arrangement was for a team from one institute to visit the other institute on an annual, alternating basis for a 3-day activity. The first dialogue took place in May 1998 in Singapore, the second in May 1999 in Brunei Darussalam, and the third and final one in July 2000 in Singapore. At these meetings, researchers from the two institutes developed collaborative projects that were to be conducted in their countries during the intervening months with frequent contacts via emails. The results were reported back at the next meeting to the whole group as well as within smaller groups. According to Sim (2000), the level of collaboration could be *minimal* (e.g., share findings on related studies conducted independently), *partial* (e.g., share research instruments), *substantial* (e.g., develop common features), or *maximal* (e.g., conduct similar national research). Over the three years, 20 projects were conducted, covering areas such as practicum supervision, psychological problems of pupils, their motivation to schooling, language education, and primary mathematics education. The IDEA agenda came to an end after three years of bilateral work because most of the key players had moved on to other work or country.

3.3 Aim and Administration of the Primary Mathematics Study

The main aim of this bilateral project was to investigate pupils' perceptions of mathematics learning in each country. The findings could stimulate discussion on how to improve pupils' learning of mathematics in each country through comparing and contrasting different factors and practices. It has been argued that even though mathematics content is universal, mathematics instruction is a cultural product and has its characteristic pedagogical flow in different countries (Schmidt et al., 1996). This study could extend knowledge of this pedagogical flow by considering two culturally very different countries: secular, technological Singapore versus Islamic, agricultural Brunei Darussalam.

The study was planned in May 1998 at the first IDEA meeting. It was conducted in November 1998. At the second IDEA meeting, the researchers discussed findings about pupils' drawings of their "best" mathematics teachers. In December 1999, a paper entitled "My best mathematics teacher: Perceptions of Singapore and Brunei pupils" was presented at the joint conference of Malaysian Educational Research Association and Educational Research Association Singapore (Kaur, Koay, Jamilah, Zaitun & Wong, 1999) and a similar paper in Malay was published by Jamilah, Zaitun, Wong, Romaizah, Kaur and Koay (2000) for the Brunei readers. Further analyses were carried out and findings were discussed at the third IDEA in 2000. A paper was distributed at the IDEA symposium held during the annual conference of the Australian Association for Research in Education (Wong, Zaitun, Jamilah, Romaizah, Kaur & Koay, 2001). These dissemination activities began from bilateral collaboration, moved to regional discussion, and ended at an international forum. This progression in steadily expanding the audience for research is a pattern of internationalisation that will occur more frequently in the coming years when educational researchers establish stronger links across countries, regions, and subject areas.

3.4 Instrument and Sample of the Primary Mathematics Study

3.4.1 Instrument

The study made use of a questionnaire adapted from the pupils' questionnaire in the KASSEL Project (Kaur & Yap, 1997), an international project coordinated by University of Exeter. Fifteen countries participated in the KASSEL Project. This illustrates how an international research project can be modified into a bilateral one like the present study. This process reduces the time and effort to create new instruments, ensures some comparability with other findings, yet allows for extension, for example, the use of pupils' drawing was not in the original KASSEL project.

The questionnaire consists of 26 items, given in English for Singapore and Malay for Brunei Darussalam. Although Primary 6 Brunei pupils also study mathematics in English, English is their second and even third language. Since the study was *not* about pupils' performance, asking the Brunei pupils to answer the questionnaire in their first language (Malay) would ensure stronger validity.

The questionnaire items deal with pupils' enjoyment of studying mathematics, use of mathematics in daily life, confidence, mathematics homework, access and use of computers and calculators, expected grades in the Primary 6 public examinations (PSLE and PCE), teaching methods, and qualities of the "best" mathematics teachers. In addition to these questions, the pupils drew their "best" mathematics teacher. This drawing technique has several advantages: it is easy and fun to use; it gives vivid snapshots of perceptions that have emotional content; it is not restricted by pre-coded responses; even young children who have problems reading questionnaire items can respond by drawing. This technique has been used to study pupils' perceptions of scientists (Chambers, 1983), mathematicians (Picker & Berry, 2000; Wong, 1995), classrooms (Burgess, 1994; Wong, 1996), and technology (Rennie & Jarvis, 1994). These studies have found the drawing technique useful in providing insights about the respective issues.

3.4.2 Sample

Primary 6 pupils were chosen for this study because they were at the end of primary schooling in both countries and this allows for easier comparison. The Singapore pupils took the survey *after* their PSLE examination in October 1998, while the Brunei pupils answered it *before* they sat for PCE in early October 1998. This difference in the times of administration in relationship to the final examinations was constrained by practical factors and should be taken into account when interpreting the findings.

The composition of the sample is shown in Table 1. In each country, the pupils were from four different schools. One of the Brunei schools was in the rural area, with the three others in urban districts. The mean age of the pupils was 12.2 years in each country. Male and female were evenly represented in the sample.

Table 1. Composition of sample by country and gender

Country	Male	Female	Total
Singapore	163	171	334
Brunei Darussalam	111	98	209
Total	274	269	543

3.5 *Findings of the Primary Mathematics Study*

3.5.1 *Enjoyment of Mathematics*

The pupils were asked whether they enjoyed mathematics when they were in the lower primary (LP) classes and at the time of administration (UP, Primary 6). The percentages answering "Always" are shown in Figure 3.

The Singapore pupils expressed less enjoyment in mathematics (overall 25% drop) as they grew older, probably because of the tremendous pressure on them to excel in mathematics at PSLE. About 37% of male and 23% of female said mathematics was the subject they liked best. On the other hand, the Brunei pupils did not express much change in their enjoyment of mathematics in the past and now, though the level (57%) was not high. About 48% of male and 54% of female said mathematics was their favourite subject, and these were higher than the Singapore percentages. The less competitive Brunei education environment may partially explain this difference.

3.5.2 *Use of Mathematics in Everyday Tasks*

For this open-ended question, 71% of the Singapore pupils and 52% of the Brunei pupils mentioned financial matters and shopping. However, 23% of the Brunei

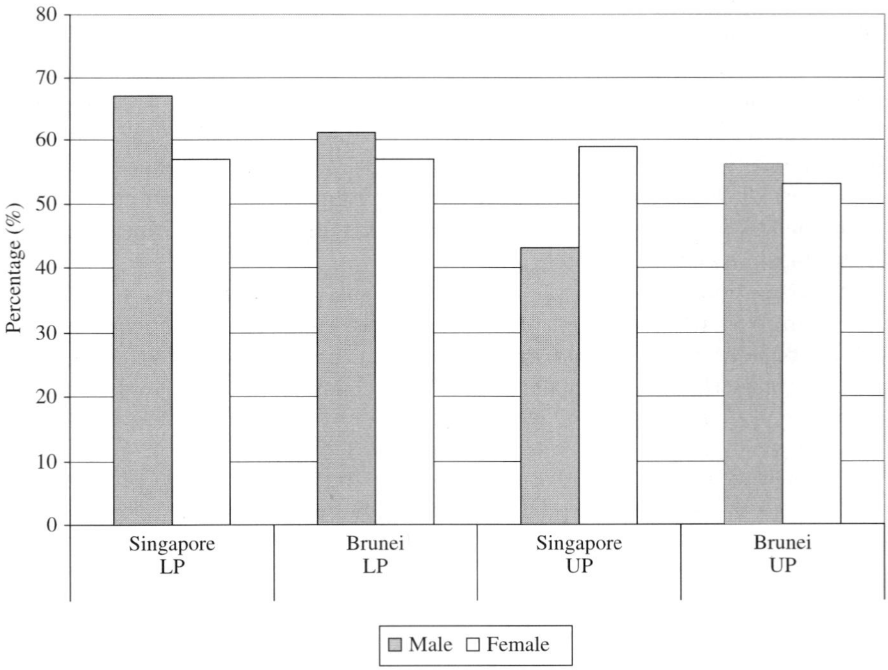

Figure 3. Percentages "Always" on Enjoyment of Mathematics

pupils did not respond compared to only 2% of the Singapore pupils. A plausible reason for this difference is that Singapore pupils have many opportunities to deal with financial matters in a highly commercialised society, whereas this is not so for Brunei pupils who live in an agricultural environment. However, in both countries, mathematics teachers should explain how mathematics can be used beyond shopping, for example, in art and craft (Metz, 1991), local games, and statistical data, because these applications were hardly mentioned by the pupils in this study.

3.5.3 Confidence in Mathematics and English

Research about academic self-concept generally agrees that this construct is multi-dimensional (e.g., Dermitzaki & Efklides, 2000). This study examined self-concept in terms of perceived confidence, and three different ways were used to assess pupils' level confidence in mathematics. These methods seem to give slightly different findings.

The first way was to compare their expressed levels of confidence in mathematics compared to English. The Brunei pupils expressed higher levels of feeling good in both mathematics and English than the Singapore pupils; see Figure 4. This was a surprising finding. One possible explanation is that the Singapore pupils answered this questionnaire *after* their PSLE, and the recent experience of taking this tough examination might have made them feel less confident about their ability. On the other hand, our experience with schools suggests that Brunei pupils tend to respond positively to the subject that they are asked about. Singapore and Brunei pupils, especially the males, felt they were better at mathematics than at English. However, the Singapore females were more confident in English than in mathematics. These findings are consistent with the common notion that boys are better in mathematics than in language and the reverse for girls.

The second way moved from general level of confidence to confidence in specific tasks, namely, mastery of multiplication tables, which is often considered, rightly or wrongly, as an indication of basic mathematics competence. Among Singapore pupils, 72% of males and 63% of females said they knew the multiplication tables very well. For the Brunei pupils, only 40% of male and 24% of female agreed. In Singapore, pupils must remember multiplication facts by heart in order to do many mental sums given in class, whereas the requirement is less demanding in Brunei. A social factor is that Singapore parents and teachers expect higher levels of mastery among the pupils compared to their Brunei counterparts.

The third way asked pupils to predict their grade in PSLE or PCE. The Singapore PSLE uses A*, A, B, C, and F (fail), whereas the Brunei PCE uses grades A, B, C, D, and F (fail). In Figure 5, we have equated Singapore A* to Brunei A and so on. This modification reflects pupils' perceptions about their "best" grade as evaluated within each country rather than absolute standards across the two countries.

Pupils in both countries held very high expectations of their mathematics performance in the public examinations, with about 85% of each group aiming to get A

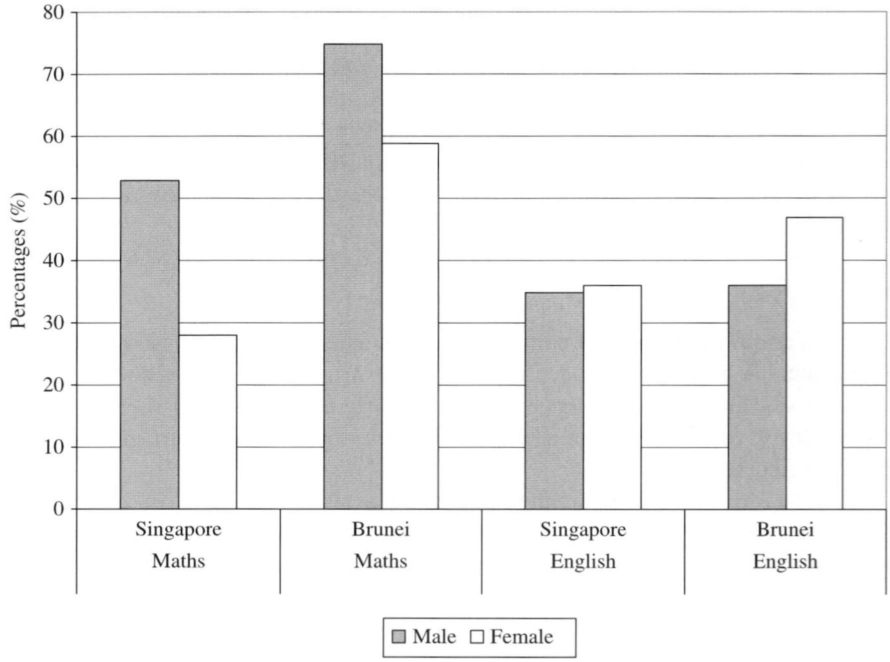

Figure 4. Percentages for "Yes" to "Being Good" in Mathematics or English

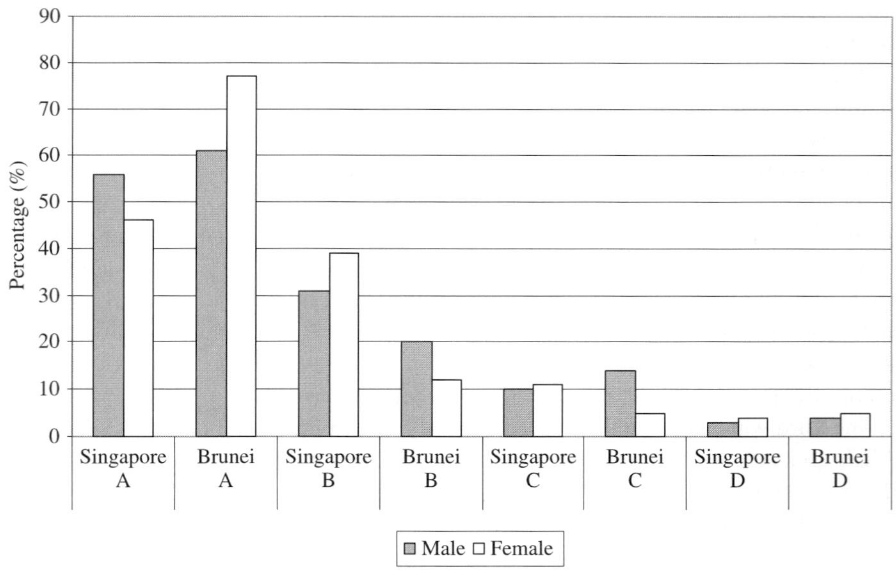

Figure 5. Percentages of Expected Grades in Public Examinations

or B grades. None of them expected to fail in mathematics. This was quite understandable though hardly realistic. Differences were also found. The Brunei pupils had higher expectations than the Singapore counterparts in obtaining the top grades. A higher percentage of Brunei females aimed for grade A than Brunei males. This expectation is not consistent with the findings about confidence in mathematics or competence in multiplication tables discussed earlier. On the other hand, the findings about Singapore pupils were more consistent across the three methods, with a lower percentage of females predicting A grade than males. Whether this difference is due to cultural expectations or the timing of the survey cannot be ascertained.

When asked about how to revise mathematics for tests or examinations, 29% of the Singapore pupils mentioned "go through their books and worksheets," 29% "did some problems from assessment books," 23% "memorise mathematical rules," and 13% "asked someone like tuition teacher or siblings to revise with them." There were no differences between males and females in the use of these revision techniques. Unfortunately, the corresponding Brunei data were missing.

3.5.4 Homework and Help in Learning Mathematics

About 71% of the Singapore pupils reported being given mathematics homework after every lesson, whereas only 33% of the Brunei pupils reported this. The pupils' responses to homework were also different. About 87% of the Singapore pupils against 63% of the Brunei pupils said they completed the homework as soon as it was given. Thus, the Singapore pupils had more homework and were more serious about completing it than the Brunei pupils. Indeed, getting pupils to complete and submit mathematics homework on time is a perennial struggle for many Brunei teachers (Wong, 1998).

When pupils need help in mathematics, about half the Singapore pupils asked their teachers and 20% asked their parents, whereas 79% of the Brunei pupils asked their teachers and few reported getting help from significant others such as parents, siblings, and friends. The Singapore parents play a more active role in helping their children in mathematics, compared to Brunei parents. Indeed, some Singapore parents have even attended workshops that train them how to coach their children in mathematics. Such workshops are unheard of in Brunei Darussalam. This difference in parental role could be a consequence of higher expectations in a competitive society like Singapore against a more relaxed one in Brunei Darussalam. Although the international mathematics education community has studied cram schools for pupils in several Asian countries, it has not paid much attention to parental workshops in mathematics and their impacts on pupil achievement.

About 49% of the Singapore pupils reported attending private tuition in mathematics over the past year. This is not as prevalent as we thought. For the Brunei sample, 67% had attended tuition in mathematics. However, this finding could be misleading because the Malay term for *tuition* used in the questionnaire, *kelas tambahan*, also means "extra classes" conducted by teachers in the afternoon. These extra classes are compulsory in some primary schools in Brunei Darussalam as well as in Singapore. This highlights the necessity of accurate translation of questionnaire items and an intimate knowledge of local situations in cross-national research.

3.5.5 Access to and Use of Computers and Calculators

About 83% of the Singapore pupils against 32% of the Brunei pupils reported that they had a computer at home. However, only 5% of the Singapore pupils and 11% of the Brunei pupils said they used computer to do homework. Such low percentages (although 11% is unexpectedly high for the Brunei sample) were not surprising because most mathematics homework in primary schools does not require the use of computers. However, this might change with recent reforms about the use of ICT being undertaken in both countries. In Singapore, 30% of curriculum time should involve the use of ICT. In Brunei Darussalam, primary school pupils should attend at least one computer-based lesson per week for each school subject. How these reforms might change the classroom and home environment for learning mathematics remains to be investigated.

When asked whether they would normally use a calculator for number work such as 390 + 30, most of the pupils (98% Singapore and 93% Brunei) responded "No". The reason is that this question is so simple that it does not require calculator. Calculators are not permitted in primary schools in both countries. This may soon change as mathematics educators in both countries wrestle with the controversial issue of calculator use in primary schools, noting the strong support for its use in several English-speaking countries and the need to reduce manipulative skills to make time for higher order thinking and ICT use. This is an interesting instance of globalisation that both countries have not followed for the past two decades. However, both countries are now re-examining the issue in the light of local necessities and global changes. Unfortunately, there is little local research to support either position, and their decision may be made on the basis of priorities and anecdotal evidence rather than systematic research.

3.5.6 Qualities of the "Best" Mathematics Teacher

Pupils were asked to write about their "best" mathematics teacher. These descriptions were classified by the research teams at the IDEA meetings into four categories. We counted how many times various words appeared under each category and ranked them by frequency. For each category, the four most frequently mentioned qualities are given in Table 2, with *lower* ranks indicating *higher* frequencies. Two examples are given below.

Pupil 196 (Brunei): Teaches us well, encourages us to learn, explains step by step so that we will understand what is being taught.

Pupil 2075 (Singapore): Able to explain clearly; caring, patient and kind; encourages pupils to ask questions; does not give too much homework; makes the lessons interesting (games/quizzes).

Most pupils gave several descriptions of teacher quality, suggesting that the "best" mathematics teacher must have a few desirable characteristics and use a repertoire of teaching techniques to help them learn. The pupils in both countries

Table 2. Ranking of teacher qualities by country

(a) Singapore

Rank	Personal Qualities	Instruction/ Pedagogy	Relationship/ Rapport	Homework/ Expectations
1	Patient	Explain clearly	Caring/Kind	Demanding/A lot of homework
2	Humorous	Ensures pupil understand	Understanding	Check/Goes through homework
3	Helpful	Good in maths/Clever	Approachable/ Encouraging/ Motivating	Constant review/ Reinforcement
4	Serious/ Dedicated	Individual help	Strict/Firm	Give difficult problems

(b) Brunei Darussalam

Rank	Personal Qualities	Instruction/ Pedagogy	Relationship/ Rapport	Homework/ Expectations
1	Calm/Good temper	Explains clearly	Answers questions	Do work in class
2	Helpful	Ensures pupils understand	Approachable/ Encouraging/ Motivating	Demanding/ A lot of homework
3	Courteous/ Pleasant/ Nice	Group work/ Discussion/ Pupils' Participation	Strict/ Firm	Encourage pupils to do more sums
4	Responsible	Simple/ Easy exposition	Supportive	Constant review /Reinforcement

reported similar qualities about their best mathematics teachers. In particular, the ability to explain things features prominently in these responses. This is consistent with earlier research into the quality of effective Singapore teachers (e.g., Lim & Wong, 1989).

3.5.7 "Best" Mathematics Teacher in Action: Pupils' Drawings

These drawings provide snapshots of the "best" mathematics teachers in action. Two sample drawings, one from each country, are shown in Figure 6. In both countries, every lesson begins with the class standing in attention to greet the teacher in unison and ends with the class saying "thank you, teacher". This public showing of respect and gratitude to the teacher is expected in Asian and Islamic cultures. Note that the Brunei drawing shows very clearly that teachers and pupils

Figure 6. Sample Drawings of “best” Mathematics Teacher

in Brunei Darussalam must be in the Islamic attire. All the four drawings show the whole-class, teacher-centred learning environment, with pupils seated in rows, teachers explaining up front, board filled with mathematical symbols, diagrams, and solutions (in 69% and 65% of Singapore and Brunei drawings respectively). About 11% of the mathematics found in the Brunei drawings was incorrect, showing that

Table 3. Percentages of pupils who gave top rank for each teaching method

Teaching Methods	Singapore	Brunei Darussalam
Teacher explanation	41.9	71.3
Class discussion	14.1	6.7
Group work	15.6	5.3
Teacher teaches pupils individually	29.9	7.2
Read textbook on my own	3.0	5.3

pupils' drawings can provide a useful source of information about their mathematics knowledge. One of the Singapore drawings in Figure 6 shows a girl doing mathematics at the board. Indeed, exposition, together with pupils sent to do work on the board, is probably common to many mathematics classes all over the world.

About 97% the drawings in each country did not show any teaching aids or computer. This suggests that the "best" teachers did not use teaching aids or the pupils were not aware of these aids. If the latter explanation were valid, then the teachers should explain to their class the purposes of the teaching aids they have used in order to raise metacognitive awareness (Bell, Crust, Shannon & Swan, 1993). However, this comment is tentative because the pupils were asked to draw their teachers and not the classroom environment.

Further support about the prevalent teacher-centred method was obtained by examining pupils' top rating of various teaching methods; see Table 3. The Brunei pupils expressed very strong preference for teacher explanation and low preference for discussion. This response has a cultural basis in the ways Brunei children are socialised from a young age to accept whatever adults (including parents and teachers) tell them and not to "answer back", which is perceived to be rude and defiant. The Singapore pupils also preferred teacher explanation, but less strongly in comparison to Brunei pupils. They also liked to have individual attention from the teachers. Group work was still infrequent, especially in Brunei. A possible reason is that the cramped classroom does not lend itself easily to effective group work. In Singapore, recent changes to the education system may have exposed its pupils to a much greater variety of teaching methods than is the case for Brunei Darussalam.

Conclusion

All four of us gained our doctorate degree from Australian universities. We have learned much about the Australian and Western perspectives of education and research, particularly from English-speaking countries. In our own teaching and research, we often cite from the corpus of English writing and research and we find ourselves educating our next generation of teachers and researchers to understand this work. The main reason for this is that there are few local studies that are significant in terms of theoretical thinking, innovative practices, or research database. Changes are taking place in Singapore where the Centre for Research in Pedagogy and Practice[7] has carried out several large scale studies to provide baseline data about Singapore pupils and teachers. Chinese mathematics educators have recently published their work in English to inform the international community of an "insider" view about Chinese practices in mathematics education (Fan, Wong, Cai & Li, 2004). The two sections above follow this trend to explain local practices through the English-speaking education community.

Aspects of internationalisation and globalisation that are evident in Singapore and Brunei mathematics education include: adapting global ideas in curriculum reviews; upgrading local publications to international status; resisting global trend of calculator use in primary schools and re-visiting earlier decisions as the educational scenes change; the prevalence of teacher-centred exposition in mathematics lessons; publication of conference proceedings; and postgraduate training in mathematics education research. The bilateral primary mathematics study has brought about greater friendship and professional links among team members as well as a deeper understanding of primary school mathematics learning and teaching in the two countries. Although these two countries differ greatly in ethnic, religious, and economic makeup, the secular or Islamic influences on mathematics education are quite minor. For both countries, there is an urgent need to develop strategies to enhance pupils' interest and confidence in mathematics leading to a greater appreciation of mathematics in daily life. This is also a concern for the international mathematics education community.

As illustrated in this chapter, comparative studies may begin with bilateral collaboration, perhaps within an inter-disciplinary setting, such as the IDEA format. The process and findings of comparative studies, whether bilateral or multi-national, can strengthen research capacity and mutual understanding through cooperation among educators from around the world. Indeed, internationalisation through comparative studies has gained great momentum in recent years. For example, Kaur and Koay have just completed another international study (Kaur, Koay & Yap, 2004), and Kaur is currently involved in the *Learners' Perspective Study*[8]. These research efforts should enrich the education community at local levels as well as contribute to knowledge creation at the international level. As Kubow & Fossum (2003) noted, "[t]here is a growing sense that the educational concerns of one nation are the concerns of other nations as well" (p. 251). Mathematics and science educators can and should contribute towards this much needed global growth, learning, and understanding for the benefits of future generations.

References

Adams, L. M., Tung, K. K., Warfield, V. M., Knaub, K., Mudavanhu, B., & Yong, D. (2000). *Middle school mathematics comparisons for Singapore mathematics, connected mathematics program, and mathematics in context (including comparisons with the NCTM Principles and Standards 2000).*

[7] http://www.crpp.nie.edu.sg/index.php, retrieved 1 April 2006.
[8] http://extranet.edfac.unimelb.edu.au/DSME/lps/, retrieved 1 April 2006.

Washington, DC: Department of Applied Mathematics, University of Washington. Retrieved July 2005 from http://www.amath.washington.edu/~adams/full.pdf

American Institutes for Research. (2005). *What the United States can learn from Singapore's world-class mathematics system (and what Singapore can learn from the United States): An exploratory study*. Washington, DC: Author. Retrieved July 2005 from http://www.air.org/news/default.aspx

Bell, A., Crust, R., Shannon, A., & Swan, M. (1993). *Awareness of learning, reflection and transfer in school mathematics. ESRC project: R000–23–2329*. Nottingham: Shell Centre for Mathematical Education, University of Nottingham.

Burgess, Y. (1994). Drawing what you know and feel. *SET: Research information for Teachers*. Hawthorn, Victoria: Australian Council for Educational Research.

Chambers, D. W. (1983). Stereotypic images of the scientist: The Draw A-Scientist test. *Science Education, 67*, 255–265.

Clements, D. H. (2002). Linking research and curriculum development. In L. D. English (Ed.), *Handbook of international research in mathematics education* (pp. 599–630). Mahwah, NJ: Lawrence Erlbaum Associates.

Cockcroft, W. H. (1982). *Mathematics counts*. London: HMSO.

Dermitzaki, I., & Efklides, A. (2000). Aspects of self-concept and their relationship to language performance and verbal reasoning ability. *American Journal of Psychology, 113*(4), 621–637.

Fan, L. H., Wong, N. Y., Cai, J., & Li, S. (Eds.). (2004). *How Chinese learn mathematics: Perspectives from insiders*. Singapore: World Scientific.

Ferrucci, B. J., Yeap, B. H., & Carter, J. A. (2003). A modeling approach for enhancing problem solving in the middle grades. *Mathematics Teaching in the Middle School, 8*(9), 470–479.

Groves, S., Jane, B., Robottom, I., & Tytler, R. (Eds.). (1998). *Contemporary approaches to research in mathematics, science, health and environmental education 1997*. Geelong: Deakin University, Centre for Studies in Mathematics, Science and Environmental Education.

Jamilah M. Y., Zaitun M. T., Wong, K. Y., Romaizah M. S., Kaur, B., & Koay, P. L. (2000). Guru matematik saya yang terbaik: Persepsi murid-murid Brunei Darussalam dan Singapura. In K. Y. Wong (Ed.), *CARE review 1999* (pp. 110–118). Bandar Seri Begawan: Centre for Applied Research in Education, Universiti Brunei Darussalam.

Kaur, B. (2003). Mathematics for all but more mathematics for some: A look at Singapore's school mathematics curriculum. In B. Clark, R. Cameron, H. Forgasz, & W. Seah (Eds.), *Making mathematicians* (pp. 440–455). Victoria: The Mathematical Association of Victoria.

Kaur, B., Edge, D., & Yeap, B. H. (Eds.). (2003). *TIMSS and comparative studies in mathematics education. Monograph number 1*. Singapore: Association of Mathematics Educators.

Kaur, B., Koay, P. L., Jamilah M. Y., Zaitun M. T., & Wong, K. Y. (1999, December). *My best mathematics teacher: Perceptions of Singapore and Brunei pupils*. Paper presented at the MERA-ERA International Conference, Malacca, Malaysia.

Kaur, B., Koay, P. L., & Yap, S. F. (2004). IPMA – Singapore's report. In D. Burghes, R. Geach, & M. Roddick (Eds.), *IPMA report: A series of international monographs on mathematics teaching worldwide – monograph 4* (pp. 175–190). Budapest: Wolterskluwer Co.

Kaur, B., & Yap, S. F. (1997). *KASSEL Project Report (NIE – Exeter Joint Study) Second Phase (October 95–June 96)*. Singapore: National Institute of Education, Division of Mathematics.

Kelly, A. E., & Lesh, R. A. (Eds.). (2000). *Handbook of research design in mathematics and science education*. Mahwah, NJ: Lawrence Erlbaum Associates.

Kilpatrick, J., Swafford, J., & Findell, B. (Eds.). (2001). *Adding it up: Helping children learn mathematics*. Washington, DC: National Academy Press.

Kubow, P. K., & Fossum, P. R. (2003). *Comparative education: Exploring issues in international context*. Upper Saddle River, NJ: Merrill Prentice Hall.

Lee, N. H. (2004, July). *Nation building initiative: Impact on school mathematics curriculum*. Invited Regular Lecture at the 10th International Congress on Mathematics Education, Copenhagen.

Lee, P. Y. (2005, November). *60 years (1945–2005) of mathematics syllabus and textbooks in Singapore*. Invited paper given at The First International Mathematics Curriculum Conference, Chicago.
Lim, S. K., & Wong, K. Y. (1989). Perspectives of an effective mathematics teacher. *Singapore Journal of Education, Special Issue*, 101–105.
Metz, M. (1991). Islamic design. In L. Jones (Ed.), *Teaching Mathematics and Art* (pp. 97–112). Cheltenham: Stanley Thornes.
Ministry of Education, Singapore. (2000). *Primary mathematics syllabus*. Singapore: Curriculum Planning and Development Division, Ministry of Education. Available from http://www.moe.gov.sg/
National Council of Teachers of Mathematics. (1980). *An agenda for action: Recommendations for school mathematics of the 1980s*. Reston, VA: Author.
Norjum, Y., & Veloo, P. K. (2004). Reforming the primary mathematics curriculum in Brunei Darussalam. *Jurnal Pendidikan Brunei Darussalam, 9*(1), 1–16.
Picker, S. H., & Berry, J. S. (2000). Investigating pupils' images of mathematicians. *Educational Studies in Mathematics, 43*, 65–94.
Polya, G. (1957). *How to solve it* (2nd ed.). New York: Doubleday & Company.
Rennie, L. J., & Jarvis, T. (1994). *Helping children understand technology*. Perth, WA: Key Centre for School Science and Mathematics.
Rowan, B. (2000). Large-scale, cross-national surveys of educational achievement: Promises, pitfalls, and possibilities. In A. C. Porter & A. Gamoran (Eds.), *Methodological advances in cross-national surveys of educational achievement* (pp. 321–349). Washington, DC: National Academy Press.
Schmidt, W. H., Jorde, D., Cogan, L. S., Barrier, E., Gonzalo, I., Moser, U., et al. (1996). *Characterizing pedagogical flow: An Investigation of mathematics and science teaching in six countries*. Dordrecht: Kluwer Academic Publishers.
Sim, W. K. (2000). Development of an IDEA. In K. Y. Wong (Ed.), *CARE review 1999* (pp. 95–99). Bandar Seri Begawan: Centre for Applied Research in Education, Universiti Brunei Darussalam.
Winter, J. (2001). Personal, spiritual, moral, social and cultural issues in teaching mathematics. In P. Gates (Ed.), *Issues in mathematics teaching* (pp. 197–213). London: RoutledgeFalmer.
Wong, K. Y. (1995). Images of mathematicians. In R. P. Hunting, G. E. Fitzsimons, P. C. Clarkson, & A. J. Bishop (Eds.), *Regional collaboration in mathematics education 1995* (pp. 785–794). Melbourne: Monash University.
Wong, K. Y. (1996). Assessing perceptions using student drawings. In M. Quigley, P. K. Veloo, & K. Y. Wong (Eds.), *Assessment and evaluation in science and mathematics education: Innovative approaches* (pp. 370–379). Brunei Darussalam: Universiti Brunei Darussalam.
Wong K. Y. (1998). Thinking hats and mathematics homework. *Science and Mathematics Education, 13*, 38–44.
Wong K. Y. (Ed.). (2000). *CARE review 1999*. Bandar Seri Begawan: Centre for Applied Research in Education, Universiti Brunei Darussalam.
Wong, K. Y. (2003). Mathematics-based national education: A framework for instruction. In S. K. S. Tan & C. B. Goh (Eds.), *Securing our future: Sourcebook for infusing national education into the primary school curriculum* (pp. 117–130). Singapore: Pearson Education Asia.
Wong, K. Y. (2005, July). *Add cultural values to mathematics instruction: A Singapore initiative*. Invited paper presented at the Fourth Asian Mathematical Conference, National University of Singapore. Electronic publication.
Wong, K. Y., Zaitun M. T., Jamilah M. Y., Romaizah M. S., Kaur, B., & Koay, P. L. (2001, December). *Perceptions of learning primary mathematics: A comparative study of Brunei and Singapore primary 6 pupils*. Paper distributed at the IDEA Symposium, Annual Conference of the Australian Association for Research in Education, Fremantle, Australia.
Wong K. Y., Zaitun M. T., & Veloo, P. (2001). Situated sociocultural mathematics education: Vignettes from Southeast Asian practices. In B. Atweh, H. Forgasz, & B. Nebres (Eds.), *Sociocultural*

research on mathematics education: An international research perspective (pp. 113–134). Mahwah, NJ: Lawrence Erlbaum.

Yang, W. C., Butler, D., Cnop, I., Isoda, M., Lee, P. Y., Stacey, K., et al. (2003). A global view of curriculum issues on mathematics with technology: A panel discussion. In W. C. Yang, S. C. Chu, T. de Alwis, & M. G. Lee (Eds.), *Proceedings of the eighth Asian technology conference in mathematics* (pp. 57–71). Blacksburg, VA: ATCM Inc.

25

LESSON STUDY (JYUGYO KENKYU) FROM JAPAN TO SOUTH AFRICA: A SCIENCE AND MATHEMATICS INTERVENTION PROGRAM FOR SECONDARY SCHOOL TEACHERS

Loyiso C. Jita, Jacobus G. Maree*, and Thembi C. Ndlalane
(University of Pretoria)

Abstract: This chapter reports on a three-way collaboration (the Japan International Cooperation Agency, the University of Pretoria, South Africa, and the Mpumalanga Department of Education). We aim to change classroom practices to "inquiry" approaches in science and mathematics, involving the learning of innovative/new kinds of knowledge, skills and attitudes and enhance opportunities for teachers to foster the kinds of learning and support required to change classroom practices in these two gateway subjects

Keywords: Lesson study; Class room practies; Inquiry approaches; Science and mathematics education; In-service education; Jyugyo kenkyu; MSSI; cluster networks; Teacher development.

Making changes in classroom practices related to inquiry approaches in science and mathematics involves, for many teachers, the learning of new kinds of knowledge, skills and attitudes in their subject areas. In South Africa, as in many other developing countries, the problem is often compounded by the inadequate provision of opportunities and resources for learning the new knowledge and skills by the practising teachers. By contrast, the Japanese education system is renowned for its unique design that provides such opportunities for teachers to participate, throughout their careers, in a school-based model of professional learning and development (Fernandez, 2002). Small groups of Japanese teachers commonly meet regularly,

*Corresponding author

B. Atweh et al. (eds.), Internationalisation and Globalisation in Mathematics and Science Education, 465–486.

to plan together and review teaching practice in what is referred to as the *jyugyo kenkyu* or commonly referred to as the "Lesson Study."

With support from the Japan International Cooperation Agency (JICA) and expertise from the University of Pretoria (UP), one Provincial Education Authority (PEA) in South Africa, the Mpumalanga Department of Education (MDE) began a three-way collaboration to explore the use and efficacy of a school-based in-service education modelled along the lines of the Japanese *jyugyo kenkyu* in the South African context. The purpose of the partnership was to develop and enhance the opportunities for secondary school science and mathematics teachers to foster the kinds of learning and support required to change classroom practices in these two subject areas.

The intervention project – known as the Mpumalanga Secondary Science Initiative (MSSI) – was launched in 2000, specifically to promote more collaborative approaches to teacher development in the province of Mpumalanga. Furthermore, based on the recommendations of the initial surveys conducted by the Japanese counterparts (Hattori, 2001; Kita, Ndlalane, Nishioka, Ono, & Paulsen, in press; Kita & Nishioka, 2001) prior to the initiation of the MSSI intervention, there was consensus among the MSSI stakeholders on the need for, and importance of pursuing some form of school-based professional development through teacher collaboration and discourse. Lesson studies thus began to take root as a joint activity between neighbouring schools and within networks or clusters of teachers in the province.

In this chapter we discuss the MSSI, as a case study on the use of one such collaborative approach to professional development, viz. the lesson study (*jyugyo kenkyu*) (Lewis; 2000; Lewis & Tsuchida , 1998a; Lewis & Tsuchida, 1998b) among groups of teachers who came together to form professional development networks, or what in South Africa is commonly referred to as teacher clusters. We explore the dilemmas and contradictions of working with this borrowed concept, i.e. *jyugyo kenkyu*, within the South African professional development context. Furthermore, we explore how the perspectives of various MSSI stakeholders influenced the development and direction of the cluster/network based professional development of science and mathematics teachers in the province.

1. The Social and Cultural Context of Mathematics Education in South Africa

South Africa is a pluralist society with a diversity of "racial" and ethnic groups. Until recently, officially since the adoption of the new constitution in 1996, this diversity of its peoples has been used to define political and socio-economic rights and privileges, with black education receiving the least priority in terms of the quantity and quality of resources for teaching and learning. In South Africa there is a vast gap between the quality of schooling and the achievement of white and African students (Maree, Claassen & Prinsloo, 1997; Saunders, 1996). Since Grade 12 marks still largely determine whether a learner will be accepted by universities (Sibaya & Sibaya, 1997), technological and scientific study in particular is, by and large, out of reach for black learners. One in 312 African students who entered school in 1980 finally qualified for university in both mathematics and physical science. The equivalent statistic for white students is one in five (Blankley, 1994). The poor participation and success rates of black learners have been more pronounced in those subjects with the highest premium in society, particularly mathematics and science. For example, between 1985 and 1993, African learners accounted for only 2.6% of the total number of students graduating in engineering and technology (Siridopoulos, 1997). It is thus not surprising that there are fewer black mathematics and science teachers within the education system (Naidoo & Lewin, 1998). These statistics would not be so alarming were it not for the fact that black people make up more than 80% of the South African population.

In its attempts to intervene in the patterns of exclusion and inequality in education, the new government of South Africa has embarked on several notable initiatives, including developing a Mathematics, Science and Technology National Strategy (Department of Education, 2001), and changing the curriculum into an Outcomes-based Education (OBE) model of learning (Department of Education, 1997). Despite the strides made to change schooling in the country, most classrooms do not seem to have been affected by this change (Jansen, 1998a, 1998b; Jita, 1999). Most teachers are yet to embrace the new curriculum discourses and practices. Rote learning and teacher talk continue to be the dominant features of teaching and learning in most classrooms. Coupled with the structural legacies of the past, most teachers in the country are lacking in their subject knowledge and pedagogical content knowledge of mathematics and science. It is against this frame of reference (which is actually still framing and defining the achievements of South African learners in mathematics and science) that an initiative, that sought to dislodge these structural and substantive disadvantages of the past, was embraced by the Provincial Education Authorities in Mpumalanga. The initiative, funded largely by the Japanese government, could therefore not have come at a better time for the South African mathematics education context.

2. The Need for Change

Although education and training are being transformed in post-apartheid South Africa, the failure rate in mathematics at school remains unacceptably high (Maree, Pretorius & Eiselen, 2003; Steyn & Maree, 2003). Only 20% of all South African university students ever graduate (Ntshwanti-Khumalo, 2003), and students in science and commerce have the highest failure rates. Many hypotheses have been investigated, including the following (Arnott Kubeka Rice & Hall, 1997; Howie, 2001; Maree & Molepo, 1999): poor socioeconomic background of learners (poor incentive to study at home), lack of appropriate supporting materials, general poor quality of teachers and teaching (especially mathematics and science teachers; including poor subject knowledge and poor motivation), language of instruction (often not the same as learner's mother tongue). In addition, much has been said about the inadequacy of the science and mathematics teachers in South Africa and other developing countries (Jita & Ndlalane, 2005). It is against this background that, when constructing a model of intervention and development assistance with the Japanese government, the Mpumalanga Department of Education (MDE) opted for the lesson study approach. The reasoning was that a more collaborative and peer-centred approach to professional development and growth would appeal more to the teachers in Mpumalanga than a top-down process that would need to be driven from an administrative centre. The resource constraints in terms of the competence and adequacy of the central office level officials who could drive the professional development processes of the science and mathematics teachers may also have motivated the choice of a much smoother approach that results in much of the responsibility to the teachers as peers in the lesson studies. Using the concept of teacher clusters or networks, the MISS partnership adopted the *jyugyo kenkyu* as an instrument for facilitating and sustaining classroom change in science and mathematics across the province of Mpumalanga.

3. Conceptual Framework

The conceptual framework for this chapter is based on the current literature on teacher communities, networks and/or teacher clusters (Desimore, Porter, Garet, Yoon, & Birman, 2002; Grossman Wineburg & Woolworth, 2001; Little, 2003; Pennel & Firestone, 1996; Supovitz, 2002). The simple argument for teacher clusters is based on the social constructivist theories of learning which assert the importance of learning in collaboration. Based on the constructivist theory assertions, groups of teachers working together in practice, constitute the best possible alternative for encouraging teacher change and classroom reconstruction. Teachers who come together to reflect on and re-examine their classroom practices and experiences have been found to create better opportunities for growth and development (Desimone et al., 2002; Supovitz, 2002). Our analyses of the experiences of the key stakeholders in the MSSI project are based on this conceptual understanding of the opportunities for teacher development and growth in the MSSI project. In the following sections, we explore further the idea of teacher communities and the kinds of opportunities they seem to have created for the science and mathematics teachers in Mpumalanga. We also discuss how lesson study was used as a vehicle to achieve benefits for the teacher communities in the MSSI project. We delve into an exploration of the lesson study process and how it is used to influence the teachers' content knowledge (CK) and pedagogical content knowledge (PCK). Shulman (1987) proposes that teachers' professional knowledge comprised by a variety of categories including what he called content knowledge (CK), pedagogical content knowledge (PCK), curriculum knowledge and knowledge of the learners. According to Shulman (1987) PCK is understood as knowledge that links the particular content and teaching practices. PCK involves how teachers blend subject matter content knowledge with teaching methods and their knowledge of students and interests. The lesson study process (see Figure 1), as adapted from the Japanese practice, enables the teachers to link the content and the teaching practice.

In the next section, we will reflect on the processes of engaging science and mathematics teachers in lesson study activities in the Mpumalanga province. That is, we reflect more on Phase 2 activities of the MSSI intervention, as opposed to Phase 1, because of the promise held by the teacher cluster approach to development relative to the cascade model that has received a fair amount of coverage in teacher professional development literature to date. Our primary sources of data are the documentary analysis of the reports submitted by the JICA expert teams, our own field notes during workshops, school/cluster meetings, and personal conversations with key participants (teachers and Curriculum Implementers) during trainings and workshops.

4. The MSSI Project

The MSSI work actually began with a small-scale baseline research survey on the knowledge, skills and attitudes of science and mathematics teachers, conducted by consultant professors from one Japanese university. The aim of the intervention was to introduce and establish a province-wide school-based In-service Education and Training (INSET) system using a cascade model, through which continuous teacher learning and teacher development would be institutionalized in the Mpumalanga schools (MSSI project documents, 2000). The major focus of the project, at its inception, was thus on building the capacity of the subject advisers (commonly referred to as Curriculum Implementers (CIs)) in the province of Mpumalanga. The reasoning, within the project, was that the "empowered" Curriculum Implementers would then help to train and develop the teachers for whom they were responsible in terms of their job descriptions. This MSSI cascade model had four layers, with the teachers at the bottom of the structure. At the top of the ladder were the experts and professors from UP who trained and developed the Curriculum Implementers, who in turn worked with the Heads of Department (HODs) in their respective schools. The HODs then worked to transfer their knowledge and skills to the rest of the teachers at the school and/or district levels. Typical of cascade models, this process of teacher development took long and convoluted turns, with the result that some critical information got lost somewhere along the long cascade chain. It was therefore not surprising that at its first evaluation conducted in 2002 (JICA,2002), one of the major drawbacks identified was the long cascade chain that largely explained the absence of wide-ranging and observable changes in many of the provincial secondary science and mathematics classrooms. The evaluation confirmed the *status quo*, where the quality of teaching mathematics and science in the province remained very mediocre. At the end of 2002 in Mpumalanga only 23 out of every 100 candidates who wrote the end-of-the-year examination, passed mathematics (De Souza, 2003). In the project terminology, the period from 2000 to 2003 was known as Phase 1 of MSSI.

In response to the findings of the evaluation team, the three partners or key stakeholders in the MSSI intervention opted for a slightly different approach to teacher development in Phase 2 of the project (April, 2003–March 2006), one that

that would bring the intervention much closer to the teachers and the classrooms. Teacher communities of learning, known as professional development clusters, or networks, were initiated with a view to cutting through the cascade model and reaching the classroom quicker than through the traditional cascade approaches. This strategy was intended to impact on the teachers' classroom practices more directly than had been the case in Phase 1 of the project. Influenced by the recent dominance of constructivist approaches to teaching, learning and development, teacher communities, networks or clusters are viewed as having the best possible chance of helping teachers to change and reconstruct their classroom practices (Adams, 2000; Guskey, 1986; Liebermann & Grolnick, 1996). This new phase of the project was called Phase 2 of MSSI. Phase 2 of MSSI introduced the concept of teacher clusters or networks as an approach that was closer to the teachers and their classroom practices. The goal was to try and impact on the teachers' classroom practices a bit more than in Phase 1. Teacher clusters or networks are generally known to have better potential of influencing the teachers' classroom practices (Adams, 2000; Guskey, 1986; Liebermann and Grolnick, 1996). Teacher networks/clusters are usually formed by a group of teachers that have a special interest in improving their subject content knowledge on a voluntary basis. These teachers meet informally to discuss and share their classroom experiences (Guskey, 1986). It is within such teacher clusters and communities that the lesson studies were adopted and used to challenge and change the classroom practices of science and mathematics teachers in Mpumalanga.

In short, MSSI intended to institutionalize peer-teacher learning, more specifically "lesson study," which is typically found in Japanese elementary and lower secondary schools.

5. Lesson Features

Many researchers who have studied lessons, like Anderson (1991) define the academic context of a lesson as "consisting of a sequence of activities occurring in the classroom during a finite time period." The following primary features of lessons were identified: purpose, activities, sequence and time period. Berliner (1983) identified ten encompassing strategies or types of activities that could take place in lessons, Stodolsky (1988) suggested fourteen, and Snyman & Kühn (1993) identified six (actualizing of existing knowledge, posing the problem, exposition of new knowledge, monitoring new content, and functionalisation). However, the South African education system has been undergoing extensive restructuring since the advent of democracy. Outcomes-based education (OBE) originated during the formation of this new democracy in South Africa. OBE asserts that all learners have the ability to succeed, and focuses on knowledge, skills, values and attitudes, unlike the traditional practice that was based on content mastery only. Lesson design has therefore been redefined according to the OBE paradigm in order to enable learners to acquire skills that will prepare them for life in general and especially in a workplace. OBE adopts a much broader approach, and Wiggins (in Warren &

Nisbet, 2001) states that lesson design and assessment should accordingly be devised to include further outcomes such as higher order thinking, reasoning, problem solving, communication skills and conceptual understanding. McNeir (1993:1) states that *these outcomes are derived from a community vision of the skills and knowledge students need to be effective adults.*

Broadly speaking, arguments on the specific number and nature of activities differ but scholars agree on broad categories that are part of every lesson. These broad categories are described as the:

a) introduction,
b) presentation, and
c) conclusion.

Numerous schemes have been suggested for classifying activities within a single lesson. Such schemes are often represented in the form of lesson observation instruments or observation mechanisms. In the Japanese system, there is a rather distinct and special way of classifying and observing lessons, especially if the aim is that of improving the teaching and learning in the classroom. It is that model of classifying and characterizing lessons that has been adopted and adapted across the province of Mpumalanga, as part of the MSSI project.

6. What is a Lesson Study?

A lesson study is "a form of action research that allows teachers to work with each other collaboratively as reflective practitioners" (Yoshida, 1999:1), representing voluntary efforts of teachers to continuously improve and develop their instructional practices. According to (Fernandez 2002:21),

> A *lesson study is a professional development process that Japanese teachers engage in to systematically examine their practice. The goal of lesson study is to improve the effectiveness of the experiences that the teachers provide to their students. The core activity in lesson study is for teachers to collaboratively work on a small number of "study lessons". These lessons are called "study" lessons because they are used to examine the teachers' practices.*

The term "Lesson Study" describes the process of studying the proceedings and the way in which the lesson is presented in the classroom (Fernandez, 2002). It is not only a study of the lesson plan but the lesson plan is used as an instrument to capture the proceedings and the activities in the classroom. The value of a lesson plan is to help the teacher to describe a specific sequence of classroom activities and to facilitate reflection on these classroom activities and events. The broad aim of a lesson study is to encourage a cohort of critical classroom practitioners to develop their own practices through open-minded dialogical communities that facilitate peer-learning opportunities (Desimone et al., 2002). In these opportunities, it is the peer-practitioners who are concerned with moving each others' understandings of practice forward by engaging in dialogue using the lesson study process. This is in contrast to the traditional forms of classroom observation where the judgment of the supervisor is often final. In the Japanese educational culture, studying and improving lessons has been established as a legitimate form of professional development (Fernandez, 2002). All beginning teachers are expected to conduct and/or participate in this form of professional development as part of their induction and formal certification process. The rationale in a lesson study is that no lesson is perfect, thus there will always be room for improvement in the classroom processes and knowledge of the teachers.

Lesson study is employed in Japanese schools to explore the practice of teachers in order to empower them. In the context of the advent of OBE in South Africa, and the accompanying curriculum innovations, which require of teachers to share, plan together, reflect and assess each others' practice within each school's development planning processes, the concept of a lesson study was naturally attractive. As described earlier, the purpose of a lesson study is to analyse how the instructional activities are presented and how these activities form the basis for both teacher and learner engagement in a classroom setting. Schematically, for the purposes of the MSSI study, a lesson plan is viewed as having nine different, but integrated stages:

The key or instrument in any lesson study is the lesson plan itself. The lesson plan serves three different and important purposes in a lesson study: it acts as a teaching instrument, a communication instrument and an observation instrument.

6.1 Lesson Plan as a Teaching Instrument

The lesson plan is used by a classroom teacher as a guide to the nature of activities that are planned for the classroom. In a sense, the lesson plan acts as a map that guides the teacher's direction on a journey to a specific destination, which is often framed as the outcome of learning in a particular classroom situation. This journey takes into consideration the participants, i.e. the learners, their learning styles, abilities and the learning resource materials available for each lesson. The lesson plan is therefore designed to be detailed, precise and meaningful for each instructional situation.

6.2 Lesson Plan as a Communication Instrument

The teacher, who is a facilitator in the classroom, facilitates learning by means of a lesson plan. The communication process between the teacher and the learner engages both of them in activities that are communicated through a lesson plan. In addition, however, it is on the basis of what is put down in a lesson plan that other teachers can collaborate and assist a colleague to achieve a meaningful outcome in their lesson.

6.3 Lesson Plan as an Observation Instrument

In a context of collaboration, a lesson plan further presents opportunities for teachers to engage with each other, provide feedback and correctives, and make suggestions

Stage 1: Collaborative planning using a variety of resources

↓

Stage 2: Collaborative formation of a lesson plan

↓

Stage 3: Simulation of a lesson with peers

↓

Stage 4: Refining and making changes

↓

Stage 5: One member teaching the actual lesson in a real classroom of learners

↓

Stage 6: Peers observing and making notes on the classroom activities and actions

↓

Stage 7: Cost-conference meeting (discussing and sharing)

↓

Stage 8: Making changes and refining the lesson plan

↓

Stage 9: Refined plan

Figure 1. The Lesson Study Process

to each other. In such cases, the lesson plan may provide a minimalist "checklist" of the activities that are planned for the lesson. As an instrument for observation and feedback, it allows for teachers to reflect together in what Guskey (1986) sees as a sharing of their success and their failures in the classroom.

7. Lesson Study as a Professional Development Exercise in MSSI Clusters

Much has been said and written about the problems of teaching science and mathematics in schools. Among others, inadequate subject knowledge of teachers is often cited as a major problem (Hattori, 2001; Kahn, 2001; Kahn, 2004; Kita et al., in press). In 1998 Howie and Rutherford reported that 42% of the science teachers in South Africa had only one year of post-school education and nearly 40% of the physical science teachers had fewer than two years of teaching experience. Therefore, the need to support the professional development of the science and mathematics teachers in Mpumalanga was hardly in dispute. Given the complexities of the teacher development task, a collaborative approach that uses the existing human resources through the integration of the experienced and the less experienced teachers was the basis for the MSSI intervention. The purpose was to promote collaboration, sharing, discussion and learning from one another. The philosophy behind this kind of INSET structure lies in its promotion of closer collaboration and sharing of ideas based on trust and commitment. The idea is to have teachers who have similar subject matter interests meeting together in a relaxed atmosphere to freely discuss their teaching practices.

8. Formation of Teacher Clusters / Networks in Mpumalanga

A group of schools in a given geographical area come together with the aim of improving the quality of teaching in science and mathematics by sharing the classroom practices as peers. These schools that are geographically situated in an area within five kilometres radius from each other form teacher clusters/networks. The clusters operate under the leadership of a selected cluster leader who is also a full-time teacher. In the case of Mpumalanga, teacher clusters are registered to operate and they meet once a month. They are expected, by the MSSI intervention, to do lesson study and to discuss content-related subject matter. The specific objectives of clustering schools, for the MSSI, were to:

a) develop a co-operative and collaborative approach to professional development;
b) develop effective approaches in the teaching of mathematics and natural sciences;
c) foster closer ties between teachers within the cluster and to encourage the sharing of expertise and resources;
d) foster dialogue and reflection amongst educators; and
e) promote peer teacher learning principles.

The lesson study was to be the major vehicle for achieving many of these specific objectives.

9. Lesson Study as Practise in MSSI Cluster Meetings

As adapted from the Japanese system, the MSSI lesson study has the following four basic components:

a) collaborative planning;
b) lesson observation;
c) reflection on the lesson; and
d) implementation of changes.

These components will now be discussed in more detail.

9.1 Collaborative Planning

At this stage the teachers meet and discuss the theme of the entire lesson. The knowledge and skills are organized into cohesive lessons which are sequenced into units. Teachers are given the opportunity to engage in lesson plans as a team in a cluster meeting.

Writing a lesson plan in a group is an instrument for building *esprit de corps* among the teachers. The teacher conference after the study lesson is a time for teachers to reflect on ways to improve the lesson and to make changes or suggestions. While planning a lesson, the teachers engage in a discussion of such issues as to how the lesson fits into the entire unit; what the intended outcomes are; how the outcomes would be achieved; what assessment instruments would be used; what types of resources were to be used; and what the time duration would be.

9.2 Lesson Observation

The focus for the lesson observation is agreed upon and communicated among the teachers so that the observers can give specific feedback to the teacher. The teacher that teaches the lesson is one of them but they all own the lesson plan. Ideally the teachers carry out a simulation lesson beforehand so that they can revise or refine the lesson plan. It is particularly helpful to examine the lesson plan from the students' perspective. Photocopies of the lesson plan are prepared to share among the teachers. The MSSI intervention has designed a classroom observation instrument that describes the activities to be reflected upon in the classroom. On the basis of this classroom observation instrument teachers that are observing the lesson write down their comments while the lesson is in progress. This process of observing becomes an experiential learning activity for teachers for the enrichment of their own classroom experiences. The learners that are participating in the actual lesson activity are made aware of the reasons why many teachers were involved in their classroom.

9.3 Reflection on the Lesson

After the lesson the teachers meet to reflect on the proceedings of the lesson. The first opportunity for comment is given to the teacher that presented the lesson and later to all the other teachers who observed the lesson. The focus of this reflection during the post-conference meeting is on what worked and why it worked, what did not work and why it did not work, and finally what should be changed and how it will be changed for the future implementation of the same lesson.

9.4 *Implementation of Changes*

This post-conference session provides the opportunity for teachers to discuss, debate and argue on what they expected to happen and/or see during the lesson. The value of this session is in the collaboration of the teachers in solving practical classroom problems. By solving these problems together, the individual teachers' practices are influenced by the ideas and contributions from others. This is an example of the application of the idea of a teacher as a reflective practitioner, which has a long and respectable history in the literature on professional development of teachers (Schon, 1987). For example, Gunstone and White (quoted by Fraser-Abder, 2002) have shown, particularly with science teachers, the role that reflection plays both in classroom practice and the phenomenon of science teaching. More recently, Cooper and Boyd (in Fraser-Abder, 2002) have described a scheme of peer and group-orientated reflection on practice and development among teachers in a New York school district which provided a systemic self-help strategy for the long-term maintenance of innovative methods in the classroom. In South Africa seven roles have been assigned to all teachers. These roles are

a) learning mediator;
b) interpreter and designer of learning programmes and materials;
c) leader, administrator and manager;
d) scholar, researcher and lifelong learner;
e) community, citizen and pastoral role
f) assessor; and
g) learning area/ subject/ discipline/ phase specialist (Department of Education, 2000).

It is evident from this literature that allowing teachers to reflect on and talk about their lessons can be a powerful instrument of change and reconstruction. The discussion allows teachers to highlight their current practices and the attempts to change them. On the other hand, the individual teacher who presented the lesson also receives personal feedback on his/her presentation style and substance of the lesson.

Let us now consider the following *vignette*, which is an example of a lesson study presentation during one of the professional development sessions at an MSSI cluster meeting.

9.4.1 *Vignette*

A grade seven teacher had invited members of the cluster from neighbouring schools to come and study her lesson. There were eight teachers that came for the lesson study on that day. The lesson itself lasted for about an hour, after which the teachers met in the staff room to study the lesson. This was a maths lesson and the teacher wanted to introduce the concept of (negative and positive) signs used in mathematics. The title of her lesson was *Direction*. After the lesson had been presented, the presenting teacher was given the first opportunity to comment on her own lesson. Another teacher from a neighbouring school, the cluster leader, acted as the chairperson and facilitator of the lesson study. This is what the presenting teacher had to say about her lesson:

"I was very nervous as there were a lots of people, educators that I had to address. I was not sure whether I exhausted all that I had planned to do. I was so engrossed in my teaching that I realised in the middle of the lesson that I had forgotten to give my learners the handout with various road maps showing the directions. It would have been easy for me, had I asked them to identify the town from the map before I started my speech and would have grabbed the full attention of the learners. I think I did a lot of talking as against my initial plans to talk less. I had a feeling that I might not find the time to have an interactive session with my learners and feared that I might not complete my lesson within the stipulated time."

The teachers that observed the lesson also commented on what they observed in this lesson. They began by mentioning the good things about the lesson and later focused on those things that needed to be changed. As one teacher commented:

"Your introduction was good and it led learners to concentrate on the story that you were telling when you got lost in a city. I never knew that story telling could be powerful in the teaching of mathematics. I thought mathematics was just figures and numbers. I also liked the way in which you developed the concepts of directions. You started with a point, dots forming a line, line joined into various directions to form angles. You then joined the angles and showed directions by using arrows. I liked that – I will try it in my own lesson. I suggest that you should explore the learners' ideas first on directions as they have a lot to contribute and you can build on their ideas by introducing new terms. Please, consider that in future."

From the comments of this teacher, it was evident that she had studied the proceedings of the lesson carefully and highlighted issues of interest that needed to be discussed. She also challenged and made suggestions on those things that needed to be changed.

This example of a segment from a lesson study in one of the MSSI clusters illustrates three major issues that were recurrent during the MSSI lesson studies and contributed to its appeal within the South African context generally. The three major issues were:

i. Firstly, the fact that the participating teachers had collective ownership of the lesson being presented, since the planning of the lesson was itself a collaborative process.

ii. Secondly, the democratic approach to sharing ideas on pedagogy, modelled around the more popular approaches to political deliberations that have great appeal since the "new democracy" of the post-apartheid era in the country.
iii. Thirdly, the direct focus of the discussions on the content knowledge and pedagogical content knowledge.

Whereas there is some anecdotal evidence from our research to suggest that the lesson studies might be contributing to improved **reflective discourse** on content and pedagogy among the practicing teachers, and therefore their classroom practice, it is still too early to suggest what these improvements in the culture and discourse patterns of the cluster teachers might mean for learner performance in the Mpumalanga province of South Africa. The efficacy and relationship of the MSSI approach to professional development to the learner's performance is the subject of an ongoing study and reflection in the project.

10. Dilemmas and Contradictions in the Perspectives of the Various Stakeholders to MSSI

Unlike many other clusters and networks documented in the research literature that are formed by teachers themselves, the clusters in Mpumalanga were formed with the "official influence" of the MDE for various reasons. The added benefit of such official sanction of clusters lay in the fact that the teacher initiatives were encouraged and recognized by their employers. This is often a morale booster in the professional development of the teachers. The obvious drawback, however, lay in the fact that the **sense of ownership and volunteerism** almost disappeared with the official endorsement of these clusters. The official incorporation of clusters into the structures of the MDE did result in some tensions and contradictions concerning the ownership and direction of the clusters themselves. It is these tensions and contradictions that we explore later in the chapter.

As official organs of the MDE, the clusters risked being bureaucratized and in fact spent a great deal of time engaging in officially assigned tasks of the department. For example, the bulk of the time in clusters was spent on discussing the new policy changes to the curriculum-2005 based assessment and evaluation, paper moderation and designing learners' projects (Rogan & Van den Acker, 2002). The first dilemma for clusters therefore was to set aside the time to engage in lesson study during their meetings while also pursuing the official agenda on the new curriculum policy issues.

11. Structure of MSSI Teacher Clusters

Teacher clusters are formed by schools that are situated close to each other and the teachers that participate teach at a high school level or what is referred to in South Africa as the Further Education and Training (FET) level. The teachers that teach the lower classes or the General Education and Training (GET) level are expected to form their own separate clusters. These teachers, at the two different levels of the education system, meet at different times and engage in different professional development activities. Each cluster has a leader selected by teachers with the approval of the official supervisor from the MDE, viz. the curriculum implementers. The cluster leader's position is not a permanent position but is supposed to change from time to time. The one other contradiction and dilemma for the teachers in this form of clustering structure is in the boundary and separation of teachers who may be from the same set of schools and teach the same set of subjects, science and mathematics, but who happen to be assigned to different grade levels. This situation tends to limit the free flow of knowledge and information among the science and mathematics teachers at the two levels of education. This artificial separation tends to create some ill-feeling among teachers at the two levels and also reinforce the discourses in regard to stereotypes of the power relations and status of teachers at the two levels of education. One teacher participating in such a cluster articulated these feelings in the following terms:

We feel that teachers from FET phase do not want to meet with us because we do not have content knowledge. As a result they have their own meetings. I think I am much better than them – the only thing is that I teach at the primary school because I could not get a teaching post from the local high schools.

From this teacher's comment it was clear that the idea of separating the teachers and creating knowledge boundaries was counter-productive. Teachers at different levels of the system can share and learn from each other. For the university participants who were expected to provide the subject expertise support to the clusters, the dilemma is how to build and strengthen the working relationship among these cluster members and across the clusters without reinforcing the borders and boundaries of bureaucracy. Knowledge respects no boundaries. All teachers can potentially learn from, and in collaboration with each other. In fact, in the collaborative planning of a lesson, every teacher can contribute and benefit from others. This is one of the dilemmas and tensions that arise in the MSSI's attempts to balance official recognition with self-initiated teacher professional development in the MSSI intervention. Failure to identify these constraints of knowledge barriers might make us lose sight of the vast resources available within the education system itself. Classroom teachers are experts in their own work and need to be assisted to reflect and draw lessons from their experiences.

We will now focus on a few crucial aspects of the cascade model in Phase 2.

12. Cascade Model in Phase 2

Teachers are classroom practitioners who are more knowledgeable about their own classes than any other person. Each lesson that the teacher presents in class brings a new set of experiences. These experiences reside with the one teacher until they are shared in a cluster meeting or outside the classroom. The training in MSSI as prescribed by MDE states that curriculum implementers should be the first group that should experience the intervention through training either by Japanese experts or experts from the University of Pretoria. This approach to training and development retains aspects of the less successful cascade system with the result that some of the benefits of the cluster approach to professional development are moderated and eroded during the cascade chain. This was another serious contradiction and dilemma in the MSSI intervention and often was a potential source of disagreement among the partnership stakeholders. Being faced with this dilemma, and to keep the partnership healthy, it was resolved to continue to cascade the training by the university experts as required by MDE, first to the curriculum implementers who then presented it, in turn to the cluster leaders. The cluster leaders then cascaded the training further down to the teachers at the school level. Although the preferred link would have been to work with the teachers directly, for sustainability considerations it was felt that the CI layer of the MDE represented a permanent structure to consolidate and further support the initiative and goals of the MSSI long after the project will have ceased as a special intervention project.

JICA, as one of the support mechanisms, annually sends a group of delegates from MDE to visit Japan for a period of 12 weeks. This visit attempts to enrich the participants with the Japanese education system, school and classroom experiences and exposure to the outside world. There is always a dilemma in selecting the participants who would be useful immediately to the project goals as opposed to the MDE's own long-term development plans. While the MDE sees this opportunity as an incentive to the curriculum implementers and other senior officials to visit and learn about the Japanese education system, the other stakeholders want to use the opportunity as a learning opportunity for teachers to meet counterparts who are facilitating lesson study and other subject-related matters in Japan. The perspective of the MDE is consistent with its cascade approach to professional development of the teachers, which begins with the senior officials and cascades down to the teachers. The approach, however, creates a dilemma in the value attached to teachers as the ultimate implementing agents.

12.1 The Time Factor

Change is a process and does not happen overnight. The training sessions on lesson study require time and commitment from all the key MSSI stakeholders, especially the teachers. Scheduling meetings of teachers for two to three hours after school hours is often not enough. In addition to this, engaging teachers during school holidays on such activities as the lesson study is also not always practical. It is inadvisable to teach a lesson in a simulated classroom without actual learners. The simulated lesson mostly comprises a group of adults who have a more advanced approach to the lesson than the learners themselves. The simulations sometimes fail to capture the learners' prior learning, the questions they ask and the general response to the presentation of the lesson. Creating time to refine lessons by studying them is therefore a real problem. Unlike in the South African context, in Japan the process of lesson study has been institutionalized and built into the school timetable as a school-based INSET programme. It is scheduled once a week and all the teachers at the school participate in the activity. Creating time for teacher learning and professional development during the school timetable is not only desirable but also necessary. Under these conditions, schools would begin to function as learning organizations.

12.2 Peer Teacher Learning

Teachers are familiar with the style of most training programmes where experts come and tell them what to do and how to do it. The philosophy behind the lesson study process is different, in that it engages teachers to explore content knowledge (CK) and the pedagogical content knowledge (PCK) on their own initiative. This is based on the assumptions that, on the one hand, teachers have vast amounts of collective knowledge, and, on the other hand, they have acquired their experiences over many years. Lesson study thus provides the opportunity for teachers to tease out those elements of content that are usually neglected. Fraser-Abder (2002) makes this point differently when he asserts that expert practitioners possess a complex personal knowledge base upon which they draw almost intuitively. Although much of this knowledge base can be acquired through training experience, most individual teachers are not able to articulate why they do what they do in their classrooms. Similarly, the teachers in MSSI were uncomfortable with the idea of sharing what they do in their classroom and the ways they go about doing it. At first they saw the demands of the lesson study as a disguised attempt to evaluate their CK and PCK, but with time spent on, and experiences of lesson studies, they became comfortable and saw the process as beneficial to their own personal and professional development. Here is what one of the teachers who participated had to say about the process and its benefits:

When the learners gave me two answers that I was not expecting, I was not sure whether to say right or wrong because I did not know. But during the post-conference meeting, the teachers helped me to clarify my own confusion. This was very helpful, and I can definitely now see the value of doing lesson study.

This is only one instance of a teacher who saw the benefits of engaging in a conversation on their teaching with colleagues. We observed and heard numerous similar sentiments from teachers about the benefits of this teacher collaboration throughout the MSSI intervention schools.

12.3 Communication

Communication plays an important role when partners are working on the same project. Since MSSI is a large provincial project, with many key stakeholders involved, communication among the stakeholders was sometimes a problem. For example, one of the perennial problems of the project was to decide on what topics to select and target for the professional development of teachers. Differing views were often expressed by the teachers who sometimes preferred topics that they were about to teach, while others, including the MDE officials, preferred to target topics in the case of which there was either evidence that students experienced difficulty or gave evidence of non-performance. In such cases, the university would also often prefer to target what researchers refer to as the "key ideas" in the disciplines, which sometimes were not aligned to what the other stakeholders preferred at the time. Resolving these differences was part of the MSSI partnership strategy. The dilemma is essentially created by the demands made by the so-called Mathematics Pacesetters (provided by the Mpumalanga Department of Education) based on the MDE programmes and the needs of the teachers. These contradictions are sometimes based on – and reflect – the power games that the various stakeholders wielded within the MSSI intervention project.

12.4 Leadership in Clusters

The MSSI cluster leaders attended a series of workshops at least three times a year. During the workshops, the leaders are exposed through the process of lesson study, monitored and coached in the lesson study processes. They are further supported at their schools by experts from UP and by the curriculum implementers. Sometimes the process of learning and development takes time as the teacher leaders find their feet and begin to own up to the innovation ideas. The problem currently is that it is MDE policy that cluster leadership should rotate every year to give as many of the teachers as possible the opportunity to lead, and to learn about leadership. The dilemma for the clusters in this, otherwise noble, idea of a rotating leadership is in the way that it works in practice. It often happens that, just when a cluster leader has finally found his/her feet and is beginning to engage as a leader of a cluster, his or her term comes to an end, thereby setting the teacher development process through clustering back to square one. A new leader then needs to be developed, and so the process repeats itself. A practical compromise in resolving this dilemma has been found by taking advantage of the fact that the policy currently does not specify how many terms a cluster leader may serve and/or does not preclude a serving cluster leader from being re-elected and/or being reappointed.

Conclusion

The cluster method of professional development of teachers has had both benefits and drawbacks as discussed above. There have been many obvious benefits in the overall agenda for teacher development in Mpumalanga. Teachers enjoy the opportunity of meeting in a cluster with colleagues from different schools to discuss their own work. Many teachers have begun to develop trust and confidence in each other as colleagues and are able to share and exchange notes on classroom practice. The challenge of having to stand in front of other teachers and expose one's own practice is fast diminishing as a barrier to these teachers' professional development. Teachers are made to think seriously about their everyday classroom experiences in preparing their lessons. The lesson preparation has challenged the teachers, at least at the levels of Content Knowledge (CK) and Pedagogical Content Knowledge

(PCK). The professional development system, operating as clusters, provides an alternative source of perceptions, beliefs and actions for its members on both CK and PCK. The lesson study process provides a fresh alternative to the traditional teacher development mechanisms. It also has the advantage of being classroom focused, as it deals with the classroom issues of content and the way the content is delivered in the classroom. It further takes into consideration the behaviour and the responses of learners as the lesson is presented in the classroom. It is thus more promising as a mechanism for classroom and teacher change.

A major challenge for many of the teachers has been to analyze the lesson and to make changes to it. Most teachers failed to describe the events as they happened during the lesson; instead they made assumptions about what the presenting teacher was doing. For example, if the teacher started the lesson by putting the cup on his or her lips, the interpretation was that the teacher was drinking. Lesson study expects the teachers to say that the teacher put the cup on his/her lips, and not jump to an interpretation of the action. During the post-conference session the teacher would be asked why the cup was put on the lips and he or she would provide the explanation and the interpretation. This explanation would clarify the reasons for the teacher putting the cup on the lips. From the teacher's explanation, other teachers would then debate, argue and challenge the lesson proceedings.

In summary, we have identified two important issues that we view as critical to the operation of the MSSI project. These two issues emanate from Phase 1 and Phase 2 of the MSSI intervention respectively. The first issue is that of *change* and the second is that of *power*. It is these two issues that we found to be critical in shaping the direction of the MSSI INSET system. Phase 1 represented and built a structure of INSET that involves the education department, the overseas agency and the university. Such a structure is not common in professional development circles. The partnership *operated through a typically top-down structure of cascading the information from the university experts to the classroom teacher*. The weakness of this structure was that it was *top-down, prescriptive, and failed to challenge the existing teachers' practices in the teaching of science and mathematics*. The existing structure failed to engage teachers who are classroom practitioners to change classroom practice. If we are serious about changing classroom practice, *the mechanisms adopted need to start at the level where teachers are*. The teachers, themselves, deserve a chance to share what they do in their classrooms and how they do it with their colleagues as a starting point for reflection and critique.

Phase 2, on the other hand, represented a change in the structure and created better opportunities for teachers to share their CK and PCK with a great deal of promise for change. There are two contradictions, however, that continue to challenge the implementation at the Phase 2 level. The first problem is *the number of opportunities* that exist in the clusters to engage in issues of CK and PCK as opposed to the regulatory issues of policy, curriculum change structures and forms. The use of cluster meetings as opportunities to distribute and fill in forms and paper work on setting, moderating and grading question papers takes away the opportunity to focus the clusters on some significant areas of practice such as the lesson study.

Because of this, the effectiveness of clusters as an informal place for teachers to carry their own activities on content knowledge is minimized. Deadlines and tasks to be done are prescribed and dictated to teachers and tend to consume more time in the clusters. Attendance registers are kept in order to monitor which schools are present and which are not. A recent study by the Japanese counterparts reveals that the cluster leaders spend 60% of their time on curriculum and policy-related issues while only 40% of their time is used giving priority to subject content-related matters and the lesson study.

Despite all these occurrences, the lesson study approach to teacher development has managed to drive the entire teacher development system in Mpumalanga to a comprehensive vision of well-planned lessons that have to be studied and reflected upon periodically. It has also established a system in which teachers have grown accustomed to relying on each other, coaching, leading discussions and exploring alternative solutions. This is perfectly aligned with the OBE (Outcomes-based Education) system currently implemented in South Africa. Since OBE is foreign to teachers in South Africa, it is imperative to work together in groups as a support mechanism and to reduce duplication of work (Le Grange & Reddy, 1998:13). In a way, the formation of clusters has broken the isolation of teachers in their classrooms and at their schools. The lesson study has further broken the separation between training courses and classroom practice that is experienced in the workshops of most traditional INSET providers. Such separation between what goes on in the classroom and what teachers do at training workshops reduces teachers' capacity to contribute to and learn from the totality of their experiences. Multiple sources of information are available within a cluster, and the use of lesson study is beginning to make these multiple resources available and significant for the benefit of large numbers of science and mathematics teachers in the province. The lesson study approach has, to date, managed to shape the direction of the INSET system in Mpumalanga, although the pace of implementation and consequences of the intervention continues to vary from cluster to cluster.

References

Adams, J. (2000). *Taking charge of the curriculum: Teacher networks and curriculum implementation.* New York: Teachers College Press.

Anderson, L. W. (1991). *Increasing teacher effectiveness.* Paris: UNESCO International Institute of Teacher Planning.

Arnott, A., Kubeka, Z., Rice, M., & Hall, G. (1997). *Mathematics and science teachers: Demand, utilisation, supply, and training in South Africa.* Craighall: EduSource.

Berliner, D. C. (1983). Developing conceptions of classroom environment: Some light on the teacher in the classroom studies of art. *Educational Psychology, 13*, 1–13.

Blankley, W. (1994) The abyss in African school education in South Africa. *South African Journal of Science, 90*, 54.

Department of Education. (1997). *Annual report, 1996.* Pretoria: Government Printers.

Department of Education. (2000). *Norms and standards for educators.* Pretoria: Government Printers.

Department of Education. (2001). *National mathematics, science and technology education strategy.* Pretoria: Government Printers.

Desimone, L., Porter, A. C., Garet, M. S., Yoon, K. S., & Birman, B. F. (2002). Effects of professional development on teachers' instruction: Results from a three-year longitudinal study. *Educational. Evaluation Policy Analysis, 24*(2), 81–112.

De Souza, C. (2003). Senior certificate exams: Plausible progress or passes below par?. *E.D.S., 39*, 1–13.

Fernandez, C. (2002). Learning from Japanese approaches to professional development: The case of lesson study. *Journal of Teacher Education, 53*(5), 393–405.

Fraser-Abder, P. (2002). *Professional development of science teachers.* New York: RoutledgeFalmer.

Grossman, P., Wineburg, S., & Woolworth, S. (2001). Toward a theory of community. *Teachers College Record, 103*(6), 942–1012.

Guskey, T. R. (1986). *Staff development.* Thousand Oaks: Corwin.

Hattori, K. (2001). *Problems of education conducted by in service mathematics teachers of secondary schools in the Mpumalanga province in the republic of South Africa as seen from the baseline survey.* Tokyo: Japan International Agency.

Howie, S. (2001). *Mathematics and science performance in grade 8 in South Africa, 1998/1999.* Pretoria: Human Sciences Research Council.

Howie, S. J., & Rutherford, M. (1998). *Producing quality science teachers: Challenges facing colleges of education in South Africa.* Paper presented at 18th annual meeting of the International Society for Teacher Education, Skukuza, 17–24 April 1998.

Jansen, J. (1998a). Curriculum reform in South Africa: A critical analysis of outcomes-based education. *Cambridge Journal of Education, 21*(4), 242–246.

Jansen, J. (1998b). "Essential alterations?" A critical analysis of the state's syllabus revision process. *Perspectives in Education, 17*(2), 1–11.

Japan International Cooperation Agency (JICA). (2002). *Evaluation report for the MSSI project in the Mpumalanga province, South Africa.* Tokyo: International Cooperation Agency.

Jita, L. C. (1999). *Transformative practices in secondary schools science classrooms: Life histories of black South African teachers.* Unpublished doctoral dissertation. East Lansing: Michigan State University.

Jita, L. C., & Ndlalane, T. C. (2005). How much science do South African teachers know?. *Proceedings of the 13th annual Conference of Southern African Association for Mathematics, Science and Technology, Pretoria*, 10–13 October 2005.

Kahn, M. (2001). Changing science and mathematics achievement: Reflections on policy and planning. *Perspectives in Education, 19*(3), 169–176.

Kahn, M. (2004). For whom the school bell tolls: Disparities in performance in senior certificate mathematics and science. *Perspectives in Education, 22*(1), 149–156.

Kita, M., & Nishioka, K. (2001). Trends and problems of natural science education in the Republic of South Africa. *Bulletin of the Research Center for School Education.* (Naruto University of Education), *16*, 99–111.

Kita, M., Ndlalane, T., Nishioka, K., Ono, Y., & Paulsen, R., (in press). Teacher development and empowerment through collaborative lesson study: The experience of MSSI in the Republic of South Africa.

Le Grange, L., & Reddy, C. (1998). *Continuous assessment. An introduction and guidelines to implementation.* Kenwyn: Juta and Company Limited.

Lewis, C. (2000). Lesson study: The core of Japanese professional development. Invited address to the special interest group on research in mathematics education, *American Educational Research Association 2000 Annual Meeting*, New Orleans.

Lewis, C., & Tsuchida, I. (1998a). A lesson is like a swiftly flowing river. *American Education,* Winter, 12–17, 50–52.

Lewis, C., & Tsuchida, I. (1998b). Planned educational change in Japan: The case of elementary science instruction. *Journal of Educational Policy, 12*(5), 13–331.

Liebermann A., & Grolnick, M. (1996). *Networks and reform in American education.* Paper presented for the National Center for Restructuring Education, Schools, and Teaching (NCREST), New York: Columbia University, Teachers College.

Little, J. W. (2003). Inside teacher community: Representations of classroom practice. *Teaching. College Record, 105*(6), 913–945.

Maree, J. G., Claassen, N. C. W., & Prinsloo, W. J. (1997). *Manual for the study orientation questionnaire in mathematics.* Pretoria: Human Sciences Research Council.

Maree, J. G., & Molepo, J. M. (1999). The role of mathematics in developing rural and tribal communities in South Africa. *S. A. J. E., 19*, 374–381.

Maree, J. G., Pretorius, A., & Eiselen, R. E. (2003). Predicting success among first-year engineering students at the Rand Afrikaans University. *Psychology Report, 93*, 399–409.

McNeir, G. (1993). *Outcomes-based education.* http://eric.uoregon.edu/publications/digests/digest 085.html checked on 24-5-2004.

MSSI. (2000). *Project documents.* Nelspruit: Department of Education.

Naidoo, P., & Lewin, K. M. (1998). Policy and planning of physical science education in South Africa: Myths and realities. *Journal of Research in Science Teaching, 35*(7), 729–744.

Ntshwanti-Khumalo, T. (2003). A tertiary update: October 2002–September 2003. *E.D.S.*, 1–20.

Pennel, J. R., & Firestone, W. A. (1996). Changing classroom practices through teacher networks: matching program features with teacher characteristics and circumstances. *Teacher College Record, 98*(1), 46–76.

Rogan, J., & Van den Acker, J. (2002). The Profile of implementation of C2005 in Mpumalanga secondary schools. *Proceedings of the 10th Annual Conference of Southern African Association for Mathematics, Science and Technology.*

Saunders, W. (1996). One lost generation after another. *Frontier of Freedom,* Fourth Quarter, 18–19.

Schon, D. (1987). *Educating the reflective practitioner.* London: Jossey-Bass.

Shulman, K. (1987). Knowledge and teaching: Foundations of new reform. *Harvard Education Review, 57*(1), 65–86.

Sibaya, P. T., & Sibaya, D. (1997). Students' performance on a teacher-made mathematics test: The interaction effects of sex, class and stream with age as a covariate. *S.A.J.P., 27*(1), 9–15.

Siridopoulos, E. (1997). *Maths and science remain at the bottom of our national priorities list.* Johannesburg: FastFacts S.A.I.R.R.

Snyman, R., & Kühn, M. J. (1993). Structured instruction and learning. In W. J. Louw (Ed.), *Classroom Practice* (pp. 36–45). Pretoria: Academica.

Steyn, T. S., & Maree, J. G. (2003). Study orientation in mathematics and thinking preferences of freshmen engineering and science students. *Perspectives in Education, 21*, 47–57.

Stodolsky, S. (1988). *The Subject matters,* Chicago: University of Chicago Press.

Supovitz J. (2002). Developing communities of instructional practice. *Teachers College, 104*(8), 1591–1626.

Warren, E., & Nisbet, S. (2001). How grades 1–7 teachers assess mathematics and how they use the assessment data. *School Science and Mathematics, 101*(7), 348–355.

Yoshida, M. (1999). *Lesson study (Jyugyo Kenkyu) in elementary school mathematics in Japan: A case study.* Paper presented at the American Educational Research Association 1999 Annual Meeting, Montreal, Canada.

26

THE POST-MAO JUNIOR SECONDARY SCHOOL CHEMISTRY CURRICULUM IN THE PEOPLE'S REPUBLIC OF CHINA: A CASE STUDY IN THE INTERNATIONALIZATION OF SCIENCE EDUCATION

Bing Wei[1] **and Gregory P. Thomas**[2]
[1]*Faculty of Education, The University of Macau, Macau, People's Republic of China*
[2]*Department of Mathematics, Science, Social Sciences and Technology, The Hong Kong Institute of Education, Hong Kong, People's Republic of China*

Abstract: Using the case of the Junior Secondary School Chemistry Curriculum (JSSCC) in the People's Republic of China, this paper addresses the issue of internationalisation of science education in a developing country. The focus was on the change of the JSSCC during the period from the late 1970s to the present. Data were collected from two sources: curriculum documents related to the JSSCC, and interviews with the individuals who were responsible for designing the JSSCC during the period under study. International influences on the four versions of JSSCC during the period were analysed in the four themes: socio-political climates, influencing factors from the abroad, the ways the international influences were exerted, and constrains by the national conditions. And finally, problems and issues involved in the international of the JSSCC were discussed

Keywords: Science Education, People's Republic of China, Curriculum change, Socio-Political Climate.

1. Introduction: Acknowledging Country-Specific Influences on Science Curriculum Development

The issue of internationalization of curriculum and pedagogy has become increasingly prominent in the field of education in recent decades as a consequence, at least in part, of the emergence of increased worldwide economic interaction and increased opportunities for the sharing of educational ideas between countries. In Asia, this

B. Atweh et al. (eds.), Internationalisation and Globalisation in Mathematics and Science Education, 487–507.

issue has been addressed by curriculum research on general school curriculum structures (e.g., Morris, McClelland, & Wong, 1997) and social science school curricula in particular (e.g., Wong, 1991). However, few studies are evident in relation to the internationalization of science school curricula. Indeed, differences between countries in relation to their science education and science curriculum development have only relatively recently been recognized and explored. For example, Ogawa (1995) examined the strong differences between Japan and the West and Loo (1996) noted that Islamic countries have their own idiosyncratic concerns. Over the past twenty years there is evidence of increasing similarities between countries in terms of their emphases in science curriculum innovations. These similarities are evident, for example, in the engagement of Science-Technology-Society (STS) orientations and constructivist frameworks for pedagogy and curriculum development. While acknowledging this tendency, Donnelly and Jenkins (2001) suggested that substantial differences between countries can be identified when investigating closely the influence of such frameworks on any innovations. They argued that, 'how these different orientations work out in practice might be expected to be influenced by national and regional intellectual and professional traditions' (p.10). Their position echoes that of Black and Atkin (1996) who offered the following suggestion for analyzing science curricula and their development in any specific country:

> At the deepest level are a nation's underlying mood and values. It is difficult to fathom precisely what impact these have on educational policy and practice in the classroom; but there is no doubt that the national spirit of any moment colours the perspectives and behaviour of all the actors who shape, and are shaped by, educational policy.
>
> (p.14)

We consider these suggestions to be salient in relation to curriculum development in the PRC. Because of China's intrinsic features and its pragmatic and predominantly utilitarian intention in relation to its reasons for learning from foreign countries, the adoption of educational ideas or approaches from overseas has never been a process of imitating simply any international trend. For instance, even in the 1950s when the former Soviet Union influenced strongly many aspects of Chinese society, including its education system, this influence was not exerted by way of simple imitation (Chen, 1981). Such a measured, context-influenced application of educational thought and practices can be further illustrated with reference to the generally guarded attitudes of Chinese curriculum scholars in relation to the potential influence of foreign curriculum thought and content on PRC curricula. Chen (1989), for instance, proposed that foreign experiences in curriculum development should be interpreted and accommodated in relation to the PRC's national conditions. In relation to adopting the chemistry curricula developed in Western countries, Liang (2002) elaborated on the same principle when stating that, 'if they [foreign curriculum thought and content] are appropriate, we follow them; if they are not appropriate, we discard them' (p.4). Ding (2001) clarifies this position well:

> The facts show that the simplistic logic of the dominant culture exerting its power does not enable us to explain or evaluate China's educational experience...we cannot underestimate the importance of external cultural influences in China's education, but we must also recognize that it was not a matter of China progressing from a lower to a higher cultural level. Rather, it was a matter of the creation of a new cultural complex, which drew on the best of foreign cultures and reached a kind of dialectic between universal and particular, an integration of diversity and unity.
>
> (p. 183)

We propose that the implication of such views is that Chinese curriculum scholars perceive that noticeable differences exist between China and foreign countries, and that foreign experiences in curriculum development should be used to meet domestic needs only after careful consideration of their relation to prevailing domestic factors. This orientation is important because it means that any interpretation of curriculum change in the PRC and influences on such change should acknowledge and account for, (a) the domestic conditions existing in the PRC at any particular time, and (b) their influence on the thinking of curriculum developers and consequently on the final curricula.

Finally, in relation to curriculum reform in the PRC we suggest it is also important to understand that the PRC is a highly centralised country in many ways. This centralised nature is evident in, amongst other things, the Communist Party's organizational structure, the government's functioning, the development of national policies and, importantly for this paper, its education system. Not surprisingly therefore, portrayals of a central, determinant role of the state in the processes of curriculum change are common. Lewin, Xu, Liitle and Zheng (1994) have generally described curriculum development in China in the following way:

> The overriding characteristic of curriculum development in China since 1949 has been the central control of a nationally unified syllabus...the main mechanisms for this control are the teaching syllabi and teaching programs.
>
> (p. 147)

Curriculum change unfolds through state sanctioned structures such as the Ministry of Education (MOE) which has the highest authority and governs all educational matters within the PRC. Under the communist system, the primary duty of the MOE is to faithfully implement educational policies formulated by the Central Committee of the Chinese Communist Party. In understanding changes to any post-Mao curriculum in the PRC it is important to be aware of such context specific top-down influences that invariably shape the nature of education and curriculum in that country and that also influence the perspectives of those individuals and groups engaged in making curriculum decisions.

2. An Historical Overview and the Need for Further Research

The structure and content of the science subjects taught in schools in China prior to the foundation of the PRC were imported from Western countries in an almost wholesale manner at the end of the nineteenth century (Reardon-Anderson, 1991). Prior to 1949, science education in China was strongly influenced by European curricula, especially in the first part of the twentieth century, and then by Japan during the periods of Japanese occupation. Not surprisingly, 1949, the founding year of the People's Republic of China, is seen by both local and international scholars as the key year in which to begin studies of the impact of international influences on the PRC curricula (Leung, 1991; Lewin, 1987; Liang, 1982; Liu, 1996). Scholars recognize generally that science education in China since 1949 has been indelibly influenced by the former Soviet Union in relation to both its curriculum content and pedagogy. The origins of the PRC's science curricula entails that international influences on science curricula in the PRC over the period of the twentieth century are essentially unavoidable. Yet these influences and the teaching that occurs in the PRC in enacting the curricula are still not well understood. Few empirical studies have explored the specific processes by which foreign science curriculum and educational ideas have influenced science curriculum in the PRC and so little is known about these processes and their outcomes. This is despite the general recognition of these pre-1949 influences. Only recently have curriculum and pedagogy within the PRC become a source of interest, especially since research and reviews such as Stevenson (1992), Stevenson and Stigler (1992) and Ma (1999) drew Western interest into trying to explain and understand the curricula and the pedagogical and social conditions that facilitated the reportedly high academic achievement of Chinese students.

In this chapter we contribute to an increased understanding of the influence of the internationalization on the process of science curriculum development. To do so we use as a case the Junior Secondary School Chemistry Curriculum (JSSCC), that is offered in the third grade of junior secondary school (the equivalent of ninth grade in the American schooling system) in the People's Republic of China. We explore the changes in the JSSCC from 1978 to 2001 in relation to international influences. Previously we identified a shifting tendency from elite to future citizenry orientations in the JSSCC during the period from 1978 to 2001 and provided sociological explanations for such a shift (Wei, 2003; Wei & Thomas, 2005). This chapter differs from our previous work as we now explore the processes by which the post-Mao JSSCC curriculum was developed with reference to domestic conditions existing since 1978 and how these conditions influenced the use of ideas from outside the PRC in the development of the curriculum by those involved in the development and revisions of the JSSCC.

3. Methodology Overview

Our study focuses on the 'written' or 'preactive' (Goodson, 1994, p.113), aspects of the JSSCC. We collected data from two sources: curriculum and curriculum-related documents related to the JSSCC and interviews with its designers during the relevant period. The documents included conference papers, discussion papers, manuscripts of syllabi and curriculum standards that had not been presented to the public, as well a range of official documents and state approved textbooks related to the JSSCC that had been released and/or approved by the Ministry of Education. The people who were involved in producing the syllabi and textbooks of the JSSCC provided some of these came from their personal collections. It is important to understand the central importance of the textbook as a curriculum document for the purposes of this study. The various versions of the chemistry textbooks of the junior secondary schools embody the intended JSSCC and strongly reflect curriculum organization and emphases. In doing so, they reflect a long tradition of textbook use in China and are therefore valuable sources of data for use in identifying changes in the JSSCC over the period of interest.

We used individuals' proximity to prime leadership roles in the development and revision of the JSSCC as the primary criterion for selecting interviewees. The first author interviewed intensively thirteen people who were appointed by the MoE to take on prominent leadership roles and this group included almost all of those who were appointed by the MoE to shoulder the major responsibility for designing the JSSCC during the period under study. Some were acquaintances of the first author, known to him through his study and work experiences in Beijing Normal University (BNU) and the People's Education Press (PEP). These acquaintances and other former colleagues then recommended other key informants for interview. Most interviewees came from BNU and the PEP as they were closest to the curriculum decision making and design processes. The PEP was established in the early 1950s and has long been authorized by the Ministry of Education to produce the syllabi and textbooks used in primary and secondary schools in China. Others were chemistry educators who were and in some cases still are responsible for teachers' training in the department of chemistry in BNU, a key teachers' university administered by the Ministry of Education.

4. Reporting Considerations and Framework

We identify four distinct phases of the JSSCC between 1978 and 2001 from our analysis of curriculum documents: 1978–1982; 1982–1992; 1992–2001, and 2001 onwards. For brevity, we have named these phases the 1978, the 1982, the 1992, and the 2001 JSSCCs respectively. Four themes relevant to the internationalization theme of this tome emerged from our analysis of the revisions of the JSSCC over this period. They were the sociopolitical climate/s within the PRC, influencing factors from abroad, the ways the international influences were manifest in the JSSCC, and constraints exerted by the national conditions within the PRC. We provide a summary of the four themes related to international influences in each of the phases of the JSSCC in provided in Table 1.

Table 1. International influences on the JSSCC from 1978 to 2001

	1978	1982	1992	2001
Sociopolitical climates	The 'Four Modernization' program was promoted;	A multi-track schooling system was established	Nine-year compulsory education was implemented	More concern was given to the quality of education
Influencing factors from the West	Science curriculum reform in the 1950/60s	'Back to the Basics' in 1970s; Science curriculum reform in the 1950/60s	'Science for all' in the 1980s	'Scientific literacy' in the 1980/90s;
The ways the international influences were exerted	Adding theoretical chemistry and decreasing descriptive chemistry	Lowering requirements of theoretical chemistry and adding descriptive chemistry; Specifying scientific processes	Paying attention to the relation of chemistry with society	'Scientific inquiry' and 'chemistry and social development' as parts of curriculum content
Constrained by national conditions	Marxist-Leninist learning theory	Confucian tradition	Lessons learnt in the past	Understanding of the concept of scientific literacy

In what follows, we present the international influences on the JSSCC during the period under study in relation to these themes. However, because these themes are interwoven in history we have chosen to present the internationalization of the JSSCC in its different historical phases rather than to structure the presentation around the four aforementioned themes. We have chosen to use a series of historical narratives that each correspond to an aforementioned phase. In each narrative we attempt to blend the historical, sociopolitical context with the personal accounts of those key players involved in the revisions of the JSSCC. In doing so we aim to provide, as much as possible, a seamless account of the 'internationalization' of the JSSCC and the factors contributing to the nature and extent of this process. We have deliberately omitted substantial detail regarding the specific changes to the curriculum and have instead decided to highlight those changes that are most relevant to the theme of internationalization. Readers interested in a specific and detailed account of the changes to the JSSCC might refer to Wei (2003) and Wei (2005). The following represents our construction of the narrative account/s of the internationalization of the JSSCC.

5. The 1978 JSSCC

The year 1976 is significant in the history of the PRC. Less than one month after the death of Chairman Mao on 9 September, the 'Gang of Four,' critical leftists in the national leadership in the latter phase of the Cultural Revolution (1966–1976), were arrested. This event marked the conclusion of this political episode. The end of the Cultural Revolution and the return of Deng Xiaoping to power in the central committee of the Chinese Communist Party (CCP) in the summer of 1977 resulted in the reinstitution of the manpower function of education and the promotion of the 'Four Modernizations' program. This programme targeted industry, agriculture, national defence, and science and technology and aimed to bring about the recovery and development of the national economy. At the direction of Deng Xiaoping the first curriculum reform in the post-Mao period was initiated in August 1977. A team known as the 'National Teaching Materials Workshop' (Wu, 2002), comprising selected staff from the PEP and other persons drawn from tertiary institutions and primary and secondary schools, was organized by the Ministry of Education within subject-specific working groups to produce new syllabi and textbooks for the various primary and secondary school subjects (Pu, 1993).

The modernization of school teaching materials, especially textbooks, was the prime objective of the curriculum reform. Deng Xiaoping himself played an overarching role in determining what was to be understood by teaching material modernization. During the last half of 1977, Deng held talks with leaders of the Ministry of Education concerning this issue. The main directives he raised in these talks included: (1) Teaching materials are crucial and they must reflect the advanced levels of modern technology and match the conditions of the PRC; (2) Much more attention needed to be paid to the teaching materials of primary and secondary schools; and (3) Foreign teaching materials should be imported, and useful things should be absorbed from them (Pu, 1993). The appointed leader of the JSSCC group, Wu Yongxing, an academic chemist as were the other members of the group, saw the key to modernizing the curriculum as, 'strengthen[ing] the basic knowledge and skills that are necessary for pupils to join in the industrial and agricultural labour production' (Wu, 1978, p. 23). In this way their aim was commensurate with Deng's vision for education as a source of manpower for the Four Modernizations programme.

In order to modernize science curricula and recognize and, where appropriate, incorporate the state of the art information from the West, a large sum of money was spent to purchase science textbooks from foreign countries under the instruction of Deng Xiaoping (Pu, 1993). More than ten titles of chemistry textbooks were imported for the chemistry group. Most of these were written in the west in the 1950/1960s with some written in Asian countries in the 1970s. When examining these chemistry texts the members of the chemistry group were astonished to find that many abstract concepts and theories traditionally covered in university in the PRC were found in the sample of chemistry texts used in foreign secondary schools. This was not only manifest in the textbooks from developed countries but also in those from developing countries of that time such as Thailand and Singapore.

The chemistry group summised that the raising of the level of theoretical chemistry was an international tendency and congruent with their concept of the 'modernization' of chemistry curriculum. Among those textbooks was *Modern Chemistry* by Metcalfe, Williams, and Castaka[1]. The chemistry group's knowledge that it was one of the most popular chemistry textbooks in the United States resulted in it becoming a main reference in defining the breadth and depth of the chemistry curriculum reform. Liang (2002) recorded this event:

> … we got some chemistry textbooks of secondary schools from overseas (particularly the United States) and felt that the level of theories in these textbooks was raised to a higher level. … When designing our own textbooks, we made much reference to the textbook *Modern Chemistry*, which was widely used in the United States.
>
> (p. 4)

Influenced by science curricula as epitomized in the sample of reviewed textbooks from western countries in the late 1950s and 1960s, a time of substantial curriculum reform in science education, much theoretical chemistry was added the 1978 JSSCC at the expense of a lot of descriptive chemistry. Wu, in further justifying this position, saw textbooks as becoming 'organically combined' (Wu, 1983, p. 37) entities when their structure was organized around basic concepts or principles.

However, prevailing Chinese traditions constrained the foreign influences in several ways. Firstly, in accordance with Marxist-Leninist learning theory, in which learning is supposed to occur in the two stages of perceptual knowing and

[1] This textbook was published in 1958 in its first version.

rational knowing that are separated by a dialectical leap (Cross & Price, 1991), descriptive chemistry was perceived strongly to provide the foundation on which the learning of theoretical chemistry was based. Therefore, the designers of the 1978 chemistry curriculum did not follow their western counterparts to substantively remove descriptive chemistry to the same extent as evident in the Western texts but deliberately kept what they considered to be an appropriate ratio of theoretical to descriptive chemistry (Xu & Liang, 1981). Secondly, laboratory work was treated as part of the subject matter and used to ratify concepts proposed in lectures and to train 'skills' of chemical experimentation (MOE, 1978). Therefore, we find it not surprising that notions of 'scientific inquiry' or 'discovery learning', which were increasingly advocated in western science curricula of the 1950/1960s, were not evident in the 1978 JSSCC.

6. The 1982 JSSCC

In December 1978, the third plenum of the 11th central committee of the Chinese Communist Party (CCP) adopted the new ideological line of 'seeking truth from facts,' which implicitly repudiated rigid adherence to Maoist principles and called formally on the party to shift its energies to the task of economic development. This plenum was a major landmark in the political and economic life of the post-Mao era. It signalled the rise of Deng as China's paramount leader and the adoption of key policy positions related to accelerating economic development and opening China to the outside world. Several policies which were thought to be beneficial for economic development prior to the Cultural Revolution were revived. In the field of education one such policy was to replace the single academic schooling system with a multi-track schooling system. The official justification for this change was the proposal that, during the Culture Revolution, the 'Gang of Four' not only tried to universalize secondary schooling prematurely but also to unify the educational system in a manner inappropriate to China's corresponding level of economic development (Zhang, 1984). The move to transform China's school system around 1980 was encapsulated by (Zhang, 1984) as follows:

> Enriching primary schools, rectifying junior secondary schools, adjusting senior secondary schools, vigorously developing vocational schools, and well handling key-point secondary schools.
>
> (p. 153)

Consequently, the Chemistry Editorial Board of the PEP decided that the 1982 JSSCC should be for those who would study in vocational and skilled senior secondary schools or get jobs in labour markets and not only for the minority of students who would continue their studies in academic senior secondary schools (Zhang, 1982).

The 1982 JSSCC was, in essence, a revised version of its 1978 predecessor. We can trace the degree of impact of international influences on the 1982 JSSCC to the personal involvement of Liang Yinghao who oversaw the process of revision. Liang, a senior editor in the PEP, and a member of the 1978 design team, was head of Chemistry Editorial Board in the early 1980s and given responsibility for designing the 1982 JSSCC. Since his participation in the design process of the 1978 JSSCC he had devoted himself to studying chemistry education literature including curricula and textbooks imported from the West, especially those from the United States and the United Kingdom. Liang recalled that around 1980 the main sources of his readings were *Journal of Chemical Education* from the United States and *Education in Chemistry* from the United Kingdom. These two academic journals which only became available in the PRC after the Cultural Revolution were windows through which Liang and other Chinese scholars learned about chemistry education beyond the PRC in the late 1970s and the early 1980s. Liang (1981) noted from his readings that, in the curricula of the late 1970s in western countries, attention was increasingly paid to information about chemical elements and compounds and their applications and to descriptive chemistry.

With the initiation of the 'open door' policy and especially with the normalization of Sino-American relations in January 1979, academic exchanges between China and the West became increasingly popular. One example of such an exchange occurred in the summer of 1980 when two chemistry teachers from the United States visited the PEP and met with the staff of the Chemistry Editorial Board. Liang (2002) recalled this event and drew particular attention to insights gained as a consequence of these talks in relation to the implemented, actual use of *Modern Chemistry,* the text that had served as primary reference for the 1978 JSSCC:

> When two chemistry teachers from Bronx Middle School, which was one of the schools that adopted this textbook most effectively, came to Beijing in 1980 we had talks with them. Surprisingly, we learnt that many abstract theories in this textbook had not been taught when this textbook was used in their classes. I was deeply impressed by this event. So, we should not rely on the written textbooks only, but should have the knowledge about the actual use of the textbooks.
>
> (p. 5)

Through such academic exchanges, Liang and his colleagues also learned of the 'Back to Basics' movement emerging in the United States and other countries in the 1970s. Consequently, their attention was drawn to the proportion of theoretical and descriptive chemistry in the curriculum. Liang (1981) was convinced that the theoretical level in the 1978 chemistry curriculum had been raised much higher than was reasonable for pupils and that some descriptive chemistry had been omitted inappropriately. However, a matter for Liang's rejoicing was that they had not followed their western counterparts to the same extent in their reduction of the amount of descriptive chemistry in the 1978 JSSCC. Even so, to respond to the call to maintain the trend to selectively integrate chemistry curricula from the West the level of theoretical chemistry was lowered while more descriptive chemistry was added to the new textbook (Liang, 1981).

An additional concern for curriculum reformers in the late 1970s in the PRC was how best to develop student's thinking abilities (Ye, 1982). This concern arose from a common perception emerging in the late 1970s that science and technology were advancing very rapidly and that the volume of scientific information was growing exponentially. However, students' contact time in school was limited. For pupils to adapt to contemporary society it was considered not enough to equip them only with scientific knowledge. Rather, it was seen as increasingly essential to cultivate their abilities (Du, 1980). Therefore, cultivating student's abilities was one of the goals of the 1978 chemistry curriculum (MOE, 1978). However, it was not clear what 'abilities' should be focused on at that time. When the 1978 JSSCC was implemented, critics argued that insufficient emphases were paid to cultivating student's abilities through the textbooks. Guo (1981), for instance, observed that, 'textbooks put more emphases on basic knowledge and skills but less on developing students' intelligence' (p.50). In response to such criticisms, 'scientific processes' that were increasingly advocated in western science curricula in the 1950/1960s, were included in the realm of training students' 'abilities' in the 1982 JSSCC. These processes were described as 'observing', 'thinking', 'experimenting', and 'learning by oneself.' In interviews Liang provided the following account for selecting the four abilities:

> In my opinion, training students' abilities was emphasized in every curriculum document in the west and the amount of these abilities were much more than ours... The reason for selecting the four was based on our understandings and the conditions that could be provided in our country at that time.

Therefore the 1982 JSSCC selectively integrated some concepts derived from the study of increasingly available foreign curricula, texts and educational literature and from increased academic exchange. However the nature and extent of the integration was in accord with the interpretation of the centrally determined objectives and intended purposes of the curriculum and in particular the interpretation of those objectives by Liang and his colleagues.

7. The 1992 JSSCC

By the middle of the 1980s the success of agricultural reform and the efficiency of the foreign and joint enterprises prompted the CCP to launch a long-term program of economic development. The ultimate goal of this program was to reach economic parity in GNP per capita with middle-income developed countries by 2049, the 100th anniversary of the founding of the PRC (Lewin et al., 1994). In order to achieve such a lofty goal, the Chinese education system was expected to become a cradle for skilled personnel. However, the goal of economic development seemed at odds with both the quantity and quality of education in the country. The 1982 national census suggested that there were 235 million illiterate or semiliterate people in the PRC and for every 10,000 rural residents there were no more than four agro-technicians

(Lewin & Xu, 1989). To make the Chinese education system more relevant to the economic goals of the country, the central committee of the CCP proclaimed the 'Decision on Reforming the Educational Structure' on 27 May 1985. As part of this 'decision,' nine-year compulsory education was proposed. In April 1986 the People's Congress legitimized nine-year compulsory education by passing the 'Law of Nine-year Compulsory Education.' Making the JSSCC and other curricula more appropriate for the majority of pupils was an overwhelming concern in the process of designing the 1992 JSSCC (Hu, 1989). It was seen as essential to help popularize nine-year schooling and so ways were sought to meet these goals for the curriculum. International influences were to play a significant role in determining the direction for the revision but once again their influence was mediated by sociopolitical factors and the principles guiding the thinking of those chosen to be primarily responsible for the revision. The two most prominent such influences were in the form of 'science for all' and in the potential for increased use of student activities.

Since the early 1980s, academic staff of the Chemistry Editorial Board of the PEP had begun to travel overseas increasingly frequently to visit educational institutes or to attend various conferences on chemistry education or science education. On their return they usually wrote papers to brief readers on what they had learnt of emerging tendencies in chemistry or science programmes and/or on textbook development (e.g., Cheng, 1986; Liang, 1983, 1984a, 1984b; Wu, 1986). Further, as a nationally designated educational press, it was becoming increasingly less difficult for the PEP to exchange textbooks and other teaching materials with its counterparts in other countries. In the mid to late 1980s, several foreign chemistry textbooks developed in western countries such as *Chemistry in Community* (ChemCom) from the United States, *Elements of Chemistry* from Australia, and other texts from Japan, Hong Kong and Taiwan, became available to the Chemistry Editorial Board in the PEP. A key message acquired from the review of these materials was that students were often engaged in activities such as 'group discussion', 'social investigation', and 'project work.' Consequently, the designers of the 1992 JSSCC in the Chemistry Editorial Board in the PEP made attempts to adopt these instructional forms, which they assumed to be helpful for developing students' abilities. However, most of academic staff in the Chemistry Editorial Board in the PEP held a reserved attitude to the context-centred, STS-oriented approach to organizing the curriculum content that was encapsulated in these texts. This reservation was evident in the debate on the implications of the move 'Science for All' that had been ongoing in the PRC since the mid- to late 1980s.

A key conference that influenced the consideration of science for all on this revision of the JSSCC occurred October 1985. It was a Science Education Conference in the Asian and Pacific Regions and was held in Istanbul. This conference focused on 'Science for All' and was attended by delegates from Beijing including Wu Yongxing who was still an influential leader in the PEP and actively involved in the production of curriculum materials. As a follow-up activity to this conference, the 'Forum of Secondary School Science Curriculum in China' was held in Beijing in September 1986. This forum, which built on what was learnt at the Istanbul conference, placed the concept of 'Science for All' before the forum audience for their consideration. The focus and tone of this forum is evident in the following passage from the opening speech delivered by Wu Yongxing:

> At present, in order to implement the Law of Compulsory Education, we are deliberating [regarding] the teaching program and syllabi of individual subjects in the nine-year compulsory education on the basis of domestic and foreign experiences... It is on this occasion that we hold this forum ... I believe that the discussion on the topic of 'Science for All' will be beneficial to the secondary school science education in China.
>
> (1986, p. 11)

In this forum, the delegates acknowledged and appreciated the spirit of 'Science for All.' They also expressed interest in the aforementioned instructional activities that seemed evident in the Western science curricula. Wu Yongxing recalled that it was commonly considered that the level of theoretical knowledge in the then junior secondary science curricula would need to be lowered if 'Science for All' was to be introduced into the science curriculum for China's compulsory science education. However, most delegates also suggested that prudent steps should be taken before considering the adoption of the use of contextual approached for organizing the teaching content as exemplified in some of the Western curricula. Rao's (1986) summary confirms this concern:

> It was commonly accepted that the spirit of 'science for all' was adaptable. However, as for the strengths and weakness of its curriculum models and that of the traditional curriculum: further studies should be undertaken. Many comrades believed that it was worthier to explore the model of combining systematic knowledge with typical practical problems.
>
> (p. 57)

This concern arose because the contextual approach evident in the sampled Western texts reminded delegates of the lessons learnt from the times of the Cultural Revolution and the 'Great Leap Forward' in 1958, another critical leftist movement in China, in which much more emphases were placed on applications of science as they related to industrial and agricultural production at the expense of subject knowledge that provided a theoretical foundation for understanding those applications. In an interview between the first author and Wu on the influence that this conference had in the 1992 JSSCC in relation to the adoption of a contextual approach, Wu commented:

> Is the concept of STS correct or incorrect? Of course, it is correct. The problem is how to combine it with teaching content. I know there are some kinds of textbooks – air, water, and other things are presented with theories implied in them but without inherent interrelation among them. I oppose [this kind

of organization]! If such an organization of teaching content were adopted in our science curricula, we would be back to the times of the 'Great Leap Forward' and the Cultural Revolution. That will be very dangerous.

8. The 2001 JSSCC

Deng Xiaoping died in February 1997. The new generation of leaders supported and took up his ambition of seeing China develop into a powerful socialist country. After Deng's death, the key points of his talks and speeches over the period during which he was paramount leader, 'Deng Xiaoping Theories,' were hailed as one of the creeds of the Party and the country. Deng's manpower function of education and his priorities for science and technology continued to be strengthened in the era of Jiang Zemin. At the 14th National Congress of the CCP held in October 1992, the notion of prioritizing education was proposed as a strategy to boost economic development. At the Second National Conference on Science and Technology held in May 1995, the Central Committee of the CCP claimed that the rejuvenation of China was dependent mainly on science and technology and on education. This strategy was encapsulated in the slogan 'rejuvenating China by science and education.' At the Fifteenth National Congress of the CCP held in October 1997, Jiang Zemin emphasized that, 'the progress of our construction of modernization is, to a great degree, dependent on the increase of national quality and development of personnel resources' (Jiang, 1997, p. 9). Jiang suggested that the fundamental reason to develop education was to raise the quality of the nation and train millions of specialized personnel to be adaptable to the modernization of China. Notably, Jiang's comment was very similar to that made by Deng in the National Education Working Conference mentioned previously.

In response to the national priority of raising the quality of the nation, the term 'quality education' which initiated from the terms 'quality of labourers' and 'quality of the nation' became discussed widely in and beyond education circles in China in the 1990s as an alternative to examination-oriented education. Examination oriented education was synonymous with, (a) too much concern for the success of the minority of pupils who would proceed to higher levels of education at the expense of the majority who would not proceed past compulsory education, (b) the ignoring of moral, physical and allround psychological development in favour of academic intellectual development, and (c) an overemphasis of information delivery at the expense of exploring and developing the processes of inquiry. In terms of activating the notion of 'quality education' the Third National Education Work Conference held in June 1999 in Beijing can be seen as a milestone. Because of the lamentable state of schooling, the need for personnel training for national economic development, and the need for continued scientific and technological development, the Central Committee of the CCP and the State Council jointly made the decision to promote without reservation 'quality education' with a focus on cultivating students' spirit of creativity and their practical abilities.

In this round of curriculum reform the PEP was no longer designated by the Ministry of Education to be the legitimate designer of the curriculum standards. Instead, a 'project system' was established by the Ministry of Education with the aim of facilitating the selection of 'qualified' designers from across the country. A group of young chemistry educators from teachers' universities united to bid successfully for the project to design the chemistry curriculum standards of the secondary schools. The membership of this group therefore varied from that of the previous groups that almost exclusively consisted of academic chemists. Prominent among this group were Wang Zuhao and Wang Lei who represented the emerging trend in science curriculum development as those with Masters degrees in chemistry and Doctorates in science education.

The Ministry of Education required that the revision project be based on four preparatory studies of, (1) The existing state of chemistry curriculum development in China; (2) trends in international chemistry curriculum development; (3) advances in contemporary psychology; and (4) the advances in modern chemistry. Consequently, much of the group's work during the preparatory studies, conducted in the first half of 2000, focused on comparative studies in which they referred to intensive and extensive readings of chemistry programs, curriculum standards of science or chemistry from western countries, and from Taiwan and Hong Kong. They analyzed texts, including *Chemistry in Community* (US), *Salter's' Chemistry Programs* (UK), *Science and Technology in Society* (UK), and *The National Curriculum* (UK) (Wang & Wang, 2000). These foreign teaching materials were not difficult to access as some of the group members specialized in the study of science and/or chemistry curriculum in western countries and had accumulated many such relevant materials. Further, few of these young chemistry educators had problems reading these English materials. The common features of these documents and programs were summarized by them as:

- A thematic approach is used to construct the curriculum contents;
- Scientific inquiry is emphasized as a main learning strategy;
- Subject knowledge is presented in various contexts;
- Social problems are heavily focused on.

(Wang & Wang, 2000)

We have suggested previously in this chapter that the above common features are often noted in curriculum documents of the 1980s and 1990s produced in western countries under the banner of scientific literacy (e.g., American Association for the Advancement of Science (AAAS), 1989; National Research Council, 1996). However, even though previous groups involved in previous three JSSCC revisions may have noted this also, support for the implementation was meagre and insufficient for such an orientation to be reflected in the organisation teaching materials and textbooks.

However, the concept of scientific literacy influenced and was manifest in the 2001 JSSCC in several ways (MOE, 2001). Firstly, as a central unifying concept, the term scientific literacy entailed the shared importance of the three dimensions of the goals of the JSSCC i.e., the development of, (a) knowledge and skills, (b) processes and methods, and (c) emotions, attitudes and values. Significantly, it lead to the explicit recognition of a set of scientific processes to be developed within the second dimension and it broadened the goals related to the third dimension. Secondly, the theme of 'chemistry and social development' was adopted as an essential directive for the selection of curriculum content for the 2001 JSSCC. As Wang (2001) noted, the selection of curriculum content was meant to, 'widen pupils' knowledge range, strengthen their sense of using knowledge, and let them experience how chemistry influences the daily lives of mankind' (p. 4). Thirdly, 'scientific inquiry' was adopted as part of curriculum content. In the textbooks of the JSSCC, personally and socially relevant content was significantly increased in comparison with that in previous textbooks (Wei, 2003). These revisions reflected the group's concern for developing curriculum that attended to what was considered 'quality education.'

Despite these changes, the 2001 JSSCC was still not in any sense a copy of its western counterparts as it featured noticeable Chinese characteristics. For example, in the purposes of the curriculum, 'skills' were still located together with 'knowledge,' both of them traditionally treated as 'double bases' in China, rather than being considered as 'processes' as is predominantly the case in western curricula. Also, 'dialectic materialism' and 'patriotism,' hallmarks of PRC curricula, remained as key elements of the JSSCC. Furthermore, attention to the technological applications of chemistry was decreased greatly in the 2001 textbooks of the JSSCC. This contrasts with the situation in western countries where the study of technological applications has been advocated widely in science curricula in recent years (e.g., AAAS, 1989; Cajas, 2001). Therefore, once again, the degree of influence of international factors was mediated by the designers' continued attention to the underlying sociopolitical foundations of the PRC.

9. Discussion

Our discussion is structured around a number of assertions which represent our distillation of the key ideas that emerge from the narratives. These assertions are not necessarily mutually exclusive. However we use them to highlight key aspects of the process by which ideas related to science education and chemistry education in particular, and emanating almost exclusively from the West, came to be incorporated and/or reflected in the JSSCC from 1978 to 2001. In particular we seek to identify common threads that seem to connect each of the narratives.

1. *Economic imperatives were important driving forces for the use of educational ideas from outside the PRC in the JSSCC revisions.*

While the sociopolitical climates within which foreign influences were incorporated varied over the period of the JSSCC a common feature of these sociopolitical climates was the persisting emphasis on the manpower function of education. It is well known that the P. R. China has devoted herself to economic development since the late 1970s. In this context education has been expected to supply sufficient well qualified and trained manpower for the so called socialist economic construction. We consider that the predominant view in the period of our study was that economic revival and development were closely linked to the enhancement of education. Therefore, we find it not difficult to understand that economic goals and the associated need for a well educated workforce constitute an important and enduring rationale for the internationalization of the JSSCC as China sought to develop an internationally competitive economy.

2. *Changes in domestic policy within the PRC enhanced opportunities for Western science and chemistry education ideas to be sourced and assessed for potential use in the JSSCC.*

As Huang (1987) observed, Chinese international exchanges with foreign countries were influenced to a great degree by China's internal policies as well as by her relationship with the West. An interesting example of such a proposition is that the international influences on the JSSCC in this study were largely dependent on the national policy of opening the door to the outside. We speculate that the impact of international educational thought would have been significantly less if there had been no open door policy as initiated by Deng. We also speculate that advances in information technology and the rise of the global marketplace in the latter two periods further hastened the opportunity for uptake and modification of ideas that directly impacted on educational policy and thought. In the first phase of the JSSCC, since this policy had not been formally instituted, the western influences were only exerted following the personal instruction of Deng Xiaoping. With the implementation of this policy increased opportunities arose to obtain information about science curricula development beyond the PRC in the 1950/1960s and the 1970s. So, it was not until the early 1980s that the western science curriculum of the previous three decades had the chance to influence the 1982 JSSCC. Increased academic exchange between China and the Western countries in the late 1980s and the 1990s resulted in trends in science curriculum development becoming timelier in terms of their availability and more comprehensive in nature. Further, we are not too surprised that prior to the designing of the 2001 JSSCC a more thorough and systematic literature review which was absent in the processes of the previous revisions was called for and conducted. It seems that at the commencement of the revisions of the JSSCC there was a gap of around about ten years between a curriculum trend from the West affecting a revision. However, with the increased access to information afforded by the open door policy and expanding opportunities for academic, information and economic exchanges we suggest that trends and ideas emanating from outside the PRC will be more quickly available for scrutiny and have the potential to more rapidly influence future revisions.

3. *The extent and nature of the accommodation and use of educational ideas and trends from outside the PRC was mediated by the individuals selected by the MoE to lead and guide the curriculum revision.*

From the narratives it is clear to us in some cases that the preferences or dislikes of particular appointed leaders determined the degree to which an international trend or idea was reflected in the revision/s they were associated with. They interpreted existing domestic social, political and cultural conditions and used these as parameters for guiding the nature and extent of the use of foreign ideas and trends. We consider this most obvious in different orientations of the 1992 and 2001 JSSCC. In the phase of the 1992 JSSCC, the thematic approach to organizing the curriculum content, and emphasis on the relation between chemistry and society were noted and appreciated by the designers. However, because of the resistance from key people in the design process these ideas were not reflected in that revision. However they became central ideas guiding the revision and organization of the curriculum content of the 2001 JSSCC because they were more appreciated as being relevant and necessary at that time by the emerging cohort of chemistry educators in universities who were selected ahead of the academic chemists to undertake the revision. Also, it is interesting for us to note that the process of delegation of authority and tasks for each revision was highly top-down in its orientation. The directives for the improvement of the education system within the PRC came from the highest levels of government to the MOE and then to individuals to manage the process and adjudicate on the appropriateness or otherwise of any aspect of a JSSCC revision.

4. *Changes in the domestic educational policy shaped the use of educational ideas and trends from outside the PRC.*

After emerging from the Cultural Revolution there was an urgent need for reform and revival of the education system in the PRC. The central importance of the examination system and the selective nature of that system have been trademarks of Chinese education systems and these characteristics were unquestioned in the PRC for the greater part of the period under study. However, the move to nine-year compulsory education, the need to harness support for that policy direction, and the move to focus on more inclusive ‘quality education’ for the majority of students rather than examination oriented education for the minority led to a reorientation of educational priorities that became reflected in the JSSCC. We propose that the conditions were right for the increased attention to ‘science for all’ in the 2001 JSSCC. Had the shift to a broader form of socially relevant science education not been in concert with the emerging national education priority then we speculate that it is unlikely that the group of young chemistry educators with their knowledge and appreciation of ‘science for all,’ and of how to construct a curriculum to reflect that orientation, would have been chosen to lead the revision. Consequently we suggest that the reorganization of the content and goals of the JSSCC would not have been in the same form as that which eventuated. Had there been no change in education policy to move to 'quality education' then in our opinion it is unlikely that the JSSCC would have used the ideas from beyond China to the same extent as what is evident, if at all.

Concluding Remarks

Western influences have always been used to serve domestic needs through the curricula of the PRC. This has been evident since the late nineteenth century when China began its modernization initiatives. The principle of substance-application (Chinese learning for substance, western learning for application) has been adopted in the practices of learning from the West. The implication of this principle is that what is learnt from the West is viewed in the PRC as to be used to compensate, improve, or consolidate the political, economic, cultural maintenance and development in the needs of China. The significance of domestic needs in determining the influence of Western education ideas has been highlighted by researchers, such as, Ding (2001) and Hayhoe (1987). We consider that foreign influence on the JSSCC's revisions have been mediated continuously by domestic conditions such as cultural traditions, a predominant Marxist dogma, and the perceptions of the designers of the JSSCC in terms of the appropriateness of the concepts and themes of contemporary science curricula. The JSSCC from 1978 to 2001 was influenced continually by inputs from the West and different phases of the JSCCC were influenced by different waves of Western science curriculum reform. Our study supports the view that economic imperatives, so often cited as driving forces for education reform, were important in determining how and to what extent the international ideas and trends in science and chemistry education from outside the PRC were used. However, while they might trigger such reforms, we should also take great care to understand that other factors may influence greatly science education reform. Our case of the factors influencing the internationalization of the JSSCC strongly supports such a view and leads us to suggest that Black and Atkin's (1996) aforementioned proposition is highly salient in considering science education reform. In our study we have tried to understand what the PRC's underlying mood and values were from 1978 to 2001 and to highlight what impact these have had on educational policy and science curriculum reform as exemplified by the JSSCC.

References

American Association for the Advancement of Science (AAAS). (1989). *Science for all Americans: A project 2061 report on goals in Science, mathematics, and technology.* Washington, DC: Author.

Black, P., & Atkin, J. M. (Eds.). (1996) . *Changing the subject: Innovations in science, mathematics and technology education.* London: Routledge.

Cajas, F. (2001). The science/technology interaction: Implications for science literacy. *Journal of Research in Science Teaching, 38*(7), 715–729.

Chen, H. T. (1981). *Chinese education since 1949: Academic and revolutionary models.* New York: Pergamon.

Chen, X. (1989). *Curriculum theories.* Beijing: People's Education Press (in Chinese).

Cheng, M. (1986). Reflection on the 8th ICCE. *Huaxue Jiaoyu, 1*, 45 (in Chinese).

Cross, R. T., & Price, R. F. (1991). Teaching chemistry in China. *Education in Chemistry, 5*, 72–74.

Ding, G. (2001). Nationalization and internationalization: Two turning points in China's education in the twentieth century. In G. Peterson, R. Hayhoe, & Y. Lu (Eds.), *Education, culture, and identity in twentieth-century China* (pp. 161–186). Hong Kong: Hong Kong University Press.

Donnelly, J.F., & Jenkins, E.W. (2001). *Science education: Policy, professionalism and change*. London: Paul Chapman Publishing Ltd.

Du, D. (1980). Some suggestions on developing student's abilities. *Guangming Daily*, January 14, 1980.

Goodson, I. (1994). *Studying curriculum*. Toronto: The OISE Press.

Guo, Z. (1981). Promoting secondary school chemistry teaching and developing pupil's abilities. *Jiaoyu Yanjiou, 6*, 49–52 (in Chinese).

Hayhoe, R. (1987). Past and present in China's educational relations with the industrialized world. In R. Hayhoe & M. Bastid (Eds.), *China's education and the industrialized world: Studies in cultural transfer* (pp. 271–290). Armonk: M.E. Sharpe.

Hu, M. (1989). Reflections on the chemistry syllabus of nine-year compulsory education. *Kecheng Jiaocai Jiaofa, 3*, 24–27 (in Chinese).

Huang, S. (1987). Contemporary educational relations with the industrialized world: A contemporary view. In R. Hayhoe & M. Bastid (Eds.), *China's education and the industrialized world: Studies in cultural transfer* (pp. 225–251). Armonk: M.E. Sharpe.

Jiang Z. (1997). A Speech at the 15th National Congress of the Chinese Communist Party in Beijing in October 1997. In Editorial Committee of the State Education Commission (Eds.), *Zhongguo Jiaoyu Nanjian 1998* (China Education Yearbook 1998) (pp. 1–9). Beijing: People' Education Press (in Chinese).

Leung, Y. M. J. (1991). Curriculum development in the people's republic of P. R. China. In C. J. Marsh & P. Morris (Eds.), *Curriculum development in Eastern Asia* (pp. 61–81). UK: Falmer.

Lewin, K. M. (1987). Science education in P. R. China: Transformation and change in the 1980's. *Comparative Education Review, 31*(3), 419–441.

Lewin, K. M., & Xu, H. (1989). Rethinking revolution: Reflections on China's 1985 educational reforms. *Comparative Education, 25*(1), 7–17.

Lewin, K. M., Xu, H., Little, A.W., & Zheng, J. (1994). *Educational innovation in China*. Harlow: Longman.

Liang, Y. (1981). Some illustrations to the outline of the revised chemistry textbooks of secondary schools (A). *Huaxue Jiaoyu, 5*, 40–44 (in Chinese).

Liang, Y. (1982). Review of chemistry education of secondary schools in China since 1949. *Huaxue Jiaoyu, 4*, 1–4 (in Chinese).

Liang, Y. (1983). Chemistry curriculum and textbooks in the United States. *Kecheng Jiaocai Jiaofa*, 1 (in Chinese).

Liang, Y. (1984a). "Chemistry in Community": A newly developed chemistry program in the United States. *Huaxue Jiaoyu, 2*, 61–63 (in Chinese).

Liang, Y. (1984b). A new tendency of chemistry education of secondary school: An important issue discussed in the 7th ICCE. *Huaxue Jiaoxue, 3*, 34–35 (in Chinese).

Liang, Y. (2002). *Reflections on chemistry education*. Beijing: People Education Press (in Chinese).

Liu, X. (1996). *Mathematics and science curriculum change in the people's republic of China*. Ceredigion, UK: The Edwin Mellen Press.

Loo, S. P. (1996). The four horsemen of Islamic science. *International Journal of Science Education, 18*, 285–294.

Ma, L. (1999). *Knowing and teaching elementary mathematics*. Mahwah, NJ: Lawrence Erlbaum Associates.

Ministry of Education. (1978). *The chemistry syllabus of secondary schools of full-time and ten-year system (trial version)*. Beijing: People's Education Press (in Chinese).

Ministry of Education. (2001). *The chemistry curriculum standard of compulsory education of full-time system*. Beijing: Beijing Normal University Press (in Chinese).

Morris, P., McClelland, G., & Wong, P. M. (1997). Explaining curriculum change: Social studies in Hong Kong. *Comparative Education Review, 41*(1), 27–43.

National Research Council (NRC). (1996). *National science education standards.* Washington, DC: Author.

Ogawa, M. (1995). Science education in a multi-science perspective. *Science Education, 79*, 583–590.

Pu, T. (1993). Enhancing teaching material development according to the strategic thought of the 'Three Orientations'. *Kecheng Jiaocai Jiaofa, 9*, 5–6 (in Chinese).

Rao, H. (1986). A summary of the symposium of science curriculum development in China, held on 14 to 24 October 1986. *Kecheng Jiaocai Jiaofa, 12*, 57 (in Chinese).

Reardon-Anderson, J. (1991). *The study of change: Chemistry in P. R. China, 1840–1949.* NY: Cambrige University Press.

Stevenson, H. W. (1992). Learning from Asian schools. *Scientific American, 267*(6), 70–76.

Stevenson, H. W., & Stigler, J. (1992). *The Learning gap: Why our schools are failing and what we can learn from Japanese and Chinese education.* New York: Summit Books.

Wang, L. (2001). On the revision of the chemistry syllabus of the nine-year compulsory education. *Xueke Jiaoyu, 4*, 3–6.

Wang, Z., & Wang, L. (2000). *A Progress report of the project of the national standard of junior secondary school chemistry curriculum.* (Internally circulated documents, in Chinese).

Wei, B. (2003). *A Study on the change of the junior secondary school chemistry curriculum: From 1978 to 2001.* Unpublished Ph.D thesis. Hong Kong: The University of Hong Kong.

Wei, B., & Thomas, P. G. (2005). Explanations for the transition of the junior secondary school chemistry curriculum in the P. R. China during the period from 1978 to 2001. *Science Education, 89*(3), 451–469.

Wong, S. Y. (1991). The evolution of social science instruction, 1900–86: A cross-national study. *Sociology of Education, 64*(1), 33–47.

Wu, Y. (1978). An introduction to the chemistry textbook of junior secondary school of full-time and ten-year schools. *Zhongxue Like Jiaoxue, 4*, 22–29 (in Chinese).

Wu, Y. (1983). Some issues involved in teaching materials of secondary school chemistry. *Kecheng Jiaocai Jiaofa, 6*, 39–48 (in Chinese).

Wu, Y. (1986). On the issues about the reform of science teaching. *Kecheng Jiaocai Jiaofa, 12*, 10–11 (in Chinese).

Wu, Y. (2002). *The reform of and exploration in chemistry teaching materials.* Beijing: People Education Press (in Chinese).

Xu, G., & Liang, Y. (1981). Thirty years of chemistry teaching materials of secondary schools in China. *Kecheng Jiaocai Jiaofa, 2*, 49–59 (in Chinese).

Ye, L. (1982). On the work of compiling and writing teaching materials of primary and secondary schools. *Renmin Jiaoyu, 1*, 16–19 (in Chinese).

Zhang, J. (1982). Some issues about revising the chemistry textbook of junior secondary schools. *Huaxue Jiaoyu, 3*, 43–47 (in Chinese).

Zhang, J. (Ed.). (1984). *Yearbook of China's education (1949–1981).* Beijing: China Encyclopedia Press (in Chinese).

27

GLOBALISATION/LOCALISATION IN MATHEMATICS EDUCATION: PERCEPTION, REALISM AND OUTCOMES OF AN AUSTRALIAN PRESENCE IN ASIA

Beth Southwell, Oudone Phanalasy and Michael Singh
University of Western Sydney, National University of Laos, University of Western Sydney

Abstract: This chapter examines the ways in which different types of international involvement by mathematics educators from Australia is an expression of, and response to the transitions in globalisation that have been occurring for the past two decades. The evidence it presents relates specifically to work in a development programme in Laos. Reference is also made to a teaching programme in Malaysia, and an independent development in the Maldives. The prevailing perceptions of communication, identity and development and renewal are examined within Delors (1996) identified tensions in learning

Keywords: Communication, identity, development, culture, renewal, curriculum

This chapter argues that the term 'globalisation' has many and varied meanings and implications for the policies, pedagogies and politics of mathematics and science education. Through consideration of the tensions in these fields of education as raised by Delors (1996), it identifies the diverse, complex and circuitous ways in which the transitions in the localising practices of globalisation are being effected through the relations between developed and less developed countries. Here we want to make the point that globalisation is a multi-faceted phenomenon that may often be thought of in the first instance as concerning relationships between nation-states (Singh, Kenway, & Apple, 2005). However, as we show through the experiences of Australians working as consultants overseas, globalisation also finds expression through and reflection in our selves. Viewed from these different facets, globalisation takes on differing orientations, some more favourable than others.

Based on the theoretical framework afforded by Delors (1996) report in which he identifies seven tensions in the field of education, the work of mathematics

B. Atweh et al. (eds.), Internationalisation and Globalisation in Mathematics and Science Education, 509–523.

and science educators from Australia in overseas projects in the Lao People's Democratic Republic, Malaysia, and the Maldives, can be examined. Despite the lack of recency in these projects, the important point is that the effects of the contemporary transitions in globalisation were already present in the signs observed a decade ago.

The aim, therefore, of this chapter is to present the reflections of consultants to three different countries in relation to the tensions identified by Delors and to draw out three issues that seem to impinge on such consultancies in this era of globalisation. In doing so, the multi-faceted nature of globalisation is emphasised rather than a set definition since many exist (Scholte, 2002). For instance, for some people 'globalisation' is quite reasonably understood as the imposition of 'western' knowledge, power and desires from developed nations on the countries of the less developed world. Consequently, much of what is included reflects the opinions of the authors rather than being an examination of the literature.

Following an account of a curriculum development project in Laos and brief accounts of two other projects, the framework of Delors (1996) tensions in education will be used to identify three perceived issues in the projects. Finally, some implications relating to these three issues will be explored.

1. The Three Case Studies

1.1 Curriculum Development for Teacher Education in Laos

Until the last two decades, the Lao People's Democratic Republic was something of an unknown for Australian mathematics educators. The French occupation had established many practices related to their own institutions, but these were disbanded during the period leading to the establishment of the Republic in 1975. At that time, there were only four upper secondary schools for the whole country with a population of approximately four million. Consequently, there were very few students who could continue to upper secondary school after finishing lower secondary school.

The Lao education system was shaped to accord with many of the practices of first the French, and later the Vietnamese. It was not until the late 1980s that the Lao government decided to open the country to the west. An educational system with Laotian characteristics was seen to be a desirable outcome.

Concerns about the shortage of teachers and the poor quality of teaching led to the review of education in Laos. With respect to mathematics it was found that many students perceived it to be a difficult subject and felt that little benefit would result from taking it as a major in their studies. There was also a shortage of textbooks in the Lao language. Those that did exist were of a poor quality. Teacher's salaries were low, so low in fact that they could not live on their teaching income and needed to take a second job such as selling goods or farming in order to support their families. A select few were able to take jobs as tutors in private colleges. Community perceptions were another factor that added to the difficulties facing Laotian education. The extremely low salaries of teachers indicated that they were not regarded very highly, and in effect were looked down on by the general public. Paradoxically, people wanted to study in good schools with good teachers but were not prepared or able to give them the necessary support or remuneration.

As the Lao People's Democratic Republic was integrated into the global market economy after the establishment of the republic in 1975, and its economy opened to the rest of the world, money became much more important to people. The economic globalisation of Laos created a situation in which many teachers resigned from their teaching position to take up the new and better paid jobs that were emerging. Locally, the restructuring incited by the agents of economic globalisation exacerbated the teacher shortage and did nothing to redress the poor quality of mathematics teaching. It was in this climate and recognising the dimensions of the problem, that the Lao government sought and was successful in gaining funds to redress the situation.

One of the early projects sought to rewrite the school curriculum and individual subject syllabuses. It was necessary then to revitalise teacher education to improve the preparation of teachers and the knowledge they needed to teach these new syllabuses. Consequently, the Curriculum Development for Teacher Education (CDTE) project was launched in 1993. The initial task for the twelve overseas consultants and their Lao counterparts was to develop a position paper that married curriculum development with teacher professional development using an 'action learning' framework. Their next task was to establish a future-oriented curriculum for the teacher education colleges. After that attention was given to individual discipline areas.

Alongside this important curriculum policy work, considerable action was taking place among the 57 teacher training colleges that had been established since 1975 to cater for Indigenous or tribal groups. Under the Education Quality Improvement Program (EQIP), the 57 existing colleges were merged to form eleven much better equipped, more centrally located colleges with better educated staff. In most cases this meant new buildings and a new curriculum for teacher education. To maintain this work, one of the main tasks of CDTE was to establish the Teacher Development Centre at Dong Dok, a village on the outskirts of the capital city where the French had previously established a university. This Teacher Development Centre became the Lao centre for professional development and resources for teacher education.

The group of consultants and counterparts as a whole determined the content and format of the teacher education curriculum. After about four months of work on the position paper and the general framework of the teacher education programme, those concerned started detailed work in the disciplines, albeit not without problems. Initially, two mathematic educators had been appointed as counterparts but one resigned to return to his university position. To enable both the foreign consultants and their Laotian counterparts to have input, the mathematics lessons were written in either English or Lao and then they were to be translated to the other language so that they could be refined. With the overseas consultant on a six months contract, the total task was not completed before she had to leave. However, she did return on several occasions, working voluntarily until the task was almost completed.

The preparations made by the consultants and counterpart team for the teacher education curriculum included materials on both mathematics content and mathematics pedagogy. A similar situation existed for science education with two Thai consultants, one for biology and one for physics and chemistry. The work was set out in lesson note format so that the staff at the teacher education colleges could work through a sequence of lessons with helpful suggestions on the method of presentation. The materials also introduced the teacher educators to action learning, and this pedagogy was made explicit throughout all the notes for the lessons. The parallel emphasis on content was necessary because many of the student-teachers in colleges only had an elementary education. Because of the educational attributes of the student-teachers as well, it was necessary to develop three different teacher preparation programs. There was a three-year program for the students with Grade 8 education; a one-year program for students with Grade 11, and also a three-year program for those who had finished Grade 11 which qualified them to teach in upper secondary school.

Towards the end of the six-month's consultancy, a conference was held in which staff from teacher education colleges were brought together at Dong Dok for a residential professional development conference. During the week of the conference, the Laotian counterparts presented some of the modules that had been completed to visiting staff from the provincial colleges. The whole mathematics teacher education course was discussed among the mathematics and science teachers present. When the interpreter was available, the overseas consultants contributed. The fruits of this week were borne out in the many subsequent changes in mathematics content and pedagogy nurtured throughout all the colleges.

1.2 The Maldives Teacher Education Project

The Maldives Teacher Education Project led to the University of Western Sydney (UWS) being invited to collaborate in upgrading the education and qualifications from diploma to bachelor degree of four senior staff from that country's Institute for Teacher Education. Of these four women, two had been teacher educators for a considerable length of time, one being the Centre's Deputy Director and the other its Director of Primary Programs. The project that was carried out used a mixed delivery mode. The four women studied University texts which were developed from already existing Australian University courses, undertook work-related assignments and participated in a fortnightly, trans-national teleconference that connected them to teacher educators at UWS. Once a year a teacher educator from UWS visited the Maldives for a two week period to orient the Maldivians to the subjects they were to study and provide offline, face-to-face assistance to meet the student's learning needs. As a consequence of this program, three of these women enrolled in Master of Education programs at UWS, investing in a year of on-campus study. The mathematics educator in the group participated in the program designed for professional development of mathematics teachers, as well as undertaking foundational units in the study of education. This proved to be a wise decision when she later became the Director of the Institute as it gave her a much broader understanding of education than if she had focussed exclusively on mathematics education.

1.3 Malaysia and the Regional Education Centre for Science and Mathematics

The South East Asian Ministers of Education Organisation (SEAMEO) established the Regional Education Centre for Science and Mathematics (RECSAM), locating it in Penang, Malaysia. From its inception until 1996, RECSAM received support from the Australian Government in that Australian consultants to RECSAM received an honorarium. Australian consultants were enlisted to spend periods of time at the Centre presenting courses to representatives of the SEAMEO countries. Australian consultants presented a range of courses covering pedagogy, resource construction and use, research methodology and education studies in mathematics and science, all of which were planned by RECSAM staff mostly as professional development courses. The courses were designed for either elementary or secondary teachers. In practice, sometimes the two groups were combined and consequently the consultants had to adjust their teaching to incorporate this range. Likewise, while the courses were also designed for either mathematics or science education, again they were often amalgamated. As a consequence, consultants found that in any one class they had elementary and secondary teachers of mathematics and science studying together. This called for them to display considerable interdisciplinary, cross-sector ingenuity to maintain a high level of engagement and ensure each participant really benefited. Countries were asked to select representatives with a working knowledge of English. This was because the consultants were mostly monolingual English speakers who were expected to deliver courses in English. Not all countries could send such teachers, because knowledge of English was not widespread. Thailand and Indonesia in particular found it difficult to comply with this.

Support for RECSAM was gradually decreased by the Australian government on the expectation that the Centre could pay its own way. It began to depend more on fee-paying courses, voluntary consultants and, contributions from other countries and commercial enterprises. The Centre still determined the subject and nature of the courses and supported consultants through free accommodation and a small allowance.

If these three situations are considered together, are they really different experiences? If they are different, in what way? What are the factors that make them different? To explore the answers to these questions, we will relate them to some of the tensions identified by Delors (1996) as a basic framework.

2. Delors' Tensions and Related Issues Identified by Consultants

The following is a brief description of the seven tensions in globalisation identified by Delors (1996).

- *global and local*

It might seem reasonable to expect that the distinctions between countries would become more blurred under the process of globalisation. The local emphases might be expected to merge with globalisation forces and even disappear. The tension arises because of the pressures placed on the teachers to not only maintain their local perspectives but to also see themselves within a global environment.

- *The universal and the individual*

A major issue of concern for countries around the world is whether to provide education for all, or just for the elite few. This might once have been, and in some cases still is, regarded as a problem for nations in the developed world also. Some countries in the developing world initiated education of a few with the explicit requirement that they would then implement universal primary education for all. Now these countries are making substantial progress in extending secondary education to all.

- *Tradition and modernity*

Consideration of the past and past methods can be helpful in making more fruitful use of the present. For those who do not want anything to change, history is not important. Technology has always been an important element in educational and economic development, whether it is in the form of pencils, the buffalo-drawn plough or the computer and electronic technology of various forms is a product of modernity.

- *Long-term and short-term considerations*

There is tension over whether the contingencies seen on the immediate horizon should limit the planning for the future that lies beyond that horizon. Sometimes the tendency is to allow short-term socio-economic expediency to dictate the provision of a lower standard of mathematics education today. Looking further ahead, over the horizon, might mean making different provision in the immediate time-frame to increase the likelihood of building firmer foundations for a different and better long-term future.

- *Need for competition and concern for equal opportunity*

Education that is directed at producing and reproducing elites can be perceived as, and usually is, very costly and very competitive. By contrast to competition, collaboration can be much more productive. The advantages of collaboration are the generation of a greater range and depth of thinking, the development of appropriate social skills and the provision of the opportunity for effective decision-making. Collaboration also leads more naturally to equal opportunity.

- *Knowledge expansion and the capacity to assimilate it*

Catastrophe and chaos theories are among the new fields of mathematics that have been developed within the last 50 years. Topology was only developed within the last one and a half centuries. Likewise, statisticians have made available new analytical and graphing procedures. It is only with the increasing availability of new technologies, that it is now possible to introduce such recent additions to mathematics into school and university curricula. The question arises as to how relevant these technological developments are to developing countries and how they can manage the increasing growth of knowledge. One possibility is to teach students how to retrieve appropriate knowledge when it is needed. This requires a solid grounding in the foundations and general principles of mathematics and science, the capacity to make connections across conceptual structures, as well as learning new strategies for learning mathematics.

- *The spiritual and the material*

No matter what religion one might follow, it is believed that deep down in everyone there is a spirit that expresses itself in the way we live and who we are. This spirit or life force shapes, if it does not decisively determine, how we live. It gives sustenance to the values we hold and ultimately what we achieve in our life. It is the force of these passions that gives us the capacity to reflect critically on everything we do and is done to us. Nevertheless, the expectation of an increase in material wealth creates tension.

Not all these tensions were observed in the three consultancies described. In keeping with the aim of this chapter, we will now determine which ones apply to which projects and consider why this might be. Reflections of the consultants in relation to these tensions identified three issues that seem to impinge on such consultancies in an era of globalisation.

3. Issues

The three issues that were identified from the personal experiences of the consultants were communication, identity and culture, and growth or renewal of knowledge.

- *Communication*

Communication difficulties arise from the tension between the global and the local and to a certain extent from that between tradition and modernity in that modernity is often related to influences from outside the country that are not fully understood. This tension between the global and the local appeared obvious to a consultant in the Lao People's Democratic Republic. The more global the world becomes, the better the communication between people and countries needs to be.

Communication was certainly a problem in the Lao project. Mathematics, with its claims of being a universal language, did not seem to present as many difficulties as other disciplines might have done. Learning fluent Lao may not have helped greatly in dealing with the technical aspects of mathematics. The absence of a common language between the consultants and their counterparts, however, may have set the project back in time due to misinterpretations and misunderstandings and the length of time required for translation. Despite this, considerable collegiality was developed among the total team and this outlasted various changes including changes in personnel. Understanding the global flows of knowledge made possible by English, the Laotians were extremely keen to learn and make this language one of their own. Part of the project was to assist them to do so. The mathematics consultant took a small part in the early morning (7.30 A.M.) lessons that were presented for English, at the same time endeavouring to learn Lao.

The consultants in the Lao project came from Canada, the United States of America, Thailand, and Australia. For the Anglophones language was a problem. The Thais could communicate relatively easily with the Laos because of the similarities in their languages. A US/American who had worked in the Peace Corps in Laos spoke the national language fluently but the rest of the consultants had no proficiency in the Lao language whatsoever. Among the consultant's counterparts from Laos, most were in the process of learning English. There was one from the mathematics team who had a fairly good grasp of English, and one other counterpart who spoke fluent English, having studied in Perth (Australia). He became the unofficial interpreter and translator for the group.

Communication was a concern due to lack of knowledge of the English language on the part of some participants and the lack of knowledge of any relevant Asian languages on the part of the consultants to RECSAM. Consultants who were regular visitors to RECSAM did learn, informally, a little Malay or Thai, but the scheme made no provision for inter-language development. Also, some courses were requested by other institutions or countries to meet a particular need. This created difficulties where a large number of participants were sent to a course from a country where the language used was not the same as that of the consultant. The permanent staff of RECSAM had appropriated English so Australian monolingual consultants had no difficulty in that regard (Singh, 2003). As the permanent RECSAM staff were from the different SEAMEO member nations, there was always someone who could communicate with the students in their own language.

There were no problems in the language of communication in the Maldivian project. All concerned spoke fluent English as most of their education had been in the English language as a result of being colonised by the British. Because some Maldivians from the country's outer islands use Divali, some of the Institute's courses were once available in that local language. This has since changed, with all education being in English only. Once again, the uneven impact of the global spread of technology access meant that there were some difficulties due to the absence of modern electronic library resources, the insufficient supply of computers and the inaccessibility of electronic networking facilities in the Maldives. Despite the challenges there were no serious problems in making telephone and email links between Sydney, a provincial state capital which is thought of as a global city, and Male, the national capital of the Maldives.

In these projects can be seen three levels of language as it contributed to effective communication among the project teams. In the Lao project, difficulties were encountered because neither the consultants not the counterparts were fluent in the other's language in general. In the Malaysian project, the situation was mixed in that some were fluent and others not. In the Maldivian project, there were no difficulties in this regard.

- *Identity*

A second predominant issue is that of identity. The tensions that give rise to this issue are those between the global and the local, between tradition and modernity and between spirituality and the material. In a provocative account, Arnett (2002, p. 774) likens the status of countries in the developing world to the development experienced by adolescents in their quest for an adult identity. He claims that teenagers sometimes struggle to gain a sense of their own identity through engagement with, and sometimes in opposition to the socio-cultural, historical and economic influences that are part of their upbringing. The suggestion is that for developing countries a similar child-like process results in a bicultural identity, or the sense of belonging to two cultures, namely, that of their heritage and another they develop for themselves. From the author's observations, this seems a reasonable point of view though it has not been tested through research. One point with which we agree with Arnett (2002) is the rejection of any view that nation-states have a fixed identity that is the product of some single, unified culture. We also agree with the implication that nation-states have multiple identities which operate with different power and force at different times in the lives of the people who identify with them. This tension is especially vivid in Laos. There the global/local flows of 'western' culture, whether it be dress, music or something else, impinge on the identity formation of young people struggling to make a life project that best positions them for a cosmopolitan world. It is in this context that mathematics and science education is positioned and is vital in assisting young people to make a success of their life.

In contrast, Green (1997) claims that there is not much evidence to indicate that nations are losing control over their educational systems. However, the proposition that we are seeing a global convergence of education raises complex issues. At this time most education systems remain grounded in a given nation-state. However, different education sectors, particularly universities, are experiencing difficulties because of market forces (Singh, 2005). However, as Adam's (2004) report of an ethno-mathematics curriculum project implemented in the Maldives demonstrates, global forces, and connections are subjected to localising pressures. In this instance, the curriculum framework was articulated with global processes in knowledge being grounded in Maldivian mathematical practices.

Most nations throughout the world are now experiencing the pluralisation of their cultures, values, economies, education systems and work. The pluralisation of societies is a cogent factor in the rejection or otherwise of traditional values and practices. This is happening at an ever-increasing rate in societies around the world, with the speed of change being possibly even greater in developed nations than others. Policies now emphasise the complementarity of the learning society, the need for lifelong learning and greater preparation for new forms of work. It is no coincidence that these changes to work are a result of labour-saving, labour-displacing technologies. Many of these developments require levels of mathematical problem solving, second language proficiency and skill with new technologies not previously considered. However, Patadia and Thomas (2002) survey in teacher education found most teacher education institutions in New Zealand had no specific policies or programs directed at these issues.

The Australian consultant believed that it was not her role to impose Australian theories and practices on Laotian education, but instead to work with the counterparts to enable them to develop their own particular ways of doing schooling. The situation was different for the consultants who were from Thailand. While Thailand and Laos share a common border and have a similar language, the contrast in relative prosperity of the two is quite marked. Thailand is an independent, sovereign nation, it was not colonised by the Europeans and has a long history of prosperity. In contrast, Laos has been subjugated by other nations for many decades at a time and is the second poorest country in the world. The Laotian counterparts' experiences in the program as reported to the authors, left them feeling as if they belonged to a subservient country and less than confident in developing their own future. Also, while it seems to be that the older members of the population desire the retention of the particularities of their Laotian culture, younger Laotians are tapping into the global flows of ideas, images, people, money, technology, power and risks. The global/local tensions created by the multitude of programs such as these, represent and reflect a small but nonetheless significant expression of a distinctively *Laotian cosmopolitan identity*.

Identity was not necessarily a problem in the courses organised through RECSAM. Each student knew the purpose they were attending RECSAM, namely, that they would return to their home country and put into practice what they had learnt and then provide a progress report on their learning-in-practice within a designated time period. Ostensibly, the teaching at RECSAM was geared to the needs of the communities from which the students came. The intention was to have the students see their learning not as something imposed by a nation from the developed world but as practical help of immediate relevance to their own situations. The courses at RECSAM were designed by agencies from within the SEAMEO nations for fellow member nations. The Australian consultants were asked to present these courses and so had to work hard to create a perception of themselves as co-workers with the mostly multicultural groups.

Problems concerning ownership or identity were not evident in the Maldives. When the project was started, primary education was provided locally by the

Maldivian government but secondary education was staffed by Sri Lankans to a great extent. Being one of the closest countries to the Maldives, Sri Lanka had the good fortune of having an excess of secondary teachers. During the two-week in-country initiation of the project, the consultant visited several primary and secondary schools around Male, observing some quite stark contrasts in the teaching strategies. Secondary schools seemed to be quite regimented. While much teaching relied on the 'chalk and talk' strategy, a chemistry lesson was observed in which the teacher engaged students in a fair amount of experimentation. Since then, education of secondary teachers has begun in the Maldives under the auspices of an Australian university. Moreover, the management and administration of the education system is now entirely in the hands of Maldivians.

Some cultures appear to have a greater reservoir of, and considerable reliance on spiritual truths and resources for their every day lives, than others. For example, in Laos, it is appropriate to celebrate births, deaths, marriages, a new year and the establishment of a new home by having Buddhist monks to bless the occasion. Likewise, the start to a new mathematics program is celebrated by conducting a similar event. The non-material aspects of life are as important to the Lao people as they are to Australia's Indigenous nations.

Materialism may have robbed some people of the possibility of seeing any relationship between spirituality and mathematics. But this is an area receiving some consideration in terms of the relationship between values and mathematics (Bishop, Clarke, Corrigan, & Gunstone (2005)), as well as science generally. This creates a local/global perspective on the connections between mathematics education and the life of the soul:

> the restoration of the soul is part of this pattern of global awakening [in relation to peace, the environment]. People are attempting to infuse soul into every part of our lives in order to revitalise and give meaning and purpose to what we do every day
>
> (Miller, 2000, p. 8).

In the perceptions of the authors, the external globalising forces impinging on identity seemed strongest in Laos and virtually non-existent in the other two projects though possibly for different reasons. This could be that Laos is a newly emerging country, whereas the other two have had a longer exposure to such globalising forces. It could also be due to other factors. For instance, the participants at RECSAM were there for a specific purpose and knew what they had to do. In the case of the Maldivian project, the leaders there were much more aware of their place in the world because of their development as a nation.

- *The expansion and/or renewal of knowledge*

This issue is related firmly to the tension between the expansion of knowledge and the capacity to cope with it. To a lesser degree, other tensions impinge on this aspect. For example, what should be the balance between global and local knowledge? Does a system plan for the immediate future or take a more long-term view of knowledge? Should universal education or education of the few be the practice?

In the three projects described above, long-term goals were obvious. It was anticipated and hoped that the outcomes of each of the projects would filter down to children in schools. These three projects, however, raise the issue of what knowledge and at what pace should educational systems progress.

The increase in the availability of electronic technologies has been quite phenomenal in the past decade (Arnett, 2002, p. 776) and is related to the coping strategies of students and teachers when faced with a vast amount of new knowledge. In East Asia, the number of televisions has risen in the period from 1975 to 1995 from 5 to 255 for every 1,000 people, while the number of telephones has risen from 4 to 49. For South Asia, the corresponding figures are 1 to 60 televisions and 2 to 16 telephones. In contrast, for countries in the developed world, the figures are 280 to 525 televisions, and 178 to 414 for telephones. Of course, since 1995 the increases in the resources consumed in the global/local production, distribution and marketing of these commodities would be even more dramatic.

The internet and other technological advances have enabled people from all over the world to communicate instantaneously and exchange information. For some people at least, there are some dangers as well as advantages in these new technologies (Koblitz, 1996). Educators receive regular warnings about the truth-claims of some web pages and the possibilities for plagiarism created by the over-enthusiastic use of the internet (Owston, 1997). Nevertheless, these new technological developments have already brought about changes in curriculum and pedagogy in some countries while signalling possible changes in others. Despite this, the availability and access to computers is quite uneven across the globe.

The Teacher Development Centre in Laos had three computers that the secretaries used and the Team Leader had a laptop. Otherwise there was no electronic technology to assist the work of the project team. This meant everything had to be written by hand and typed by the secretaries for consideration by the team. Not surprisingly, very little equipment of any kind existed in the many rural schools of Laos. For instance, in the school at Vang Vieng, a tourist village situated less that 100 kilometres from the capital city, there were only small slates to be used jointly by two or more children. That this community is now a thriving tourist resort reflects and expresses the uneven integration of education into the global flows of people and money. The school buildings had mud floors and over 50 children occupied the 30 seats in the typical classroom. While the conditions of schools in the towns were better, even there, books and pencils were scarce or absent in a few.

Increasingly, new information and communication technologies became available to RECSAM, although not immediately to the students. Several computers were installed in the library which was well stocked with English language publications, although not necessarily the most recent. Because most of the SEAMEO member nations reflected considerable uneven development internally, the focus at

RECSAM was on the use of technologies that were appropriate to their local school economies such as pencils and natural materials.

In the Maldivian project, participants and their colleagues were keen to learn more of the knowledge that was becoming available. This can be seen in the learning undertaken by the participants as they studied towards continuously improving their qualifications. All completed master's degrees (three in Australia and one in England) and one proceeded to candidature towards a PhD. Initially they were hindered by the distance between them and the providing university but were able to use some electronic technology to assist them.

All three projects suffered from a lack of resources, both print, personnel and electronic, to capture new knowledge. This, to a certain extent, justifies the use of external consultants. Once again, however, the degree to which the three projects experienced difficulties in this regard varied with Laos having the most difficulty.

4. Implications and Conclusions

From the reflections of the authors and their experiences in the three reported projects, and using Delors's seven tensions of learning as a framework, three issues have been identified as operating within the environment of globalisation. While the three issues are not experienced in the same way in each project, consideration of them serves to contrast the differences and highlight the reasons for these differences.

The emphasis on problem solving, the sustainability of linguistic diversity and information communication technologies are contributing in this area. While there is a difference in degree, there are no substantial differences in the tension-ridden changes confronting the countries of both developed and developing worlds. In the aftermath of the 26 December tsunami that swept across the Indian Ocean from Malaysia to well beyond the Maldives, we can see that these two worlds need to join together in sharing and collaboration.

Mathematics is a discipline that is in great demand in both developed and less developed worlds. There is a shortage of mathematics teachers for secondary schools. In many instances the teaching of mathematics at the elementary level is relatively poor. It is to be expected then that major efforts need to be made to upgrade and develop teachers and their knowledge of mathematics content, pedagogical skills and the learning characteristics of students. This applies to most countries but especially to countries in the less developed world where access to higher education, particularly teacher education, has not been readily available for various reasons. Having conceptualised the problem as redressing this lack, initiatives have been mobilised as if they were framed variously as a single international project; or sometimes in relation to science, and sometimes as part of a total curriculum reform. This last approach to re-presenting the problem was the situation in the Lao People's Democratic Republic.

Globalisation cannot be ignored, not even in mathematics education. Perhaps all too belatedly we might ask, what might be affected through mathematics education in response to the differing orientations to globalisation present in our selves, our national societies, the international organisations to which we belong and the humanity of which we are a part. This chapter has presented a consideration of the effects on mathematics education of globalisation as perceived by people in consultancy roles. It has hinted at a basis for elaborating possibilities for 'both-ways education' by suggesting warranted transformations in cultural ethos, the communication of desires and narratives of life trajectories.

There are many questions that need to be asked about the consequences of the contemporary transitions in the ideological, historical and localising practices of globalisation. However, there are not always people who are willing to give reasoned answers. Atweh (2004) is an exception in this regard. He has provided a model for social justice in mathematics education that may be appropriated and reworked for use in many projects and movements grappling with global/local disjunctions in inequitable power relations. Often, these are expressed in the every day tensions and harmonies between experts or consultants and their indigenous counterparts.

So how can Australian mathematics educators, or mathematics educators from any other developed country, help nations cope with the effects of globalisation? Certainly, if they are to do this, they will need to take into consideration the three issues suggested above. That is not to say, however, that in other countries there may not be other issues more prevalent than these three. It seems reasonable to assume, however, that a language of communication is an issue for most consultants. Also, the expansion of knowledge is part of education in any country. The issue of identity may not be so relevant for developed countries as it is for developing countries.

The language of communication needs particular attention. Appropriate foreign languages need to be taught in Australian schools and prospective consultants need to undertake a course of instruction in the language of the country to which they are going, unless, of course, they know that they will be able to communicate effectively with the participants in their projects.

A deep knowledge of the educational system and people concerned and the philosophical basis of the culture is critical for consultants. This is most effective when supplied by co-participants in the project but much can be learnt beforehand. Developing problem solving skills and learning how to learn are essential characteristics to pass on to participants in projects as well as in the home country.

The authors have sought to briefly describe three projects that have involved Australian mathematics educators, set these within the framework of the Delors (1996) report and suggest three issues with implications for Australian consultants. The 1996 Delors report remains a useful analytical tool for making explicit the tensions present in the project of globalising educational pedagogies, policies and politics. That report also set out four pillars for learning in the world. The first of these, *learning to know*, focuses on educating learners towards a greater understanding of nutrition, health and self-sustainability. *Learning to do* is designed to provide the literacy and numeracy education required to lift people out of poverty. *Learning to be* emphasises the role of all types of education in the formation of students' multi-dimensional identities as citizens of nation-states and the world. *Learning to live together* is concerned with teaching students ways of making hope, justice and peace practical. As demonstrated in this chapter, mathematics and science educators have much to contribute to this type of world education, not the least of which is their will to share.

References

Adam, S. (2004). Ethnomathematical ideas in the curriculum. *Mathematics Education Research Journal, 16*(2), 49–68.

Arnett, J. J. (2002). The psychology of globalization. *American Psychologist, 57*(10), 774–783.

Atweh, B. (2004). Towards a model of social justice in mathematics education and its application to critique of international collaborations. In I. Putt, R. Faragher, & M. McLean (Eds.), *Mathematics education for the third millennium: Towards 2010* (pp. 47–54). Melbourne: Mathematics Education Research Group of Australasia.

Bishop, A., Clarke, B., Corrigan, D., & Gunstone, D. (2005). Teacher's preferences and practices regarding values in teaching mathematics and science. In P. Clarkson, A. Downton, D. Gronn, M. Horne, A. McDonough, R. Pierce, & A. Roche (Eds.), *Building connections: Research, theory and practice*, (pp. 153–160). Melbourne: Mathematics Education Research Group of Australasia.

Delors, J. (1996). *Learning: The treasure within.* Report to UNESCO of the International Commission on Education for the Twenty-First Century. Paris: UNESCO.

Green, A. (1997). *Education, globalization and the nation-state.* London: Macmillan.

Koblitz, N. (1996). The case against computers in K-13 Math Education (Kindergarten through calculus). *The Mathematical Intelligencer, 18*(1), 9–16.

Miller, J. P. (2000). *Education and the soul.* New York: State University of New York.

Owston, R. D. (1997). The World Wide Web: A technology to enhance teaching and learning. *Educational Researcher, 26*(2), 27–33.

Patadia, H., & Thomas, M. (2002). Multicultural aspects of mathematics education. *Mathematics Teacher Education and Development, 4*, 57–68.

Scholte, J. A. (2002). Definitions of globalization. In M. K. Smith (Ed.), Globalization and the incorporation of education, *The Encyclopedia of informal education Infed.* www.infed.org/biblio/ globalization.htm (Retrieved November 1, 2005).

Singh, M. (2003). Appropriating English: Reinventing the project of globalising English to help sustain the bio-linguistic diversity of humanity. In A. Pandian, G. Chakavarthy, & P. Kell (Eds.), *New literacies, new practices, new time.* Serdang: Universtie Putra Malaysia Press.

Singh, M. (2005). Enabling transnational learning communities: Policies, pedagogies and politics of educational power. In P. Ninnes & M. Hellsten (Eds.), *Internationalizing higher education: Critical explorations of pedagogy and policy.* Hong Kong: University of Hong Kong.

Singh, M., Kenway, J., & Apple, M. (2005). Globalizing education: Perspectives from above and below. In M. Apple, J. Kenway, & M. Singh (Eds.), *Globalizing education: Policies, pedagogies and politics.* New York: Peter Lang.

AUTHOR INDEX

Adam, S., 215, 518
Adams, J., 470
Adams, L. M., 444
Adams, R. K., 408
Adams, T., 251, 259
Agudelo, C., 431
Ahmad, A., 116
Ahonen, S., 365
Aikenhead, G. S., 90, 164
Alagia, H., 394
Alangui, W., 215
Alexander, R., 168
Ali, T., 116
Allen, R. L., 296
Alrø, H., 13
Altbach, P., 412
Anderson, L. W., 470
Ang, G., 250
Annan, K., 321
Anyon, J., 286
Appelbaum, Peter, 65
Appiah, A., 219
Apple, M. W., 4, 82, 89, 104, 152, 428, 429, 509
Archibugi, D., 7
Aristotle, 23, 24, 25
Arnaudyan, A., 299
Arnett, J. J., 517, 520
Arnott, A., 467
Aronowitz, S., 299
Ascher, M., 27, 213, 236, 244
Ascher, R., 213, 236, 244
Astiz, M. F., 82
Atkin, J. M., 156, 488, 505
Atweh, B., 96, 101, 102, 103, 104, 106, 235, 249, 255, 259, 261, 280, 321, 324, 326, 330, 331, 333, 335, 336, 337, 340, 341, 375, 379, 387, 522
Baird, M., 82, 88
Baker, D. P., 82, 156, 165, 168
Baker, Frank., 63
Ball, D. L., 271
Ball, S. J., 83
Barnes, B., 118, 120
Baroody, A. J., 281
Barton, A. C., 131
Barton, B., 215, 217, 218, 220
Bar-Yam, Y., 367
Baudrillard, J., 80, 194
Bauman, Z., 6, 114, 116, 117, 118, 119, 120, 121, 123
Bazerman, Charles, 65
Bbenkele, Edwin C., 72
Beaton, A. E., 303
Beck, U., 5, 80, 404
Becker, B., 176
Becker, J. P., 271
Becker, J.R., 215
Bell, A., 459
Bell, D., 7
Bentley, P. O., 32
Bereday, G. Z. F., 153
Berliner, D. C., 470
Bernstein, B. B., 174, 196, 344, 348, 349, 351
Berry, J. S., 451
Berryman, S.E., 367, 370
Biggs, J., 251
Birman, B. F., 468
Bishop, A. J., 103, 113, 114, 167, 249, 303, 304, 345, 519
Black, P., 156, 488, 505
Blank, R. K., 155
Blankley, W., 467

Blat-Gimeno, J., 406
Blocker, H. G., 322
Bogue, Ronald, 73
Borba, M., 215, 231, 234, 383, 391, 392, 393, 394, 396, 399, 400
Borgesen, D. S., 288
Boss, Judith E., 63
Bourdieu, P., 33, 123, 356
Bowers, C.A., 47, 356, 357
Boyer, C. B., 248
Bracey, G. W., 155
Brah, A., 80
Braidotti, Rosi, 74
Bray, E., 137
Broomes, D., 34
Brousseau, G., 432, 433
Brown, A. J., 28, 189, 191
Brown, P., 126
Bryman, A., 191
Bryson, M., 355, 356, 357
Buenfil, R. N., 409
Buenfil-Burgos, R., 252
Burbules, N. C., 195, 344
Burchell, G., 82
Burgess, Y., 451
Burke, P., 288
Burman, E., 310
Burstein, L., 303
Burton, L., 96, 97, 264
Butler, D., 445

Cai, J., 270, 271, 272, 274, 276, 277, 281, 304, 460
Cain, C., 218, 222
Cajas, F., 502
Cajete, G. A., 90
Cajori, F., 229
Callahan, C. M., 155
Cambridge, J., 343, 345, 346
Cannon, J. R., 287
Cao, Z., 304, 307
Caprio, M. W., 288
Carnochan, S., 296
Carnoy, M., 79, 82, 85, 213, 403
Carranza, J. A., 409
Carrington, W. J., 411
Carson, R., 215
Carter, J. A., 444
Carter, L., 81, 88
Castells, M., 5, 79, 395, 398, 425, 427
Castro, C de Moura, 403
Castro, M., 431
Chalmers, D., 252
Chambers, D. W., 451
Chang, H. J., 405
Chavels, S. G., 406
Chen, H. T., 488
Chen, X., 488
Cheng, M., 498
Cheng, Y. C., 157
Cheung, W. M., 157
Chisholm, L., 138, 140
Chomsky, N., 116, 120
Chrispeels, J., 288
Christiansen, I.M., 10
Christie, P., 139
Cisneros-Cohernour, E. J., 403, 406, 412, 413, 414, 415
Claassen, N. C. W., 466
Clarke, B., 519
Clarke, D. J., 105, 107
Clarkson, P., 96, 102, 106, 235, 249, 255, 259, 261, 280, 321, 331, 335, 340, 341, 375, 379, 387
Clegg, S., 345
Clements, D. H., 444
Clements, K., 345
Clements, M. A., 345
Cnop, I., 445
Cobb, Paul, 46, 437
Cobern, William W., 41, 42, 43, 44, 45, 46, 48, 53, 62
Cockcroft, W. H., 443
Codd, J., 117
Code, Lorraine, 50, 237
Coe, M. D., 237, 241
Cogan, L. S., 165
Colebrook, Claire, 62
Collins, Harry, 65
Collins, J., 405
Colosio, L. D., 409
Constanza, A., 103
Cook, S. W., 305
Coombes, A. E., 80
Cooper, B., 175
Cooper, M., 310
Cooper, Viv, 52
Corrigan, D., 519
Cortesao, L., 82
Cotton, T., 98
Cox, T. H., 97
Cozma, T., 293
Cronin, C., 212
Cross, R. T., 495
Crotty, M., 178

Crowther, D. T., 287
Crust, R., 459
Crystal, D., 177
Cucos, C., 292, 293, 297
Cummins, J., 220
Cunningham, G. K., 308
Curro, G., 251
Cvetkovich, A., 213

D'Ambrosio, U., 27, 35, 199–208, 213, 217, 219, 221, 232, 233, 234, 235, 236, 248, 345, 391, 392, 396
Dale, R., 35
Daniels, H., 348, 351
Dash, R. C., 322
Daun, H., 82
Davis, P. J., 25
Dawson, G., 375, 377
De Greiff, P. D., 212
de Laeter, J., 83
De Souza, C., 469
Dear, M., 79
Deer, K., 358
Dekkers, J., 83
Delanty, G., 79, 80, 82
Deleuze, G., 60, 61, 62, 64, 66, 70, 73, 138
Delors, J., 509, 510, 513, 514, 522
delos Santos, A. G., 330
Denny, P., 220
Dermitzaki, I., 453
DeSalle, Rob, 63
Desimone, L., 468, 471
Detragiache, E., 411
Deuss, K., 240
Diaz de Castillo, B., 239
Díaz, M. A., 410, 411
Diaz, R. P., 229, 237, 238, 243
Dimitriades, G., 82
Ding, G., 489, 505
Donaldson, M., 189
Donnelly, J.F., 488
Dori, Y. J., 292
Dowling, P. C., 28, 102, 173, 174, 175, 176, 177, 181, 186, 187, 188, 189, 191, 192, 193, 196
Dreyfus, T., 271, 272, 281
Driscoll, A., 220
Driver, Rosalind, 220
Drori, G. S., 83, 84, 88
Drucker, P. F., 367
Du Plessis, W., 139
Du, D., 139, 497
Dubeck, Leroy W., 63
Duggan, S., 89
Dunwoody, Sharon, 67
Duthilleul, Y., 378

Easley, J., 152
Eckstein, M. A., 152, 168
Eco, Umberto, 60
Edge, D., 445
Edwards, R., 357
Efklides, A., 453
Eglash, R., 214, 216, 217, 221
Eiselen, R. E., 467
Eisenberg, T., 271, 272, 281
Ell, F. R., 327
Ellerton, N. F., 345
Emanuelson, J., 107
Engeström, Y., 351, 352, 353
Enslin, P., 132, 138
Eraut, M., 353, 354
Ernest, P., 19, 20, 21, 22, 23, 25, 34, 36, 432
Esteley, C., 385, 388, 389, 390, 394
Evans, T., 144

Fallshaw, E., 250
Fan, L. H., 460
Fauvel, J., 358
Featherstone, K., 228
Fensham, Peter J., 49, 85, 90, 151
Fernandez, C., 465, 471, 472
Ferrer, B., 327, 329
Ferrucci, B. J., 444
Fillmore, L.W., 221
Findell, B., 444
Finn, P. J., 286
Fiorentini, D., 391
Firestone, W. A., 468
Fischman, G., 81, 91
Fish, S., 178
Fisher, D., 158
FitzSimons, G. E., 343, 345, 348, 349, 350, 355, 358, 359
Flecha, R., 429
Flude, M., 176
Flusty, S., 79
Forgasz, H. J., 303, 307
Fossum, P. R., 460
Foucault, M., 7, 25, 30, 114, 174, 190, 217
Fox, D., 155
Fradd, S. H., 90
Frankenstein, M., 14, 34, 35, 105, 213, 214, 215, 234, 248
Fraser, N., 100

Fraser-Abder, P., 476, 481
Freire, P., 4, 15, 35, 47, 286, 295, 299, 300, 387, 390, 399
Freudenthal, H., 126
Friedman, J., 80
Fuchs, H.-W., 156
Fullan, M., 286, 291, 367

Gabriel, Peter, 61, 62
Galbraith, P., 356
Gallard, A. J., 286
García, G., 431
Garden, R. A., 303
Gardiner, H., 90
Gardner, H., 367
Garet, M. S., 287, 292, 298, 468
Garner, R., 366
Gates, P., 337
Gee, J. P., 196
Geiger, V., 356
Geldenhuys, J. L., 137, 141
Gerdes, P., 27, 28, 35, 215, 216, 233
Gertzog, W. A., 295
Gess-Newsome, J., 287, 292, 293, 298
Gewirtz, S., 96, 99, 100
Ghosh, Amitav, 70, 71, 72
Giardina, M. D., 82
Gibbons, M., 14, 24
Gibbs, W. W., 155
Giddens, A., 116, 117, 118, 119, 120, 122, 209, 213
Gilmer, G.F., 215, 216
Gilmer, P. J., 288, 298
Gintis, H., 152
Girard, R., 241, 242
Giroux, H. A., 286, 299
Godden, G. L., 350
Goldin, G. A., 271
Goldsmith, E., 80
Goldsmith, L. T., 89
Gómez, P., 431
Gonzalez-Cantu, R., 409
González-Torres, E., 409
Goodrum, D., 84, 88
Goodson, I., 491
Goos, M., 356
Gorman, Christine, 68
Gough, A., 129, 131, 132, 141
Gough, Noel, 39, 53, 57, 60, 62, 63, 64, 66, 67, 84, 138
Gouws, A., 132, 137, 138
Govett, A. L., 287
Grattan-Guinness, I., 230, 237, 238
Green, A., 517
Grolnick, M., 470
Groome, M., 131
Gross, Paul R., 41, 407
Grossman, P., 468
Groves, S., 448
Guattari, F., 60, 61, 62, 66, 73, 138
Guba, E., 392, 395
Gummer, E., 288
Gunstone, D., 476, 519
Gunstone, R. F., 48, 49, 50, 288
Guo, Z., 497
Guskey, T. R., 287, 291, 470, 473
Gutstein, E., 4, 14

Habermas, J., 24, 209, 210, 211, 212, 213, 221
Hackling, M., 84
Haigh, M. J., 252, 259
Håkansson, A.-K., 358
Hall, G., 467
Hammer, D., 287
Hammer, M., 176
Hancock, S.J.C., 215, 216
Hanvey, R. G., 296
Haraway, Donna J., 41, 60, 63, 65, 66
Harding, Sandra, 40, 41, 64, 65, 84, 130, 143, 145
Hardt, M., 5, 7, 14, 211, 212
Hardy, G. H., 8
Harewood, S. J., 82
Hargreaves, A., 104
Harré, Rom, 45
Harris, M., 215
Hart, L., 97
Hartshorne, K., 139
Harvey, D., 80
Hasan, R., 196
Hassenpflug, W., 406
Hattori, K., 466, 474
Hayhoe, R., 153, 505
Hayles, N. Katherine, 41, 48, 49, 50, 193
Heath, S. B., 196
Hellmundt, S., 252, 259
Hemler, D., 287
Henning, E., 356
Hersh, R., 21, 25
Hewson, P. W., 295
Hiebert, J., 166, 270
Ho, A., 263
Hobsbawm, E., 117
Hofstede, G., 412
Holland, D., 218, 222, 222

Holland, E. W., 174
Holliday, B. W., 155
Holliday, W. G., 155
Hoshannisyan, A., 297
House, J. D., 156
Hovhannisyan, G., 299
Howie, S. J., 467, 474
Howson, A. G., 32
Høyrup, J., 26
Hsu, F. L. K., 310
Hu, M., 498
Huang, S., 503
Hudson, A., 345
Hunter, B., 327, 329
Hurd, P. D., 85, 88, 89
Husén, T., 303

Ifrah, G., 229, 248
Ihde, D., 5
Ilon, L., 371
Irwin, K. C., 321, 327, 329
Ismael, A., 254
Isoda, M., 445

Jablonka, E., 107
Jackson, M., 259
Jacobsen, E., 106
Jameson, F., 80
Jami, Catherine, 41
Jamilah, M. Y., 441, 450
Jane, B., 448
Jansen, J., 467
Jansen, Sue Curry, 47
Janvier, C., 272
Jarvis, T., 451
Jegede, O. J., 90
Jenkins, E. W., 152, 488
Jesson, J., 84
Jiang, Z., 500
Jickling, B., 257
Jita, L. C., 465, 467
Jones D. M., 405
Jones, P. W., 344
Jones, R., 155
Joseph, G. G., 27, 113, 214, 215, 217, 248
Joubert, R., 137
Joyce, M., 180

Kahn, M., 474
Kaminski, D. M., 130
Kaplan, S. N., 155
Kaptelinin, V., 356
Kaur, B., 445, 450, 460
Keitel, C., 95, 105, 107, 156, 304
Kelleher, J., 286
Keller, E. F., 143, 144
Kellner, D., 213
Kelly, A. E., 448
Kelly, D. L., 164, 165
Kelly, M., 263
Kember, D., 254
Kent, P., 349
Kenway, J., 131, 133, 141, 142, 143, 509
Kerckhove, D., 395
Keys, W., 157
Kilpatrick, J., 105, 107, 156, 304, 437, 444
Kinnear, Judith, 69
Kita, M., 466, 474
Kitchen, R.S., 215
Kitschelt, H.P., 373
Klug, F. J., 406
Knaub, K., 444
Knijnik, G., 216
Knudtson, Peter, 41
Knuver, A., 157
Koay, P. L., 441, 450, 460
Koblitz, N., 520
Kogan, M., 177
Köller, O., 156
Kosma, T., 365, 378
Kraak, A., 132, 139, 140
Kreinberg, N., 141, 142, 143, 144
Krell, G., 97
Krockover, G. H., 131, 144
Krugman, P., 405
Kubeka, Z., 467
Kubow, P. K., 460
Küchemann, D. W., 189
Kühn, M. J., 470
Kuhn, T. S., 375
Kuiper, W., 157
Kuku, A., 103, 104
Kung, H., 122
Kuperus, P. K., 34
Kuutti, K., 351, 353, 354, 355, 356

Lachicotte, W., 218, 222
Lakatos, I., 21, 25, 202
Lam, C., 307
Lambek, J., 27
Larkin, J. H., 272
Lash, S., 79, 80, 81, 86, 87, 88, 89, 90
Latour, Bruno, 65, 71
Lauder, H., 125, 126

Lave, J., 27
Law, John, 58, 59, 60
Layton, D., 152
Le Grange, L., 131, 143, 484
Le Métais, J., 153
Lear, J.M., 215
Leask, B., 251
Lederman, N. G., 287, 288, 293, 298
Lee, N. H., 445
Lee, O., 90
Lee, P. Y., 445
Leinhardt, G., 271
Lemke, Jay L., 65
Lesh, R. A., 448
Leung, F. K. S., 304, 307
Leung, S. S., 271
Leung, Y. M. J., 490
Levin, B., 115
Levin, H. M., 82
Levinas, E., 120
Levitt, Norman, 41
Lévy, P., 393, 395
Lewin, K. M., 467, 489, 490, 497, 498
Lewis, C., 165, 466
Lewis, S., 141, 142, 143, 144
Li, H.-L., 82
Li, S., 460
Liang, Y., 488, 490, 494, 495, 496, 497, 498
Liebermann, A., 470
Lim, S. K., 457
Lincoln, Y., 392, 395
Lindley, David, 63
Lindsey, R. B., 231
Ling, L., 310
Lingard, B., 82
Linn, M., 165, 166
Lister, I., 35
Little, J. W., 468
Liu, J., 358
Liu, X., 490
Loden, M., 97
Londoño, M., 428, 429, 430
Lonning, R. A., 287
Loo, S. P., 488
Losee, J., 25
Louden, W., 287
Lundvall, B.-Å., 7
Lyotard, J. F., 25

Ma, L., 304, 490
Mabandla, B., 132, 137
Mander, J., 80, 119
Mangena, M., 130, 132, 137, 142
Manthorpe, C. A., 130
Marcuse, H., 29
Maree, J. G., 466, 467
Martín Izquierdo, H., 425
Martin, M. O., 160, 373
Martin, Marjory, 69
Martínez Miranda, R., 386, 387
Martin-Kniep, G., 286
Marton, F., 253, 254, 255
Mason, M., 137
Masschelein, J., 429
Masters, G., 134
Matang, R., 254
Matsushita, K., 352
Maxwell-Jolly, J., 220
Maybury-Lewis, David., 41
Mayer, R. E., 304
Mayer, Victor J., 62
Mayson, S. E., 343, 345, 346, 348
McCarthy, C., 82
McClain, K., 437
McClelland, G., 488
McClure, M. W., 370
McInerney, P., 99
McIntyre, D., 254
McKinley, E., 84
McKinney, C., 132, 137
McKnight, C. C., 155, 165
McLaren, P., 81, 91
McNeir, G., 471
McNulty, B., 298
McTaggart, R., 251
Mehrtens, H., 10
Meintjes, S., 137
Mellin-Olsen, S., 35
Merick, L. C., 229
Merzbach, U. C., 248
Messner, R., 156
Metz, M., 453
Michie, M., 90
Miller, D., 322, 324, 336, 340
Miller, J. P., 519
Millroy, W.L., 215
Mitchell, C., 137
Mitchell, Timothy, 71
Mithaug, D. E., 323
Mlcek, S., 349
Mok, I. A., 107, 307
Mok, K., 432
Molepo, J. M., 467
Möller, J., 156
Monkman, K., 82, 88

Moon, B., 175
Morales, L., 239
Moreno Mínguez, A., 425
Morovic, J., 365
Morris, M., 288
Morris, P., 488
Morrow, R. A., 82
Moscovici, H., 285, 298
Moshier, Suzanne E., 63
Mosimege, M., 254
Moss, D. M., 287
Motaung, E., 356
Moulin, Anne Marie, 41
Mtetwa, D. K., 234
Mudavanhu, B., 444
Murillo, G., 428

N'Dour, Youssou, 61, 62
Naidoo, P., 465, 466, 467
Ndlalane, T., 465
Nebres, B., 96, 341, 387
Negri, A., 5, 7, 211
Nelson, Diane, 71
Neubrand, J., 304
Neyland, J., 113, 115
Nichols, D., 237, 241
Nisbet, S., 471
Nishioka, K., 466
Niss, M., 29, 31
Noah, H. J., 152, 168
Norjum, Yusop, 446
Norris, Stephen P., 63, 67
Noss, R., 176, 177
Nowotny, H., 14
Nozic, R., 99
Nozick, R., 325, 336
Ntshwanti-Khumalo, T., 467
Nunes, T., 27
Nunnally, J. O., 308

O'Mahony, P., 80
O'Neill, A.-M., 117
O'Riley, Patricia A., 65, 66
Ocampo, J. A., 425, 427
Odora-Hoppers, C., 144
Ogawa, M., 91, 153, 167, 488
Ohanova, B., 299
Oldknow, A., 254
Olssen, M., 117
Ono, Y., 466
Orey, D. C., 234, 235
Ostergaard Johansen, L., 35
Oweiss, I.M., 230
Owston, R. D., 520

Pandor, N., 132, 136, 137
Paolini, A., 80
Park, J.-K., 82
Park, K., 302
Parsons, Malcolm, 51, 52
Pasquale, M. M., 89
Passeron, J. C., 33
Patadia, H., 518
Paulsen, R., 466
Peat, David, F., 40
Pendlebury, S., 132, 138
Pennel, J. R., 468
Pepper, S. C., 29
Perold, H., 132
Perrett, G., 263
Perry, M., 304
Perry, P., 431, 437
Pescador-Osuna, J. A., 409
Peters, Michael, 22, 60
Petitjean, Patrick, 41
Petocz, P., 247, 254, 255, 257, 263
Pevsner, D., 156
Pfannkuch, M., 254
Phillips, Linda M., 63, 66, 67, 68, 69
Picker, S. H., 451
Pieterse, G., 137, 141
Pilger, J., 120
Pimm, D., 271
Pinar, W. F., 130
Pinch, Trevor, 65
Pinxten, R., 216
Plomp, T., 157
Plummer, D. L., 97
Polesel, J., 135
Polonyi, T., 365, 378
Polya, G., 443
Popper, Karl, 50
Porter, A. C., 468
Porush, David, 64
Posner, G. J., 295
Powell, A. B., 234, 248
Powell, A., 34, 104, 213, 214, 215
Pretorius, A., 467
Price, R. F., 495
Prinsloo, W. J., 466
Pu, T., 493, 494
Puk, T., 156, 165
Pyle, E. J., 287

Rabinow, Paul, 71
Radaelli, C., 228
Raizen, S. A., 155, 164
Rakow, S. J., 155
Rao, H., 499
Raplay, J., 425
Rawls, J., 98, 99, 322, 323, 324
Reardon-Anderson, J., 490
Recinos, A., 228, 235, 239, 243
Reddy, C., 132, 141, 484
Rees, W., 121
Reich, R. B., 87
Reid, A., 247, 252, 254, 257, 259, 263
Reis, S. N., 155
Rennie, L. J., 451
Renshaw, P., 356
Restivo, S., 25, 29, 221
Reynolds, W. M., 130
Rhoten, D., 79, 85
Rice, M., 467
Richardson, Laurel, 64
Riley R. W., 155
Rivera, F., 211
Rivet, A., 131
Rizvi, F., 82, 251
Robins, K. N., 231
Robitaille, D. F., 105, 154, 303
Robottom, I., 448
Rodrik, D., 405
Rogan, J., 478
Rogers, L., 105
Romaizah, M. S., 450
Rorty, R., 25
Rorty, Richard, 44, 64
Rosa, M., 231, 232, 234, 235
Rosener, J., 97
Ross, Andrew, 63
Rouse, Joseph, 65
Rowan, B., 448
Rowe, A. P., 241
Rowlands, S., 215
Roy, Kaustuv, 70
Rubenson, K., 367, 368
Rupert, M., 425
Russell, T., 288
Rutherford, M., 474
Ryan, J., 252, 259
Ryle, G., 23

Sagan, C., 248
Samson, M., 132, 135, 139, 140, 141
Sardar, Ziauddin, 41
Saunders, W., 466
Sawada, T., 271
Sawiran, M., 104
Scantlebury, K., 84
Schaar, J. H., 323, 324, 331
Schapper, J. M., 343, 345, 346, 348
Scheyvens, R., 250
Schmandt-Besserat, D., 26
Schmidt, W. H., 155, 165, 450
Scholte, J. A., 80, 510
Schon, D., 476
Schoorman, D., 347
Schuetze, H. G., 367, 368
Schuster, M., 141
Scott, J., 82
Secada, W., 97
Secada. W. G., 216
Sekiguchi, Y., 280
Selltiz, C., 305
Sen, A., 230
Sennett, R., 79, 120, 122
Sepheri, P., 97
Sergiovanni, T. J., 286
Serrant, T. D., 370
Serres, Michel, 43
Setati, M., 97
Shah, Sonia, 72
Shannon, A., 459
Shapin, Steven, 65
Shayer, M., 189
Sheldon, M., 327, 329
Shephardson, D. P., 131, 144
Shimizu, Y., 107, 271
Shulman, K., 468
Shulman, L. S., 283, 287, 298
Sibaya, D., 466
Sibaya, P. T., 466
Silver, E. A., 271
Sim, W. K., 449
Simon, Paul, 39
Simon, R., 286, 299
Sims, V., 304
Singer, F.M., 366, 377
Singh, M., 509, 516, 517
Siridopoulos, E., 467
Sjøberg, S., 153
Skinner, D., 218, 222
Skovsmose, O., 3, 29, 34, 102, 215, 216, 338, 392, 399, 425, 426, 433, 434
Skyrme, D., 22
Slattery, Luke, 42
Slattery, P., 130
Slaus, I., 365

Slaus-Kotovic, A., 365
Slayton, J., 82, 296
Smith, E. H., 322
Smith, S. P., 281
Smyth, J., 115
Snow, C., 221
Snyders, A., 254
Snyman, R., 470
Sokal, A., 192, 194
Songer, N. B., 165
Sosnoski, J., 180
Sowder, J., 318
Spunde, W., 253
Stacey, K., 445
Stake, R., 152
Steel, J., 345
Stevenson, H. W., 490
Steyn, J., 138, 139
Steyn, T. S., 467
Stigler, J. W., 166, 270
Stodolsky, S., 470
Stoer, S. R., 82
Strike, K. A., 295
Stromquist, N. P., 82
Supovitz J., 468
Sutton, Clive, 46
Suzuki, David, 41
Swafford, J., 444
Swan, M., 459
Swanton, C., 322
Symonds, W.C., 177

Tajika, H., 304
Tan, E., 131
Taubman, P., 130
Taylor, C., 120, 123
Taylor, J. A., 345
Teese, R., 135
Terrel, R. D., 231
Thomas, J., 340
Thomas, M. O. J., 330
Thomas, P. G., 330, 490
Thomas, R. R., 97
Thompson, C., 327, 329, 346
Thompson, J., 343, 345, 346
Thompson, T., 288
Tikhomirov, O. K., 393
Tippins, D. J., 286
Tobin, Kenneth, 46
Toffler, A., 227
Tomlinson, C. A., 155
Tomlinson, J., 80
Tomlinson, M., 7
Torres, C. A., 82, 344
Travers, K. J., 152, 303
Triandis, H. C., 252
Trouche, L., 355
Tsuchida, I., 165, 466
Tung, K. K., 444
Turmo, A., 164
Turnbull, David, 42, 70, 71
Tymoczko, T., 21, 25
Tytler, R., 448

Unterhalter, E., 132, 135, 139, 140, 141
Usher, R., 357

Vakoch, D. A., 248
Valero, P., 338, 387, 421, 424, 425, 426, 428, 430, 431, 433, 434, 437
Valverde, G. A., 165
Van den Acker, J., 478
van der Westhuizen, D., 356
Van Dyne, S., 141
van Maanen, J., 358
Vardumyan, S., 297
Vári, P., 373
Varrella, G. F., 285, 297, 298, 299
Veloo, P. K., 446
Verran, Helen, 72
Vidal, R., 410, 411
Villa-Lever, L., 407
Villarreal, M., 383, 385, 389, 390, 392, 393, 394, 396, 400
Vistro-Yu, C. P., 321, 327
Vistro-Yu, C., 327, 329, 337
Vitela, N., 411
Vithal, R., 102, 215, 216, 424
Vlasceanu, L., 366
Volet, S. E., 250
Von Bleyleben, K. A., 406
Vulliamy, G., 347

Wackernagel, M., 121
Wagner, D., 97
Wagner, Jon, 53
Wake, G., 349
Walkerdine, V., 215
Wallace, J., 287
Wals, A., 257
Walters, D., 251, 259
Wang, H., 165
Wang, J., 155
Wang, L., 501, 502

Wang, Z., 501
Warfield, V. M., 444
Waring, M., 152
Warren, E., 470
Watkins, D., 263
Weaver, John A., 64
Weber, M., 175, 191
Webster, B., 158
Wedege, T., 11, 35
Wehner, J., 176
Wei, B., 490, 493, 502
Weinstein, Matthew, 65
Wells, A. S., 82, 296
Welsch, W., 80
Westbury, I., 303
White, G., 288
White, Richard T., 49
Whyte, W. H., 115, 126
Wihlborg, M., 259
Wilkinson, D., 237
Williams, J., 349
Williams, P., 412
Willinsky, J., 114
Willis, S., 133, 143
Wilson, L. D., 155
Win, E., 137
Wineburg, S., 468
Winter, J., 443
Wise, A., 115
Wiseman, A. W., 82
Wittgenstein, L., 25
Wolf, L., 177
Wolfe, R. G., 164
Wolff, L., 403
Wong K. Y., 307, 450, 457
Wong, N. Y., 460
Wong, P. M., 488
Wong, S. Y., 488
Wood, L. N., 255, 263
Woolgar, Steve, 65
Woolworth, S., 468
Wotley, S. E., 347
Wrathall, J. P., 358
Wright, R., 116, 117, 121, 122
Wu, Y., 493, 494, 498, 499
Wylam, H., 189
Wynne, Brian, 43

Xu, G., 495
Xu, H., 489, 498

Yackel, E., 437
Yager, R. E., 286
Yang, W.C., 445
Yap, S.F., 450, 460
Ye, L., 497
Yeap, B. H., 444, 445
Yong, D., 444
Yoon, K. S., 468
Yoshida, M., 471
Young, D., 73
Young, I. M., 100, 101, 102, 103, 104, 324, 336, 337
Young, M. F. D., 152
Young, R. M., 29

Zach, K., 155
Zaitun, M.T., 446, 450
Zaslavsky, C., 27
Zhang, J., 495
Zheng, J., 409
Zilcosky, John, 74
Zohrabyan, A., 299

SUBJECT INDEX

Academic exchanges, between Argentina and Brazil
analysis, 398–400
development of collaboration in Argentina, 388–390
development of collaboration in Brazil, 391–394
phases, 395–398
shifts and changes, 394–395
Academic exchanges, between New Zealand and the Philippines
advantages and disadvantages, 338–340
background, 325–327
benefits to partners, 329–332
costs to partners, 332–335
funding issues, 336–338
inequalities, 340
issue of the process of selecting the partner institution, 335
Actor-network theory (ANT), 60
Adobe Creative Suite, 176
Adult education, in mathematics
learning technologies in, 355–357
phenomena of globalisation and internationalisation of education, 344–347
return to study, 350–351
sociocultural activity theory, 351–355
tensions and recommendations, 357–359
vertical/horizontal discourses and adult numeracy, 348–350
Adult numeracy, 348–350
Andrade, F., 97
Anglo-European views, of mathematics and mathematics education, 104
Arab renaissance, in mathematics, 214
Armed globalisation, 6
Ash, 296
Attitudes Scale, 305
Authorisation, notion of, 11–13
Autonomy, 3
Awards, in teaching, 415

Bacon, F., 366
Barton's QRS system, 217–218
Basic Mediational Triangle Expanded, 352
Bilateral primary mathematics study
aim, 450
findings, 452–459
instrument and sample, 450–452
multi-national *vs* bilateral comparative studies, 448–449
research agenda, 449–450
Blackfoot physics, 40, 42
Blank spots, 53
Blind spots, 53
Block Pattern Problem, 272
Bourdieu's modern mode of domination, 123
"brain drain," phenomenon of, 101
Brazilian Indians, 14
Brunei Darussalam situation, mathematics education, 445–448
access to and use of computers and calculators, 456
confidence in mathematics and english, 453–455
enjoyment of mathematics, 452
homework and help in learning mathematics, 455–456
qualities of the "best" mathematics teacher, 456–459
use of mathematics in everyday tasks, 452–453

Capitalism, 5
Centro Nacional para la Evaluación de la Educación Superior (CENEVAL), 410
César Gaviria era and education, 427–428
Chapman, D., 51–52
Chinese and U.S. students' representations, differences, 271–274
Chinese and U.S. teachers' representations, differences, 274–276
Chinese Communist Party (CCP), 493, 495, 500
Chinese curriculum scholars, 488
CivEd homepage, 193–194
Classroom practices, changing face of, *see* Mpumalanga Secondary Science Initiative (MSSI)
Closed systems, 44
Collectivism, 252
Colombian General Law of Education, 429–430
Colombian mathematics educators, 104
Columbian situation, in education
  internationalization, 436–438
  mathematic education in Esperanza Secondary School, 434–436
  mathematic education policy during 1990s, 430–434
  reforms in 1990s, 427–430
Conceptualisations of social justice, in mathematical education
  distributive model, 99
  market model, 98–99
  recognition model, 100
Congruence, 50
Conscientizaçao, 4
Consejo Nacional de Ciencia y Tecnología (CONACYT), 410–411
Constrained constructivism, 50
Constructivism and Western science, 45–53
Constructivist learning model, 286
Constructors, in mathematical literacy, 8–11
Consumerism, 122–123
Consumers, 13–14
Contact (Sagan, Carl), 248
Contemporaneity, in science education, 79–81
Cosmopolitanism, 80
Cristina, B. E., 394–398
Critical mathematical literacy, 4
Critical Theory, 24
Cross-national studies and mathematics education
Chinese and U.S. students' representations, 271–274
Chinese and U.S. teachers' representations, 274–276
  importance of cross-cultural differences, 280–282
  pedagogical representations in Chinese and U.S. classrooms, 277–280
Cryptography, mathematics-based, 9
CSF science curriculum, 89
Cueto, S., 97
Cultural capital, 164
Cultural deficit, 41
Cultural imperialism, in education, 104–105
  effects of globalisation, 116–120, 123–126
  impact of social stratification, 121–122
  mathematical education, 113–114
  and scientific management, 115–116
  trends, 122–123
Cultural Revolution, in China, 495, 499
Curricular Guidelines for School Mathematics, in Columbia, 432–433
Curriculum
  development for teacher education, in Laos, 510–512
  division of labour among teachers, 175–176
  and mode of authority action, 175–176, 190–193
  modes of interaction, 193–196
  as technology, 174–175
Third International Maths and Science Study (TIMSS), 177–189
  in UK, 177
Curriculum and StandardsFramework (CSF), 83
Curriculum 2005 documents, 132, 140
Curriculum Implementers (CIs), 469
Curriculum reforms, in Romania
  centralized administrative and bureaucratic political thoughts, 376–379
  international academic competitions, 373–376
  mass school of the industrial era, 366–367
  from the perspective of a globalised knowledge society, 368–370
Project RO-3724, 370–373
Cybernetics, 212

D'Ambrosio (Professor), interview with; *see also* Ethnomathematics
  early years of, 200
  globalization and education, 207–208
  on his academic work, 207
  mathematics in the context of power, 203
  research and practice in math education, 200–201
  role of parents, teachers and youth, 206

role of technology in mathematics and science, 203–204
teacher education and curricula in Brazil, 204–206
on Western mathematics, 202
DeBoar, 84
De Castell, S., 355–356
Decentralization, of education, 82
Determinism, 5
Disaggregation, of tasks, 10
Discourse theory, 8
Disembedding modes, 119
Disempowerment, 3
Disposables, 14–15
Distance education, 350
Distributive model, of social justice, 99
Diversity concept, in mathematical education, 97
Division of labour, between mathematicians and school mathematics teachers, 175–176
Doctoral Program at UNESP – Rio Claro, 391
Do-it-yourself system, 12
Don Quixote's science
windmills in Cervantes' Europe, 173–174
Dresden Codex, 237

Economic globalization, 20
Empire, 211–212
Empowerment, 3
English Language Development Institute in Algebra (ELDI-A), 220, 222
Enigma-machinery, 9
Entropy, concept of, 44
Equality, in classrooms, 98
Equality, in science education, 141–144
Equity concept, in mathematical education, 97
Escuela Neuva system, 103
Esperanza Secondary School, 421–423, 434–436
Ethnomathematics, 27, 391, 396; *see also* D'Ambrosio (Professor), interview with; Mayan civilization
and anthropology, 232–233
in classroom, 215–216
critiques on, 214–215
D'Ambrosio's views, 221–223, 232–235
definitions and interpretations, 213–214
and globalisation, 231–235
global order of things, 211–213
hybrid set of altered practices, 219–221
issues with practices, 218–219
issues with theory, 217–218
as QRS system, 217
research on, 216
Eurocentrism, of scientists, 40, 47, 62
European aristocratic male, 215
Europeanization, 228

Fact File, 51
Feminism, in South Africa, 135
Flexible capitalism, 118
For-profit multinational corporations, 20
Forum of Secondary School Science Curriculum in China, 498
Free-growing capitalism, 5
Free-market economists, 63
Freyle, Juan Diaz, 230
Functional mathematical literacy, 4

Gender agenda, education
in Australia, 133–135
discussions of globalisation, 139–141
2005 Kenton Conference, 137–138
research on girls and science education, 130–131
in South Africa, 135–136, 138–139
Gender Equity Task Team (GETT), 140
Geometric-numeric patterns, of Mayan diamond, 237–239
Geophilosophy, 60–61
Ghettoised people, 6
Global information culture and science education, 86–91
Global interconnectedness, 118
Globalisation; *see also* Mayan civilization
case studies, 510–513
concept, 228
defined, 5
effects on education in Mexico, *see* Mexican situation in education
ethnomathematic perspective, 231–235
and growth of capitalism, 5
identified by Delors, 514–515
and international professional development, *see* Professional development (PD), as a form of globalization
issues related, 515–521
of mathematical knowledge, 228–231
and movement of goods, 5–6
predictability of, 5
processes of, 5
role in poverty, 6
and war, 6–7
Globalised capitalism, 14
Globalization and internationalization, of mathematics education

in Brunei Darussalam, 445–448, 457
concept of globalization, 20
dimensions of impact, 22
issues with knowledge, 23
as mess, 57–60
multi-national and bilateral comparative studies, 448–459
in Singapore, 442–445, 457
Global shifts, 296
Global village, 20
GPIMEM, 392–395
Graduate Program in Mathematics Education (GPME), Brazil, 390, 392
"Great Architect" of the universe, 235–236

Hegemonic homogenisation, 80
Higher education, in Mexico, 407
High school education, in Mexico, 406
Hindu-Arabic numeration system, 229
Hindu discoveries, in mathematics, 229
Hofstede's framework of socioculture, 412
Hypertext, 189
Hypothetical reasoning, mathematics-based, 9–10

ICME10, 387
Incas community, 236
Individualism, 252
Individual rights, 98
Industrial high schools, 370
Informational knowledge, 87
Information and Communications Technology (ICT), in education, 357
based mathematics lessons, 445, 456
In-service Education and Training (INSET) system, 469
INSET programme, 480
Institute Dialogue on Educational Advances (IDEA), 449–450
Instructional tasks, 278
Intergovernmental Panel on Climate Change (IPCC), 43
International academic competitions, 373–376
International collaborations, in mathematical education
cultural imperialism, 104–105
as exploitative, 100–102
marginalisation of, 102–103
powerlessness, 103–104
violence in, 106–107
International Congress on Mathematics Education (ICME), 249
Internationalisation, of mathematics
conferences, 249
construction of learning environments, 261–264
organisations, 249
perspectives of learning and teaching by students and lecturers, 253–261
as a subject, 248–249
as a value for learning, 249–253, 261–264
International Journal of Mathematical Education in Science and Technology (iJMEST), 249
International Mathematical Olympiad, 373
Internet, 386
Israel model, of PD, 289–291, 293, 295

Japanese education system, 465–466
Japanese school curriculum, 156, 166–167
Johannesburg Earth Summit, 255–257
Junior Secondary School Chemistry Curriculum (JSSCC) in China, a study
discussion and analysis, 502–505
historical overview and need for research, 490
international influences, 492
methodology, 491
reporting considerations and framework, 491–493
year 1978, 493–495
year 1982, 495–497
year 1992, 497–500
year 2001, 500–502

KASSEL Project, 450–451
K-12 classroom, 287, 297
Kenton Conference 2005, 137
Khaldun, I., 230
Kindergarten education, in Mexico, 406
Knowledge, globalization and internationalization of, 21
mathematical knowledge and society, 25–35
types and production, 23–25
Knowledge economy *vs.* traditional economy, 22

Lao education system, 510–512
Lather, 91
Leadership, in clusters, 482
Learners Perspective Study project, case study, 107–109
Learning and teaching experiences, by students and lecturers
collaborative work, 254
conceptions in mathematics, 255–256

on internationalization and teaching, 259–261
notions of teaching, 254
student views, 255
teaching and sustainability, 257–259
views of diversity, 253–254
Learning for democracy, 1
Learning technologies, in education, 355–357
Lesson study
as a communication instrument, 472
defined, 471
as observation instrument, 472–473
as part of MSSI project, 475–478
as professional development exercise, 474
as teaching instrument, 472
Lester, F., 271–272, 274, 277
Likert scales, 308–309

Mainstream participation, of women in African society, 136
Malaria outbreaks, 70–72
Malaysia and Regional Education Centre for Science and Mathematics, 513
Maldives Teacher Education Project, 512–513
Marcelo, C. B., 391
Marginalised knowledges, 62
Market model, of social justice, 98–99
Marxist-Leninist learning theory, 494–495
Mass school, of the industrial era, 366–367
Mathematical gaze, 28
Mathematical knowledge and society
in developing countries, 33
ethnomathematics and proto-mathematics, 27–29
in industrialized society, 29–30
mathematics as academic discipline, 26
modern research mathematics, 26–27
role of audio-visual media, 33
teaching methods of industrial nations, 31–32
Mathematical literacy
as an operator, 11–13
for consumers, 13–14
cross-national studies, *see* Cross-national studies and mathematics education
in education of the 'disposables,' 14–15
knowledge-power issue, 7–8
as a part processes of construction, 9–11
social constructivist philosophies, of mathematics, 21
social justice of, *see* Social justice, in mathematical education
a tool of cultural imperialism, *see* Cultural imperialism, in education
types, 4
uncertainty, 16–17
Mathematical practices, as diverse cultural forms of knowledge, 233
Mathematic education
in Esperanza Secondary School, 434–436
in South Africa, *see* South Africa situation, in mathematics education
Mathematicoscience, 190
Mathematics education
in Brunei Darussalam, 445–448, 457
multi-national and bilateral comparative studies on international collaboration, 448–459
research in the context of globalisation, 423–426
in Singapore, 442–445, 457
Mathematics Education Library, 396
Mathematics Pacesetters, 482
Mayan civilization
decoding of messages, 241–242
general discussion, 236
number system of the divine creation, 229, 242–243
numerology, 244
process of globalization, 236–237
sacred geometric-numeric patterns, 237–238
sacred mats, 239–241
Mayan people, 236–237
Mediterranean-based mathematics, 234
Merchant, B., 413
Mexican General Law of Education, 410
Mexican Indian students, *see* Mexican situation, in education
Mexican situation, in education
characteristics of Mexican educational system, 406–408
context, 405–406
globalisation effects, 408–416
Mexican universities, 407
Micronesian navigators, knowledge level of, 43
Mode 1 knowledge production, 24
Mode 2 knowledge production, 24–25
Modernism, concept of, 23
Mosquito rhizomatics, 66–73
Mpumalanga Secondary Science Initiative (MSSI), 469–471
academic context of a lesson, 470–471
aspects of the cascade model in Phase 2, 479–482
dilemma and contradictions for clusters, 478
lesson study, 471–474
structure of clusters, 478–479
teacher clusters/networks, 474–478

Multicultural societies, 210

National economic well-being, 156
Nation-states, history of, 212
Nature-as-an-object-of-knowledge, notion of, 40–45
Navajo notions of space, 216
N'Dour, Y., 62
Neoconservatism, 82, 90
Neoliberalism, 82, 85
New Right biologism, 63
Newtonian mechanics, 47, 52–53
Newton's Law of Gravitation, 49
Nigerian worldview, 46
Non-European mathematics, 113
Nordkvelle, 356
North America Free Trade Agreement (NAFTA), 408
Numerology, of Mayan civilization, 244

OECD's Programme for International Student Assessment (OECD/PISA), 404, 411
 and culture, 164–168
 findings of influence, 160–164
 responses, 155–160
OECD's Programme for International Student Assessment (OECD/PISA), 84
 contextual constructs and their relations with science achievement, 154–155
 survey methodology, 153–154
Operational matrix, 190
Oral proto-mathematics, 27
Ornellas, 409
Outcomes based Education (OBE) model of learning, 467, 470

Paradox of citizenship, 426
Paradox of inclusion, 426
Parents Influence scale, 305
Passat, 116
Pedagogical content knowledge (PCK), 468, 481
Pedagogical representations, in Chinese and U.S. classrooms, 277–280
Peers Influence scale, 305
Peer teacher learning, 481
Pentagon Framework curriculum, 443
Personal constructs, 46–47
Photosynthesis phenomenon, 43–44
Popular media, in science education, 63
Postcolonialism, 72
Post-Fordism, 64, 79, 89
Postmodernism, concept of, 23, 25
Power and mathematics, 7–8
Powerlessness concept, in education, 103–104
Practical wisdom, 23
Printing press, invention of, 228
Privatised curriculum space, 126
Professional development (PD), as a form of globalization
 contributions to globalization, 296–300
 essentials of, 287–288
 models of, 288–295
 theoretical referents, 286–287
Program of Incentive for Teachers-Researchers, 385
Progressive educators, 300
Project 2061: Science for All Americans, 83–85
Project RO-3724, 370–373
Pseudo-capitalism, 116
Public curriculum space, 115, 125
Public domain texts, 187
Pythagorean Theorem, 217

Quiché codex, 242

Ramesier, S. J., 164
Ramphele, 139–140
 concept of social justice, 322–325
Recognition model, of social justice, 100
Reflective discourse, 478
Reflexive reshaping, of knowledge, 118–119
Regional Education Centre for Science and Mathematics, in Malaysia, 513
Rhizome and nomad tools, in Western science education
 mosquito rhizomatics, 66–73
 nomadic and sedentric point of view, 73–74
Rhizome space, 60–61
Romania model, of PD, 288–291, 293, 295
Roman numeral system, 229
Routinisation, notion of, 11–13

Science education
 areas of research, 88–89
 and contemporaneity, 79–81
 educational reforms, impact of, 81–86
 and global information culture, 86–91
Science Education Conference in the Asian and Pacific Regions, 498–499
Science fiction (SF), 63, 70, 72
Science textbooks, 67
Scientific Literacy and Cultural Studies Project (SLCSP), 42

Scientific management, of mathematics education, 115
Scientific media reports, 67–70
Seabrook, 121, 123
Seminar of Mathematics and Mathematics Education (SMME), 392
Singapore situation, mathematics education, 442–445
  access to and use of computers and calculators, 456
  confidence in mathematics and english, 453–455
  enjoyment of mathematics, 452
  homework and help in learning mathematics, 455–456
  qualities of the "best" mathematics teacher, 456–459
  use of mathematics in everyday tasks, 452–453
SMILE scheme, 176
Social constructivism, 286
Social constructivist philosophies, of mathematics, 21
Social constructs, 46
Social justice, in mathematical education
  academic exchanges between New Zealand and the Philippines, 325–335
  advantages and disadvantages, 338–340
  agenda, 96–98
  concept and issues of fairness, equality and equity, 322–325, 335–338
  conceptualisation of, 98–100
  and international collaborations, 100–107
  Learners Perspective Study project, case study, 107–109
Social mathematics, 28
Social theorising, 7
Sociocultural activity theory, 351–355
Socio-cultural knowledge, 261
South Africa situation, in mathematics education
  academic context of a lesson, 470–471
  aspects of the cascade model in Phase 2, 479–482
  dilemma and contradictions for clusters, 478
  lesson study, 471–474
  MSSI Project, 469–470
  need for change, 467–468
  social and cultural context, 466–467
  structure of clusters, 478–479
  teacher clusters/networks, 474–478
South East Asian Ministers of Education Organisation (SEAMEO), 513
Specially Designed Academic Instruction in English (SDAIE), 220–221
Speech act theory, 8
Standardized assessment examinations, 414
Statistics Education Research Journal (SERJ), 249
Students' attitudes towards mathematics in China and Australia, case study
  aim and methodology, 305
  discussions, 315–318
  issues in design, 305–310
  process and result, 310–315
Sultan Hassanal Bolkiah Institute of Education (SHBIE), 447
Super imperialism, 116
Survey research, 153–154
  case study, 305–318
Swedish student nurses' conceptions, of internationalisation, 259–261
Symbolic-analytical workers, 87
Symbolic-analytic workers, 90
Symbolism, 282

Task Influence scale, 305
Teacher clusters/networks, 474–478
Teacher Development Centre, in Laos, 520
Teacher Influence Scale, 305
Teacher-learners, 286
Teachers' content knowledge (CK), 468, 481
Teknologirådet, 13
Test Curriculum Matching Analysis (TCMA), 158
Theoretical referents, to professional development (PD), 286–287
Third International Maths and Science Study (TIMSS), 84, 249, 373–375, 410, 445
  analysis of website, 179–189
  Australian report of, 163
    contextual constructs and their relations with science achievement, 154–155
    and culture, 164–168
    findings of influence, 160–164
    responses, 155–160
    survey methodology, 153–154
Third World countries, 71
Tocqueville, 123
Top-down research-development-dissemination (RDD) model, of curriculum, 119
Torres, C. A., 344
Transnational capitalism, 116
Trap-door function, 9
Tribal wisdom, 41
Trigwell, K., 253

Unceratinty, in mathematical literacy, 16–17
UN Girls' Education Initiative (UNGEI), 136
United Nations Decade on Education for Sustainable Development, 136
United Nations Literacy Decade, 136
United Nations' Millennium Development Goal, 135
Universalisation, of education projects, 119–120
Universal mathematics, 234
Universiti Brunei Darussalam (UBD), 447

Vertical/horizontal discourses and adult numeracy, 348–350
Violence, in education, 106–107

Western consumerist lifestyles, 33, 210–211
Western mathematics, 234
Western science, globalization of
  general discussion, 61–66
  global reach of European and American imperialism, 40
  issue of constructivism, 45–53
  nature-as-an-object-of-knowledge, notion of, 40–45
  nomadic and sedentric point of view, 73–74
  rhizomes and nomads as tools, 61–66
  scepticism of, 41
  universality of, 41–42
Western-style market economy, 368
Whalley, 348
Wikipedia, 20–21
Williams, P., 412
Willinsky, J., 114
World Bank's mathematics education programs, 106, 370–373
World Education Forum Education for All Dakar Framework of Action, 136

Xiaoping, D., 493–494

Zero, symbol for, 229